Sources and Studies in the History of Mathematics and Physical Sciences

Springer
New York
Berlin
Heidelberg
Barcelona
Hong Kong
London
Milan
Paris
Singapore
Tokyo

Sources and Studies in the History of Mathematics and Physical Sciences

K. Andersen
Brook Taylor's Work on Linear Perspective

H.J.M. Bos
Redefining Geometrical Exactness: Descartes' Transformation of the Early Modern Concept of Construction

J. Cannon/S. Dostrovsky
The Evolution of Dynamics: Vibration Theory from 1687 to 1742

B. Chandler/W. Magnus
The History of Combinatorial Group Theory

A.I. Dale
A History of Inverse Probability: From Thomas Bayes to Karl Pearson, Second Edition

A.I. Dale
Pierre-Simon Laplace, Philosophical Essay on Probabilities, Translated from the fifth French edition of 1825, with Notes by the Translator

P.J. Federico
Descartes on Polyhedra: A Study of the *De Solidorum Elementis*

B.R. Goldstein
The Astronomy of Levi ben Gerson (1288–1344)

H.H. Goldstine
A History of Numerical Analysis from the 16th through the 19th Century

H.H. Goldstine
A History of the Calculus of Variations from the 17th through the 19th Century

G. Graßhoff
The History of Ptolemy's Star Catalogue

A.W. Grootendorst
Jan de Witt's *Elementa Curvarum Linearum, Liber Primus*

T. Hawkins
Emergence of the Theory of Lie Groups: An Essay in the History of Mathematics 1869–1926

A. Hermann, K. von Meyenn, V.F. Weisskopf (Eds.)
Wolfgang Pauli: Scientific Correspondence I: 1919–1929

C.C. Heyde/E. Seneta
I.J. Bienaymé: Statistical Theory Anticipated

J.P. Hogendijk
Ibn Al-Haytham's *Completion of the Conics*

Continued after Index

Jens Høyrup

Lengths, Widths, Surfaces

A Portrait of Old Babylonian Algebra and Its Kin

With 89 Illustrations

Springer

Jens Høyrup
Section for Philosophy and Science Studies
University of Roskilde
P.O. Box 260
DK-4000 Roskilde
Denmark
jensh@ruc.dk

Sources and Studies Editor:
Gerald J. Toomer
2800 South Ocean Boulevard, 21F
Boca Raton, FL 33432
USA

Library of Congress Cataloging-in-Publication Data
Høyrup, Jens.
Lengths, widths, surfaces : a portrait of old Babylonian algebra and its kin / Jens Høyrup.
p. cm. — (Studies and sources in the history of mathematics and physical sciences)
Includes bibliographical references and index.
ISBN 0-387-95303-5 (alk. paper)
1. Mathematics, Babylonian. 2. Algebra. I. Title. II. Sources and studies in the history of mathematics and physical sciences.
QA22 .H83 2001
510'.935—dc21 2001032839

Printed on acid-free paper.

Production managed by Timothy Taylor; manufacturing supervised by Erica Bresler.
Photocomposed copy prepared from the author's TEX files.
Printed and bound by Maple-Vail Book Manufacturing Group, York, PA.
Printed in the United States of America.

9 8 7 6 5 4 3 2 1

ISBN 0-387-95303-5 SPIN 10838641

Springer-Verlag New York Berlin Heidelberg
A member of BertelsmannSpringer Science+Business Media GmbH

To all the Assyriologist-friends in Copenhagen,
Leningrad, Illinois, and Germany East and West
who never refused assistance;

to Peter, Jöran, and Jim;

and in memory of O. Neugebauer

ogni approfondimento di ricerca rivela una complessità di elementi dei quali precedenti sintesi non avevan tenuto sufficiente conto, e, se li avevano presi in considerazione, non era sembrato che infirmassero una tesi di primaria importanza, di solito condizionata dal gusto imperante al tempo in cui venne formulata quella sintesi.

Mario Praz, *Gusto neoclassico*

Preface

"[...] it is through wonder that men now begin and originally began to philosophize" – thus Aristotle's *Metaphysica* 982^{b}12 [trans. Tredennick 1933: I, 13]. Some 25 years ago *I* started wondering when reading the secondary literature about the early history of mathematics: what could be the reasons that induced the Babylonians to work on second-degree equations, as it was said they did? Obviously not practical applicability – nor, however, it appeared, that kind of curiosity which made the ancient Greeks create mathematical *theory*.

In parallel with many other questions, I pursued the matter until I believed – around 1980 – to have arrived at least at a rough explanation. Looking at what I wrote back then I can still recognize the inception of my present ideas about the historical sociology of Babylonian mathematical knowledge; but beyond the general Assyriological literature, my basis consisted of translated sources whose interpretation had been commonly accepted since the 1930s.

In 1982 I gave a guest lecture in Berlin on my sociological interpretation, after which a member of the audience asked me *what* this Babylonian algebra looked like. I answered in agreement with what I had understood on the basis of the translations, and thus gave a picture close to the rhetorical algebra of the Middle Ages. Peter Damerow, who had organized the session, at that moment asked me why I was so sure, and showed a geometrical interpretation which Evert Bruins had proposed for a particular text; I recognized the diagram from one of the geometrical proofs from al-Khwārizmī's *Algebra* (it is shown below in Figure 88, p. 413), which made me curious. I got hold of a grammar and a dictionary and soon realized that the diagram was totally irrelevant in the context where Bruins had used it; but I also discovered that the current interpretation of the Babylonian "algebraic" texts was made to fit the numbers but did not agree with what followed from a careful reading of the words between the numbers.

For the outsider, Assyriology comes close to being an occult science, and it took some years before I was able to publish a decent detailed account of my arguments and my results [Høyrup 1990]. That I got so far was largely due to the support I got from Bendt Alster, Mogens Trolle Larsen, and Aage Westenholz of the Carsten Niebuhr Istitute, University of Copenhagen, and to the discussions I had with the participants in the "Workshops on Concept Development in Babylonian Mathematics" organized in Berlin in 1983, 1984, 1985, and 1988 – especially with Peter Damerow, Robert Englund, Jöran Friberg, Hans Nissen, Marvin Powell, Johannes Renger, and Jim Ritter. I also got precious advice and patient encouragement from Wolfram von Soden, even though it took me years to convince him that I might be on the right track.

That I got further is thanks to the colleagues who prevented me from concentrating all my scholarly energies on Mesopotamia, and seduced me into pursuing parallel work on ancient Greek, Islamic, and Latin medieval mathematics. Though I started in the likeness of Columbus, hitting land on a course I had initially chosen for the wrong reasons, I continued rather like Odysseus, visiting many unfamiliar countries, staying long with Circe and with Calypso. I also lost some experienced companions and masters on the way whom I think of with much regret – first Kilian Butz and Kurt Vogel, more recently Wolfram von Soden and Wilbur Knorr; I even visited the realm of the dead and learned immensely from the shadows of Thureau-Dangin and Neugebauer.

I was never left alone on the shore of Ithaca (if that is where I am now), but the possessions on my shelves are no less precious for me than the gifts of the Phaeacians for Odysseus: books, articles, letters from colleagues, and my own notes and writings on many intersecting themes. The pages that follow build on these riches, synthesized as far as I can at the present stage. The core of the argument is an analysis of the techniques and conceptualizations of Babylonian "algebraic" and related mathematics from the "Old Babylonian" earlier second millennium BCE (the "golden age" of Babylonian mathematics), based on texts in transliteration and "conformal translation"; on this foundation, a global portrait of the mathematical type in question is delineated.

These are the topics of Chapters I–VII. They deal with a moment in the history of mathematics, but the approach is not historical: it is synchronous and does not ask about the development nor, *a fortiori*, about the forces that shaped this development. The rest of the book (Chapters VIII–XI) is devoted to history proper: the historical shaping of Old Babylonian mathematics itself, the detailed geographical and chronological pattern; the origins and transformation; and, finally, kinship and historical influence.

I shall abstain from reformulating in prose what may just as well be read from the table of contents, and close this *prolegomenon* with three technical remarks:

All translations into English in the following – both from the sources and from modern publications – are mine, if no other translator is identified.

References mainly follow the author/editor-date system (with alphabetization after first author in the bibliography, pp. 418*ff*). However, standard editions of Babylonian texts and Assyriological reference works are referred to by the customary abbreviations, which are also listed in the bibliography.

Babylonian tablets are referred to by habitual museum or publication numbers. The "Index of Tablets" (pp. 426*ff*) inventories all tablets referred to in the text and refers to the publications from which I have taken the single texts. It also lists the references to each text in the preceding pages.

Jöran Friberg read and commented valuably on part of the first draft and Eleanor Robson on the second version, for which I thank both sincerely. I hardly need to point out that I remain responsible for everything, both where I have followed their suggestions more or less faithfully and where I have decided differently.

Contents

Chapter I
Introduction

The Discovery of Babylonian "Algebra"

Since the early days of cuneiform studies it was known – and hardly considered amazing, given among other things the importance of measure and number in the Old Testament – that the Babylonians were in possession of numbers and metrology; since the later nineteenth century the existence of a place value system with base 60 (the "sexagesimal" system) and its use in Late Babylonian mathematical astronomy were also well-known facts. During the following few decades, finally, a number of Babylonian and Sumerian mensuration texts were deciphered. By the end of the 1920s it was thus accepted that Babylonian mathematics could be spoken of on an equal footing with Egyptian mathematics, as indirectly acknowledged by Raymond C. Archibald when he added a section on Babylonian mathematics to his exhaustive bibliography on ancient Egyptian mathematics.[1]

Nonetheless it came as an immense surprise in the late 1920s when Babylonian solutions of second-degree equations were discovered at Neugebauer's seminar in Göttingen.[2] Until then, systematic treatment of

---

1 The main part of the bibliography is in [Chace et al. 1927: 121–206]; an unpaginated supplement with the section on Babylonian mathematics is in [Chace et al. 1929].

The state of the art of the early twentieth century is illustrated by the treatment of the Babylonians in [Cantor 1907: 19–51]; a thorough coverage of publications from the period 1854–1929 that somehow deal with the topic (even when as a secondary theme only) can be found in [Friberg 1982: 1–36] – a recommendable annotated bibliography also for later decades, from which the (much less extensive but still annotated) chapter on the subject in [Dauben 1984: 37–51] is drawn.

2 In 1985 I was told by Kurt Vogel about the intense amazement with which the

second-degree algebra was believed either to begin with the Indian mathematicians and then to have been borrowed by the Arabs; with Diophantos; or, in geometric disguise, with Euclid (*Elements* II) and Apollonios.[3] To a large extent, of course, the disagreement hinged upon the understanding of the term "algebra".

The first publications relating the discovery appeared in Neugebauer's newly founded journal *Quellen und Studien zur Geschichte der Mathematik, Astronomie und Physik*. In an analysis of a text concerned with the partition of trapezia (which Carl Frank [1928] had already tried to penetrate without understanding much), Neugebauer [1929: 79*f*] concluded that

> One may legitimately say that the present text confronts us with a piece of Babylonian mathematics that enriches our all too meagre knowledge of this field with essential features. Even if we forget about the use of formulae for triangle and trapezium, we see that complex linear equation *systems* were drawn up and solved, and that the Babylonians drew up systematically problems of *quadratic* character and certainly also knew to solve them – all of it with a computational technique that is wholly equivalent to ours. If this was the situation already in Old Babylonian times, hereafter even the later development will have to be looked at with different eyes.

In the following (second) fascicle, Neugebauer and Schuster each had an article dealing (in Neugebauer's case among other things) with Old Babylonian and Seleucid solutions of quadratic equations, respectively.[4] In the first fascicle, Neugebauer and Struve [1929a] had already investigated the Babylonian way of dealing with circles, circular segments, and truncated cones.

To what extent these publications mark a watershed is revealed by a slightly ironic remark made in 1935 by Neugebauer in the preface to the first volume of his monumental edition and analysis of Babylonian mathematical texts [MKT I: v]:

> When saying my aim to have been since the very beginning to prepare an edition of all available mathematical cuneiform texts, then it is meant that the work certainly has not changed its fundamental nature but all the more its scope. The first manuscript, supposedly "print ready" already in 1929, contained only the c. two dozens table texts from Hilprecht's publication BE 20,1, the three London texts BM 85194 and BM 85210 from CT IX, BM 15285 from RA **19** (Gadd), the two Paris texts AO 6456, AO 6484 (TU 31 and 33), and finally the six texts from

discovery was received. Vogel also informed me that the first to make the discovery was H. S. Schuster.

The whole development of the historiography of Babylonian mathematics since the late 1920s is analyzed in [Høyrup 1996].

3 See, among other publications, [Nesselmann 1842]; [Rodet 1878]; and [Zeuthen 1886].

4 The Old Babylonian period lasts from c. 2000 BCE to 1600 BCE; the mathematical texts appear to date from its second half. The Seleucid epoch goes from 312 BCE to 64 BCE. A brief overview of the periodization of Mesopotamian history that can be read independently of the intervening chapters is found on p. 309.

Frank SKT. That was less than the half of the present Chapters I to III and Chapter V.

Apart from the corpus of tables, even this early list consists of texts that had not been interpreted successfully before – among which are those which Schuster, Struve, and Neugebauer dealt with in *Quellen und Studien* in 1929-30. That is, the supposedly "print-ready" manuscript from 1929 was already a decisive leap forward – yet the impetus created by this initial breakthrough made the leap look like a quite modest step from the perspective of 1935.

Most important for this change of perspective was precisely the new understanding of what had come to be identified as "Babylonian algebra". The credit for making the breakthrough indubitably belongs to Neugebauer and his collaborators, all of them primarily mathematicians and historians of mathematics. As pointed out by Schuster [1930: 194], he was not the first to try his hand at the tablet which he analyzed; yet when publishing it in 1922, the eminent Assyriologist F. Thureau-Dangin had seen nothing but unidentified "arithmetical operations", in spite of his longstanding interest in Babylonian metrology and computational techniques. Once the breakthrough had been made, however, Thureau-Dangin was able to contribute with his outstanding philological competence, and a tense but efficient and always polite race began. By the end of the thirties, the shared efforts of Neugebauer and Thureau-Dangin (supported by contributions from Kurt Vogel, Solomon Gandz, and a few others) had produced a fairly detailed picture of Babylonian mathematics, in which "algebraic" problems (in particular problems of the second degree) occupied a prominent place. In particular it was known that most mathematical texts came from the Old Babylonian period, and apparently from its second half; with extremely few exceptions (only one of which was really certain), the other texts known at the time belonged to the Seleucid era. What little was known about practical computation in the Sumerian third millennium BCE disappeared from view, probably being considered as "not really mathematics". Although conspicuous changes in the terminology from one period to the other had been pointed out at an early stage by Schuster [1930: 194] as well as Neugebauer [1932a: 6], Babylonian mathematics came to be regarded as essentially the same in the two periods.

The Standard Interpretation

With minor disagreements, Neugebauer and Thureau-Dangin had also produced *an interpretation* of the "algebraic" part of the Babylonian corpus that soon gained global acceptance. For decades, this interpretation – which claimed it to *be* an algebra – and the translations based on it constituted the basis for all further work within the field.

In order to understand that *interpretation* in a strong sense is really involved, we shall have to look at the difficulties which the early workers encountered.

The language of the texts is Babylonian, a dialect of the Akkadian tongue (a Semitic language not too far from classical Hebrew and classical Arabic; the other main dialect is the Assyrian). They are written on clay tablets in cuneiform, partly in syllabic writing, partly by means of logograms (word signs) – almost always of Sumerian origin (in which case we shall speak of them as "Sumerograms") but mostly read in Akkadian (as "*viz.*" is read in English as "namely", and not with the original Latin value "*videlicet*"). The first difficulty arises because the same signs function both as syllabic signs and as logograms; even worse, many signs possess several logographic meanings (not necessarily semantically related) together with one or more *groups* of phonologically related syllabic readings.

The sign  may serve as an example, and at the same time serve to introduce some of the subtleties of the transliteration of cuneiform.[5] Its conventional sign name found in the lexical lists is KAŠ (sign names, used when the actual reading is uncertain, are written as small capitals[6]). It may stand for Sumerian k a š, "beer" (spaced writing is used for Sumerian words), and for the Sumerian possessive and demonstrative suffix .b i; the latter reading is used in Sumerian as an approximate phonetic representation of the compound b+e > b e, "it is said". These three uses have given rise to logographic use in Akkadian for the corresponding words *šikarum*, "beer" (Akkadian readings are italicized), *-šū/šā*, "its", *šū/šūatum*, "this", and *qabûm*, "to say". In the Old Babylonian period it will be found with the phonetic values *bi*, *bé*, *pí*, and *pé* (accents and subscript numbers are used to distinguish different writings of the same syllable; *be*, *bé*, and *bè* thus correspond to three different signs but to the same phonemic string); in later periods, it can have the additional phonetic values *gaš*, *kaš*, and *kás*. To this can be added the role in a number of composite sign groups used logographically: different sorts of

5 Assyriologists distinguish "transliteration", where syllabic writing is rendered syllable for syllable, and logograms are rendered as such, from the "transcription", in which logograms are translated into phonetic (not syllabic) Akkadian.

6 Many recent publications introduce a further distinction and write sign names in large capitals when used in isolation or within genuine Sumerian texts, and use small capitals for all logograms used within Akkadian texts, irrespective of whether the corresponding Sumerian value is identified or not. The reason for this convention is that this Sumerian value will only in exceptional cases have corresponded to the intended pronunciation of the words; in the context of the present argument, however, I have found it more adequate to facilitate the identification of terms that possess a Sumerian interpretation – not because the use of Sumerian terms implies general continuity with third millennium mathematics, but in order to provide a basis for distinguishing cases where the evidence suggests continuity from those where the use of Sumerian appears to be an Old Babylonian construction.

On the pronunciation of *š* and other non-standard characters, see the "Note on Phonetics" on p. 43.

beer; innkeeper; etc. Finally, the sign may represent twice the area unit ešè, written .

As we see, some uses belong only to specific periods. Moreover, specific text types have their particular usages, which further reduces the ambiguity – but they do not eliminate it, and they only reduce it when the characteristic usages of a particular text group have been discovered. The same applies to the use of semantic determinatives and of phonetic complements to logograms[7] – aids of which the Babylonian scribes themselves made use in order to evade ambiguity, but again in ways that change from one text group to the other.

On top of this is the difficulty of understanding the terminology itself, once it is determined how to transliterate the texts into syllabic writing and logograms. Like every technical terminology, that of Babylonian mathematics was ultimately derived from daily language – but often technical meanings cannot be guessed from general meanings, even when these are known. Once we have analyzed the term "perpendicular", it is easy to see how a pending plumb line suggests the idea of the vertical and hence – via its relation to the horizontal plane – of the right angle. Yet etymology alone could never tell us whether verticality or orthogonality is the technical significance; worse, the use of phrases like "raising the perpendicular" support the wrong hypothesis of verticality. Ultimately, a technical terminology has to be understood from its technical uses, and the interpretation can at most be suggested and checked by, but never derived from, everyday meanings. Technical uses, however, may be difficult to understand as long as we do not understand the terminology in which they are expressed.

As an example we may consider the sentence

30 *a-na* 7 *ta-na-ši-ma* 210

(for the moment I translate the Babylonian numerical notation into Arabic numerals; number writing presents separate problems to which we shall return). *ana* is a proposition meaning "to" (etc.), and *tanašši* is the second person singular, present tense, of the verb *našûm*, "to raise", "to carry". The

---

7 A semantic determinative is a sign that is not meant to be pronounced but which indicates that the preceding or following sign stands for (e.g.) a divine or geographical name, a profession, an artefact made of wood or of metal, etc. Originally determinatives may have indicated in which lexical list the sign was to be found (not everybody shares this interpretation). Thus [ĝiš]apin stands for a real plough, apin, namely, made of wood (ĝiš) and found in the list of wooden artefacts; [mul]apin stands for a celestial constellation called "the plough" (mul = "star"); with determinative lú ("man") it becomes "peasant", with Sumerian reading engar.

Phonetic complements may give the pronunciation of the first or last syllable of a word, thus identifying both the Akkadian interpretation of a logogram and the grammatical form. In a.šà[*lam*] the complement thus shows that a.šà, "field", stands for the Akkadian word *eqlum* but in the accusative form (*eqlam*).

enclitic particle *-ma*, finally, can be translated "and then" or "and thus" or simply as ":". In partial translation, the sentence thus becomes

30 to 7 you raise: 210.

Similarly, "raising" 2 to 20 gives 40, and 17 to 30 gives 510. The obvious conclusion seems to be that "raising" is nothing but multiplication. On the present evidence, however, this conclusion is not warranted, and the meaning of our sentence might as well be

you make 30 perpendicular to 7 [as sides f a rectangle]: [the area is] 210.

Only investigation of the further contexts in which the operation occurs allows us to reject the geometrical interpretation – when prices and amounts of grain occur as "factors", rectangular construction can at most be a frozen metaphor for the process involved;[8] technically, multiplication (though not necessarily any kind of multiplication) must be meant.

In other cases, ambiguity is even harder to kill. The sentence

10 *it-ti* 10 *šu-ta-ki-il-ma* 100

is an appropriate example. *itti* is a preposition meaning "together with", and *šutakil* is the imperative of one of the verbs *šutakūlum* and *šutākulum*, meaning "to make hold each other" and "to make eat each other", respectively.[9] A literal translation will hence be

make 10 and 10 hold/eat each other: 100.

Even mutual "holding"/"eating" thus seems to be a multiplication, apparently nothing but a synonym for "raising". Mutual "holding"/"eating", however, is only used when the "factors" involved are the measures of line segments.[10] In this case, rectangular construction is therefore at least as good an interpretation as multiplication (not least since it is not used when the areas of triangles or trapezia are calculated). If nonetheless we believe that the "lengths" and "widths" of rectangles that our text make "hold" or "eat" each other are merely names for unknown numbers, and the corresponding "surfaces" nothing but a frozen metaphor for their products, we are led to a numerical interpretation of the operation, which confirms our initial numerical hypothesis. But if we take the texts at their words, assuming them to speak about measurable and measured rectangular sides and areas, the geometrical

---

8 As we shall see (p. 22), a frozen geometrical metaphor (though stereometrical and not plane) is in fact involved.

9 It is generally not possible to determine the length of vowels from the syllabic writing; whether one or the other spelling is correct thus depends on the root from which the word is derived.

As we shall see later, the reading *šutakūlum* (possibly *šutakūllum*) is to be preferred, corresponding to a derivation from *kullum*, "to hold".

10 As we shall see, the "only" of this statement is in need of slight qualifications; the discussion of these will have to be postponed until a later stage in the argument.

interpretation results, in its turn confirming the geometric reading. At this stage, nothing but an act of faith allows us to choose between the two readings.

The interpreters of the thirties did choose. Their faith was numerical, and their gospel was accepted. Goetze, an eminent philologist with interest in the mathematical texts, says that a certain text at a certain point

> turns from the square with equal sides (*mitḫartum*) to the rectangle in which at least two sides are of different measurements. This must *of course* be interpreted arithmetically [...].[11]

Even when suggesting that fundamental algebraic identities "like $(a-b)(a+b) = a^2-b^2$" may have been found by means of geometric diagrams, van der Waerden [1962: 71*f*] maintains that

> we must guard against being led astray by the geometric terminology. The thought processes of the Babylonians were chiefly algebraic [i.e., numerical – JH]. It is true that they illustrated unknown numbers by means of lines and areas, but they always remained numbers. This is shown at once in the first example [of the preceding], in which the area xy and the segment $x-y$ are calmly added, geometrically nonsensical.

In exactly the same vein, Neugebauer had argued [MKT II, 63*f*] from the proliferation of "completely nonsensical [*sinnlosen*] inhomogeneous problems" that the main emphasis of Babylonian mathematics was on algebraic, not geometrical relations" – "algebraic" being again understood as "numerical".

If only numbers were involved, however, then the different "multiplicative" terms had to be synonyms – there *is* only one multiplication, as argued by Thureau-Dangin – and it was legitimate to translate all of them into the same operation. Similarly, if uš and sag̃ – the logograms meaning "length" and "width" – were only meant as names for unknown numbers, there seemed to be no problem in understanding them as numerical dummies in the style of the "thing" of medieval Arabic and Italian algebra or in replacing them by letter symbols x and y. Thureau-Dangin's view was that the logograms were read as Akkadian words, and that the texts functioned as the rhetorical algebra of the Middle Ages. Neugebauer argued that the logograms might have been meant as a non-verbal representation, much in the style of modern symbolic algebra. Thureau-Dangin's Babylonian algebra was thus strikingly medieval in character, while the Babylonian algebra which most historians of mathematics found in Neugebauer's works[12] looked

11 [Goetze 1951: 148*f*]; emphasis added.

12 Actually, this was not the "algebra" Neugebauer had put into them; but most readers understood as *interpretations* the symbolic computations by means of which Neugebauer had established the correctness of the Babylonian solutions. His conjecture about the possible function of logograms notwithstanding, Neugebauer remained an agnostic as to the interpretation of the mathematical thinking of the Babylonians, and argued explicitly for agnosticism. That he shared the numerical understanding wholeheartedly and without hesitation illustrates how natural it was

astonishingly modern and similar to ours.

The Texts, the Genre, and the Problems

It is the purpose of the present book to replace this standard interpretation by a less modernizing reading (and to draw the consequences that follow). A first condition for doing so, however, is to know a bit more about the genre which we shall investigate.

The mathematical texts are school texts. They contain no theorems and no theoretical investigations, and to speak of their authors as "Babylonian mathematicians" is therefore misleading unless we take care to remain *very* aware of this difference. They were teachers of computation, at times teachers of pure, unapplicable computation, and plausibly specialists in this branch of scribal education; but they remained teachers, teachers of scribe school students who were later to end up applying mathematics to engineering, managerial, accounting, or notarial tasks (all to be understood within the conditions of Babylonian social and technological practice). We shall return to the impact of this professional situation on the character of Babylonian mathematics and to a sense in which it may still be legitimate to speak of the teachers as mathematicians (see p. 384); for the moment we may restrict ourselves to the observation that all such applications of mathematics as they taught aimed at *finding the right number*.

This preoccupation with numbers is also characteristic of the mathematical texts. A first categorization divides them into *table texts*, *problem texts*, and (recently recognized as a group by Eleanor Robson) *calculations and rough work*. The category of table texts encompasses tables of reciprocals, tables of multiplication, tables of squares, etc., together with tables of technical "constant coefficients" and metrological conversions, to whose use we shall return. "Problem texts" contain mathematical problems; even in the few cases where geometrical figures (regular polygons, squares or other figures inscribed into squares, etc.) are treated, the aim is always to find the measure of certain lines or areas when other dimensions are given.

The problem text category can be subdivided in several (mutually intersecting) ways.[13] One subcategory consists of *procedure texts*. These are texts which start by stating a problem and next explain how to proceed in order to find the solution. Other texts list sequences of questions, perhaps stating also the solution but not the steps that lead to it.

Some procedure texts contain only one or a few problems – in the latter

felt to be at the time.

13 For several of these categories and names I am indebted to Jöran Friberg – after two decades of discussion I am not sure exactly which, but most will be his. The notion of "series texts" goes back to Neugebauer.

case normally variations on a common theme. Other tablets contain many problems. The latter type may either be *anthology texts*, containing problems with scarce mutual connection (neither as concerns the objects dealt with nor the methods used); or they may be systematic *theme texts*, where longer sequences of problems are closely related in one or the other way. Theme texts may be either *catalogue texts* listing only problem statements, or procedure texts. A particular kind of theme texts are the *series texts*, texts that occupy several tablets which together make up a series – mostly written in utterly elliptic logographic writing.

Though not always as elliptic as the series texts, the statements of catalogue texts are generally compact, and offer little possibility for a finely structured format. Procedure texts, in contrast, are often (not invariably) given in such a format, determining among other things a systematic shift between the past and the present tense and between the first, second, and third grammatical person.

Formally, all problems deal with matters which scribes could be expected to encounter in their professional life: the dimensions of fields; rent paid by tenants; profits from trade; the digging of irrigation canals and the building of siege ramps; etc. The questions, however, are often of a kind that could never arise in practice; when would you know, for instance, the profits arising from a commercial transaction without knowing the prices at which you bought and at which you sold?

Some problems certainly correspond to real practice; since these are normally not "algebraic" in character, they do not occur in the following. The others fall in three groups (with some intermediate cases).

Some problems have been derived from practical questions by the inversion of known and unknown entities; if one knows the quantity of oil involved and the buying and selling "rates" (the inverse prices, i.e., the quantity of oil corresponding to 1 shekel of silver), to find the total profit from a transaction is straightforward but no mathematical challenge; finding the rates from their difference and the profit, as in TMS XIII (below, p. 206), is demanding on two accounts: firstly, the problem is of the second degree, and asks for the solution of a second-degree standard "equation"; secondly, seeing without the use of symbolic algebra that the problem belongs to this species and finding the appropriate equation is a challenge in itself.

These problems may deal with entities of any kind; the second group – by far the largest if the number of extant problems is concerned – only deals with rectangular and quadratic fields. These problems are so uniform that one soon forgets their connection to the real world and sees only abstract geometrical problems about rectangles and squares; by invariably using Sumerograms for the lengths and widths of these abstract configurations[14] the authors of the

[14] Only three published texts and one unpublished specimen which I know about are exceptions to this rule – see below, pp. 320 and 324.

texts demonstrate that they saw them in the same way. These problems are, indeed, *functionally abstract*, and the *standard representation* to which other problems are reduced (we shall return to this notion of a standard representation in more detail below, see p. 34). Where *we* would solve the above oil problem by calling the rates x and y and treating them as pure numbers, the Babylonian calculator would treat them as the length and the width of a rectangle, whose dimensions could be determined from the total quantity of oil involved, the difference between the rates, and the profit. In this sense, Neugebauer was right in considering the uš and sag̃ as equivalents of the symbols of modern algebra.

The third group consists of properly geometric problems of a particular kind, of which we shall see a ṇumber of examples in Chapter VI.

Chapter II
A New Reading

An Example

So far everything has been fairly abstract. It is time to look at a real specimen of Old Babylonian "algebra", for which purpose the simplest of all mixed second-degree problems may serve:[15]

1. a.šà$^{l[am]}$ *ù mi-it-ḫ̮ar-ti ak-m[ur-m]a* 45.e 1 PI-*ṣi-tam*
2. *ta-ša-ka-an ba-ma-at* 1 *te-ḫ̮e-pe* [3]0 *ù* 30 *tu-uš-ta-kal*
3. 15 *a-na* 45 *tu-ṣa-ab-ma* 1-⌐e] 1 íb.si$_8$ 30 *ša tu-uš-ta-ki-lu*
4. *lìb-ba* 1 *ta-na-sà-aḫ̮-ma* 30 *mi-it-ḫ̮ar-tum*

Thureau-Dangin [1936a: 31] was the first to publish and translate the problem:

> J'ai additionné la surface et (le côté de) mon carré: 45´.
>
> Tu poseras 1°, l'*unité*. Tu fractionneras en deux 1°: 30´. Tu multiplieras (entre eux) [30´] et 30´: 15´. Tu ajouteras 15´ à 45´: 1°. 1° est le carré de 1°. 30´, que tu as multiplié (avec lui-même), de 1° tu soustrairas: 30´ est le (côté du) carré

and to interpret it:

> On donne: $x^2+x = 45'$
> d'après le scribe,
>
> $$x = -30' + \sqrt{30'^2 + 45} = 30'$$
>
> Il ne fait qu'appliquer la formule type indiquée ci-dessus, à savoir

---

15 BM 13901, obv. I 1–4. BM 13901 is the museum number in the British Museum, conventionally used to identify the tablet. As are many other tablets, it is inscribed on the obverse as well as the reverse; the present problem is found in column I of the obverse, lines 1–4. (For publication data, etc., see the list of tablets, pp. 426*ff*; for the value of non-standard characters, see the "Note on Phonetics" on p. 43).

$$x = \frac{-30' b \pm \sqrt{(30' b)^2 + ac}}{a}$$

Neugebauer, the following year, translated as follows in [MKT III, 5]:

1. Die Fläche und (die Seite) des Quadrates habe ich addi[ert] und 0;45 ist es. 1, den koeffizienten

2. nimmst Du. Die Hälfte (von) 1 brichst du ab. [0;3]0 und 0;30 multiplizierst du.

3. 0;15 zu 0;45 fügst du hinzu und 1 hat 1 als Quadratwurzel. 0;30, das Du (mit sich) multipliziert hast,

4. von 1 subtrahierst Du und 0;30 ist das Quadrat.

The first observation to make concerns the numerical notation. The mathematical texts make use of a place value system with base 60 and no indication of the absolute order of magnitude. The single "digits" are written by means of signs for the numbers 1 through 9 and the decades 10 through 50 (fixed patterns of the wedges meaning *1* and *10*, respectively). The 45 of our text can thus mean $45 \cdot 60^n$, where n can be any integer. In translations it is convenient to indicate the order of magnitude (or, if this cannot be determined with certainty, to choose a coherent, plausible order of magnitude). For this purpose, Neugebauer separated the integer from the fractional part of a number by ";", and other digits by a comma; Thureau-Dangin preferred to generalize the familiar notation for angle measurement, marking "order zero" by ° and indicating decreasing and increasing orders of sexagesimal magnitudes, respectively, by ´, ˝, ‴, ... and by `, ``, ```, ...

In the following I shall follow Thureau-Dangin's notation, omitting, however, the sign ° when it is not needed as a separator. 1`40 thus stands for 100, 1``40` for 6000, 1°40 for $1\frac{2}{3}$, and 1´40˝ for $\frac{1}{36}$. In order to keep as close as possible to the situation of the Babylonian calculator, the reader should pronounce it in all cases as "one-forty", keeping the order of magnitude as silent knowledge.

As regards the mathematical interpretation, the two authors agreed; both took care to translate differently the two verbs *kamārum* (*ak-mur-ma* of line 1) and *waṣābum* (*tu-ṣa-ab-ma* of line 3), but both also considered the two terms as mere synonyms for the same addition of numbers. Similarly, none of them asked whether anything distinguishes the *bāmtum* (*ba-ma-at*) of line 2 from the normal "half" (*mišlum*) occurring elsewhere in the tablet. Both saw that the term *mitḫartum* (lines 1 and 4), which in other contexts stands for the geometric square configuration, must refer to the side of the square. Both saw that the Sumerian expression íb.si$_8$ is a finite verb with root si$_8$, "to be equal". Neugebauer translated it "to have as square root", whereas Thureau-Dangin, believing it to be a logogram for *mitḫartum*, chose "to be the square of".

On one point they disagreed: the PI-*ṣi-tam* of line 1. Thureau-Dangin suggested that the reading of PI might be *wa*. He did not mention the meaning of the word that results (*wāṣītum*, something that sticks out or protrudes, e.g., from a building), which indeed seemed wholly irrelevant. Instead he interpreted very tentatively from what might seem to be the technical function of the word: a specification of 1 as 1° and neither 1\` nor 1´ (etc.). Neugebauer objected that the term is present in all but one of the problems of the tablet with only one unknown, and only there, and suggested hesitatingly that it might refer to the second-degree coefficient of the normalized equation.[16] The occurrence of the word *lìb-ba* ("inside") in line 4 was wholly neglected by both, as was the question of why the half is found in the specific situation of line 2 by an operation of "breaking" (*ḫepûm*).

We may say that the received interpretation made sense of the numbers occurring in the text. But it obliterated the distinction made in the texts between terms which after all need not be synonymous unless the arithmetical interpretation is taken for granted; in some of the more complicated texts it contradicted the order in which operations are performed; and it had to dismiss some phrases as irrelevant (e.g., *libba*) or to explain them by gratuitous *ad-hoc* hypotheses (e.g., *waṣītum*).

Alternatively, one might try the hypothesis that the text is meant to say what it says, that is, that it deals with a square with unknown but measurable side and area, and not with the sum of an unknown number and its second power. We may also take advantage of the fact that the "multiplication" *šutakūlum* is only used when the "factors" are line segments, and deduce that the operation is likely to correspond somehow to the construction of a rectangle – which means that not only the problem but also the procedure is geometric. If we do so, coining furthermore a word "confrontation" corresponding to the etymology of the word *mitḫartum* ("a confrontation of equals", *viz.*, the square configuration parametrized by its side) and introducing the "moiety" as the translation of the anomalous term for the half, we are led to something like the following:

1. The surface and my confrontation I have accumulated: 45´ is it. 1, the projection,
2. you posit. The moiety of 1 you break, 30´ and 30´ you make hold each other.
3. 15´ to 45´ you append: by 1, 1 is the equalside. 30´ which you have made hold
4. in the inside of 1 you tear out: 30´ the confrontation.

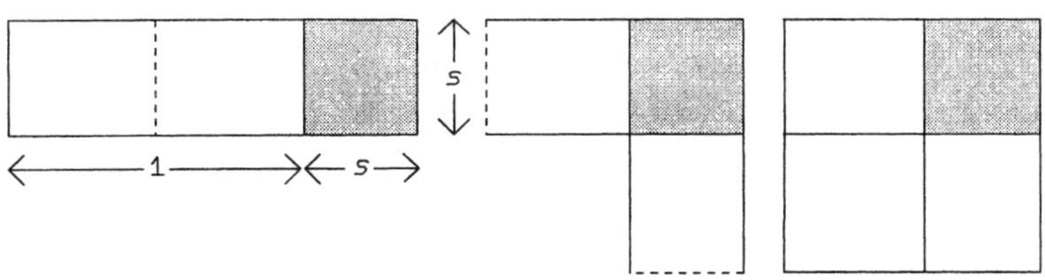

Figure 1. A possible geometric procedure for BM 13901 #1; distorted proportions.

---

16 Unfortunately, the term is always used about the coefficient of the first-degree term, except in one case which is corrupt anyhow.

A possible interpretation of these lines is shown in Figure 1 in slightly distorted proportions: The square □(*s*) represents the surface; from this a projecting line 1 is drawn ("posited"); together with the side this "projection" contains a rectangle ⊏⊐(1,*s*), whose surface is evidently equal to the side *s*.[17] According to line 1, the total surface of square and rectangle is thus 45′. "Breaking" the "projection" into two parts 30′ and 30′ and making these "hold" each other as sides of a rectangle (indeed a square) produces a completed gnomon, whose surface is 45′+30′×30′ = 45′+15′ = 1, which is flanked by the "equalside" 1. "Tearing out" that 30′ which was moved around in order to "hold" leaves 1–30′ = 30′ as the (vertical) side *s* of the original square.

So far this is nothing but a *conceivable* interpretation, a possible alternative to the received numerico-algebraic interpretation, which, however, has the advantage of following the present text more closely. In the following we shall see that it fits the whole Old Babylonian text corpus, not only resolving the recognized anomalies but also explaining phenomena which had never been detected because of faith in the traditional reading of the texts. Since the late texts have distinct characteristics, we shall concentrate for a while on the Old Babylonian material, postponing the treatment of the Late Babylonian and Seleucid texts to Chapter X.

Structural Analysis and Close Reading

Before we get that far, we shall have to introduce two principles of interpretation. One is that of "structural analysis", the other that of "close reading". Both may be exemplified by the problem just discussed.

The principle of "structural analysis" consists of observing not only what is done by each particular operation – for example that "raising" 30′ to 7 brings forth 3°30′, i.e., 7·30′ – but also registering in which situations the operation in question is used in the total corpus *and in which not*. This kind of analysis was rarely made in the early years, but it can be exemplified by Neugebauer's argument against Thureau-Dangin's conjectural interpretations of the *waṣītum*. The observation that mutual "holding" always involves line segments is another application of the same principle

The rule of "close reading" consists of taking the texts and not only the numbers seriously and in being attentive to their details, to the variable contexts in which each term occurs, and to the organization of procedures compared to alternative possibilities which are *not* used;[18] if 30′ is "torn

---

17 Here and in the following, □(*s*) designates the square with side *s*, and ⊏⊐(*l*,*w*) the rectangle with length *l* and width *w*. Numerical multiplications which correspond to a rectangle construction will be written with the sign ×.

18 A similar principle has been advocated by Karine Chemla [1991] as a tool for analyzing the methods of ancient Chinese mathematics.

out", not simply from "1" but "from the inside of 1", the first hypothesis should be that this peculiar expression is used with purpose and carries a meaning or at least is meant to suggest a connotation, *not* (as actually presupposed in the two translations quoted above) that it is empty talk.

Properly speaking, such a reading can of course only be achieved on the transliterated original text; if the analysis refers to a translation, even approximate compliance with the precept puts strong constraints on the way the translation has to be made. Neugebauer [MKT III, 5 n.20] explained his choice of what he considered as "*substantially* adequate" ("*sachlich* adäquaten") instead of literal translations by the sarcastic observation that "who intends to study the history of terminology by means of a *translation*, *he* is anyway beyond salvation". As a first approximation to the texts – an expression used elsewhere by Neugebauer – this may have been a sound step, but the consequence was that general historians of mathematics, believing that terminology and mathematical contents could be separated, based their understanding on translations from which almost everything that might contradict the established interpretation had been omitted as "substantially irrelevant".

Structural analysis and close reading affect the interpretation but should of course not change the text itself. Because of the ambiguities of cuneiform writing (and because tablets are often broken or damaged, which may make the reading difficult and require that lost signs, words, or passages be restored), it might, however, touch the transliteration. It shows the extrème finesse of workers like Neugebauer, Thureau-Dangin, and Sachs that their failure to recognize explicitly the conceptual distinctions which are revealed by the structural analysis never made them err and choose the wrong operation when restoring a lost word. In what follows, my transliterations of texts originally published in [MKT], [TMB], and [MCT] are thus almost identical with what is found in these wonderful volumes; what is new belongs at the level of interpretation, and results from explicitation of what the founding fathers often seem to have known intuitively without *knowing that* or *precisely why* they knew it, and which was therefore lost when the following generation of historians used their work.

Numbers and Measures

The numbers of the mathematical text are mostly written in the sexagesimal place value notation, which was already presented above.[19] In certain cases,

19 Actually, the Old Babylonian calculators seem to have seen it rather as a mixed decimal-seximal system. Two texts from Susa (TMS XII and XIV) indicate missing digits by means of a special sign, a so-called "intermediate zero"; but these do not indicate a missing sexagesimal place but missing tens or ones. Even

however, phonetically written number words occur; in others numerals are provided with phonetic complements indicating which number word in which grammatical form (e.g., the genitive of an ordinal number) is intended. The translation of the place value numbers has already been explained (p. 12). Number words will be translated as number words, and mixed writings in a similar mixed writing (e.g, "its 7th").

For practical use, other number notations were in use – neither economic texts nor legal documents could employ a system which did not distinguish 8 g (1 shekel) of silver from 30 kg (1`` shekel = 1 talent). Evidence exists that practical reckoners used the place value system for intermediate calculations, much the same way as late medieval reckoners used the Arabic numerals for computations but entered the results in Roman numerals – less easily tampered with – in the ledgers. The mathematical texts make little use of the notations by which practical reckoning eliminated numerical ambiguity; after all, they mainly served as *aides-mêmoire* and not as independent information. In one of the texts analyzed below (VAT 7532, see p. 209), however, we shall see that 60 is written not as 1 (meaning 1`) but as "1 *šūši*", meaning "1 sixty".

A few simple ("natural") fractions – $\frac{1}{2}$, $\frac{2}{3}$, $\frac{1}{3}$ – are often written by means of special logograms; they will be translated in modern fractional notation, as $\frac{1}{2}$, etc. When written syllabically, they are translated "one half", etc. Higher aliquot parts may be expressed by means of the corresponding ordinal numbers, as indeed with us; depending on whether the ordinal is written in full syllabic writing or as a numeral followed by a complement, such expressions will be translated (e.g.) as "the seventh" or "the 7th".[20]

All kinds of metrological notations turn up in the mathematical, and even in the "algebraic" texts.[21] Some of them, however, only turn up rarely in the latter group – capacity measures, for instance, we shall encounter in the

---

without a separation sign, 30 16 could never be read as 46, nor 30 41 as 71 or 14 3 as 17 – the texts that insert zeroes thus do so not in order to avoid erroneous readings but for the sake of system – and that system is obviously decimal-seximal and not sexagesimal *stricto sensu*.

20 A few mathematical texts include multiplicative or additive-multiplicative combined fractions, of the types "parts of parts" ($\frac{1}{p}$ of r, where r itself is an aliquot part or $\frac{2}{3}$) or "ascending continued fractions" (r, and q of r, where also q may be an aliquot part or $\frac{2}{3}$); the text TMS V contains a section where parts of parts are explored systematically, elsewhere both kinds of composite fractions seem mostly to be used when normal notations fail or become too clumsy (but in YBC 4714, as we shall see, as a way to express $\frac{1}{6}$ by means of "natural" fractions, as "half of the 3rd part"). The scattered occurrences of these expressions combined with their importance in Arabic mathematics suggests that they will have been part of general Akkadian parlance without being granted franchise in the school – cf. [Høyrup 1990c].

21 [Powell 1990] is a detailed survey of Mesopotamian metrologies from the late fourth through the first millennium BCE; a convenient survey of most of those metrologies that are important in the mathematical texts can be found in [MCT, 4–6].

problem dealing with the buying and selling of oil. Such metrologies are best presented in the context where they turn up.

The metrologies belonging with the "standard representation" (see p. 10), on the other hand, deserve a general presentation.

The basic measure of horizontal distance is the nindan ("rod"), equal to c. 6 m. Mostly, this unit is not written but remains implicit. The nindan is subdivided into 12 kùš ("cubits") of c. 50 cm, and the kùš into 30 šu.ši ("fingers").

In practical agricultural computation, the most important area unit was the bùr, some 6 ha. The basic unit of the mathematical texts, however, – the unit tacitly understood when no unit is written – is the square nindan or sar (that is, 1 sar ≈ 36 m^2). The bùr equals 30\` sar.[22]

For vertical extensions, the basic measure is the kùš. The corresponding measure for volumes is 1 nindan^2·1 kùš, that is, an area of 1 sar provided with a standard thickness of 1 kùš. This volume, like the area forming its base, carries the name 1 sar.[23] For the sake of comprehensibility it may at times be necessary to refer to it as a "[volume] sar", but as a rule it is preferable to keep as close to the original text as possible in all cases where the context makes clear which unit is meant.

It should be kept in mind that our very concept of a metrological "unit" does not map the Babylonian usage too well. Often, it is true, a number followed by the name of a unit stands for that number of the unit in question. As already stated, however, the so-called "basic units" are often omitted, the user of the text being supposed to know that lengths when nothing else is indicated are measured in nindan, etc. In order to avoid the ambiguity arising from the indefinite absolute order of magnitude of sexagesimally written numbers, the name of a unit may be used much as *semantic determinatives* are used to eliminate the ambiguities of the cuneiform script. In the texts below we shall encounter, e.g., "10 nindan" meaning "10′, [of the order of magnitude of the] nindan";[24] "5 kùš", to be understood as "5′, [nindan, i.e., 1] kùš"; and "5 1 kùš" with the equivalent interpretation "5′, [nindan, i.e.,] 1 kùš".[25] In the last example, kùš occurs as a unit in our sense, in the other two the units function as determinatives.

The system of "basic units" belongs together with the use of the sexagesimal system in technical computation. Its function derives from the fact

---

22 Intermediate area units are the iku = 1\`40 sar, to be understood as □(10 nindan); and the ešè = 10\` sar = ▭(10 nindan, 1\` nindan). Units above the bùr also exist.

23 The sar also serves as a brick measure (12\` bricks, the number of bricks that, if they were of a certain type, would fill a volume sar). Since no brick problems are examined in the following, this need not concern us here.

24 "10′ nindan" might seem a more straightforward interpretation, but this would be written simply as "10".

25 The phrases quote the tablets BM 13901 (rev. II 16) and BM 85200 (rev. I 16, and rev. II 14).

that the metrological sequences were not always arranged sexagesimally: thus, as we have seen, the nindan is subdivided into 12 kùš and the kùš into 30 šu.ši – no factor 60 occurs. If, for instance, a platform had to be built to a certain height and covered by bricks and bitumen, a "metrological table" had to be used to transform the different units of length into sexagesimal multiples of the nindan and kùš, allowing the determination of the surface and the volume in the basic units sar and [volume] sar. A list of "constant coefficients" (igi.gub) would give the amount of earth carried by a worker in a day over a particular distance, the number of bricks to an area or volume unit, and the volume of bitumen needed per area unit – all expressed in basic units (if no transformation into basic units had taken place, different coefficients for the bitumen would have had to be used for small platforms whose dimensions were measured in kùš and for large ones measured in nindan). With these values at hand the number of bricks and the amount of bitumen as well as the number of man-days required for the construction could be found by means of sexagesimal multiplications and divisions – once again facilitated by recourse to tables, this time tables of multiplication and of reciprocal values. Finally, renewed use of metrological tables would allow the calculator to translate the results of the calculations into the units used in technical practice.[26]

Below we shall encounter this kind of transformation in the tablets VAT 8389 and 8391.

Mathematical Operations

The results of the close reading can only be adequately demonstrated on actual texts; moreover, as was argued, only on texts in the original language or in a translation which maps the structure of the original text very closely – a "conformal translation". The gross outcome of the structural analysis, on the other hand, can be summed up on its own, and even has to be if the principles of the translations used in the following are to be set out in advance.

The text corpus embraces texts from a couple of centuries and from several localities and schools. The overall features of the terminology and organization of the texts are shared, but the finer details vary. In order not to make the picture too perplexing, discussions of the details are printed in brevier or (if brief) relegated to the footnotes. Such passages can be skipped during a first reading.

---

26 But as Eleanor Robson comments on this last sentence, "in reality, the scribe should be able to do at least the metrological conversions without recourse to tables, having learned them ad nauseam at school".

Additive Operations

The terms traditionally understood as synonymous names for addition turn out to fall into two distinct groups, and thus to cover two different operations. The most obvious distinguishing feature is that only terms from one of the groups will occur when a quadratic completion is performed. The other group dominates when lengths and areas or areas and volumes are added; however, in specific text groups and under specific circumstances to which we shall return, terms from the first group may occur.

The first group contains the Akkadian term *waṣābum* (AHw "hinzufügen") and the Sumerogram d a ḫ, which functioned as a logogram for the Akkadian word. Both will be rendered "to append" in the conformal translation (see below, p. 40). The operation is asymmetric, one entity being always "appended *to*" (*ana*) another; and it is only used about concretely meaningful "additions", where one of the "addends" is absorbed into the other, which so to speak conserves its identity while increasing in magnitude.[27] As a conceptual model one may think of the addition of interest to a bank account, which remains *my* bank account in the process. The example is no anachronism; the Akkadian term for "interest", indeed, is precisely *sibtum*, derived from *waṣābum* and meaning "the appended".

No particular term for the corresponding "sum" seems to exist; nor is there any obvious need for it, in view of the "identity-conserving" character of the operation.

The other operation is symmetric, and does not presuppose the addition to be concretely meaningful. It adds the measuring numbers of two or more addends, connecting them with the word *u* ("and"), and may thus be regarded as a genuine arithmetical operation. The Akkadian main term is *kamārum* (AHw "schichten, häufen"), to which correspond the Sumerian term g̃ a r . g̃ a r and the unexplained logogram UL.GAR. No text contains the slightest hint that these were not simply used as word signs for *kamārum*, and I shall therefore use the same translation for all three, *viz.*, "to accumulate".

With this operation, the sum has a name: *kumurrûm* (derived from *kamārum*), with the logograms g̃ a r . g̃ a r and UL.GAR. I shall translate all three as "the accumulation".

Occasionally, other terms derived from *kamārum* turn up: one (used in VAT 8520) is *nakmartum*, which we may translate "the accumulated", since the grammatical form

---

[27] Cases where a linear extension is "appended" to an area might seem to contradict this concretely meaningful character of the operation. The explanation is that the surveyors' tradition from which the scribe school had borrowed its inspiration operated with a notion of "broad lines", lines provided with a virtual standard breadth of 1; cf. below, note 76 and p. 291.

may involve the idea of a process – that of accumulating – seen from the end point. Another is *kimrātum*, which offers the peculiarity of being a plural (found in AO 8862, see below, p. 162). It seems to refer to the sum as compounded from still identifiable constituents; I shall use the translation "the things accumulated".

In a few texts (AO 6770, YBC 4714, and BM 85200+VAT 6599, below, pp. 179, 111, and 137), *u* alone is used now and then for the process of addition or for the sum ("length and width, 50′"). Moreover, in one of these texts (BM 85200+VAT 6599), an abbreviated form of the term used for the sum total in accounting is employed twice (nigin, translation "total"); in both cases, two *numbers* – *viz.*, a pair belonging together in the table of reciprocals – are added.

Subtractive Operations

Even "subtractions" are of two kinds, "removal" and "comparison". Removal is always concrete, and is the inverse of *waṣābum*, "appending". The main term is *nasāḫum* (AHw "ausreißen"), "to tear out", with the Sumerogram zi. Like *waṣābum*, *nasāḫum* is identity-conserving in the same sense as my bank account remains *my* bank account when a payment has been made, only with a reduced balance. It can be used only when the subtrahend is really *part* of the entity from which it is subtracted. What is left may be spoken of as *šapiltum*, "the remainder", derived from the verb *šapālum*, "to be(come) low, deep, small", which agrees well with this conservation of identity.

Speaking of *nasāḫum*/zi as the "main term" implies that others exist, which is indeed the case.[28] Depending on special circumstances (through connotations rather than in precisely defined ways), the texts may use *ḫaraṣum*, "to cut off"; *tabālum*, "to withdraw"; and *šutbûm*, "to make leave".

Some texts reveal a tendency to "cut off" from linear entities and to "tear out" from areas; others may "tear out" from a line and "tear out inside (*libbi*)" an area; most texts, however, use the *libbi* (or the similar phrases *libba* or *ina libbi*, "from the inside") indiscriminately in both cases or do not use it, and *ḫarāṣum* is not in common use at all.

tabālum, "to withdraw", has a general connotation of reclamation by legal action. In one case (TMS V) it occurs in a "dressed" problem dealing precisely with such a situation, and in another (YBC 4608) to describe the removal of one side from what is already known to be the sum of two opposing sides of a quadrangle – that is, so to speak, removal of what can be "justly" removed.

In extra-mathematical contexts, *šutbum* is often used when you remove something or make somebody go away that is somehow due to be removed or go away: making workers go out for work; removing guilt, demons, or garbage; taking a statue from its pedestal for use in a procession. In a few mathematical texts it is used within arguments by "single false position", describing the removal of the due fraction from the model magnitude (thus VAT 7532, see p. 209).

---

28 A detailed treatment of the use of these and other "subtractive" terms with precise reference to texts is contained in [Høyrup 1993b].

Comparison is also a concrete operation, and used to say how much one magnitude *A* exceeds another magnitude *B which it does not contain*. It is thus *no* inverse of *kamārum*, "accumulating", and cannot be the reversal of any addition (since the sum always contains the addends).[29] The phrase in use states that *A eli B d itter/īter*, "*A* over *B*, *d* it goes/went beyond" (from *eli* ... *watārum*, "go beyond", "be(come)/make greater than"),[30] with the Sumerographic equivalent *A* ugu *B d* dirig. In some texts dirig is also used as a logogram for the excess, that is, for that amount *d* by which *A* "goes beyond" *B*.

In a few cases, comparisons are made the other way round, saying not by how much *A* exceeds *B* but instead by how much *B* falls short of *A*, using the verb *maṭûm*, "to be(come) small(er)" (Sumerogram lal). This possibility is used, either in order to obtain one of the preferred relative differences (e.g., $\frac{1}{7}$ instead of $\frac{1}{6}$, cf. p. 59)[31] or because constraints of format require the smaller magnitude to be mentioned first. The widespread legend that the Babylonians made use of negative numbers comes from misreading of Neugebauer's treatment of the topic.[32]

When translating these operations into symbols, I shall render "*q* torn out from *p*" (with synonyms and other grammatical forms) as $p-q$; that "*p* over *q*, *d* it goes beyond" I shall render $p = q+d$; that *p* falls short of *q* by *d* will become $p = q-d$.

"Multiplications"

Four different groups of terms have traditionally been understood as "multiplications".

The group which most clearly deserves the name contains only one member, the Sumerogram a.rá (from RÁ,[33] "to go"). This is the term used

---

29 Because of the symmetric character of the accumulation operation, its actual inverse is the *splitting into* or *singling out of* components (*bêrum*, cf. p. 407).

30 I use this somewhat clumsy translation in order to be able to render the grammatical structure of the phrase (including case and the use of prepositions) precisely. The meaning would be just as well served by the translation "*A* exceeds *B* by *d*".

31 On the preference for certain factors and relative differences, see [Høyrup 1993a], and below, p. 287.

32 Neugebauer never spoke of Babylonian negative numbers. What he did was to render as precisely as possible the distinction between "*A* over *B*, *d* it went beyond" and "*B* (compared) to *A*, *d* it was smaller" when translating the phrases into algebraic formulae, as "$A-B = d$" and "$B-A = -d$", respectively. Cf. below, p. 294, and [Høyrup 1993b: 55–58].

33 Actually, the verb takes on a number of different forms depending on grammatical number and aspect [SLa, §268]: g̃en and du (singular perfective and durative),

in the tables of multiplications, that is, for the multiplication of number by number. The phrase is *a* a.rá *b*, meaning "*a* steps of *b*".

This grammatical interpretation follows from the usage of the Seleucid tablet BM 34568, where a question is asked repeatedly which in modernized translation becomes "what times what shall I take in order to get *A*?" It always has the former factor in the nominative and the second in the genitive case.

The evidence is very late, but a.rá belongs to that part of the terminology which remained in use from the Old Babylonian through the Seleucid epoch. Moreover, the metaphor of "repeated going" expressed in Akkadian (*alākum*) and used in a general way (for multiplication as well as repeated "appending") is found in various Old Babylonian texts from Susa.

The core term in the next group is *našûm*, "to raise", with the Sumerogram íl. Another Sumerogram used in the same function is nim, logographically connected to *elûm* and *šaqûm* and their various derivations (both "to be/become/make high")"; I shall translate it "to lift"; in one mathematical text (Str 368) it is used alternatingly with *našûm*.[34]

These terms designate the determination of a concrete magnitude by means of a multiplication. They are used for all multiplications by technical constants and metrological conversions; for the calculation of volumes from base and height; and for the determination of areas when this is not implied by the construction of a rectangle (that is, for triangles, for trapezia and trapezoids, and even for rectangles which *are already there*).

The original use of the term turns out to be connected to the determination of volumes. In these, indeed, the base is invariably "raised" to the height; in all other cases, the order of the factors is random from a mathematical point of view, and depends first of all on stylistic criteria – as a rule, it is the quantity that has been computed in the preceding sentence that is "raised" to the other factor, irrespective of its meaning or role in the computation – cf. [Høyrup 1992: 351*f*].

As we have already seen, an area *B* sar was understood as the carrier of a "virtual height" of 1 kùš and thus to represent also a volume of *B* [volume] sar. A prismatic volume with base *B* and height *h* could therefore be understood as resulting from the "raising" of the virtual height of the base to the real height. The original usage hence seems to reflect a very obvious imagery. From there, the term will have been transferred to similar functions by analogy – first perhaps to the determination of areas, which might be understood as composed from unit strips, similar to the unit slices from which volumes were composed, but eventually to all multiplications referring to considerations of proportionality. Although it is not obvious from *our* way to conceive of lengths, areas, and volumes, "raising" is thus a *category-conserving* multiplication.

It cannot be excluded that the first use of the metaphor did not concern volume computation in general but the particular case of brickwork calculations: *ullûm*, D-stem of *elûm*, is precisely the term used when a building is made higher and when

re$_7$ and su$_8$.b (plural ditto). Since both g̃en and du are written DU = RÁ, it will be convenient to write the verb as RÁ in order to keep present the relation with the term a.rá (which should certainly be pronounced this, since it gives rise to the Akkadian loanword *arûm<*ara-um*.

34 As observed above, *našûm* means "to carry" as well as "to raise". The equivalence with nim, however, leaves no doubt about the correct interpretation.

brickwork is elevated *n* brick layers (AHw, 208b); similarly *šuqqûm*, D-stem of *šaqûm*, is used about the raising of buildings (AHw, 1180b–81a). But as Eleanor Robson asks me, "how else would you express it in Akkadian?"

The third group is constituted by the term *eṣēpum*, "to double", with the Sumerographic equivalent tab (general meaning "to be/make double", "to clasp to", "to duplicate"), and with some affinity to the Susa use of *alākum*, "to go". The operation is one of concrete repetition, often said to be *ana n*, "until *n*" ($2 \leq n \leq 9$), for which reason I shall translate it "to repeat"/"to repeat until *n*".

The fourth group of supposed multiplications is better dealt with under a different heading.

Rectangularization, Squaring, and "Square Root"

This group, indeed, represents no multiplication at all, neither of number by number nor of [measured] quantity by number or quantity; what it refers to is the geometric operation of *rectangularization*. The central term is *šutakūlum*, "to make [two segments *a* and *b*] hold each other", *viz.*, as the sides of a rectangle ⊏⊐(*a*,*b*) – at times grammatical constructions are used which rather imply that *a* together with *b* contain or "hold" the rectangle, for which reason I shall use the shorter translation "to make hold". Occasionally, when "*a* and *b* I have made hold", the next line says that "a surface I have built".

By most workers (Thureau-Dangin being the chief exception), the term has traditionally been interpreted instead as *šutākulum*, "to make [*a* and *b*] eat each other". The cuneiform writing does not allow us to distinguish, but the logographic use of the Sumerogram ì.gu_7.gu_7[35] (in certain texts abbreviated gu_7, "to eat") has been taken to decide the question. A stronger argument points in the opposite direction, however. In the text quoted above (p. 11), one of the sides was provided with the epithet *ša tu-uš-ta-ki-lu*, "which you have made hold/eat". Other texts use a noun *takīltum* instead of the relative clause but with precisely the same function; this term can only derive from *kullum* and mean "which is made hold". For brevity, I shall translate it as "the made-hold".

This might seem to make the use of ì.gu_7.gu_7 look strange, as a pun based on the phonetic near-identity between *šutākulum* and *šutakūlum*. The use of such "puns" or rebus-writings, however, is one of the key principles on which the cuneiform writing is based (above, p. 4, we have encountered the writing of .bi, "its", and be, "it is said", by the same sign). The logographic use of gu_7 is hence no weighty counterargument against the derivation from *kullum*.

---

35 gu_7 seems to be a better reading than kú, which is used in the classical editions of the mathematical texts – cf. [Borger 1967].

As long as the operation was interpreted as a mere multiplication of numbers, the two metaphors were equally strange. "Holding" only becomes an image near at hand when it is understood that the "building" or setting-out of a rectangular surface is involved.[36]

Beyond ì.gu$_7$.gu$_7$ and the abbreviation i.gu$_7$, several other logograms for this operation are in use: UL.UL (probably to be read du$_7$.du$_7$ for *nitkupum*, "to butt each other"; misidentified as ZUR.ZUR in [MKT]); UR.UR; and NIGIN, which can be interpreted as a contracted LAGAB.LAGAB. In all four cases, the reduplication is probably not to be understood as a genuine Sumerian grammatical form but rather as a way to render in pseudo-Sumerian the reciprocity of the process that the Akkadian language presents by means of the Št and Gt-stems (*šutakūlum* and *nitkupum*, respectively).

It may not be warranted at all to speak of UR.UR, UL.UL, and NIGIN as logograms in the proper sense – their use varies slightly from that of related Akkadian terms, and they are at times embedded in somewhat different phrase structures. It may be better to see them as genuine *ideograms*, in the likeness of "+" and "=". Ideograms will (and would) of course be read in words when a written text is rendered orally; but their semantic range does not coincide with a particular spoken word, and the process is indeed one of translation; whether "+" is translated "and", "plus", or "added to" depends on context, stylistic feeling (and accident), and it is not possible to declare one translation correct and the others erroneous.

Since the relation between the terms is not fully transparent, I shall translate ì.gu$_7$.gu$_7$ (which is an indubitable logogram for *šutakūlum*) "to make hold"; UR.UR will be translated "to oppose", and UL.UL/gu$_7$.gu$_7$ "to make encounter". NIGIN, finally, I shall render "to make hold", because certain texts refer backwards to a NIGIN in the statement with a syllabic *šutakūlum* – see below, note 220.[37]

It may seem a challenge to the interpretation of these terms as rectangle constructions that they are also used when a circular perimeter is multiplied by itself (and next "raised" to 5′ in order to give the area). Most likely, the origin of this procedure is a traditional practitioners' rule of thumb, according to which the cross-section of a cylinder (a log, etc.) was found as $\frac{1}{12}$ of the square whose side was the string that went around the cylinder; in any case, when the area of a semi-circle is found from the product of arc and diameter ($A = \frac{1}{4}a \cdot d$), "raising" is used for both multiplications, as one might expect in a generalization of the computation of a triangular area. The origin of the former rule in a non-school environment is also suggested by the way the perimeter is found from the diameter in cases where the latter magnitude is given; in these instances, the perimeter is always "tripled" (*šullušum*), never "raised to 3". This idiosyncratic formulation follows certain practitioners' traditions until the late Gothic master builders, and is probably conserved because it corresponds to a concrete operation of repetition – see [Høyrup 1997: 87].

---

36 In one text, gu$_7$ as well as *šutakūlum* are used to describe the construction of non-rectangular quadrangles (YBC 4675, obv. 1, rev. 15; below, p. 244). Here, any interpretation as a multiplication is wholly out of the question. The first line explains indeed that the quadrangle is held by "length and (different) length", i.e. that the quadrangle is irregular.

37 LAGAB – then to be read nig̃in – can serve as a Sumerogram for *lawûm*, "to surround", which might suggest the reading of NIGIN – a contracted LAGAB.LAGAB – as nig̃in, "to make surround", which gives some sense. In the context of rectangularization and squaring, however, LAGAB is likely to be iconic and no Sumerogram, the sign *being* nothing but a square □.

Several tablets play with repeated multiplication, but no specific term for (and, we may infer, concept of) powers *of numbers* exists. *Squaring*, as a specific process, is *geometric* squaring.

As already stated (p. 12), the term for the square configuration is *mitḫartum*, a term which refers to a confrontation of equals (*viz.*, equal sides) and which when expressed as a number coincides with the side of the square. This has caused some bewilderment, and even some disdain for Babylonian mathematical though: apparently, the Babylonians did not know, or were unable to express, the difference between a square and a square root.

Bewilderment and disdain are ill founded: our own concept is just as ambiguous, only different. To us, in accordance with Euclid's *Elements* (I, def. 14 and 22), a square is a "figure", i.e., something which is "encompassed by some boundary or boundaries". As a consequence, the square *is* an area and *has* a side. A Babylonian *mitḫartum*, on the other hand, understood as a confrontation of equal sides and thus as the square frame, *is* the side and *has* an area.

mitḫartum derives from the verb *maḫārum*, "to confront" – often in the sense of confronting a peer, an equivalent, or a "counterpart" (*meḫrum*, from the same root); the Št-stem (causative-reflexive) of the same verb, *šutamḫurum*, "to make *s* confront itself", designates the construction of a square with side *s*.[38] Often, however, one of the terms for rectangularization is used instead, and the texts ask, e.g., "to make *a* and *a* hold", "to make *a* hold together with itself" or, more rarely, simply "to make *a* hold".

Some texts use LAGAB and NIGIN (a single and a repeated square, we remember) as ideograms (not necessarily logograms) in the same functions as *mitḫartum* and/or *šutamḫurum*. Some series texts use íb.si$_8$ in the same roles;[39] however, the normal use of this important term is different.

Originally it is a Sumerian finite verb form, and it often occurs as a verb in phrases "*Q*.e *s* íb.si$_8$", where Q and s are numbers, $s = \sqrt{Q}$. .e is either the "ergative" or the "locative-terminative" suffix",[40] si$_8$ a verb stem meaning

---

38 These terms and their close mutual connection are not restricted to mathematical discourse; in the Gilgameš epic, Enkidu is spoken of as the *meḫrum* of Gilgameš, and the predicted fighting between the two peers in strength is spoken of with a term (*ittamḫaru*, "they confronted each other") close to *šutamḫurum* (Old Babylonian version, second tablet, V 27 and VI 11, respectively [ed. Thompson 1930: 23*f*]).

39 A single procedure text, moreover (YBC 6504), uses íb.si$_8$ in parallel with du$_7$.du$_7$, in a construction where it might function logographically for *šutamḫurum*. In a catalogue text dealing with the division of squares into other geometric figures (BM 15285), it alternates with *mitḫartum*.

40 Sumerian is an "ergative" language, that is, the subject of intransitive verbs and the object of transitive verbs appear in the same (unmarked) case, whereas the agent of transitive verbs appears in a marked agentive or "ergative" case – the marker being the suffix .e. In a first approximation, an ergative language can be understood as one in which the unmarked voice is the passive – cf. the sentences "I sleep" and "I am beaten", where the agent of the latter may be identified

"being equal"; íb combines a mark of finiteness /ĩ/ with the "inanimate pronominal element" /b/ – *s*, indeed, is no person. In total, the phrase therefore has to be translated "alongside *Q*, *s* is equal", or, as I shall do in the following in agreement with the inherent ambiguity of the suffix, "by *Q*, *s* is equalside" (employing a term which allows to put emphasis both on the idea of *side* and on the total configuration of *equal* sides[41], and leaving out the article in order to intimate that what is a noun in English corresponds to an "adjectival verb" in the original). Many texts demonstrate that this underlying Sumerian phrase was well remembered: when they ask for *s*, the interrogative pronoun *mīnûm* occurs in the accusative form (*mīnâm*).[42] The meaning is that when the area *Q* is laid out as a square, it is flanked by *s* as side of this square (equal of course to the other sides). Other texts have left behind the etymology, and use íb.si$_8$ as a noun, the name for this side (to be translated in the following as "the equalside", with inclusion of the article) – so to speak the geometrical equivalent of a square root.[43] As we shall see in Chapter IX, this difference may be one of the keys that allow us to reconstruct the process in which Old Babylonian mathematics was shaped.

---

through an additional "by him".

In all likelihood, this use of .e is derived from its use as an indicator of close proximity, as a marker of the "locative-terminative" case [SLa § 170] – meaning approximately "close to"/"near by" (the underlying idea being, exactly like in the English agentive use of "by", that the one who is close to an action is probably responsible).

41 The translation "equalside" was proposed by Jöran Friberg. In earlier publications I have used instead "equilateral" (even this originally proposed by Friberg), which, however, is less apt to render the inherent ambivalence of the concept involved.

In earlier publications I have also blindly accepted the interpretation of .e as an agentive suffix. This was originally proposed by Thureau-Dangin, and was indeed the only possibility within the arithmetical reading – *the number* 9 cannot meaningfully be claimed to be close to *the number* 81.

42 This shows that the Old Babylonian calculators had reinterpreted the phrase from the point of view of Akkadian grammar. Akkadian, in contrast to Sumerian, is an "accusative" language like English, and treats all subjects alike, having the object in a marked accusative case – cf. "I sleep" and "he beats me". Akkadian scribes would therefore understand a noun carrying the suffix .e as a subject, and the unmarked noun as an object, and therefore understand its case as an accusative.

That this is an error with regard to the original meaning follows from the usage of a couple of texts from Ur, probably from the nineteenth century BCE (UET V, 859 and 864,): they both ask en.na(m) ba.si$_8$.e, but the former uses a.na.àm for the accusative (namely, for the object of the verb g̃ar). This distinction only gives meaning if the author made a different grammatical reconstruction (even this one from the perspective of Akkadian grammar), identifying the cases of transitive and intransitive subjects, and seeing the resulting number as an intransitive subject.

43 In texts where the use of the undeclined Sumerogram en.nam does not allow us to distinguish cases, the word order shows whether a verb or a noun is meant, both Sumerian and Akkadian being verb-final. en.nam íb.si$_8$ thus means "what is equalside?", whereas íb.si$_8$ en.nam means "the equalside, what?"

Quite a few of the texts that use the term as a noun employ the homophonic (unorthographic) writing íb.si. Other texts use ba.si or ba.si$_8$ (the latter form also occurs as a verb); since a distinction between the different writings turns out to be informative, I shall translate the ba-forms as "(be) equilateral". The shift si$_8$>si simply shows that the etymology is no longer thought of, but the pronunciation still Sumerian (Thureau-Dangin, in contrast, believed íb.si$_8$ to be a logogram that was pronounced *mitḫartum*; in a few cases this is indeed the case, but mostly not); ba is an alternative prefix which probably contains a locative element /a/, indicating that the process takes place "(out) there".[44] At times, the noun appears as an Akkadianized loanword, *basûm*, further proof that the pronunciation remained Sumerian.[45]

The function of the "equalside" was not restricted to the case of squares. That it was also used about the side of a cube should not be astonishing – after all, even a cube is flanked by equal sides. But true generalizations also occur, somewhat similar to our reference to the "root(s)" of an equation, derived in twisted ways from the concept of a square root. We shall encounter some examples below.

When the term refers to the side of a square, the íb.si$_8$ is usually spoken of in the singular; when a reference to two sides (meeting in a common corner) is needed, the texts speak about the íb.si$_8$ and its "counterpart" (*meḫrum*). A few texts, however, refer to "each" (ta.àm) of the íb.si$_8$ (YBC 4607 and 5037) or the *mitḫartum* (NBC 7934).

Division, Parts, and the igi

In our view, division is both *a problem* – to solve the equation $bx = a$ – and an arithmetical operation. The familiar assertion that division does not exist in Babylonian mathematics obviously refers to the absence of division as an arithmetical operation of its own.

The way a division *problem* is dealt with depends on whether the divisor b is "regular" or not. A "regular" number (our term) is a number whose reciprocal has a finite expression in the sexagesimal system (that is, a number which divides some power of 60), and which evidently also possesses a finite

44 The prefix ba. is often used regularly with the identity-conserving subtraction zi, "to tear out"; in contrast, the identity-conserving addition daḫ, "to append", is often preceded by bi., which suggests a "here", a closer contact.

In a group of early texts from the Eshnunna-region (close to modern Baghdad), some texts use the ba- and some the íb-form ("equilateral" and "equalside", respectively) for the quadratic case; the ba-form is used as a noun. ba.si$_8$.e is also used in the quadratic case (but as a verb) in two early texts from Ur (UET V, 859 and 864). Elsewhere, the ba-form is only used for the cubic and for generalized cases.

45 Some publications use the transliteration íb.sá instead of íb.si$_8$. Later lexical lists, indeed, render the pronunciation of the Sumerian verb as *sa-a*; the preference of most editions of mathematical texts for si$_8$ is supported by the occasional homophonic shifts to íb.si and ib.sí, which, however, might as well be íb.se and ib.sé. Since we also find the syllabic writing ba.se.e (which can *not* be ba.si.i) for *ba.si$_8$* (IM 52301 rev. 7, 9, below, p. 213), the real phonetic value might be between *-se* and *-sä*.

expression itself; regular numbers are thus numbers of the form $2^p \cdot 3^q \cdot 5^r$, where p, q, and r are integers. The reciprocal of such a number n is called igi n g̃ál.bi ("[of 1,] its igi n g̃ál",[46] cf. above, p. 4), often shortened to igi n g̃ál or igi n. A standard table containing the reciprocals of the regular numbers from 2 to 1`21 was in use; when a division by such a number n is to be performed, our texts ask for igi n to be "detached" (*patārum*/du$_8$),[47] after which the dividend is "raised to" the igi. In modern terms, division by n is thus performed as multiplication by $^1/_n$.

The literal meaning of igi n g̃ál is not clear, but Old Sumerian texts from Lagaš (c. 2400 BCE) speak of $^1/_3$, $^1/_4$ and $^1/_6$ shekel of silver and of $^1/_4$ sar of land in this way [Bauer 1967: 508–511; Lambert 1953: 60, 105*f*, 108, 110; Allotte de la Fuÿe 1915: 132], long before the creation of the sexagesimal place value system and the tables of reciprocals (see p. 314); the abbreviated form igi 4 is in a table of square areas that may go back to 2450 BCE (OIP 14,70, see [Edzard 1969: 101]). This early use of the term appears to exclude an explanation of the term proposed by E. M. Bruins ([1971: 240], and elsewhere), apparently in agreement with a belief that was already held in Old Babylonian times: igi, originally a picture of an eye, is used logographically for *īnum*, "eye"; for *amārum*, "to see"; and for *pānum*, "face", "front side". In agreement with the latter use, some Old Babylonian mathematical texts (e.g., Haddad 104) replace the writing igi with *pa-ni*, "in front of", which suggests a reference to the table of reciprocals, where igi n is indeed present "in front of" n – a folk etymology close at hand, not least because g̃ál means "to be/place (somewhere)", "to be at disposition".[48] But since the explanation is ruled out by chronology (and does not fit the enclosing phrase "of 1, its ..."), all we can say is that an n'th was designated igi n

---

46 Strictly speaking, early tables of reciprocals (some certainly from Ur III, some perhaps from the early Old Babylonian period) show that the meaning is "[of 1`,] its igi $n$ gal". These tables, indeed, list $^2/_3$, $^1/_2$, $^1/_3$, $^1/_4$, etc., of *sixty* = 1` – see Steinkeller [1979: 187]. In later times, genuine reciprocals in our sense will have been thought of.

47 Thureau-Dangin [1936: 56] suggested an interpretation of *paṭārum* in analogy with the modern idea of "solving" a problem. In the mathematical text Str 367, obv. 15 (below, p. 239), however, a triangle is "detached" from a trapezium of which it is part; in rev. 10 of the same text a similar subtractive use occurs; in both cases, the next line contains the "detachment" of an igi. This leaves no reasonable doubt that this "detachment" is really understood as the splitting of the whole into n parts, one of which is then taken out from the bundle. Compare also presently the parallel use of du$_8$ and zi.

If Thureau-Dangin's conjecture had been correct, we may add, one might also expect the term to be used when other problems are solved, e.g., when a nominal "equalside" is to be found. Yet the íb.si$_8$ may be "made come up" (*šūlûm*<*elûm*), be "taken" (*leqûm*), or it may be asked "what" (*mīnûm*) it is; it is never "detached". CBS 19761, obv. II 10, it is true, should probably read *i-na* íb.si$_8$.e *pu-ṭu-ur-ma* [Robson 2000: 37]. However, the construction shows that íb.si$_8$ is here a logogram for *mitḫartum* (here it is a noun, but when related to the equalside in line 12 it is a verb); the meaning must therefore be that those "flanks" (*pāt*) of which the next line speaks should be separated as entities from that square figure which they delimit.

48 Similarly, an Akkadian *pa-nu* is found as an explanatory interlinear gloss preceding igi in BM 96957, obv. I 7 [ed. Robson 1996: 183, cf. 186].

g̃ál for n = 3, 4, and 6 no later than the early 24th century BCE; that igi ... g̃ál is a composite verb in Sumerian that may mean "to look upon" (literally, to place the eye unto) – but that this need not be its exact function in the actual context.

An alternative reference of the expression could indeed be the "eye" or circular pattern in which 2, 3, 4, 5, and 6 impressions of the stylus were arranged in signs for $\frac{1}{2}$, $\frac{1}{3}$, $\frac{1}{4}$, $\frac{1}{5}$, and $\frac{1}{6}$ of the smallest unit in the grain metrology of the late fourth millennium. This explanation – thus that igi n g̃al means something like "n (dots) placed in eye (i.e., circle)" – was first proposed as a possibility by Jöran Friberg (neither he nor I remember where). As observed by Eleanor Robson, the latest evidence for this notation and the earliest for the use of igi for fractions are separated by at least half a millennium; as far as I am aware, however, no other way of w̰riting fractions is attested in documents from this long period. In the absence of supplementary evidence – unlikely to turn up, given the early origin of the term and the character of the early documents – we cannot decide.

If the divisor is not regular (a recurrent situation in mathematical procedure texts), the text takes note of the non-existence of the igi and then formulates the division as a problem – "what may I posit to b which gives me a" – and states the answer immediately, "posit p; raise p to b, a it gives to you" (or some similar construction). Since mathematical problems were always constructed backwards from the solution (the safest and often the only way to ensure that a reasonable solution exists, as any writer of mathematics textbooks will know), this could always be done without difficulty.

In "real-life" computation, answers would of course not have been known in advance, but irregular divisors would occur only rarely – technical "constant coefficients" (be it the number of bricks a worker could carry 1` nindan per day, be it the amount of bitumen that would cover a square cubit) were chosen as regular numbers.[49] A couple of tables containing approximate reciprocals of irregular numbers have been found;[50] they may possibly have served in the cases where

---

49 This may seem an arbitrary procedure, but is in fact nothing but a rounding: our computational aids demand that technical coefficients contain a limited number of significant digits, and thus we round to (e.g.) the closest four-place decimal fraction; instead, the Babylonian methods ask for "rounding" to a neighbouring regular number.

The grid of regular numbers found in or easily derived from the Old Babylonian standard table of reciprocals is evidently coarse, but probably not coarser than the precision of the technologies for which it was used; whether the amount of bitumen to be spread on 1 square cubit of coarse mud bricks should be 14′ or 15′ silà was probably not precisely predictable. In Late Babylonian astronomy, the need for precision was much higher; this explains the elaboration of multi-place tables of reciprocals in the Seleucid era.

A few non-regular coefficients can be found in the Old Babylonian tables of constant coefficients – the areas of certain complicated geometrical figures, and some coefficients connected to metals and obviously artificially constructed so as to constitute an arithmetical progression [Robson 1999: 56, 127]. None of them are likely to have played any role in practical calculation.

50 YBC 10529 records the reciprocals of regular as well as irregular numbers from 56 to 1`20; M 10, John F. Lewis Collection, Free Libr. Philadelphia, lists the reciprocals of 7, 11, 13, 14, and 17.

irregular divisors turned up anyhow; at least they have no function within the mathematical problem texts. Most likely, however, their preparation was in itself a mathematical exercise.

In some cases, a text will state that "igi *n* I do not know" or "the igi is not detached" even though the divisor is regular; this happens when the divisor does not occur in the standard table of reciprocals and cannot easily be derived from it (a trained calculator will immediately have known that 22°30′ is half of 45, and the corresponding igi hence the double of igi 45 = 1′20″). In the Old Babylonian epoch, the concept of igi *n* is thus first of all associated with the practice of tables, and it should be identified only indirectly or even obliquely (cf. note 46) with the abstract mathematical idea of a "reciprocal number"; that the term was first of all understood in relation to the table and the use of the values as factors is also suggested by the name igi.gub, "fixed igi", for the technical constant coefficients – these coefficients are numbers used in calculations in the same way as are the igi, and they are listed in tables, but they have not the least thing to do with reciprocals.[51] In consequence I shall not translate the term igi but use the term as a loanword.

igi *n* is, however, also used in the traditional sense in our texts, namely, as the *n*th *part of something*.[52] In this function I shall translate the word as "part". The use of two different translations may seem to contrast with the principle of conformal translation, but is indeed called for precisely by this principle. In texts in which both uses occur together, various devices are indeed often used to keep them apart. The devices vary, but their presence shows that the authors of the texts took care to keep apart concepts which their standard terminology conflated; our translation should therefore also make a difference.

The most common device is the use of ellipsis. A number of tablets (Str 367, VAT 7532, VAT 7535, etc.) speak of the *n*'th part of an entity (even of 1 if this number represents an unknown length in an argument of false position) by a full phrase igi *n* ĝál; the number facing *n* in the table of reciprocals, for its part, is simply igi *n*. In the tablet BM 85210 the same distinction is made, but supplemented by the use of different verbs: as always, the reciprocal is "detached" (du_8); the *n*th part of *m*, however, is "torn out" (zi). BM 85194 uses the short form for both concepts, and makes the distinction exclusively by the choice of verb.

---

51 The primary association with the table of reciprocals is also reflected in a couple of Sumerian loanwords in Akkadian, *igûm* and *igibûm*, "the igi and its igi", referring to a couple of numbers from the table, not necessarily in orders of magnitude that make them mutual reciprocals – 5 and 12, 1°30′ and 40′, 1°20′ and 45′, etc. We shall meet them in several problems below.

52 Related to this (perhaps via generalization, perhaps as another aspect of an original broader spectrum of meanings) is the use of igi in the sense of the "share" of somebody in a partition (thus AO 8862 #4).

Bisection

When dividing by a number which, so to speak, is 2 only "by accident", our texts find "igi 2" = 30′ and "raises" to that number. Beyond that, a particular sign (which I shall translate ½) and corresponding words (*mišlum*/šu.ri.a, both to be translated "half") for the half exist. They are used, for instance, when the width of a rectangle is said to be half its length; if one entity is said to exceed another by its half; or to indicate a measure ("half a barleycorn"). Such relations and measures are accidental; they might as well have been different.

Besides this conventional half, however, a different, "natural" or "necessary" half (of something) is used in the texts, the *bāmtum* (no Sumerogram seems to exist[53]); I shall translate it "moiety". It occurs in places where something is "broken" (*ḫepûm*/gaz) into two necessarily equal parts: *viz.*, when the base of a triangle is bisected for the purpose of an area calculation; when the same is done to the sum of opposing sides in a trapezium; and when the radius is found from a circular diameter. In the above text example (p. 13), it was used when (expressed arithmetically) the "coefficient of the first-degree term" was bisected as a preparation to the quadratic completion of the equation. Below (pp. 162 and 239) we shall discuss the tablets AO 8862 and Str 367, texts where the distinction between the coincidental and the necessary half is very visible.

The general references of *bāmtum* fit its particular use as a mathematical term: a side of ribs; one of two opposing mountain slopes; etc.

As stated above, the "moiety" is invariably found by "breaking". This verb, on the other hand, has no other function in the mathematical texts; it always goes together with *bāmtum* (or with ½ or šu.ri.a in the texts where

---

53 Non-presence of a term in the text material at our disposal is in general a weak argument. But absence from texts where it *should* have occurred is strong. Such texts are Str 366, 367, 368, all of which insist on writing everything with Sumerograms (except grammatical particles that do not exist in Sumerian), although in clearly Akkadian sentence structures: they eschew the word (Str 366), or they use the number sign ½. This is also done in VAT 7532 (below, p. 209) and VAT 7535. The tablet YBC 6504 (which was also exceptional in its use of íb.si$_8$ in the function of *šutamḫurum*, "to make confront itself") has recourse to šu.ri.a.

BA and BA.A, used in the function of *bāmtum*, are taken in [MKT] to be Sumerograms; actually, we are confronted with elliptic writings or, more likely, with irregular assimilations to the pronominal suffix *-šu*, *bāššu* < **bāmat-šu*; the noun **bûm* [MCT, 161], constructed backwards from similar forms, should also be *bāmtum*.

The writing of *bāmtum* as $^{\text{ba-a}}$bà in a lexical list should probably be understood as $^{\text{ba-a}}$30: the sign BÀ coincides with the writing of 30; the phonetic complement tells us that the "30′" in question starts with *ba*.

these are used as logo-/Sumerograms for *bāmtum*). In the mathematical texts (but only here), "breaking" is thus the same as *bisection*.

Mathematical Organization and Metalanguage

The mathematical operations discussed above are used within texts that are organized with a particular logical structure – hypothetical-deductive arrangements, relation between "true" entity and representative – and which make use of what we may call a metalanguage – names for unknown quantities, terms that indicate equality, terms that delimit expressions (corresponding to the parentheses of our algebraic expressions), terms that announce results, etc. It is advantageous to describe these structures and terms, no less than the operations themselves, before we approach the texts.

The Standard Format of Problems

This aspect of the organization was already mentioned briefly on p. 9: the problems of the procedure texts consist of a statement followed by a prescription, a description of the procedure to be followed. Some of the early texts from Eshnunna in Northern Babylonia start by setting the stage explicitly as one of riddle-solving: "If somebody asks you thus", and then go on with the statement formulated in the first person singular, preterite tense: "I have done so and so ..."[54] (with the striking exception that the excess of one entity over another, even if it is the excess of the length over the width of a rectangle which the speaker has laid out, is reported in the neutral third person singular). Most texts, however, start directly with the statement in the first person singular, past tense, as if the master himself is speaking. After the statement follows the prescription, formulated in the second person singular, present tense, "you do so and so", or the imperative, "do so and so". The prescription may but need not be introduced by a phrase "You, by your proceeding" and closed by a corresponding "The procedure".[55] At times a

---

54 It should be observed that the Akkadian preterite tense often – and not least in this case – corresponds to an English perfect, and not to the mostly imperfective English past tense; in Old Babylonian times, the use of the Akkadian so-called "perfect" was restricted to particular situations, "*now* I have ...", "then I did", etc. (see GAG, § 80).

55 "Proceeding" translates *epēšum*, "to make" or "making" (logogram dù, originally "to build", "to erect on the ground"); "procedure" is the translation of *nēpešum*, derived from the same root; it is mostly used to close the description of the proceeding (in post-Old Babylonian medical texts it also closes the description of pharmaceutical prescriptions); in the text Haddad 104, however, problem

particular step is justified by a quotation from the statement, "because he has said ...", as if the one to explain the procedure is not the master (or the "somebody") but an instructor (appearing as the "big brother" in texts describing the school).

Tense and person are unambiguous in texts written in syllabic Akkadian, to which the above rules apply with high regularity. It is a fair guess that texts written with logograms (giving no indications as to grammatical form) should be understood in the same way, at least those which open the procedure prescription with a Sumerian phrase za.e ..., "you, ..." In one text from Susa (TMS XVI, see below, p. 88), however, a logogram zi in the statement is quoted in the prescription as an infinitive (the usual lexical form and the standard way to render Sumerian roots in Akkadian grammar teaching and lexical lists [Jacobsen 1988: 218]), *na-sà-ḫu*, "to tear out". It may therefore be imagined that the logograms of the more compactly written texts were, and should be, understood in this abstract (i.e., de-personalized and de-temporalized) sense.[56] But the use of the infinitive may also imitate the manner of the lexical lists. In any case, as Eleanor Robson insists, Susa scribes "were almost certainly Elamite, not Akkadian speakers", and their Akkadian texts only evidence of what non-native speakers would do.

Standard Names and Standard Representation

Central to every calculational technique that somehow can be characterized as "algebraic" is the *analytical approach*, that is, "the assumption of what is searched for as if it were given, and then from the consequences of this to arrive at the truly given", in Viète's words.[57] This, however, cannot be done unless the magnitude in question is at least identifiable – as a line (etc.) in a diagram in Greek geometric analysis, as a letter in Modern symbolic algebra, as a verbal name in medieval "rhetorical" algebra. If the technique deserves the name of a *discipline*, this identification will have to be standardized.

---

statements announce that what follows is the *nēpēšum* for a particular object – or, if variants, start by *šumma*, "if (however)".

56 In this respect, Neugebauer ([1932: 222]; [MKT I, viii]) might thus be right that the logograms function "precisely as mathematical symbols". In other respects, as we shall see (note 141), the formulation is misleading.

57 *In artem analyticen isagoge*, Ch. I [ed. Hofmann 1970: 1]. Or, as formulated by Pappos (*Collection* VII.1 [ed., trans. Jones 1986: 82]), "in analysis we assume what is sought as if it has been achieved, and look for the thing from which it follows, and again what comes before that, until by regressing in this way we come upon some one of the things that are already known, or that occupy the rank of a first principle".

There is an ongoing and widely branched discussion whether Pappos (and hence Viète) has understood or misunderstood the concept of the ancient geometers; this is irrelevant for the present purpose, everywhere in the following the term is taken in Pappos's and Viète's sense.

Our Babylonian texts are verbal. They identify unknown quantities by words, and mostly do so in highly standardized ways. Moreover, they share another characteristic with medieval and modern algebraic computation: unknown quantities may be of any kind – weights, prices, distances, ...; but this unruly diversity is *represented* by a specific class of quantities, and the well-known structure in which these participate is used to map the relationships in which the unknown quantities are involved. In contemporary algebraic computation, this *standard representation* is numerical, and the structure used to map the relationships between physical entities, prices, or whatever we deal with, is the arithmetical organization of the class of (mostly real) numbers. The Old Babylonian standard representation, instead, was geometric – more precisely, it was constituted by the *geometry of measurable segments and surfaces*. This is seen both when the product of two numbers is referred to as a.šà, "the surface" (YBC 6967, below, p. 55); in the picking of a particular term among several possible synonyms according to its range of connotations (e.g., the preference of certain texts for "cutting off" from linear entities and "tearing out" from areas, see above, p. 20); and in the distribution of such terms as *libbi* (ibid.), none of which make sense outside a geometric representation, and which – apart from the use of a.šà – could not remain in consistent use once an original geometric mode of thought had been forgotten, governed as their appearance is by ineffable criteria beyond the control by rules.

Familiarity with the standard representation was developed by means of standard problems dealing with rectangles and squares. Dominant in the text material (which on this account is large enough to be considered representative) are the rectangle problems.

Rectangles are understood as figures which, without further qualification, are determined by two dimensions, the uš and the sag̃.[58] uš functions (*inter alia*) as a logogram for *šiddum*, "flank", "long side"; sag̃, originally "head" in Sumerian, is used (*inter alia*) for *pūtum*, "front" – including the "front" of a field, the narrow side turned toward the irrigation channel.

However, with the exception of a couple of early texts from the Eshnunna region (cf. below, p. 319), the Akkadian terms are never used as standard

58 At times, procedure texts start by stating which kind of geometrical figure is dealt with. In the case of a rectangle, the corresponding phrase is simply uš sag̃. If a triangle is the topic – another figure determined by uš and sag̃, since a triangle with no further specification is supposed to be "practically right" – the initial information is sag̃.dù, a logogram for *santakkum*, "wedge" or "triangle" (perhaps to be read sag̃.KAK, KAK *being* a triangle, cf. note 37 – see [Robson 1999: 43]). In general, Babylonian mathematical texts should be understood "by default": the information which is offered is supposed to be sufficient, and the configuration dealt with can thus be assumed to be the simplest possible given this information – "simplest possible" of course as understood by the Babylonians, for whom the rectangle was more fundamental than the triangle.

As to the notion of "practical perpendicularity," cf. below, p. 228.

"variables", not even in texts which write virtually everything in syllabic writing, and which use, e.g., *šiddum* for the distance bricks are carried; outside Eshnunna they are invariably written in Sumerian, as are (with very rare exceptions) igi and íb.si$_8$.[59] As in the case of igi/part, the explicitly differentiated use calls for different translations. uš and sag̃, used in connection with the rectangles of the abstract standard representation, I shall translate "length" and "width", terms whose abstract-geometrical connotations fit the Babylonian use. *šiddum*/uš and *pūtum*/sag̃ when used about supposedly *real* structures (fields, irrigation channels, etc.) will be translated "flank" and "front".[60] When no configuration is involved (e.g., when a carrying distance is meant), *šiddum*/uš can adequately be translated as "distance". sag̃ when used as a logogram for *rēšum*, "head", will be translated accordingly.

A variant of the rectangle is the pair *igûm-igibûm*, "the igi and its igi" (cf. above, note 51). In principle, these are of course *numbers*[61] (a possible origin of the table of reciprocals as fractions of 1` nindan is probably immaterial for the understanding of the entities in mature Old Babylonian mathematics); but their product (1 or 1`) may be referred to as a.šà, "surface" (YBC 6967, see p. 55), their difference may measured in nindan (YBC 4668, rev. II 32) – and they may actually be the length and the width of an excavation (BM 85200+VAT 6599, *passim*, see p. 158). *igûm-igibûm* thus represents a *standard problem* which may be varied in more or less intricate ways; but it refers back to the conventional *standard representation* of measurable line segments.

The other standard configuration was the square. We have already discussed the central term – *mitḫartum*, translated "confrontation" – at some length; in the present connection it should suffice to recall that the "confrontation" designates the square configuration of equal sides (thus the frame, not the contents) and is parametrized by the length of the side.

The area of both rectangles and squares (and other figures as well) is spoken of as a.šà, outside Eshnunna always written in this Sumerian form but often with phonetic complements that indicate an Akkadian pronunciation *eqlum*. The general meaning of this word is "field" or "open area"; the consistent use of the Sumerogram indicates a technical use in the mathematical

59 The newly published text from Nippur CBS 19761, obv. II 11 contains an exception to this rule, but within a highly unusual construction – cf. note 47.

60 Even numerically, representing and "real" fields are kept apart. When the order of magnitude can be determined, the dimensions of the standard "abstract" rectangle mostly turn out to be 20′ nindan×30′ nindan, i.e., c. 2×3 m^2. These are "fields" that could be easily drawn to scale in the school yard. Texts that emphasize their reference to "real" fields by explicit use of a practical area unit often have the unhandy dimensions 20 nindan×30 nindan, c. 120×180 m^2. In TMS XVI (below, p. 87) we shall encounter a distinction between "width" and "true width" which is likely to refer to this distinction between the real field and the model field.

61 Cf. also their summation by the accounting term nig̃in/"total" in one text, see page 20.

texts (as a rule, and in order to make clear the difference, these will make use of other, semantically less adequate words when they want to speak about the divisions of a *real* field); for this reason, I shall translate it as "surface", halfway between the concrete "field" and the fully abstract "area".

In certain contexts, a.šà may also designate a volume, in agreement with the common metrology for areas and volumes (and with the underlying idea of areas being provided with a standard thickness of 1 kùš). More often, when real three-dimensional structures are dealt with – either prismatic excavations in the ground or constructions filled with earth (including bricks, made from mud) – the texts speak of saḫar, "earth" or "dirt" (I shall use the former translation).

Prismatic excavations themselves are designated túl, túl.sağ, or ki.lá; túl (perhaps even túl.sağ?) may be used logographically for *būrtum*, "cistern", "well", and túl as well as ki.lá (and túl.sağ?) for *kalakkum*, "excavation", "cellar", "silo"; there is little doubt that the same Akkadian word is meant in all the mathematical texts – and thus *kalakkum*. I shall translate it as "excavation". The horizontal dimensions of the excavation are designated uš and sağ, and the depth GAM or bùr, the latter (and probably the former) a logogram for *šuplum*.[62] The base of the excavation is designated *qaqqarum* (mostly written with the logogram KI, probably to be read gagar[63]), "floor" or "ground" (I shall use the latter translation). The reason may be that a.šà could be interpreted "by default" as "volume" in connection with a three-dimensional object – but also that *qaqqarum* happened to be the conventional name in connection with this object or configuration.[64]

One might perhaps expect the túl.sağ to be both standard representation and standard configuration, as is the uš-sağ-rectangle – the rectangular prism is, after all, the most elementary model on which to construct third-degree problems. As a matter of fact it iṣ only a standard configuration, and never used for representation (that is, no other problems are reduced explicitly or implicitly to this structure and solved accordingly). The probable explanation is that the Old Babylonian calculators knew no general method of solving irreducible third-degree problems; they could solve some by means of factorization (see below, p. 151), but they mostly used the "excavation" as the basis for first- and second-degree problems.

A number of other recurrent problems that do not serve as standard representations can be found in the text material. Below, we shall encounter several examples: the "broken reed" (VAT 7532, p. 209), and the "unfinished ramp" (p. BM 85194, #25–26, p. 217) – cf. in general pp. 286*f*.

In *our* symbolic algebra, change of variable is a customary technique. Even though the geometric representation imposes different conditions, something analogous is found in many Old Babylonian mathematical texts, often marked by the use of the epithets *kīnum*/gi.na, "true", or *sarrum*/lul,

---

62 The height, in structures that go upwards instead of downwards (walls, brick piles, etc.) is designated sukud (at times the corresponding syllabic *mūlûm* is used).

63 In lexical lists, Akkadian *qaqqarum* has the equivalent $^{ga.gar}$KI [AHw, 900b].

64 One text from early Old Babylonian Ur (UET V, 859) speaks of the volume as saḫar and refers to the base not simply as a.šà but as a.šà.ga, ga being apparently an abbreviation for gagar, the usual Sumerian equivalent for *qaqqarum*, "ground".

"false". The length and area of a figure measured in terms of a measuring reed of unknown length may be termed "false", while the measures in nindan and sar are "true" (Str 368, VAT 7535); the "false grain" of a field may be the grain (in case, grain to be paid in rent) per unit area (that is, the rent to be paid *had* the area of the field been 1 sar) (VAT 8389, VAT 8391, pp. 77 and 82); if the width of a rectangle has been augmented during the procedure, the original width will be "true" (AO 8862, p. 162). But the "true length" of a triangle may also be the length which comes closest to being perpendicular to the width (YBC 8633, p. 254).

In Greek geometric analysis, as we know, it is not necessary to identify one or two particular line segments as *the* unknown(s) – all elements of a diagram can be treated on a par – all that is needed is a way to identify them ("the line *AB*", "the square *AC*", etc.). The Babylonian texts contain no drawings apart from such as are used to illustrate problem statements, but some texts make use of similar devices, identifying, for instance, a length that has been prolonged not as the "false length" but more precisely as "the length of the surface 2" (TMS IX, below, p. 89).[65] Whether such expressions are to be regarded as "changes of variable" depends, if not on taste alone, on whether we want to emphasize the affinity with Modern (that is, post-Viète) algebra or the difference.

Structuration

The gross arrangement of procedure texts into statement and prescription and the corresponding distribution of grammatical person and tense were dealt with above (p. 32). But the logical structure of the texts and the way to compose fairly unambiguous mathematical expressions is also standardized, though not uniformly in all text groups (a question to which we shall return, see Chapter IX).

Occasionally, the hypothetical-deductive structure of the problem is made explicit by an introductory *šumma*, "if" – also familiar as the opening of the protasis of omina ("if the liver looked so and so", etc.). Since it is found in most of those texts from Eshnunna that do not carry the full "if somebody has asked you thus" (and in this function almost exclusively in texts stemming from the same northern part of Babylonia), the presence of the term in this place may originally have been a vestige of this phrase. In one text from Eshnunna, however, as in certain other texts, *šumma* is used to introduce

65 Supplementary identifiers – be they numerical, be they references to the role the entity in question has already played in the procedure – serve both when the same concrete entity (e.g., the width of a rectangle) is modified during the process (in which case the analogy with a change of variable is at hand) and when several unconnected entities share the character of (e.g.) a "length" (where the analogy is certainly misleading).

variations of an exemplar, and thus in the meaning "if (instead the situation is as follows:)". It may also serve to introduce a smaller piece of deductive reasoning inside the prescription from already established foundations ("if [as you have now established] ..."), or to open the proof; it can be found in either function both in texts from Eshnunna and other localities in the periphery and in texts from the former Sumerian heartland.[66]

inūma, "as", is used in a couple of texts from the periphery to mark a piece of deductive reasoning on already established foundations. *aššum*, "since", has the same function in some texts from the periphery and in some from the core (in the periphery specimens it mainly serves as the opening phrase of the procedure prescription[67]); it may also be used in connection with quotations from the statement.

The pertinency of single steps in the procedure may indeed be argued from the words of the statement. This is done by an accurate quotation, followed by the word *iqbû* (< *qabûm*), "he has said", or (more often) the whole quotation is contained in the phrase *aššum ... qabûku*, "since it is said to you".

In certain cases, the statement falls into two sections, the first of which contains general information – the value of a technical constant to be used, the rent to be paid per bùr of a field, etc. – whereas the second presents the actual problem. The second section may then be introduced by the word *inanna*, "now" (see note 395).

Even the prescription may be divided explicitly into subsections. Such divisions may be marked by the verbs *saḫārum* (AHw "sich wenden, herumgehen", etc.), "to turn around", *târum* (AHw "sich umwenden, zurückkehren; (wieder) werden zu"), "to turn back", and nigin(.na), which in general may function as a logogram for either (in the mathematical texts, the value is likely to be *târum*). At least in the text AO 8862, *saḫārum* appears to be used in the statement about a quite concrete walk around a field which has just been marked out, and *târum* about a return to the starting point. *târum* is used in a similar way in texts from several text groups; it seems no unreasonable guess that this concrete usage may have been the origin of the abstract use as a textual delimiter in the prescriptions.

Many problems – not least those of "algebraic" character, thus also the example quoted on p. 11 – are shaped as *equations,* that is, as statements that (the measure of) a more or less complex quantity equals a number; also present though less common are equations that declare that (the measure of)

66 A division of the text corpus into a "northern" and a "southern" group was originally undertaken by Goetze, who based himself on orthographic criteria [MCT, 146–151] (almost all mathematical texts known in 1945 came from illegal diggings and were thus of unknown provenance). A more detailed analysis will be found below in Chapter IX, where I shall propose to replace Goetze's division by a distinction between the (former) "Sumerian core" and its "periphery".

67 Not, however, in texts beginning "if somebody ...".

one quantity equals (the measure of) another quantity. In the former case, the equality is mostly implied by the enclitic particle *-ma* on the verb, which I shall render by the sign ":".[68] In the latter case, the term *kīma* (AHw "wie; als, wenn, daß") may be found; I shall translate it "as much as".

A term for equality that may serve as a kind of bracket when complex quantities are constructed verbally is *mala* (AHw "entsprechend (wie), gemäß"), "so much as". It is found in the expression "so much as *a* over *b* goes beyond", meaning $(a–b)$.

The numerical value of a quantity *Q* may be asked for in two ways, either by the question *mīnûm Q* (AHw "was", standard translation "what", logographic equivalent en.nam), or by the question *Q kī maṣi* (*kī*, AHw "wie, als, daß", *maṣûm* "entsprechen, genügen, ausreichen"), which I shall translate "corresponding to what *Q*".[69] Collective questions for each of several values may be asked for with the question *kiyā* (AHw "wieviel"), which I shall translate "how much each" (cf. note 369).

Recording

Numbers occur as data in the statement, as intermediate outcome of calculations, and as final results. Several terms and phrases may be used in this connection.

Of particular importance is the verb *šakānum* (AHw "hinstellen, (ein)setzen, anlegen; versehen mit"), with logogram g̃ar, which I shall translate "to posit". The term appears to designate various kinds of material recording – "putting down" in a computational scheme, writing the value of a length or an area into a drawing, etc. Its is mainly used in two functions: to take note of data in the beginning of the prescription, and thus to prepare their use; and in the formulation of the division problem, "what may I posit to *b* which gives me *a*", with the answer "posit *p*; raise *p* to *b*, *a* it gives to you" (etc., cf. p. 29). In the latter, insertion into a computational scheme is likely to be meant (cf. p. 89).[70] Some texts also "posit" an "equalside" and its "counterpart", i.e., two sides of a square; the same process is indicated in other

68 When following after a noun, the function of the particle is one of identification, and I shall translate *X-ma* as "that *X*".

69 In the division question, the accusative *mīnâm* is found, "what may I posit to ...". Similarly when a geometric square root (an "equalside" or an "equilateral") is asked for by the verbal construction, "by *S*, what is equalside/equilateral". When the "equalside" is regarded as a noun, it may be asked for by *mīnûm*, or the student is asked to "make the equilateral of *S* come up", *S basâšu šuli*, or to "take" (*laqûm*) it.

70 In the text YBC 6504, g̃ar is used to take note of both intermediate and final results. We shall return to its generally unconventional and probably experimental terminology on p. 343.

texts by the verbs *lapātum* (AHw "eingreifen in, anfassen; schreiben"), "to inscribe", or *nadûm* (AHw "werfen, hin-, niederlegen"), "to lay down". In one geometric text (BM 15285) there is no doubt that the actual sense of the latter term is to lay down *in drawing*; most likely, its general meaning in mathematical texts is "to lay down in writing or drawing". *lapātum* is also used regularly about numbers that afterwards serve in additive and subtractive operations; since these seem to have been performed on a counting board and not in clay (see below, p. 73 and note 222), it may also have referred to recording on such a device, which is still in agreement with the general meaning of the term, "to grasp/take hold of"; possibly, even "positing" includes this operation.

A specific phrase for recording an (invariably intermediate) result is *rēška likīl*, "may your head hold (it)" (from *rēšum*, AHw "Kopf, Haupt; Anfang", and *kullum*, AHw "(fest)halten", also the root of *šutakūlum*, "to make hold"). It seems to be reserved for numbers that are not to be inserted in a fixed scheme and therefore are not "posited".

The appearance of a result may be announced in several ways. It may be said that a number "comes up for you" (standard translation of *illiakkum*, from *elûm*, AHw "auf-, emporsteigen"), or that a calculation "gives" a certain result (*nadānum*, AHw "geben", Sumerogram sum); alternatively, the text may state that "you see" the result (*tammar*, from *amārum*, AHw "sehen"), or that it is "given" to you/me (*nadānum*). Very often, the calculation is simply followed by the enclitic particle *-ma* (translated ":") and the number, or by nothing but the naked number. As we shall see in Chapter IX, the format for the announcement of results is one of the most striking ways in which text groups differ.

The "Conformal Translation"

In some measure, any translation is a mistranslation, for the simple reason that different languages are structured differently on all levels from syntax to conceptual structures and to the network of connotations in which terms participate. Any trained translator knows that he has to choose, that rendering one level of the text adequately entails that other levels are betrayed, and that the purpose of the translation determines which level is to be treated faithfully, and which levels are to be betrayed – or which kind of compromise between different considerations is to be made, determining in which way "faith unfaithful [is to keep] him falsely true", in Tennyson's words.

The betrayal inherent in any translation increases if the languages are as far from each other as Akkadian and English, and if even the "language game" – in the original Wittgensteinian sense (as opposed to the way the phrase is used by postmodernists) of an irreducible complex of extra-linguistic practice, concepts, and usage – of the original is structurally different from any language game with which the user of the translation can be expected to be familiar. Neugebauer's and Thureau-Dangin's choice of "substantially adequate" translations presupposed that the Babylonian "mathematical" texts were so well described by this epithet that their "substance" could be identified with the "game" of post-Renaissance mathematics.

This is obviously only a rough approximation, which fails as soon as we try to understand not only the numbers but also the actual words and the underlying practice of the original texts; a translation which tries to render faithfully the pattern of the original game must therefore by necessity betray both idiomatic English and current mathematical usage.[71]

The texts on which the rest of the book is based are bilingual, in Akkadian and in English translation. The Akkadian text is "transliterated", that is, syllabic writing is rendered syllable for syllable, and logograms are rendered as such, either as Sumerograms or with sign names.[72] The translation is meant, firstly, to be a possible basis for discussion of the text without reference to the original; secondly, to serve as support for the reader who wants to be able to follow what goes on in the original without knowing more than the rudiments of its language. For both reasons, the translation should be "conformal", that is, conserve the structure of the original, rendering always a given expression by the same English expression, rendering different expressions differently (with the only exception that certified logographic equivalence is rendered by coinciding translation). To the extent it does not entail exorbitant clumsiness, the translation is *de verbo ad verbum*, and word order is conserved – but the absence of articles in Akkadian and the posterior position of the adjective limits this principle strictly. To the extent I have found it possible, terms of different word class but derived from the same root are rendered by derivations from the same "English" (actually often Latin) root. These are the standard translations which were given above for each

---

71 Repeatedly, I have encountered the objection that one has to respect and accept current language, either as a matter of principle or because deviations from this rule result in illegible translations. I ask myself whether those who formulate the objection would also translate eighteenth-century chemical texts based on the phlogiston theory into the English vocabulary of oxygen chemistry.

72 In the transliteration, we remember, syllabic writing appears as *italics*, Sumerograms are rendered in s p a c e d w r i t i n g, and sign names as SMALL CAPS.

A transliteration is already an interpretation, a decision whether the single sign is to be read as a syllable (and as *which* syllable, also a decision) or as a logogram (and, again, as *which* logogram); but by means of a sign list (e.g., [MEA] or [ABZ]) it permits identification of the original sign, and thus also modification of the syllabic/logographic interpretation.

There remains the purely palaeographic interpretation, the identification of the sign. As with any handwriting, cuneiform may sometimes be hard to read even for the trained Assyriological eye. On this point the transliteration offers no assistance, and a first recourse is the published photograph or the hand copy (the drawing of the tablet); ultimately, a collation with the tablet in the museum may be required. (On damage, see below).

term.

In many cases, the extra- and intra-mathematical uses of a term are intimately connected (some examples are discussed below, on p. 301) via shared connotations; in order not to make this connection disappear from view in the translation, the conformal principle requires that the English word chosen to render a particular Babylonian term should not only be distinct from the translations of other terms but also inasfar as possible share the everyday connotations of the original term.

The result undeniably *is* a betrayal of decent English style – and the acceptance of what seemed to me to be necessary deviations from the principle of conformality has certainly made me no less "falsely true" than Tennyson's Lancelot. But experience shows me that it *is* possible for those who do not know the original languages to work in and with this artificial code.

A list of all Akkadian terms and logograms with appurtenant standard translation is found in Table 1, p. 43, and a reverse list of standard translations with Akkadian and logographic equivalents in Table 2, p. 46. It may be convenient to make a photocopy of the pages in question in order to have easy access to these lists while working on the texts.

An index locating references to single terms and key phrases in the discussion is found on p. 430.

Clay is less exposed to the vicissitudes of time than paper and palm leaves, but nevertheless cuneiform tablets are rarely intact. Fortunately, the mathematical texts are very repetitive, which often allows the reconstruction of lost or damaged passages. Such reconstructions are put in square brackets [...], in agreement with usual conventions; if part of the sign or passage in question is conserved, this may be indicated by high or low writing of the brackets. Reconstructions which I consider hypothetical are marked ¿...?.

At times, the scribe seems to have forgotten a sign, a word, or a passage while copying. Such assumed omissions are indicated in angle brackets ⟨...⟩. In other cases, he has inserted a superfluous sign, word, or passage – most often a repetition. Such passages are put into braces {...}. In the translation, passages in {...} are omitted or their presence indicated without the contents if there is no particular reason to preserve it; if angle brackets indicate that the scribe has abbreviated, they are left out in the translation.

Brackets and braces are found in both the transliterations and the translations – in the latter located so as to indicate which part of the meaning is affected (a missing Akkadian prefix may thus be rendered as a missing suffix in English, if the functions of the two are similar). The translations also make use of round brackets (...), used to insert explanatory words which have no counterpart in the original.

In the translations, numbers written in the place-value system are rendered in agreement with Thureau-Dangin's extended degree-minute-second convention (see p. 12). In the transliteration, the single sexagesimal "digits" are separated by commas, and no indication of absolute magnitude is given. On the translation of number words and fractions, see p. 16.

A NOTE ON PHONETICS

Both Akkadian and Sumerian employed a number of speech sounds which are not familiar in English or which are transcribed by means of non-standard characters.

In Sumerian, as pointed out, accents and subscript numbers have *no* phonological implications. /g̃/ *may* perhaps be similar to /ŋ/, but this is by no means certain. /ḫ/ is probably similar to Akkadian /*ḫ*/ (see imminently), but how closely can be questioned. /š/ is likely to be close to Akkadian /*š*/

In Akkadian, dashes over vowels indicate that they are long. /*š*/ corresponds to /sh/ in English "shine". Dots under *s* and *t* indicate that these are "emphatic"; one may get an approximate idea of the pronunciation by trying to pronounce the consonant in question together with a (French/German) back-tongue /r/; /*q*/ is a similar "emphatic" *k*. /*ḫ*/ corresponds to /ch/ in Scottish "loch" or German "Bach".

Table 1: Akkadian Terms and Logograms with Appurtenant Standard Translation.[73]

a.na (~*mala*): so much as
a.na.àm: what
a.rá: steps of / step
a.šà (~*eqlum*): surface
adi: until
aḫertum: rest
alākum (~RÁ): go
amārum: see
ammatum (~kùš): cubit
an(.ta/na) (~*elûm*): upper
ana (~.ra): to
annikîam: here
aššum (LB ~mu): since
atta (*ina epēšika*) (~za.e (kìd.da/ ta.zu.dè)): you (by your proceeding)
ba.si$_8$/si (~*basûm*): (is) equilateral
bal: conversion
bāmtum: moiety
bandûm: bandûm
banûm: build
bar.NUN (~*ṣiliptum*): diagonal
basûm (~ba.si): equilateral
bêrum: single out
bùr (~*šuplum*): depth
bùr[iku] (~*būrum*): bur
būrum (~bùr[iku]): bur
daḫ (~*waṣābum*): append

[73] It should be kept in mind that some of the logographic equivalences listed in the following are rare or at least restricted to specific text groups.

dagal: breadth
dakāšum: thrust forward
dal (~*tallum*): crossbeam
dirig (~*watartum*): going-beyond
dirig, ugu ... (~*eli* ... *watārum*): go beyond, over ...
du$_7$.du$_7$: make encounter
du$_8$(.a) (~*paṭārum*): detach
dug$_4$ (~*qabûm*): say, saying
edûm (~zu): know
eṣēpum (~tab): repeat
elēnu: over-going
eli (~ugu): over
elûm (~an): upper
elûm: come up (as a result)
en.nam (~*minûm*): what
eperum (~saḫar): earth
epēšum (~kìd): proceed/proceeding
eqlum (~a.šà): surface
éš.gàr: task
ezebum (~tag$_4$): leave
gaba(.ri) (~*meḫrum*): counterpart
GAM (~*šuplum*): depth
GAM meaning circle, see gúr
g̃ar (~*šakānum*): posit
g̃ar.g̃ar (~*kamārum*): accumulate/ accumulation
garim (~*tawirtum*): meadow
gaz (~*ḫepûm*): break
gi (~*qanûm*): reed
gi.na (~*kīnum*): true
gíd.da: long(er)
gim(.nam), see gin$_7$.nam
gín (~*šiqlum*): shekel
gin$_7$.nam (~*kī*(*ma*)): as much as
gu.la (~*rabûm*): great(er)
gu$_7$(.gu$_7$) (~*šutakūlum*): make hold
gúr (~*kippatum*): circle
gur: gur
ḫarāṣum (~kud?): cut off
ḫaṣābum (~kud): break off
ḫepûm (~gaz): break
ḫi.a: various (things)
íb.si$_8$ (~*mitḫartum*): confrontation
íb.si$_8$: (is) equalside
íb.tag$_4$ (~*šapiltum*): remainder
íd (~*nārum*): river
igi (~*igûm*): *igûm*
igi *n* (g̃ál.bi): igi *n* / *n*th part
igi.bi (~*igibûm*): *igibûm*
igi.du$_8$ (~*tammar*)): you see
igi.te.en (~*igitēnnum*): fraction
igibûm (~igi.bi): *igibûm*
igitēnnum (~igi.te.en): fraction
igûm (~igi): *igûm*
íl (~*našûm*): raise
imtaḫḫar (<*maḫārum*): confronts itself
ina (~.ta): from/by
inanna: now
inūma: as
ištēn ... *ištēn*: one ... one
ištēn ...*šanûm*: the first ... the second
ištēnum: one
ištu: out from
itti (~ki): together with
kamārum (~g̃ar.g̃ar, UL.GAR): accumulate
kašāṭum: cut away
ki (~*itti*): together with
KI (~*qaqqarum*): ground
kìd (~*epēšum*): proceed/proceeding
kī maṣi: corresponding to what
ki.2, ki.3, etc.: nº 2, nº 3, etc.
ki.lá (~*kalakkum*?): excavation
ki(.ta) (~*šaplûm*): lower
kī(*ma*) (~gin$_7$(.nam)): as much as
kiyā: how much each
kīam: thus
kimrātum (<*kamārum*): things accumulated
kīnum (~gi.na): true
kippatum (~gúr): circle
kú, see gu$_7$
kud (~*nakāsum*): cut down
kud (~*ḫaṣābum*): break off
kullum: hold

kumurrûm (<*kamārum*) (~g̃ar.g̃ar): accumulation
kúr (~*nukkurum*): alter, be altered; alternate
kùš (~*ammatum*): cubit
la (~nu): not
LAGAB (~*mitḫartum*): confrontation
lal (~*maṭûm*): be(come) small(er) / diminish
lapātum: inscribe
leqûm: take
letûm: split
libbā: in the inside (*ina libbā*: from the inside)
libbum (~šà): inside
lugúd.da: shorter
lul (~*sarrum*): false
-ma: after verbs ":", after a noun *X* "that X"
ma.na (~*manûm*): mina
maḫārum: confront
makṣarum: bundling
makāsum: collect (taxes, rent)
mala (~a.na): so much as
manātum: contribution
manûm (~šid): count
manûm: number
manûm (~ma.na): mina
maṭûm (~lal (tur)): be(come) small(er)
meḫrum (<*maḫārum*; ~gaba(.ri)): counterpart
mindatum: measure
mīnûm (~en.nam): what
mišlum (~šu.ri.a): half
mitḫartum (<*maḫārum*; ~LAGAB; ~íb.si$_8$): confrontation
mu (~*aššum*): since
murub$_4$ (~*qablûm*): middle
mūšarum (~sar): sar
muttarittum: descendant
muttatum: half-part
nadānum (~sum): give
nadûm: lay down
nakāsum (~kud): cut down
nakmartum (<*kamārum*): accumulated
nārum (~íd): river
nasāḫum (~zi): tear out
nāsḫum (<*nasāḫum*): the tearing-out
našûm (~íl): raise
nēmelum: profit
nēpešum (<*epēšum*): procedure
nesûm: move away
níg.šid: account
NIGIN (~*šutakūlum*): make hold
nig̃in, see šu.nig̃in
nig̃ín(.na) (~*târum*): turn around
nim (~*elûm*, ~*našûm*): lift
nindan: nindan
nu (~*la*, *ul*(*a*)): not
pàd: see
paṭārum (~du$_8$): detach
parāsum (*warkatam*) (~kud): distinguish (what is below)
pirkum: bar
pūtum: front (Eshnunna also width)
qa (~silà): silà
qablum (~murub$_4$): middle
qabûm (~dug$_4$): say
qanûm (~gi): reed
qaqqarum (~KI): ground
qātum: hand
ra (~*ana*; at times ~*ina*): to (at times "from")
RÁ (~*alākum*): go
rabûm (~gal, ~GU.LA): great, be(come) great(er)
ramānišu: itself
redûm, see *ruddûm*
rēška likīl: may your head retain
rēšum (~sag̃): head
riāḫu (Late Babylonian): remain
ruddûm (~daḫ according to OB lexical lists): to attach (additively)
saḫar (~*eperum*): earth
saḫārum: turn around
sag̃ (~*rēšum*): head

sağ.dù (or sağ.KAK, ~*santakkum*): triangle

sağ.ki.gud: trapezium

sağ(.ki): width

santakkum (~sağ.dù/sağ.KAK): triangle

sar (~*mūšarum*): sar

sarrum (~lul): false

SI (abbreviation for dirig=SI.A; ~*watārum*): go beyond

silà (~*qa*): silà

sukud: height

sum (~*nadānum*): give

ṣeḫērum (~tur): be(come) small(er)

sehrum (~tur): small

ṣiliptum (~bar.NUN): diagonal

ša: which / that of

šà (~*libbum*): inside

šakānum (~ğar): posit

šâlum: ask

šám: buy/buying

šanûm: second

šapiltum (~íb.tag_4): remainder

šaplum (~ki.ta): lower

še (~*še'um*): grain

še'um (~še): grain

šid (~*manûm*): count

šiddum: flank (Eshnunna also "length"; by carrying "distance")

šiqlum (~gín): shekel

šittatum: left-over

ŠU.BA.AN.TU (~*banûm*?):

šu.nigin / šu.nigín: total

šu.ri.a (~*mišlum*): half

šulmum: integrity

šumma kīam išâlka umma šūma: If ⟨somebody⟩ asks you thus:

šumma: if

šumum: name

šuplum (~BÙR): depth

šūšum: threescore

šutakūlum (<*kullum*; ~gu_7): make hold

šutamḫurum (<*maḫārum*): make confront itself

.ta (~*ina*): from / by

ta.àm: each

tab (Late Babylonian ~*tepû*): join

tab (~*eṣēpum*): repeat

tabālum: take away

tag_4 (~*ezēbum*): leave

takīltum (<*kullum*): made-hold

takkirtum (<*nakārum*): change

tallum (~dal): crossbeam

tammar (<*amārum*; ~igi.du_8/pa(d)): you see

târum (~nigín): turn back

tawirtum (~garim): meadow

tebûm: go away

túl.sağ (~*kurpum*? / *būrtum*? / *kalakkum*?): excavation

tur (~*ṣeḫrum*, ~*ṣeḫērum*): small

tur (~*maṭûm*): be(come) small

ṭepû (Seleucid ~tab): join

ugu (~*eli*): over

u: and

UL.GAR (~*kamārum*): accumulate / accumulation

UL.UL, see du_7.du_7

ul(a) (~nu): not

UR.UR: oppose

uš: length

UŠ (= 1` nindan): UŠ

wasābum (~daḫ): append

wabālum: bring

wāṣbum (<*waṣābum*): the appending

wāsītum: projection

watārum (~dirig): go beyond

wāṣûm: project(ing)

wuṣubbûm (<*waṣābum*): appended

za.e (kìd.da/ta.zu.dè) (~*atta* ...): You (by your proceeding)

zâzum: divide

zi (~*nasāḫum*): tear out

zu (~*edûm*): know

Table 2: The Standard Translations with Akkadian and Logographic Equivalents

account: níg.šid
accumulate: *kamārum*/ UL.GAR / g̃ar.g̃ar
accumulated: *nakmartum* (<*kamārum*)
accumulation: *kumurrûm* / g̃ar.g̃ar / UL.GAR
alter, be altered; alternate: *nukkurum* / kúr
and: *u*
append: *waṣābum* / daḫ
appended: *wuṣubbûm*
appending, the: *wāṣbum*
as: *inūma*
as much as: *kī(ma)* / gin$_7$(.nam)
ask: *šâlum*
attach (additively): *ruddûm*
bandûm: *bandûm*
bar: *pirkum*
be(come) small(er): *ṣeḫērum* / tur / lal / *maṭûm*
breadth: dagal
break: *ḫepûm* / gaz
break off: *ḫaṣābum*
bring: *wabālum*
build: *banûm*
bundling: *maksarum*
bur: *būrum* / bùr^{iku}
buy/buying: šám
change: *takkirtum*
circle: *kippatum* / gúr
collect (taxes, rent): *makāsum*
come up (as a result): *elûm*
confront: *maḫārum*
confrontation: *mitḫartum* (rarely: LAGAB / íb.si$_8$)
confronts itself: *imtaḫḫar*
contribution: *manātum*
conversion: bal
corresponding to what: *kī maṣi*
count: šid / *manûm*
counterpart: *meḫrum* / gaba(.ri)
crossbeam: dal / *tallum*
cubit: kùš / *ammatum*
cut away: *kašātum*
cut down: *nakāsum* / kud
cut off: *ḫarāṣum*
thrust forward: *dakāšum*
depth: *šuplum* / bùr / GAM
descendant: *muttarittum*
detach: *paṭārum* / du$_8$
diagonal: *ṣiliptum* / bar.NUN
diminish: lal
distinguish (what is below): *parāsum* (*warkatam*)
divide: *zâzum*
each: ta.àm
earth: *eperum* / saḫar
equalside: íb.si$_8$
equilateral: ba.si$_8$/si / *basûm*
excavation: ki.lá / túl.sag̃
false: *sarrum* / lul
first, (the) ... the second: *ištēn* ...*šanûm*
flank: *šiddum*
fraction: igi.te.en / *igitēnnum*
front: *pūtum*
give: *nadānum* / sum
go: RÁ / *alākum*
go away: *tebûm*
go beyond: dirig / *watārum*
go beyond, over ...: *eli* ... *watārum* / ugu ... dirig
going-beyond: *watartum* / dirig
grain: *še'um* / še
great, be(come) great(er): *rabûm* / gal / GU.LA
ground: *qaqqarum* / KI
gur: gur
half: *mišlum* / šu.ri.a

half-part: *muttatum*
hand: *qātum*
head: *rēšum* / sag̃
height: sukud
here: *annikīʾam*
hold: *kullum*
how much each: *kiyā*
if: *šumma*
if ⟨somebody⟩ asks you thus: *šumma kīam išâlka umma šūma*
igi: igi (g̃ál.bi)
igibûm: *igibûm* / igi.bi
igûm: *igûm* / igi
inscribe: *lapātum*
inside: *libbum* / šà
integrity: *šulmum*
itself: *ramānišu*
join: *ṭepû* / tab (Late Babylonian)
know: *edûm* / zu
lay down: *nadûm*
leave: *ezēbum* / tag_4
left-over: *šittatum*
length: uš
lift: *elûm* / nim
longer: gíd.da
lower: *šaplum* / ki(.ta)
made-hold: *takīltum*
make confront itself: *šutamḫurum*
make encounter: $du_7.du_7$
make hold: *šutakūlum* / NIGIN / $gu_7(.gu_7)$
may your head retain: *rēška likīl*
meadow: garim / *tawirtum*
measure: *mindatum*
middle: *qablum* / $murub_4$
mina: *manûm*
moiety: *bāmtum*
move away: *nesûm*
name: *šumum*
nindan: nindan
not: *la* / *ul*(*a*) / nu
now: *inanna*
number: *manûm*
nº 2, etc: ki.2, ...
one: *ištēnum*
one ... one: *ištēn* ... *ištēn*
oppose: UR.UR
out from: *ištu*
over-going: *elēnu*
over: *eli* / ugu
posit: *šakānum* / g̃ar
part, *n*th: igi *n* (g̃ál.bi)
procedure: *nēpešum*
proceed/proceeding: *epēšum* / kìd
profit: *nēmelum*
project(ing): *wāṣûm*
projection: *wāṣītum*
raise: *našûm* / íl
reed: *qanûm* / gi
remain: *riāḫu* (Late Babylonian)
remainder: *šapiltum* / $íb.tag_4$
repeat: *eṣēpum* / tab
rest: *aḫertum*
river: *nārum*/ íd
sar: sar / *mūšarum*
say, saying: *qabûm* / dug_4
second: *šanûm*
see: *amārum* / (Sargonic, Ur) pàd
shorter: lugúd.da
silà: *qa* / silà
since: *aššum* / (LB) MU
single out: *bêrum*
small: tur / lal / *ṣeḫrum* / *ṣeḫērum* / *maṭûm*
so much as: *mala* / a.na
steps of / step: a.rá
surface: *eqlum* / a.šà
shekel: *šiqlum* / gín
take: *leqûm*
take away: *tabālum*
task: éš.gàr
tear out: *nasāḫum* / zi
things accumulated: *kimrātum*
threescore: *šūšum*
thus: *kīam*
split: *letûm*
tearing-out, the: *nāsiḫum*
together with: *itti* / ki
total: šu.nig̃in / šu.nig̃ín
trapezium: sag̃.ki.gud

triangle: *santakkum* / sağ.dù / sağ.KAK
true: *kīnum* / gi.na
turn around: *saḫārum*
turn back: *târum* / niğín(.na)
until: adi
upper: *elûm* / an(.ta/na)
UŠ: UŠ (= 1` nindan)
various (things): ḫi.a
what: *mīnûm* / en.nam / a.na.àm
width: sağ(.ki)
you (by your proceeding): *atta* (*ina epēšika*) / za.e (kìd.da/ta.zu.dè)
you see: *tammar* / igi.du$_8$

Chapter III
Select Textual Examples

The present chapter contains a sequence of problems in which the basic techniques and methods of Old Babylonian "algebra" are used. At first we shall have another look at our introductory example in bilingual format, and afterwards turn to problems #2 and #3 from the same theme text on squares.

BM 13901 #1[74]

Obv. I

1. The surfa[ce] and my confrontation I have accu[mulated]: 45´ is it. 1, the projection,
a.šà$^{!}$[am] *ù mi-it-ḫar-ti ak-m*[*ur-m*]*a* 45.e 1 *wa-ṣi-tam*

2. you posit. The moiety of 1 you break, [3]0´ and 30´ you make hold.
ta-ša-ka-an ba-ma-at 1 *te-ḫe-pe* [3]0 *ù* 30 *tu-uš-ta-kal*

3. 15´ to 45´ you append: ⌊by] 1, 1 is equalside. 30´ which you have made hold
15 *a-na* 45 *tu-ṣa-ab-ma* 1-⌊e] 1 íb.si$_8$ 30 *ša tu-uš-ta-ki-lu*

4. in the inside of 1 you tear out: 30´ the confrontation.
lìb-ba[75] 1 *ta-na-sà-aḫ-ma* 30 *mi-it-ḫar-tum*

---

74 Based on the transliteration in [MKT III, 1].

75 It has been proposed (Mogens Trolle Larsen, Jöran Friberg and others) that the presumed *lìb-ba* should be read as a Sumerogram, as šà.ba < *šà-bi-a, šà being the logogram for *libbum*, /bi/ the locative suffix and /a/ the locative suffix (cf. [SLa, §182]); the meaning, however, would then be a grammatically rather

The statement of the problem, we observe, "accumulates" the area and the side of the square (the "surface" and the "confrontation", respectively), which allows the author to add the measuring numbers for the area (the "surface") and the side (the "confrontation") without asking whether this addition has any concrete meaning. In order to obtain such a concrete interpretation the text has to "posit" the "projection" *1* (protruding toward the left in Figure 2). This allows it to interpret the measure *s* of the side concretely as a rectangular surface ⊏⊐(1,*s*), which together with the surface □(*s*) of the square becomes 45´. The "projection"/*wāṣītum*, is a verbal noun derived from *waṣûm*, "to go out"; as indicated above, it designates *inter alia* something protruding or projecting, e.g., from a building. In mathematical texts, its value is always *1*, and its designates the breadth that transforms a "Euclidean line" into a "broad line".[76]

In line 2, the outer "moiety" of the "projection" – its "necessary" or "natural" half, we remember – is broken off and "made hold", i.e., the two "moieties" (with adjacent semi-rectangles) are made the sides of a rectangle (actually a square) with area 30´×30´ = 15´. In the same process, the original composite rectangle □(*s*)+⊏⊐(1,*s*) is transformed into a gnomon of the same area 45´.

In line 3, the square 15´ is "appended" (quite palpably, we see) to the gnomon 45´, producing a new square of area 45´+15´ = 1, "by" or alongside which 1 is said to "be equalside". From this latter 1 that "moiety" 30´ "which was made hold" is "torn out" (another quite palpable process), leaving us with the original (vertical) side of the square, which must hence be 1–30´ = 30´.

The translation of *libba* as "in the inside" may be too weak; originally, *libbum* means "heart", "bowels", or "womb", or (metaphorically) the contents, e.g., of a letter.

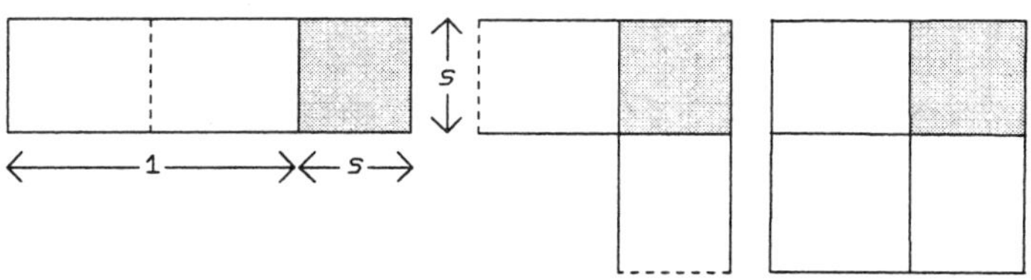

Figure 2. The procedure of BM 13901 #1, still in slightly distorted proportions.

impossible "in its inside 1" (the expanded šà.ba *ša* 1, "in its inside, that of 1", would be possible, but still not in agreement with the style of the present text). A further objection to the Sumerographic reading is the presence of the expression *i-na lib-ba* in other texts (below, we shall encounter it in BM 85210+VAT 6599 and in YBC 6504), in which the locative suffix of the Sumerographic reading would be a repetition of the preposition *ina*. I prefer to stick to Thureau-Dangin's explanation of the phrase as an Akkadian locative accusative ([1936a: 31 n.4], cf. [GAG, §146]).

76 On this understanding of lines as provided with an implicit breadth equal to fundamental unit (which turns out to be quite widespread in pre-Modern practical geometries), see [Høyrup 1995a].

In the Old Babylonian age, the word begins also to be used quasi-prepositionally as in the present translation, but always in situations where an interior or body is involved. That this is also the case for the mathematical texts follows from the situations where the word does *not* occur: The texts may "append to q" or "append to the inside of q", but they will never "raise to the inside of q".

The translation of the suffix .e in line 1 as "is it" assumes that this sign (employed here and in other places of the tablet in cases when the outcome of an operation is not used in the next operation but is followed by another number) is a borrowing of the Sumerian ergative suffix (but distorted or misunderstood by the Akkadian scribe, cf. note 42.). This assumption is hypothetical – but see p. 66.

We observe that the procedure is wholly analytical, in the sense that the unknown "confrontation" is treated as if it were given, until it can be extricated from the complex relationship in which is was initially buried. The numerical steps are exactly those of a modern algebraic solution of the equation $x^2+1\cdot x = \frac{3}{4}$.

BM 13901 #2[77]

Obv. I

5. My confrontation inside the surface [I] have torn out: 14`30 is it. 1, the projection,
mi-it-ḫar-ti lìb-bi a.šà [*a*]*s-sú-uḫ-ma* 14,30.e 1 *wa-ṣi-tam*

6. you posit. The moiety of 1 you break, 30′ and 30′ you make hold,
ta-ša-ka-an ba-ma-at 1 *te-ḫe-pe* 30 *ù* 30 *tu-uš-ta-kal*

7. 15′ t⌈o 14`30 you appe]nd: by 14`30°15′, 29°30′ is equalside.
15 *a-*⌈*na* 14,30 *tu-ṣa-*]*ab-ma* 14,30,15.e 29,30 íb.si$_8$

8. 30′ which you have made hold to 29°30′ you append: 30 the confrontation.
30 *ša tu-uš-ta-ki-lu a-na* 29,30 *tu-ṣa-ab-ma* 30 *mi-it-ḫar-tum*

#2 is the obvious subtractive companion piece to #1. However, the parallel fails in two respects. Firstly, the "confrontation" is assumed to be 30, not 30′; the reason is quite simple, namely, that the "confrontation" has to be less than the surface in order to allow subtraction. Secondly, the subtraction is the inverse operation of "appending", namely, "to tear out". This shows that no exact subtractive reversal of "accumulation" existed (even subtraction by comparison requires that the two entities involved *can* be compared; as observed, the inversion of an accumulation is a splitting into components). The statement therefore jumps to the geometric interpretation, assuming that the "confrontation" that is subtracted is a "broad line" provided with a "projec-

---

77 Based on the transliteration in [MKT III, 1].

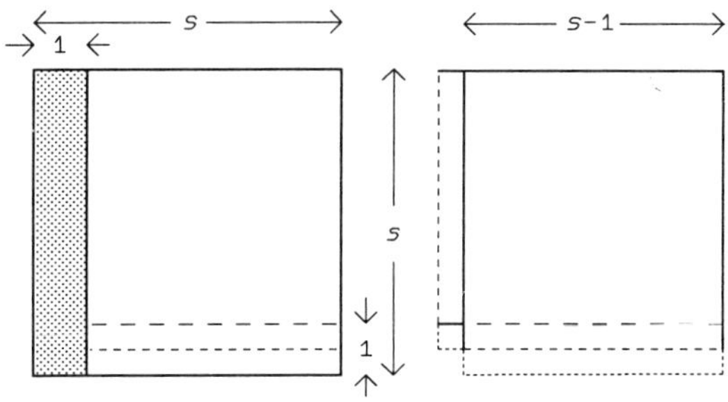

Figure 3. The procedure of BM 13901 #2.

tion" *1*.

Again, a rectangular surface results, with sides *s* and *s*–1 (see Figure 3). The excess of length over width is broken, the outer "moiety" is moved together with the adjacent semi-rectangle in order to obtain a gnomon, which makes it "hold" a square with area 30′×30′ = 15′. Again, this complementary square is "appended" to the gnomon, which results in a new square with the surface 14`30+15′ = 14`30°15′. Alongside this magnitude, 29°30′ "is equalside", i.e., the new square has the side 29°30′. To this, the "moiety" 30′ which "was made hold" is "appended" – i.e, it is put back to its original position, reestablishing the vertical side of the original square as 29°30′+30′ = 30.

BM 13901 #3[78]

Obv. I

9. The third of the surface I have torn out. The third of the confrontation to the inside
ša-lu-uš-ti a.šà *as-sú⟨-uḫ-ma⟩ ša-lu-uš-ti mi-it-ḫar-tim a-na lìb-bi*

10. of the surface I have appended: 20′ is it. 1, the projection, you posit,
a.šà*lim* *ú-ṣi-ib-ma* 20.e 1 *wa-ṣi-tam ta-ša-ka-an*

11. the third of 1 {the projec[tion}, 20′, you tear out:] 40′ to 20′ you raise,
ša-lu-uš-ti 1 {*wa-ṣi*[*-tim*} 20 *ta-na-sà-aḫ-ma*] 40 *a-na* 20 *ta-na-ši*

12. 13′20″ you inscribe. [The moiety of 20′, of the th]ird which you have {torn out}⟨appended⟩,[79]
13,20 *ta-la-pa-at* [*ba-ma-at* 20 *ša-l*]*u-uš-tim ša ta-sú-ḫu*

78 Based on the transliteration in [MKT III, 1].

79 See the commentary below concerning the passages in { } and ⟨ ⟩ in lines 11 and 12.

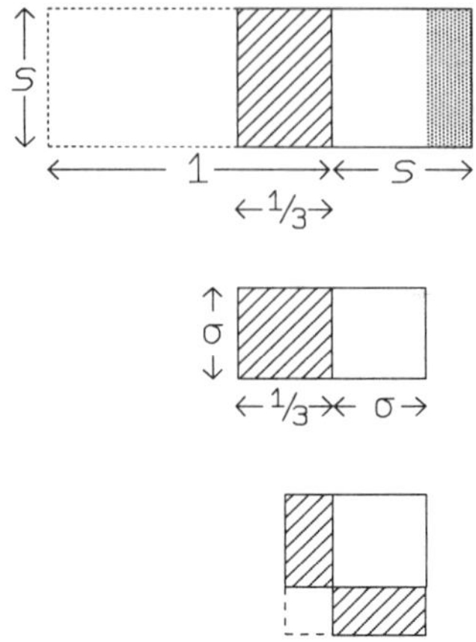

Figure 4. The procedure of BM 13901 #3.

13. you break, 10′ ⌈and 10′ you make hold, 1′40″] to 13′20″ you append,
te-ḫe-pe 10 ⌈*ù* 10 *tu-uš-ta-kal* 1,40] *a-na 13,20 tu-ṣa-ab*

14. by 15′, 30′ [is equalside. 10′ which you have made hold in the inside of 30′] you tear out: 20′.
15.e 30 [íb.si$_8$ 10 *ša tu-uš-ta-ki-lu lìb-ba* 30] *ta-na-sà-aḫ-ma* 20

15. Igi 40′ [, 1°30′, to 20′ you raise, 30′,] the confrontation.
igi 40 ĝál.b[i 1,30 *a-na* 20 *ta-na-ši-ma* 30] *mi-it-ḫar-tum*

#3 introduces two complications at a time (originally, #16 or something similar may have belonged between #2 and #3, introducing only one of them): firstly, not the whole area of the square is involved but only ⅔ of it; the problem is thus *non-normalized*. Secondly, not the whole side but only ⅓ of it is added. The presence of two different thirds has caused some confusion; only knowledge of similar problems (from the present and from other tablets) therefore allows us to decipher the procedure.

The statement is interesting in itself. At first one third of the surface is "torn out"; as observed in connection with #2, this implies that we jump from measuring numbers to their geometrical interpretation. The jump has the striking and singular effect that even the third of the "confrontation" is "*appended*"; once we are in the geometric interpretation we stay there.

As a first step in the solution, the "projection" is "posited", as we would expect. That its third is taken is also regular (see Figure 4). Then, however, the third of the surface which is to be "torn out" is also identified with the third of the "projection". The remaining part of the surface is found, correctly, to be 40′ = ⅔ [of the original surface]. The geometrical configuration possessing the total area 20′ is thus composed from two rectangles, ⊏⊐(40′*s*,*s*) and ⊏⊐(20′,*s*), which makes immediate application of the trick of #1 and #2 impossible.

At the end of line 11, this area 20′ is therefore multiplied by 40′. The aim is to find the total area of a new configuration composed from the square on 40′*s* and 20′ times the side of this same square.[80] In the second step of Figure 4 this is shown as a change of vertical scale by a factor 20′. The resulting configuration is similar to that of #1, and it is cut, pasted and completed in the same way, as seen in the third step; however, in line 12 the third of the original surface which was removed is again mistaken for the third of the "projection" which sticks out, and which is actually the entity to be bisected (this unmistakeable confounding of "tearing-out" and "appending" in line 12 supports the assumption that a parallel mix-up has occurred in line 11). The result of the process is that the side [40′*s*] of the new square is 20′, whence the original "confrontation" is found to be 30′. Both the original scaling and its reversion are made by "raising".

YBC 6967[81]

More common than "square problems" in the Old Babylonian corpus are "rectangle problems". Very often, complex problems are meant to be reduced to one of the simpler cases of finding the sides of a rectangle of which the area and either the sum of the two sides or their difference are given.

The present problem is already of the latter type, apart from the fact that it does not deal with geometry but with two numbers belonging together in the table of reciprocals – i.e., two numbers whose product is 1 or (in the actual case, cf. note 46) 1` = 60. The problem is thus a representative of the class of *igûm-igibûm* problems ("the igi and its igi", see p. 35), and thereby an adequate exemplification of the relation between actual problem and standard representation (see p. 34).

Obv.

1. [The *igib*]*ûm* over the *igûm*, 7 it goes beyond
[igi.b]i *e-li* igi 7 *i-ter*

2. [*igûm*] and *igibûm* what?
[igi] *ù* igi.bi *mi-nu-um*

3. Yo[u], 7 which the *igibûm*
a[*t-t*]*a* 7 *ša* igi.bi

---

80 That this is the aim of the multiplication 40′·20′ is of course difficult to see in the present problem, where so many numbers possess the value 20′. This is the point where we have to make use of the numerous parallel cases, several of which are found in the same tablet.

81 Based on the transliteration in [MCT, 129].

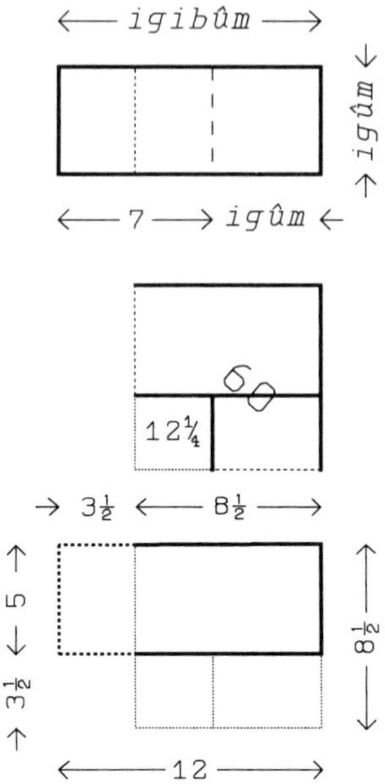

Figure 5. The procedure of YBC 6967.

4. over the *igûm* goes beyond
ugu igi *i-te-ru*

5. to two break: 3° 30′;
a-na ši-na ḫe-pé-ma 3,30

6. 3° 30′ together with 3° 30′
3,30 *it-ti* 3,30

7. make hold: 12°15′.
šu-ta-ki-il-ma 12,15

8. To 12°15′ which comes up for you
a-na 12,15 *ša i-li-kum*

9. [1` the surf]ace append: 1`12°15′.
[1 a.ša$^{!}$]$^{a\text{-}am}$ *ṣí-ib-ma* 1,12,15

10. [The equalside of 1`]12°15′ what? 8° 30′.
[íb.si$_8$ 1],12,15 *mi-nu-um* 8,30

11. [8° 30′ and] 8° 30′, its counterpart, lay down.
[8,30 *ù*] 8,30 *me-ḫe-er-šu i-di-ma*

Rev.

1. 3° 30′, the made-hold,
3,30 *ta-ki-il-tam*

2. from one tear out,
i-na iš-te-en ú-su-uḫ

3. to one append.
a-na iš-te-en ṣí-ib

4. The first is 12, the second is 5.
iš-te-en 12 ša-nu-um 5

5. 12 is the *igibûm*, 5 is the *igûm*.
12 igi.bi 5 *i-gu-um*

If the *igibûm* is x and the *igûm* is y (these symbols, which we habitually see as unknown *numbers*, are adequate in the present case), the problem is

$$x–y = 7 , \quad x·y = 60 .$$

Whereas *we* are accustomed to represent (e.g.) the length and width of a rectangle by unknown numbers, the Old Babylonian calculator represents his unknown numbers by his standard instruments – measurable line segments – and speaks in line 9 of the product as a "surface". The procedure is already familiar (see Figure 5, and compare with Figure 2 and Figure 3): We know that the length (*igûm* $= x$) of the rectangle exceeds its width (*igibûm* $= y$) by 7. This excess is bisected and the rectangle transformed into a gnomon, still of area 60, which is "appended" to the completing square $\square(^7/_2) = 12\,^1/_4$. The side of the completed square – its "equalside" – must then be $8\,^1/_2$, which is "laid down" together with "its counterpart", as two sides of the completed square. "Tearing out" that part of the excess which was moved in order to "hold" the complement (termed *takīltum*, "the made-hold") we get the *igûm;* putting it back to its original position we restore the *igibûm*.

Even though familiar on the general level, the procedure and wording differs from what we have seen above in several respects. Firstly, the gnomon is "appended" to the completing square, not *vice versa*. This switching of roles is possible because both entities are already in place – in cases where one entity stays in place and the other is moved according to the geometric reading of the texts, it is invariably the entity that is moved which is "appended". Secondly, the "equalside" appears as a noun. Thirdly, the relative phrase "which you have made hold" (*ša tuštakīlu*) is replaced by a verbal noun, the *takīltum* (translated "the made-hold"), in a way that leaves no doubt that the two expressions are equivalent. This, we remember (see p. 23), was the decisive grammatical argument that the verb for rectangularization must be read *šutakūlum*, "to make hold (each other)", and not as *šutākulum*, "to make eat (each other)".

Problems about square areas and sides will only have to "append" the "made-hold" (if the sum of area and side is given), or only to "tear it out" (if the given quantity is the difference). In the present problem, both processes take place; it is noteworthy that the "tearing-out" takes place before the "appending", even though the Babylonians would normally treat addition before subtraction, exactly as we do (for which reason BM 13901 has the sum of area and sides before the difference). The explanation is that it is *the same* entity which is "torn out" and "appended"; evidently, it cannot be "appended" before it is at disposition, which presupposes the "tearing-out". When this is not the case, "appending" comes before "tearing-out" – as we shall see time and again in what follows. This principle of concrete meaningfulness turns out not to be respected in all text groups but to be a norm which is chosen deliberately (see below, p. 383). I shall therefore refer to it in the following as the "norm of concreteness".

The text illustrates that the Old Babylonian operation with lines and areas was really *an algebra*, if this be understood as analytical procedures in which unknown quantities are *represented* by functionally abstract entities – numbers in our algebra, measurable line segments and areas in the Old Babylonian technique.

BM 13901 #10[82]

The following problem introduces a technique which is needed for the majority of complex problems, and which I shall speak of as the "accounting technique". The ancient texts appear to have no name for it but only for its outcome – cf. the ensuing commentary.

Obv. II

11. The surfaces of my two confrontations I have accumulated: 21°15´.
a.šà *ši-ta mi-it-ḫa-ra-ti-ia ak-mur-ma* 21,15

12. Confrontation (compared) to confrontation, the seventh it has become smaller.
mi-it-ḫar-tum a-na mi-it-ḫar-tim si-bi-a-tim im-ṭi

13. 7 and 6 you inscribe. 7 and 7 you make hold, 49.
7 *ù* 6 *ta-la-pa-at* 7 *ù* 7 *tu-uš-ta-kal* 49

14. 6 and 6 you make hold, 36 and 49 you accumulate:
6 *ù* 6 *tu-uš-ta-kal* 36 *ù* 49 *ta-ka-mar-ma*

15. 1`25. Igi 1`25 is not detached. What to 1`25
1,25 igi 1,25 *ú-la ip-pa-ṭa-ar mi-nam a-na* 1,25

16. may I posit which 21°15´ gives me? By 15´, 30´ is equalside.
lu-uš-ku-un ša 21,15 *i-na-di-nam* 15.e 30 íb.si$_8$

17. 30´ to 7 you raise: 3° 30´ the first confrontation.
30 *a-na* 7 *ta-na-ši-ma* 3,30 *mi-it-ḫar-tum iš-ti-a-at*

18. 30´ to 6 you raise: 3 the second confrontation.
30 *a-na* 6 *ta-na-ši-ma* 3 *mi-it-ḫar-tum ša-ni-tum*

The problem, as we see, deals with two squares, and gives the sum of their areas. In this as well as in the analogous #11, the sides differ by $\frac{1}{7}$ – but in

82 Based on the transliteration in [MKT III, 2*f*].

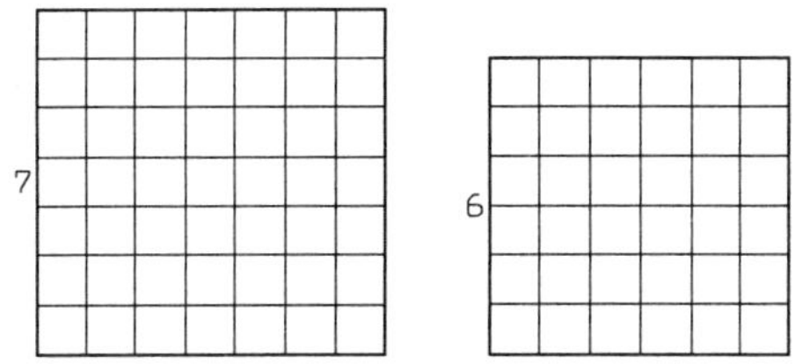

Figure 6. The two squares of BM 13901 #10.

the present problem, the second side is obtained by decreasing the first by this fraction, whereas the second side exceeds the first by the same amount in #11. #11 uses the familiar verb *watārum*, "go beyond", while #10 employs *maṭûm*, "to be(come) small(er)". In combination the two texts show that the difference between the two verbs has nothing to do with any difference between positive and negative numbers (as sometimes asserted) but at most with a distinction between additive and subtractive roles for numbers[83] – which in the present case amounts to nothing but a way to say whether $^1/_7$ is to be taken of the larger or the smaller number.

In #10 (#11 is solved by the same method), the first step is either that of a "single false position", or at least close to it in spirit: 7 is chosen as a number from which $^1/_7$ can easily be removed. Doing this leaves 6 (the single steps in this argument are left implicit, but they are spelled clearly out elsewhere, e.g., in VAT 7532, rev. *6f* – below, p. 209).

These numbers are "inscribed": taking again VAT 7532 as a model (together with the drawing in BM 15285 #24, below, p. 60), we may understand this as shown in Figure 6: Two lines of length 7 and 6 are drawn[84] and "made hold", which produces squares of area 49 and 36, and thus with a total area equal to 1`25.

At this point two interpretations become possible. Either the two squares are imagined to *be* those spoken of in the statement. Then the area 1`25 represents the total area measured in an undetermined unit equal to the area of the small square, and the subsequent steps (dividing and determining the square root) find the value of the side of the small square to be 30′ [nindan]. Alternatively, the squares 6×6 and 7×7 are thought of as having really the

83 An explicit distinction of this kind is strongly suggested by other texts – thus by TMS XVI, lines 5 and 23, see below, p. 85.

The problems of subtractive and of supposedly negative numbers in Old Babylonian mathematical texts is dealt with in detail in [Høyrup 1993b]; cf. below, p. 294.

84 Maybe only imagined. Whether in the actual case "to inscribe" means "to draw" (as when we "inscribe until twice") or tells us that the number is really written alongside the line (as done in VAT 7532) is not clear. The presence of a line, however, whether drawn or imagined, and whether spoken of implicitly or explicitly, follows from the use of the construction verb *šutakūlum*, "to make hold".

sides 7 [nindan] and 6 [nindan], in which case the division and the taking of the square root yield first the quadratic and next the linear proportion between the original and the new set of squares. Reasons of didactic simplicity might suggest the former interpretation – it seems to be easier to keep one set of squares present to the minds of students than two sets; strictly speaking, this method is not that of a false position but rather the "bundling" of which the texts sometimes speak explicitly (see p. 66). However, indubitable false positions are sometimes used (thus several times in VAT 7532, below, p. 209); moreover, the incontrovertible subdivisions of rectangles into squares in BM 8390 and TMS VIII (below, pp. 61 and 188, respectively) determine the number of such squares by means of a "raising" (the sides being *already there*, which makes construction superfluous), and not by "holding". All in all, the genuine false position seems the better explanation of the present text where 7 and 6 are "made hold".

The text gives no descriptive name to the number 1`25, i.e., to the number of small squares – a number which in translations into symbolic algebra would present itself as a coefficient.[85] Elsewhere, however, a "coefficient" *is* spoken of explicitly – the Susa text TMS XVI, l. 9–10, refers to the "coefficients" of length and width in a rectangle problem as *kīma* uš/sag̃, "as much as [there is] of lengths/widths" (below, p. 87, cf. p. 101). This term is one reason that I speak of the computation of a coefficient as an "accounting" procedure – the other reason is that what goes on *is* in fact a kind of accounting, computing additive and (in other cases) subtractive contributions, and finding the total.

BM 15285 #24[86]

The brief problem that follows comes from a theme text about the subdivision of a square with side 1 UŠ = 1` nindan.[87] For each problem, a diagram shows the division.

1. [1 UŠ the co]nfrontation.
[1 UŠ *mi-i*]*t-ḫa-ar-tum*

85 E.g., representing the first "confrontation" by $7z$ and the second by $6z$, we get the equation $(7\times7+6\times6)z^2 = 21°15'$, and thus $1`25\cdot z^2 = 21°15'$.

86 This enumeration corresponds to the complete edition of the text in [Robson 1999: 208–217]. In [MKT I, 138], the problem is #10.

87 The use of the same sign has led all workers on the text to read UŠ as uš, "length", without caring that this reading is grammatically impossible. The reading of the sign as the length unit UŠ is the only possibility if we believe the scribe to have understood the language in which he was writing – see [Høyrup 2000a: 157].

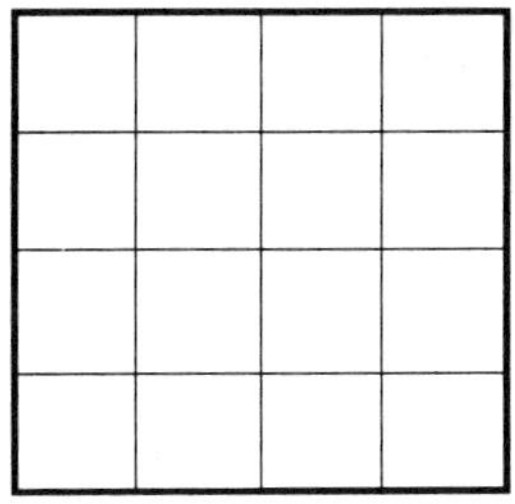

Figure 7. The diagram of BM 15285 #24.

2. Inside, 16 confrontations
· *lìb-ba* 16 *mi-it-ḫa-ra-tim*

3. I have laid down. Their surface what?
ad-di a.šà.bi en.nam

The problem is included here because of its evident similarity with the previous (and the next) problem. However, the text may also give occasion to a few terminological remarks.

Firstly, it demonstrates that the "confrontation" is both the (measure of the) side of the square (quite explicit in line 1) and the quadratic configuration as a whole (no less obvious in line 2, both because the "confrontation" of line 1 possesses an inside, and because no counting of sides of the smaller squares would find 16). Other problems of the text use the Sumerogram íb.si$_8$ instead in either function.

Secondly, the text leaves no doubt that "laying down" (*nadûm*) may mean *laying down in a drawing*.

VAT 8390 #1[88]

A third variation on the theme of the square subdivided in smaller squares is this:

Obv. I

1. [Length and width] I have made hold: 10` the surface.
[uš *ù* sağ] *uš-ta-ki-il-ma* 10 a.šà

2. [The length t]o itself I have made hold:
[uš *a*]-*na ra-ma-ni-šu uš-ta-ki-il-ma*

---

88 Based on the transliteration in [MKT I, 335*f*]. #2 of the same tablet is a close parallel, replacing only the condition $\square(l) = 9\square(l\text{–}w)$ by $\square(w) = 4\square(l\text{–}w)$; all restitutions of damaged passages are therefore beyond doubt.

3. [a surface] I have built.
[a.šà] *ab-ni*

4. [So] much as the length over the width went beyond
[*ma*]-*la* uš ugu sağ *i-te-ru*

5. I have made hold, to 9 I have repeated:
uš-ta-ki-il a-na 9 *e-ṣi-im-ma*

6. as much as that surface which the length by itself
ki-ma a.šà-*ma ša* uš *i-na ra-ma-ni-šu*

7. was [ma]de hold.
uš-t[*a*]-*ki-lu*

8. The length and the width what?
uš *ù* sağ en.nam

9. 10` the surface posit,
10 a.šà ğar.ra

10. and 9 (to) which he has repeated posit:
ù 9 *ša i-ṣi-pu* ğar.ra-*ma*

11. The equalside of 9 (to) which he has repeated what? 3.
íb.si$_8$ 9 *ša i-ṣi-pu* en.nam 3

12. 3 to the length posit
3 *a-na* uš ğar.ra

13. 3 t[o the w]idth posit.
3 *a-n*[*a* s]ağ ğar.ra

14. Since "so [much as the length] over the width went beyond
aš-šum ma-[*la* uš] ugu sağ *i-te-ru*

15. I have made hold", he has said
uš-ta-k[*i-il*] *iq-bu-ú*

16. 1 from ⌊3 which t]o the width you have posited
1 *i-na* ⌊3 *ša a-n*]*a* sağ *ta-aš-ku-nu*

17. tea[r out:] 2 you leave.
ú-[*sú-uḫ-m*]*a* 2 *te-zi-ib*

18. 2 which yo[u have l]eft to the width posit.
2 *ša t*[*e-z*]*i-bu a-na* sağ ğar.ra

19. 3 which to the length you have posited
3 *ša a-na* uš *ta-aš-ku-nu*

20. to 2 which ⟨to⟩ the width you have posited raise, 6.
a-na 2 *ša* ⟨*a-na*⟩ sağ *ta-aš-ku-nu* íl 6

21. Igi 6 detach: 10´.
igi 6 *pu-ṭur-ma* 10

22. 10´ to 10` the surface raise, 1`40.
10 *a-na* 10 a.šà íl 1,40

23. The equalside of 1`40 what? 10.
íb.si$_8$ 1,40 en.nam 10

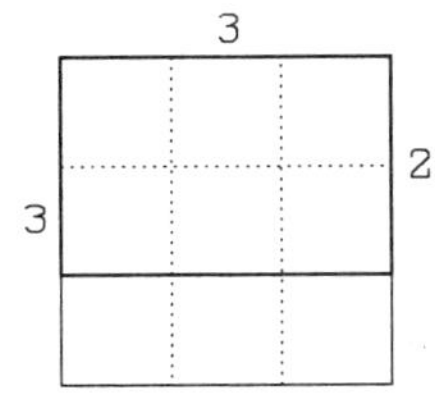

Figure 8. The configuration dealt with in VAT 8390 #1.

Obv. II

1. 10 to 3 wh[ich to the length you have posited]
10 *a-na* 3 *š[a a-na* uš *ta-aš-ku-nu]*

2. raise, 30 the length.
íl 30 uš

3. 10 to 2 which to the width you have po[sited]
10 *a-na* 2 *ša a-na* sağ *ta-aš[ku-nu]*

4. raise, 20 the width.
íl 20 sağ

5. If 30 the length, 20 the width,
šum-ma 30 uš 20 sağ

6. the surface what?
a.šà en.nam

7. 30 the length to 20 the width raise, 10` the surface.
30 uš *a-na* 20 sağ íl 10 a.šà

8. 30 the length together with 30 make hold: 15`.
30 uš *it-ti* 30 *šu-ta-ki-il-ma* 15

9. 30 the length over 20 the width what goes beyond? 10 it goes beyond.
30 uš ugu 20 sağ *mi-nam i-ter* 10 *i-ter*

10. 10 together with [10 ma]ke hold: 1`40.
10 *it-ti* [10 *šu*]-*ta-ki-il-ma* 1,40

11. 1`40 to 9 repeat: 15` the surface.
1,40 *a-na* 9 *e-ṣi-im-ma* 15 a.šà

12. 15` the surface, as much as 15` the surface which the length
15 a.šà *ki-ma* 15 a.šà *ša* uš

13. by itself was made hold.
i-na ra-ma-ni-šu uš-ta-ki-la

The problem, still homogeneous, is more complex in structure than the previous two specimens. Even though it should not be difficult to follow the reasoning in Figure 8,[89] it is therefore illuminative in several respects.

---

89 The calculation may be mapped as follows in symbolic algebra: Since $l^2 = 9\cdot(l-w)^2$, $l = 3\cdot(l-w)$, whence $w = l-(l-w) = 3\cdot(l-w)-(l-w) = 2\cdot(l-w)$. Since

Firstly, this is one of the texts which state explicitly that "a surface I have built" when two lines have been "made hold" – probably meaning that "I have marked out a field". This is the major reason that I have chosen the integer order of magnitude instead of minutes (the other is a historically and hermeneutically irrelevant concern for transparency – if the length is 30´ and the width is 20´, both the excess and the surface they "hold" will be 10´).

We notice that many entities are termed "surfaces". Since the original users of the text had no indication of absolute order of magnitude to help them keep the different surfaces apart, this explains why the text constantly identifies them with reference to how they were brought forth.

The text shows us several of the devices which allowed the Old Babylonian calculators to formulate complex problems. In line I 4 we have the "bracket" *mala*, "so much as", and in I 6 the "sign of equality" *kīma*, "as much as", between (the measures of) different entities.

"Positing" occurs in several functions. In I 10, 9 is "posited", after which its "equalside" is found; Eleanor Robson's recent work [1999: 176–179, 245–277] makes it virtually certain that this means inscription on a clay tablet for rough work. When numbers are "posited to" the length and the width, the most likely interpretation is that they are to be inserted in a drawing as in Figure 8 (this is how numbers are inserted in the diagrams of VAT 7532 and related texts, see below, p. 209).

A striking feature of the text is the use of "raising" in I 20 and II 7. The former seems to concern the determination of a rectangular area in terms of the square on l–w, and the latter beyond doubt finds the area of a rectangle. In both cases, however, the rectangle in question *is already there*, there is no construction ("building") to be performed. This distribution of "holding" and "raising" is certainly not accidental – it is repeated in #2 of the tablet without the slightest variation. The squares on l and l–w, in contrast, are hypothetical entities constructed only (mentally?) for the purpose of formulating the problem; they have to be constructed anew in the proof. Most likely, the full Figure 8 thus does not correspond to what the Old Babylonian calculator would trace; all he would "lay down" as a support for his reflections might well be the bold line – and 2 and 3 would be then be "posited" consecutively to *the same width*, as also suggested by the lack of distinctive epithets like those given to the different "surfaces".

$A = l{\times}w =$ 10`, we therefore have $2{\cdot}3{\cdot}(l–w)^2 = 6{\cdot}(l–w)^2$, whence $(l–w)^2 =$ 10`/6 = 1`40, $l–w = 10$, $l = 3{\cdot}(l–w) = 30$, $w = 2{\cdot}(l–w) = 20$.

YBC 6295[90]

This small text (no problem but an instruction) shows how homogeneous considerations similar to those of the three preceding texts might be used even in three dimensions, and informs us about the name that might be given to the method. The issue is how to find the cube root of a number – or, actually, the side of a cubic volume – which the table "has not given to you", that is, which is not listed in the standard table of cube roots.

1. The bundl[ing] of a (cubic) equilateral.
[*ma-*]*ak-ṣa-ru-um ša* ba.si

2. The equilateral of 3° 22′ 30″ what?
ba.si 3,22,30 en.nam

3. Since (for) 3° 22′ 30″ the equilateral it has not given to you,[91]
aš-šum 3,22,30 ba.si *la id-di-nu-kum*

4. 7′ 30″ (for) which the equilateral it gives you
7,30 *ša* ba.si *i-na-di-nu-kum*

5. below 3° 22′ 30″ posit:
ša-pa-al 3,22,30 ĝar.ra.ma

6. 3° 22′ 30″ 7′ 30″.
3,22,30 7,30

7. The equilateral of 7′ 30″ what? 30′.
ba.si 7,30 en.nam 30

8. Igi 7′ 30″ detach: 8.
igi 7,30 *pu-ṭur-ma* 8

9. 8 to ⟨3°⟩22′ 30″ raise 27.
8 *a-na* ⟨3,⟩22,30 íl 27

10. The equilateral of 27 what? 3
ba.si 27 en.nam 3

11. 3 the equilateral to 30′ the second equilateral
3 ba.si *a-na* 30 ba.si *ša-ni-im*

---

90 Based on the transliteration in [MCT, 42].

91 Neugebauer and Sachs write "they" for the subject; however, since both occurrences of the verb (*id-di-nu-kum*, *i-na-di-nu-kum*) are in the subjunctive, a third person singular is equally possible and seems to me more plausible than a sudden appearance of Basil Bernstein's "restricted code". When the calculator tries to "take" an "equalside" from the table it may obviously happen that the table does not "give" it to him.

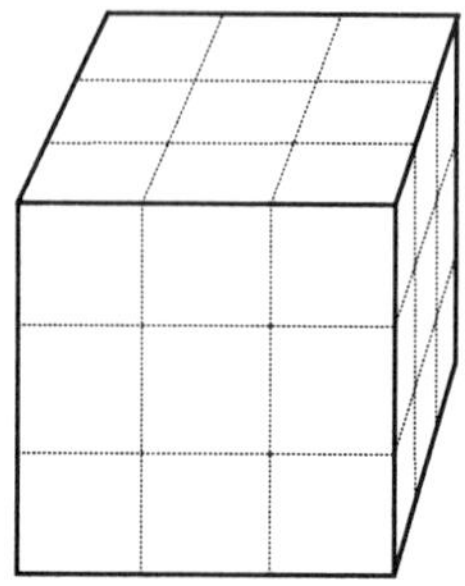

Figure 9. The division of the cube into a bundle of smaller cubes.

12. raise, 1°30.
íl 1,30

13. The equilateral of 3°22′30″ is 1°30.
ba.si 3,22,30.e 1,30

The trick, we see, consists in dividing the volume into a bundle of smaller cubic volumes which *are* given in the table, as shown in Figure 9. Such a volume is 7′30″, whose (cubic) "equilateral" is 30″; multiplication by igi 7′30″ shows that the original volume consists of 27 = 3^3 of these smaller cubes; therefore, the (cubic) "equilateral" of this original volume is 3·30′ = 1°30′.

makṣarum, "bundling", is derived from the verb *kaṣārum*, "to bind together"; below, we shall encounter the term in two-dimensional use in the text YBC 8633 (p. 254). In generalized version and without the name, we shall also encounter it in the "excavation text" BM 85200 + VAT 6599, where it is used to solve certain inhomogeneous third-degree problems.

The suffix .e in line 13 is an obvious borrowing from the phrase 3,22,30.e 1,30 ba.si$_8$, "by 3°22′30″, 1°30′ is equilateral", conserved even though ba.si$_8$ (written ba.si) has been reinterpreted as a noun, probably because it serves to separate the two numbers. Here there is little doubt that it is understood as an ergative suffix, interpreted as a nominative marker (cf. p. 52 and note 42).

BM 13901 #8–9[92]

These two problems constitute a couple, the second member of which allowed Neugebauer to reconstruct with high certainty the heavily damaged first

92 Based on the transliteration in [MKT III, 2].

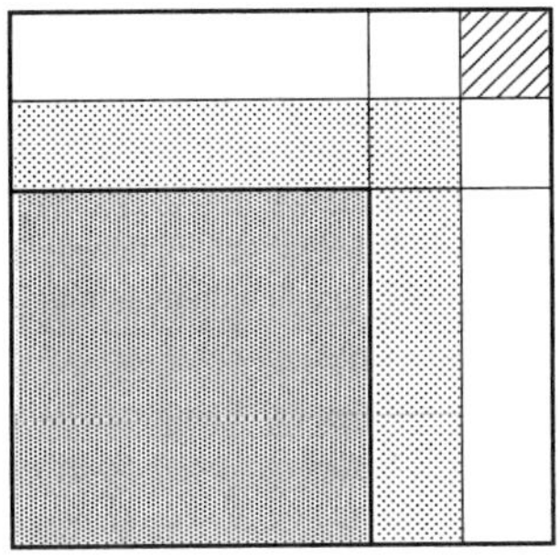

Figure 10. A possible basis for the operations of BM 13901 #11.

member.[93] In both, the sum of two square areas is given; in the first, the sum of the sides will also have been given, while the second states their difference.

Obv. I

#8

43. [The surfaces of my two confrontations I have accumulated:] 21′40″,
[a-šà *ši-ta mi-it-ḫa-ra-ti-ia ak-mur-ma*] 21,40

44. [and my confrontations I have accumulated: 50′. The moiety of 21′]40″ you break,
[*ù mi-it-ḫa-ra-ti-ia ak-mur-ma* 50 *ba-ma-at* 21,]40 *te-ḫe-pe*

45. [10′50″ you inscribe. The moiety of 50′ you break, 25′ and 25′ you m]ake hold.
[10,50 *ta-la-pa-at ba-ma-at* 50 *te-ḫe-pe* 25 *ù* 25 *tu-u*]*š-ta-kal*

46. [10′25″ inside 10′50″ you tear out: by 25″, 5′] is equalside.
[10,25 *lib-bi* 10,50 *ta-na-sà-aḫ-ma* 25.e 5] íb.si$_8$

Obv. II

1. 5′ to the first 25′ you append[: 30′ the first confrontation].
5 *a-na* 25 *iš-te-en tu-ṣa-ab*[*-ma* 30 *mi-it-ḫar-tum iš-ti-a-at*]

2. 5′ inside the second 25′ you tear out: [20′ the second confrontation.]
5 *lib-bi* 25 *ša-ni-im ta-na-sà-aḫ-m*[*a* 20 *mi-it-ḫar-tum ša-ni-tum*]

---

93 Thureau-Dangin [1936a] had proposed a different construction, which, however, makes the two problems identical, with the only variation that #8 expresses the difference by means of *maṭûm*, "to be(come) small(er)", whereas #9 uses *watārum*, "to go beyond". With hindsight this can be seen to be impossible, *maṭûm* being exclusively used in order to obtain one of the favourite *relative* differences (as in #10, see p. 59) or when the particular format of the series texts enforces the corresponding order of the members – see [Høyrup 1993b: 55–56], and below, p. 295.

#9

3. The surfaces of my two confrontations I have accumulated: 21´4[0″].
a-šà *ši-ta mi-it-ḫa-ra-ti-ia ak-mur-ma* 21,4[0]

4. Confrontation over confrontation, 10´ it went beyond.
mi-it-ḫar-tum ugù *mi-it-ḫar-tim* 10 *i-te-er*

5. The moiety of 21´40″ you break: 10´50″ you inscribe.
ba-ma-at 21,40 *te-ḫe-pe-ma* 10,50 *ta-la-pa-at*

6. The moiety of 10´ you break: 5´ and 5´ you make hold.
ba-ma-at 10 *te-ḫe-pe-ma* 5 *ù* 5 *tu-uš-ta-kal*

7. 25″ inside 10´50″ you tear out: by 10´25″, 25´ is equalside.
25 *lìb-bi* 10,50 *ta-na-sà-aḫ-ma* 10,25.e 25 íb.si$_8$

8. 25´ until twice you inscribe. 5´ which you have made hold
25 *a-di ši-ni-šu ta-la-pa-at* 5 *ša tu-uš-ta-ki-lu*

9. to the first 25´ you append: 30´ the confrontation.
a-na 25 *iš-te-en tu-ṣa-ab-ma 30 mi-it-ḫar-tum*

10. 5´ inside the second 25´ you tear out: 20´ the second confrontation.
5 *lìb-bi* 25 *ša-ni-im ta-na-sà-aḫ-ma* 20 *mi-it-ḫar-tum ša-ni-tum*

Since the reconstruction of #8 builds on the presupposition that it can be copied *mutatis mutandis* from #9, we shall first look at this latter problem.

In contrast to what we have seen until now it is not possible to read the present prescription as the description of a construction (the very first operation is the bisection of an area, which presupposes that it is already there). Instead it appears to refer to a *standard diagram*, in which the two squares can be located, and within which the "breakings", etc., take place.

The prescription is compatible with several such diagrams; all we can say is that the "moiety" of the sum of the surfaces and of the excess of one "confrontation" over the other must be easily identifiable. One possibility is shown in Figure 10. Here it is presupposed that the smaller square is drawn inside the larger square. The average surface – the "moiety" of 21´40″ – is the total dotted and hatched area, as found without difficulty by counting the various pieces involved. The square on the "moiety" of the difference between the "confrontations" (the 25″ found in lines 6–7) is the hatched area in the

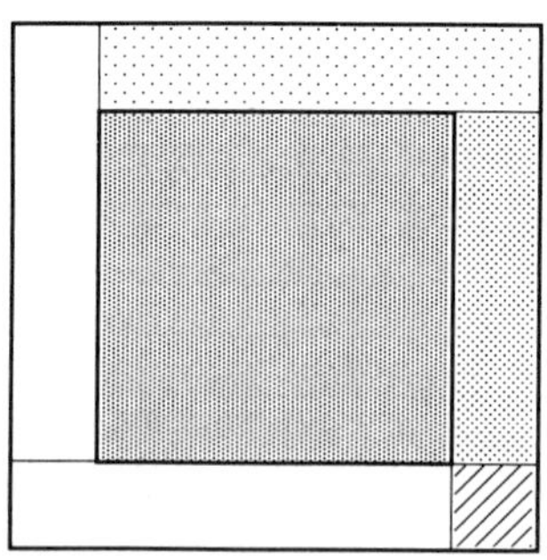

Figure 11. An alternative diagram for BM 13901 #9.

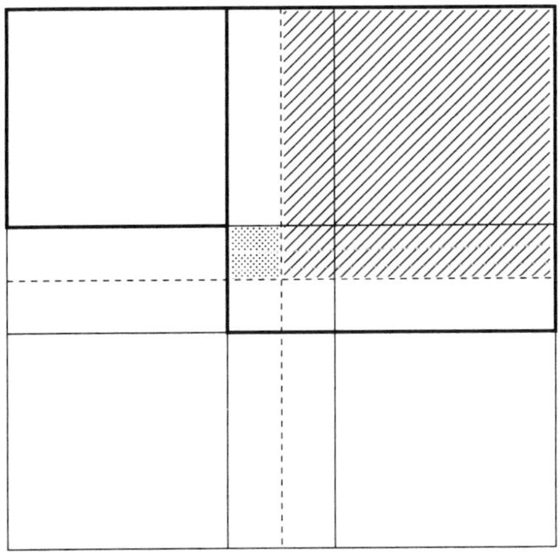

Figure 12. The diagram on which BM 13901 #8 is likely to be based.

upper right corner. When it is "torn out", we are left with the dotted area, which is the square on the average side, found to be 25′ and "inscribed" twice. "Appending" the semi-difference yields the larger "confrontation", "tearing it out" leaves the smaller one. Since the quantity that is to be "appended" and "torn out" is independent of the two sides of the dotted square (it is the "moiety" of a neutral excess, not of *a part* of the larger "confrontation"), "appending" precedes "tearing-out" without problems.

This configuration was copied from *Elements* II.8. The Babylonians are likely to have preferred a variant where the smaller square is located concentrically within the larger one, as shown in Figure 11.[94] The operations are very similar, the only difference being that the "breaking" of the difference between the "confrontations" is less manifest, since it is already split into two; in both cases, however, the "breaking" of the sum of the surfaces is more abstract than what we have seen so far.

In principle, it is possible to follow the operations of #8 on either Figure 10 or Figure 11, if only we accept that the "breaking" of the difference between the two "confrontations" be equally abstract. For several reasons, however, it is likely that a different configuration was employed, namely, the square of the sum of the two "confrontations" – see Figure 12. Firstly, this configuration turns up repeatedly in the record as a standard configur-

94 This is general a favourite configuration [Friberg 1990, § 5.4.1], which is also treated in the geometric text BM 15285; problems about two squares in the catalogue text TMS V make explicit the concentric positioning. Cf. also below on UET V, 864, p. 250.

ation;[95] secondly, the parallel offered by rectangular problems where the area and the sum of the sides is given suggest concrete representation of this sum.

The steps of #8 are easily followed in the diagram. The "moiety" of 21´40″ – that is, the average surface – is the sum of the hatched and the dotted areas. The square on the "moiety" of 50´ – that is, on the average side – is the hatched area alone. "Tearing out" the latter from the former leaves us with the dotted area (25″), which is the square on the semi-difference between the sides. This semi-difference is therefore 5´. "Appending" it to the first copy of the average side gives us the first "confrontation"; "tearing it out" from the second copy provides us with the second.

Even the procedure of #9 is followed without difficulty in Figure 12. Since *the problem* and *the procedure* need not coincide, it is therefore not to be excluded that #9 was *solved* with reference to this type of diagram, whether or not the problem as such was conceptualized in agreement with Figure 11. This would avoid the overlapping of areas, and thereby make the "breaking" of the area somewhat less abstract. However, the use of a configuration in the solution which is not already given by the statement would normally call for some constructional step; both for this reason and because of comparison with the solution of rectangle problems I therefore tend to regard this possibility as less plausible than the use of a diagram in the which corresponds to the conceptualization of the problem – either that of Figure 11 or (somewhat less likely) that of Figure 12.

The configurations of Figure 11 and Figure 12 can be used for a variety of purposes, but in connection with the present two problems they represent algebraic dead ends – as soon as we try to change the coefficients of the problems the diagrams are of no use, and other methods have to be used (as shown, for instance, in BM 13901 #14, see below, p. 73). All the more interesting is their reappearance (as identities, evidently not as problems) in *Elements* II.9-10 – not least because they remain dead ends even in the Greek context and serve nowhere else in the *Elements* nor in the rest of the corpus of Greek geometry – cf. [Mueller 1981: 286*f*, 301*f*] and [Herz-Fischler 1987: 86]. This is a question which we shall have reasons to return to in Chapter XI.

95 Below, we shall encounter it in the analysis of Db_2-146 (p. 257), YBC 6504 #2 (p. 174), and AO 8863 #4 (p. 162); it also underlies the solution of BM 13901 #19.

BM 13901 #12[96]

This problem with its solution constitutes another algebraic dead end within the Old Babylonian context, but in spite of that is highly interesting both in itself and because of its apparent historical repercussions.

Obv. II

27. The surfaces of my two confrontations I have accumulated: 21´40˝.
a.šà *ši-ta mi-it-ḫa⟨-ra⟩-ti-ia ak-mur-ma* 21,40

28. My confrontations I have made hold: 10´.
mi-it-ḫa-ra-ti-ia uš-ta-ki-il$_5$*-ma* 10

29. The moiety of 21´40˝ you break: 10´50˝ and 10´50˝ you make hold,
ba-ma-at 21,40 *te-ḫe-pe-ma* 10,50 *ù* 10,50 *tu-uš-ta-kal*

30. 1´57˝21{+25}‴40⁗[97] is it. 10´ and 10´ you make hold, 1´40˝
1,57,21{+25},40.e 10 *ù* 10 *tu-uš-ta-kal* 1,40

31. inside 1´57˝21{+25}‴40⁗ you tear out: by 17˝21{+25}‴40⁗, 4´10˝ is equalside.
lib-bi 1,57,21{+25},40 *ta-na-sà-aḫ-ma* 17,21{+25},40.e 4,10[98] íb.si$_8$

32. 4´10˝ to one 10´50˝ you append: by 15´, 30´ is equalside.
4,10 *a-na* 10,50 *iš-te-en tu-ṣa-ab-ma* 15.e 30 íb.si$_8$

33. 30´ the first confrontation.
30 *mi-it-ḫar-tum iš-ti-a-at*

34. 4´10˝ inside the second 10´50˝ you tear out: by 6´40˝, 20´ is equalside.
4,10 *lib-bi* 10,50 *ša-ni-im ta-na-sà-aḫ-ma* 6,40.e 20 íb.si$_8$

35. 20´ the second confrontation.
20 *mi-it-ḫar-tum ša-ni-tum*

Formally, the problem is of the fourth degree, but it is easily solved as a biquadratic in several ways. One way (which is not followed) would be to use Figure 12. If the two "confrontations" are called s_1 and s_2, respectively, it is easily seen on the diagram that

$$\square(s_1+s_2) = [\square(s_1)+\square(s_2)]+2\sqsubset\!\sqsupset(s_1,s_2) = 21'40''+2\cdot10' = 41'40'' = \square(50'),$$

and that

$$\square(s_1-s_2) = [\square(s_1)+\square(s_2)]-2\sqsubset\!\sqsupset(s_1,s_2) = 1'40'' = \square(10').$$

96 Based on the transliteration in [MKT III, 3].

97 That is, «1´57˝46‴40⁗» erroneously for «1´57˝21‴40⁗».

98 This number is correct but *not* the square-root of 17,46,40.

Instead, the text chooses to represent the *surfaces* $\square(s_1)$ and $\square(s_2)$ by the *sides* of a rectangle (say, L and W), whose surface $\sqsubset\!\sqsupset(L,W)$ is then found as $\square(\sqsubset\!\sqsupset(s_1,s_2)) = \square(10') = 1'40''$, while the sum $L+W$ of the sides is known to be 21′40″ – see Figure 13. This is a standard problem which by accident we have not encountered yet, and solved according to the standard, as shown in the lower part of the diagram: we may imagine the rectangle to be prolonged with the width, in such a way that the total length equals the known magnitude $L+W =$ 21′40″. This segment is bisected and "made hold", which produces a square with side 10′50″ and surface 1′57″21‴40⁗. Part of this square is identical with the original rectangle $\sqsubset\!\sqsupset(L,W)$, which is "torn out". The dotted remainder is a square with surface 17″21‴40⁗ and hence side 4′10″. Adding this to the side of the square gives us the length L of the rectangle; "tearing it out" from the other side gives W. s_1 and s_2 are then found as the "equalsides" of L and W.

Once again we notice that "appending" precedes "tearing-out", as soon as the entity to be "appended" and "torn out" has not arisen as a *part* of the larger "confrontation".

According to the palaeographic judgment of both Thureau-Dangin and Neugebauer, the text belongs to the early phase of Old Babylonian mathematics. We shall return to evidence that it is already the product of a process of maturation and canonization (below, p. 349) – but since we do not know the earlier steps of this process, the present problem may still be one of the earliest extant examples of *representation*, and thus one of the first examples in the historical record of genuine algebra. At this level, it is certainly no dead end, however much the actual trick it employs is so.

The fact that the error in obv. II 31 follows from a previous error shows that it is not a mere result of careless copying but is due to an error committed by the scribe who made the calculation (whether the scribe of the present tablet or of an original which he copied). The appearance of a correct value for the "equalside" shows that it was calculated from the known end result –

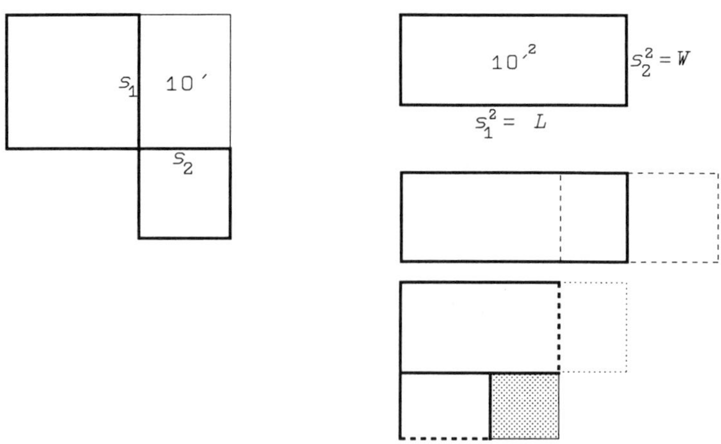

Figure 13. The procedure of BM 13901 #12.

certainly more convenient than to extract the square root of 17,21,40 by arithmetical methods.

The error itself casts some light on the Old Babylonian techniques for computation. The only simple way in which an extra contribution of 25‴ can arise is from the determination of $(50'')^2$ as $25\cdot(10'')^2 = 25\cdot(1'''\,40'''')$ = 25‴+16‴40⁗; firstly, this suggests that the calculator knew by heart the value of 10^2 but neither $10\cdot 50$ = 8`20 nor $50^2$ = 41`40. Secondly, by error 25‴ has then been inserted twice into the calculational scheme or device instead of once – which cannot easily happen unless addends disappear from view once they are inserted, as in the medieval use of the dust abacus or in a counter abacus or some similar tool. This agrees well with the absence of computations of complex products from the tablets for rough work investigated by Eleanor Robson: these will only have served to "posit" results obtained on a separate device (cf. below, notes 222–225, on errors from which similar consequences can be drawn).

BM 13901 #14[99]

The following problem about two squares returns us to the domain of widely applicable "algebraic" procedures, and introduces us to more advanced applications of the accounting technique.

As can be seen from the brackets, the text of the problem is badly damaged; however, a perfect parallel is found as #24 of the tablet, only dealing with three instead of two squares, which allows very certain restitution of all damaged passages.

Obv. II

44. The surfaces of my two confrontations I have accumulated: ⌈25′]25″.
a-šà *ši-ta mi-it-ḫa-ra-ti-ia ak-mur-ma* ⌈25,]25

45. The confrontation, two-thirds of the confrontation [and 5′, nind]an.
mi-it-ḫar-tum ši-ni-pa-at mi-it-ḫar-tim [*ù* 5 nind]an

46. 1 and 40′ and 5′ [over-going 4]0′ you inscribe.
1 *ù* 40 *ù* 5 [*e-le-nu* 4]0 *ta-la-pa-at*

47. 5′ and 5′ [you make hold, 25″ inside 25′25″ you tear out:]
5 *ù* 5 [*tu-uš-ta-kal 25 lib-bi 25,25 ta-na-sà-aḫ-ma*]

Rev. I

1. [25′ you inscribe. 1 and 1 you make hold, 1. 40′ and 40′ you make hold,]
[25 *ta-la-pa-at* 1 *ù* 1 *tu-uš-ta-kal* 1 40 *ù* 40 *tu-uš-ta-kal*]

99 Based on the transliteration in [MKT III, 3].

2. [26′40″ to 1 you append: 1°26′40″ to 25′ you raise:]
[26,40 *a-na* 1 *tu-ṣa-ab-ma* 1,26,40 *a-na* 25 *ta-na-ši-ma*]

3. [36′6″40‴ you inscribe. 5′ to 4]0′ yo[u raise: 3′20″]
[36,6,40 *ta-la-pa-at* 5 *a-na* 4]0 *t*[*a-na-ši-ma* 3,20]

4. [and 3′20″ you make hold, 11″6‴40⁗] to [36′]6″40‴ [you append:]
[*ù* 3,20 *tu-uš-ta-kal* 11,6,40] *a-na* 3[6,]6,40 [*tu-ṣa-ab-ma*]

5. [by 36′17″46‴40⁗, 46′40″ is equalside. 3′]20″ which you have made hold
[36,17,46,40.e 46,40 íb.si$_8$ 3,]20 *ša tu-uš-ta-ki*[*-lu*]

6. [inside 46′40″ you tear out]: 43′20″ you inscrib[e].
[*lìb-bi* 46,40 *ta-na-sà-aḫ-*]*ma* 43,20 *ta-la-pa-a*[*t*]

7. [igi 1°26′40″ is not det]ached. What to 1°2[6′4]0″
[igi 1,26,40 *ú-la ip-pa-t*]*a-ar mi-nam a-na* 1,2[6,4]0

8. [may I posit which 43′20″ g]ives me? 30′ its bandûm.[100]
[*lu-uš-ku-un ša* 43,20 *i-n*]*a-di-nam* 30 *ba-an-da-šu*

9. [30′ to 1 you raise: 30′] the first confrontation.
[30 *a-na* 1 *ta-na-ši-ma* 30] *mi-it-ḫar-tum iš-ti-a-at*

10. [30′ to 40′ you raise: 20′], and 5′ you append:
[30 *a-na* 40 *ta-na-ši-ma* 20] *ù* 5 *tu-ṣa-ab-ma*

11. [25′] the second [confrontat]ion
[25 *mi-it-ḫar-t*]*um ša-ni-tum*

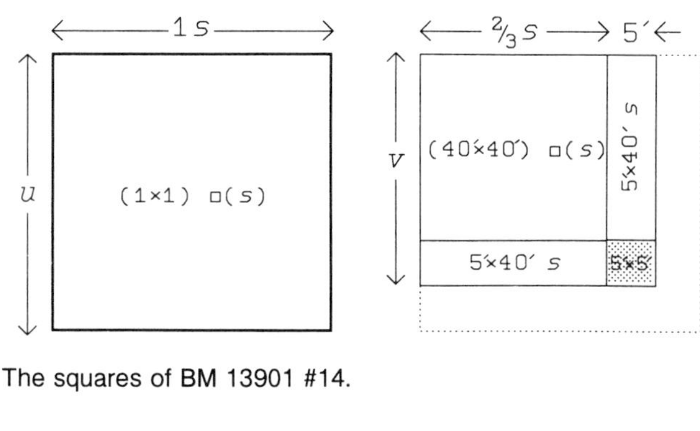

Figure 14. The squares of BM 13901 #14.

100 Probably a Sumerian loanword. The term recurs in #17 of the tablet, where its numerical value is $\frac{1}{4}$. In both cases, its mathematical function is clear, the number which (in the present case) should be "raised" to 1°16′40″ in order to give 43′20″ (functionally, it is thus simply a quotient). The etymological meaning of the term is unclear, but it might be "that which is to be given together with" (ba, "to allot" etc., .da, comitative suffix < da, "side"). Though rare in the text extant material, the term must have been quite important: in the early part of the first millennium BCE, a calculator was termed a "scribe of the *bandûm*".

For convenience we may label the sides of the two squares u and v. We are told that □(u)+□(v) = 25′25″, and that $v = \frac{2}{3}u+5′$. At first 1, 40′ (= $\frac{2}{3}$) and "5′ over-going 40′" are "inscribed". Then the accounting process begins, which eventually is going to transform the problem into the type known from #3 of the tablet (above, p. 53), "square areas plus sides" – in symbols

$$□(as)+2⊏⊐(b,as) = c$$

where $s = u$ (that a *new* square □(s) is really thought of will be apparent below). We may follow the argument on Figure 14, where □(u) and □(v) are shown in relation to s. First the shaded area in the corner of □(v) is found to be 25″, which is "torn out" from the sum; this leaves us with the non-shaded areas, which in total must equal 25′25″–25″ = 25′. Next, since $u = 1·s$, the large square is found to be 1×1·□(s) (since 1 and 1 are "made hold", we must imagine s as the unit by which the sides of the squares are measured). The upper left square from □(v) is similarly found to be 40′×40′ □(s) = 26′40″ □(s), which are "appended" to the 1 □(s). With a = 1°26′40″, this gives us a total of a squares □(s) or a rectangle ⊏⊐(s,as), as shown in the upper part of Figure 15. This can be changed into a square on as by blowing up the figure vertically by a scaling factor a = 1°26′40″, which will change the total area into a·25′ = 1°26′40″·25′ = 36′6″40‴ (lines 2–3).

Up to this point, the two "wings" of □(v) have been neglected. They are now found to be each 5·40′ s – that is, to be strips ⊏⊐(5·40′,s). This computation is performed as a "raising" – as is to be expected, since we start

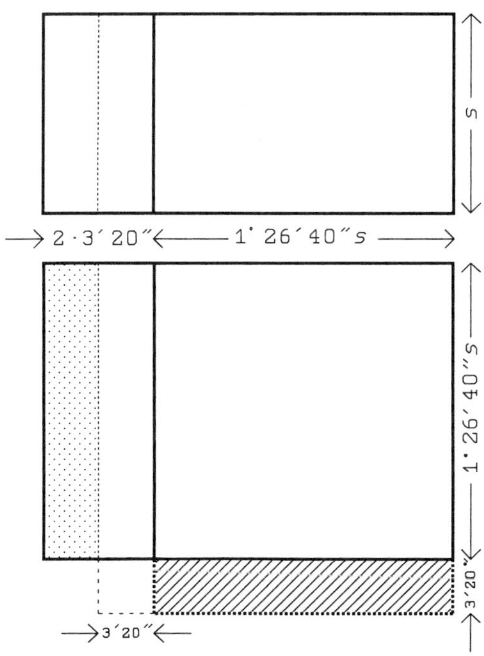

Figure 15. The final transformations of BM 13901 #14.

out from strips of width 5′, the lengths of which, however, are 40′·*s* and not *s*: thus, a factor of proportionality 40′ has to be applied.[101] Next, of course, they participate in the scaling, which gives them the adequate length *as*, but this involves no computation, whence $b = 40' \cdot 5' = 3'20''$.

As in #3 we are thus left with a problem of the type "square area and sides" (with side *as*). This is solved as we have seen a number of times (with omission of the "breaking", since the two wings are already separate), and *as* is found to be 43°20′ (line 8). From this, *s* is found (by the customary method for division by an irregular divisor) to be 30′. *u* follows when *s* is "raised" to 1, and *v* when it is "raised" to 40′, and 5′ is "appended".

It is noteworthy that the total number of sides (*b*) is never found, since this would have to be done by a doubling and would be cancelled by an ensuing "breaking". This implies that the calculator did not work mechanically in order to obtain a standard situation or problem.[102]

The need to introduce a third square, *viz.*, on the unit *s*, follows from lines 1 and 9. If the total area had been expressed as a multiple of the first area plus a multiple of its side (in symbols: as $a \cdot \square(u)+bu$), there would have been no occasion to "make 1 and 1 hold" – the area of the first square *is* 1 times this area (in #18 of the tablet, where no new square intervenes, the scribe takes advantage of this observation – see p. 108). Similarly, the multiplication in line 9 would make no sense. What was only a disputable and after all improbable conjecture in #10 – that the total area was thought of as measured in terms of the square on an undetermined unit (equal to $\frac{1}{7}u$ and to $\frac{1}{6}v$, see p. 60) – is inescapable in the present case.[103]

The computation of 1×1 and 1·*s* in lines 1 and 9 has further implications. Even if measurement in terms of $\square(s)$ and *s* was thought of, the author of the text must have been aware of the possibility of making shortcuts in the formulation and omitting these steps. The explicit presence in the text of operations with no computational effect demonstrates that shortcuts were deliberately avoided. That the two "wings" of $\square(v)$ are only calculated

---

101 If made explicit, the underlying argument is thus that ⊏⊐(*αp*,*q*) = ⊏⊐(*p*,*αq*). This transformation is used in many texts.

102 A description of the organization of the procedures as a "nesting of algorithms" [Ritter, forthcoming], perspicacious though it is, should therefore not be taken to the literal extreme; the nesting is mitigated by higher-level insights which allow adequate coordinated modification of the nesting and the nested algorithm together. The same tacit obliteration of a "repetition" to two in the nesting and a "breaking" in the nested algorithm is found elsewhere (Str 363, same problem type, and BM 85194 #25, different type, see below, p. 221) but not in all cases where it would be possible (thus not in VAT 7532, below, p. 209); we may therefore exclude, *both* that the present calculator went outside the normal canon on his own, *and* that what he did was simply to follow a more complex mechanical fixed routine.

103 The possibility of using a "model figure" with side 1 nindan is excluded by the inhomogeneous nature of the problem. The recurrent claim that the Babylonians would use the number *1* to represent the unknown in mixed second-degree problems is only true if we take it to mean that they represent it by 1·*s*.

separately and not added up to a total of 6´40´´ s will therefore be no mere shortcut induced by the knowledge that this number will soon have to be broken again: it must mean, instead, that the two wings are really thought of during the ongoing process as separate entities.

The use of a reference square □(s) which is distinct from the first square □(u) in itself calls for a commentary: it may possibly have to do with that distinction between "real field" and "model" or "school yard" field that was mentioned in note 60, with the consequences that u = 1`$s$ and $v$ = 40$s$+5 (and thus $u$ = 30 nindan, $v$ = 25 nindan, $s$ = 30´ nindan). In that case the statement should be translated "The surfaces of my two confrontations I have accumulated: 25`25. The confrontation, two-thirds of the confrontation and 5 nindan". The insertion of the unit nindan precisely here (and in #24, which presents us with the same dilemma) might even be considered evidence in favour of this reading. On the other hand, as we shall see (p. 222), #23 also refers to "10 nindan" in a situation where it can only mean "10´ nindan". Without rejecting the alternative, I have therefore chosen to ascribe not only s but also u and v to the preferred order of magnitude.

A final observation follows from the formulation of this problem, namely, from its careful distinction between "raising" and "making hold". Whatever the balance between mental and tangible geometry there is no doubt that the categories of the geometrical conceptualization are present – abstraction or metaphorization (or whatever description of the process attested to in #12 is most fitting, with its rectangle held by square areas) does not entail a shift to a numerical understanding.

The tablet Str 363 [MKT I, 244*f*] contains three variants of the present problem (v = 40´u–10´; u = s+10´, v = 40´s+5´; u = s+20´, v = 40´s+5´). Even though it was obviously written in a different school (and probably at a different time), as demonstrated by a heavily Sumerographic style and a dissimilar terminological canon, its problems agree with what we have seen here both by making use of a third square and in the arrangement of operations.

VAT 8389 #1[104]

The following problem comes from one of two twin tablets containing a sequence of exercises about the rent of two plots or parts of a field.[105] The

104 Based on the transliteration in [MKT I, 317*f*], with corrections in [MKT III, 58], first proposed in [Thureau-Dangin 1936].

105 "Plot" translates garim, Sumerogram for *tawirtum*, AHw "(Feld)flur, Umland, Umgebung"; also elsewhere in mathematical texts it is used to designate partial fields – probably in order to avoid the normal term for a field *eqlum*, which (in Sumerographic writing but Akkadian pronunciation) was used in the sense of

plots and the rents remain the same throughout all problems; what varies are the actual data.

All problems are of the first degree, and none of them are solved by properly algebraic means. There are three reasons to include two of them, this character notwithstanding:

- They illuminate by which heuristic (and in part "almost-algebraic") means the Old Babylonian calculator might attack not quite straightforward first-degree problems;
- they show us how some of the operations which are familiar from the "algebraic" texts were used outside this context;
- finally, they show how the measures of practical transactions were translated into the "basic" units on which sexagesimal calculation was based.

The specific rents are 4 gur per bùr. The bùr, we remember, is an area unit equal to 30` sar (1 sar = 1 nindan$^2$). The gur is a capacity measure, equal to 300 silà, the silà (approximately 1 liter) being the basic unit for sexagesimal calculation with capacity measures. In the present problem, we are told that the total rent of the first field exceeds that of the second field by 8`20, and that their total area is 30`. Here, no units are given, because these quantities are already expressed in the basic units silà and sar.

Obv. I

1. From 1 bùr 4 gur of grain I have collected,
i-na bùr[iku] 4 še.gur *am-ku-us*

2. from 1 seco[nd] bùr 3 gur of grain I have col[lected].
i-na bùr[iku] *ša-ni*[*-im*] 3 še.gur *am-*[*ku-us*]

3. grain over grain, 8`20 it went beyond
še-um ugu *še-im* 8,20 *i-ter*

4. My plots I have accumulated: 30`.
garim[ia] ĝar.ĝar-*ma* 30

5. My plots what?
garim[ú-a] en.nam

6. 30`, the bùr, posit. 20`, the grain which he has collected, posit.
30 *bu-ra-am* ĝar.ra 20 *še-am ša im-ku-sú* ĝar.ra

7. 30`, the second bù[r], posit.
30 *bu-r*[*a-a*]*m ša-ni-am* ĝar.ra

8. [1]5`, the gr[ain wh]ich he has collected,
[1]5 *š*[*e-am š*]*a im-ku-sú*

9. [8`]20 [wh]ich the grain over the grain went beyond,
[8],20 [*š*]*a še-um* ugu *še-im i-te-ru* ĝar.ra

(measurable) surface or area.

10. and 30` the accumulation of the surfaces of the plots posit:
ù 30 *ku-mur-ri* a.šà garim.meš ğar.ra-*ma*

11. 30` the accumulation of the surfaces of the plots
30 *ku-mur-ri* a.šà garim.meš

12. to two break: 15`.
a-na ši-na ḫe-pé-ma 15

13. 15` and 15` until twice posit:
15 *ù* 15 *a-di si-ni-šu* ğar.ra-*ma*

14. Igi 30`, of the bù[r, d]etach: 2″.
igi 30 *bu-ri-i[m p]u-ṭur-ma* 2

15. 2″ to 20`, the gra[in wh]ich he has collected,
2 *a-na* 20 š[e *š*]*a im-ku-su*

16. raise, 40′ the fa[lse] grain; to 15` [wh]ich unt[il] twice
íl 40 *še-um* l[ul] *a-na* 15 [*š*]*a a-d*[*i*] *ši-ni-šu*

16a. you have posited,
ta-aš-ku-nu

17. raise, 10` [ma]y your head hold!
íl 10 *re-eš-ka* [*l*]*i-ki-il*

18. Igi 30, of the secon[d] bùr, detach, 2″.
igi 30 *bu-ri-im ša-ni-i*[*m*] *pu-ṭur-ma* 2

19. 2″ to 15`, the grain which he has collected,
2 *a-na* 15 *še-im ša im-ku-sú*

20. raise, 30′ the false grain; to 15 which until twice
íl 30 *še-um* lul *a-na* 15 *ša a-di ši-ni-šu*

20a. you have posited, raise, 7`30.
ta-aš-ku-nu íl 7,30

21. 10` which your head holds
10 *ša re-eš-ka ú-ka-lu*

22. over 7`30 what goes beyond? 2`30 it goes beyond.
ugu 7,30 *mi-nam i-ter* 2,30 *i-ter*

23. 2`30 which it goes beyond, from 8`20
2,30 *ša i-te-ru i-na* 8,20

24. which the grain over the grain goes beyond,
ša še-um ugu *še-im i-te-ru*

Obv. II

1. tear out: 5`50 you leave.
ú-sú-uḫ-ma 5,50 *te-zi-ib*

2. 5`50 which you have left
5,50 *ša te-zi-bu*

3. may your head hold!
re-eš-ka li-ki-il

4. 40′, the ch[ange], and 30′, [the chan⸢ge,
40 *ta-ki-i[r*†*-tam]*[106] *ù* 30 *[ta-ki-ir*†⸢*-tam*

5. accumulate: 1°10′. The igi [I do not know].
g̃ar.g̃ar-*ma* 1,10 *i-gi-a-a[m ú-ul i-de]*

6. What to 1°10′ may I posi[t]
mi-nam a-na 1,10 *lu-uš-ku-[un]*

7. which 5`50 which your head holds gives me?
ša 5,50 *ša re-eš-ka ú-ka-lu i-na-di-nam*

8. 5` posit. 5` to 1°10 raise.
5 g̃ar.ra 5 *a-na* 1,10 íl

9. 5`50 [i]t gives to [y]ou.
5,50 *[i]t-ta-di[-k]um*

10. 5` which [you have p]osited, from 15` which ⸢until⸣ twice
5 *ša [ta-aš]-ku-nu i-na* 15 *ša* ⸢*a-di*⸣ *ši-ni-šu*

11. you have posited, from [o]ne tear out,
ta-aš-ku-nu i-na i[š]-te-en ú-sú-uḫ

12. to one append:
a-na iš-te-en ṣí-im-ma

13. The first is 20`, the second is 10`.
iš-te-en 20 *ša-nu-um* 10

14. 20` (is) the surface of the first plot, 10` (is) the surface of the second plot.
20 a.šà garim *iš-te-at* 10 a.šà garim *ša-ni-tim*

---

106 The superscript dagger† indicates that a reading is changed with respect to the publication of the text which I refer to. The present damaged word was restored by Neugebauer and Thureau-Dangin as *ša/ta-ki-il-tum*, at a moment when this word was supposed to mean "factor"; since the numbers in question are not involved in any operation of rectangularization (nor "held" in any other way), this interpretation now seems impossible.

From what remains of the signs, the readings *ta-ki-il-tum* and *ta-ki-ir-tum* seem equally possible (this has been confirmed to me by Aage Westenholz, though strongly doubted by Eleanor Robson). *takkirtum* is a nominal derivative from *nakārum* and means "change"; as we shall see, this might fit the way the numbers 40′ and 30′ are actually used in the following. Apart from its derivation from *kullum*, "to hold", *takīltum* might arise as a *taprīst*-type nominal from the D-stem of *nakālum*, meaning "to shape artfully/artificially", which could have to do with the characterization of the two numbers as "falses grain". I have chosen the reading *takkirtum* as the more plausible. In the actual form, none of the terms are indeed known from other mathematical texts (*takīltum* with this derivation seems to be wholly absent from the cuneiform record, but this might be because all occurrences have been automatically related to *kullum*); however, YBC 4714, rev. III 5 uses the logogram kúr corresponding to *nakārum* when referring to a modified width.

15. If 20` (is) the surface of the first plot,
šum-ma 20 a.šà garim *iš-te-at*

16. 10` the surface of the second plot, thei[r] grains what?
10 a.šà garim *ša-ni-tim še-ú-ši-n[a]* en.nam

17. Igi 30`, of the bùr, detach: 2″.
igi 30 *bu-ri-im pu-ṭur-ma* 2

18. 2″ to 20`, the grain which he has collecte[d],
2 *a-na* 20 *še-im ša im-ku-s[ú]*

19. raise, 40′. To 20`, the surface of the f[irst] plot,
íl 40 *a-na* 20 a.šà garim *i[š-te-at]*

20. raise, 13`20 the grain of 20`, [the surface of the meadow].
íl 13,20 *še-um ša* 20 [a.šà garim]

21. Igi 30`, of the seco[nd] bùr, [detach:] 2″.
igi 30 *bu-ri-im ša-ni[-im pu-ṭur-m]a* 2

22. 2″ to 15`, the gr[ain which he has collected, ra]ise, 30′.
2 *a-na* 15 *še[-im ša im-ku-sú* í]l 30

23. 30′ to 10`, the s[urface of the second plot]
30 *a-na* 10 a[.šà garim *ša-ni-tim*]

24. raise, ⌜5⌝ the gra[i]n [of the surface of the second plot].
íl ⌜5⌝ *še-[u]m* [*ša* 10 a.šà garim *ša-ni-tim*]

25. 13`30 [the grain of ⌜the surface? of the first plot]
13,20 [*še-um* ⌜*ša*/a.šà? garim *iš-te-at*]

26. over [5] the gr[ain of ⌜the surface? of the second plot]
ugu [5] *še[-im* ⌜*ša*/a.šà? garim *ša-ni-tim*]

27. what goes beyond? [8`20 it goes beyond].
mi-nam i-ter [8,20 *i-ter*]

The prescription, we observe, starts by taking note of the data. The value of the bùr in sar (30`) is "posited" once for each field; in the present context this will mean that it is written on a tablet for rough work;[107] then the rent per bùr expressed in silà is "posited", that is, written next to it. No calculation is needed for this conversion, since it could be taken from a metrological table. Finally, the difference 8`20 between the areas and the total area 30` are "posited".

The calculation begins by "breaking" the total area and "positing" each half on its own – certainly together with the data for each plot. Then, for each plot, the igi of the bùr is "detached" and "raised" to the specific rent in silà – which of course yields the rent in silà per sar, termed "the false grain"

107 For this interpretation I rely on Eleanor Robson's discovery and analysis of such tablets, as referred to above – when analyzing the text in [Høyrup 1990: 295] I was only able to conclude that the two copies of 15` were "written down or represented in some other way in two different calculation schemes or concrete representations of the two fields".

(presumably to be understood as the grain to be collected from the field under the falses assumption that its surface be 1 sar). "Raising" this to that surface 15` which belongs with the plot means *finding the rent under the presupposition that the two fields are equally large*. For the first plot this amount is 10` (I 17), for the second it is 7`30 (I 20a). The former is to be kept in the head while the second is found; then the difference between the total rents *under the same presupposition* is found (I 22) to be 2`30.

I 23 to II 1 finds this difference to fall 5`50 short of what it should be – again a number to be held in the head.[108] Therefore we have to find how many sar are to be transferred from the second to the first plot in order to obtain the given difference. The idea is that transferring 1 sar will increase the rent of the first plot by 40′ silà and decrease the rent of the latter by 30′ silà – the two "changes", if that is how the passage is to be read. All in all, the difference will therefore increase by 40′+30′ = 1°10′ (II 5). Dividing 5`50 by 1°10′ shows that 5` sar have to be transferred – and since they are to be transferred, they are "torn out" from the second plot before they can be "appended" to the first plot, in agreement with the "norm of concreteness" (p. 58).

The whole calculation is followed by a proof.

The problem is of a type which in later epochs would often be solved by means of a double false position. The underlying idea of that method is to make two guesses on the magnitude of the plots, to find the corresponding differences, and to find the true value by linear interpolation; however, it makes use of a rather opaque calculation scheme where intermediate results have no concrete interpretation and the heuristics of the interpolation is therefore lost. The Old Babylonian solution, particular as it is, is certainly of greater pedagogical value.

VAT 8391 #3[109]

This problem comes from the second of the twin tablets. As can be seen, it is very similar to the preceding one, having as data the difference between the areas instead of the sum, and the sum of the rents from the two plots instead of their difference.

108 We observe that all numbers to be used in sexagesimal multiplication (including igi-values) are "posited"; the numbers that are to be "held in the head" are those that serve in or result from additive/subtractive operations (and which, as far as the present number is concerned, shall later *result* from a multiplication).

109 Based on the transliteration in [MKT I, 321*f*].

Rev. I

3. If from 1 bùr of surface 4 gur of grain I have [collected,]
šum-ma i-na bùr^{iku} a.[šà] 4 še.gur *[am-ku-us]*

4. from 1 bùr of surface 3 gur of grain [I have collected,]
i-na bùr^{iku} a.šà 3 še.gur *am-[ku-us]*

5. now, 2 plots. Plot over plot, 10` it went beyond.
i-na-an-na 2 garim garim ugu garim 10` *i-ter*

6. Their grain I have accumulated: 18`20.
še-e-ši-na ĝar.ĝar-*ma* 18,20

7. My plots what?
garim$^{ú-a}$ en.nam

8. 30`, the bùr, posit. 20`, the grain which he has collected, posit
30 *bu-ra-am* ĝar.ra 20 *še-am ša im-ku-sú* ĝar.ra

9. 30`, the second bùr, posit. 15`, the grain which he has collected,
30 *bu-ra-am ša-ni-am* ĝar.ra 15 *še-am ša im-ku-sú*

9a. posit.
ĝar.ra

10. [10` wh]ich plot over plot went beyond, posit.
1[0 *š*]*a* garim ugu garim *i-te-ru* ĝar.ra

11. [18`20, the accu]mulation of the grain, posit.
[18,20 *ku-*]*mur-ri še-im* ĝar.ra

12. [1, project]ing, posit:
[1 *wa-ṣi*]*-am* ĝar.ra-*ma*

13. Igi 3[0`, of the bùr, detach:] 2˝; to the grain which he has collected
igi 3[0 *bu-ri-im pu-ṭur-m*]*a* 2 *a-na še-im ša im-ku-sú*

14. raise, 40´, the fa[lse] grain; [to 1]0` wh[ich plot] o[ver plot goes beyo]nd
íl 40 *še-um* l[ul *a-na* 1]0 *š*[*a* garim] u[gù garim *i-te-r*]*u*

15. raise, 6`40; from 18`20, the accumulation of the grain,
íl 6,40 *i-na* 18,20 *ku-mur-ri še-im*

16. tear out: 11`40 you leave.
ú-sú-uḫ-ma 11,40 *te-zi-ib*

17. 11`40 which you have left, may your head hold!
11,40 *ša te-zi-bu re-eš-ka li-ki-il*

18. 1, projecting, to two break: 30´.
1 *wa-ṣi-am a-na ši-na ḫe-pé-ma* 30

19. 30´ and 30´ until twice posit:
30 *ù* 30 *a-di ši-ni-šu* ĝar.ra.ma

20. Igi 30´, of the bùr, detach: 2˝; to 20`, the grain which he has collected,
igi 30 *bu-ri-im pu-ṭur-ma* 2 *a-na* 20 *še-im ša im-ku-sú*

21. raise, 40′; to 30′ which until twice you have posited
íl 40 *a-na* 30 *ša a-di ši-ni-šu ta-aš-ku-nu*

22. raise, 20′; may your head hold!
íl 20 *re-eš-ka li-ki-il*

23. Igi 30`, of the second bùr, detach: 2″.
igi 30 *bu-ri-im ša-ni-im pu-ṭur-ma* 2

24. 2″ to 15`, the grain which he has collected,
2 *a-na* 15 *še-im ša im-ku-sú*

25. raise, 30′; to the seco[n]d 30′ which you have posited, raise, 15′.
íl 30 *a-na* 30 *ša-ni-[i]m ša ta-aš-ku-nu* íl 15

26. 15′ and 20′, which your head holds,
15 *ù* 20 *ša re-eš-ka ú-ka-lu*

27. accumulate: 35′; the igi I do not know.
ĝar.ĝar-*ma* 35 *i-gi-am ú-ul i-di*

28. What to 35′ may I posit
mi-nam a-na 35 *lu-uš-ku-un*

29. which 11`40 which your h[ea]d holds gives me?
ša 11,40 *ša r[e-e]š-ka ú-ka-lu i-na-di-nam*

30. 20` posit. 20` t[o] 35′ raise, 11`40 it gives to you.
20 ĝar.ra 20 *a-[na]* 35 íl 11,40 *it-ta-di-kum*

31. 20` which you have posi[ted (is) the sur]face of the first plot;
20 *ša ta-aš-ka-[nu* a.]šà garim *iš-te-at*

32. from 20`, the surface of the plot, 1[0` which] surface over surface went beyond,
i-na 20 a.šà garim 1[0 *ša*] garim ugu garim *i-t[e]-ru*

33. tear out, 10` [the surface you] leave.
ú-sú-uḫ-ma 10 [a.šà *te-*]*zi-ib*

(Followed by a proof, rev. II 1–9)

This time, there is no need for counterfactual assumptions. What goes on is quite straightforward, though the idea seems to have escaped earlier workers on the text: if the first plot is A and the second B, A may be split into $B+(A-B) = B+10$`. The rent of 10` sar of A is found in I 15 to be 6`40 silà. If this is "torn out" from the total rent 18`40 silà, the remainder 11`40 silà must come from equal surfaces from $A$ and $B$. Therefore the yield of an "average sar" is found in I 27 to be 35′ silà. Division shows the area that yields the 11`40 silà to be 20` sar. By error (due of course to the numerical coincidence) this is identified with $A$, and $B$ is found by "tearing out" 10` silà; instead, 20` should evidently be "broken", one "moiety" identified with $B$, and the other, augmented by the excess 10`, with A.

The way in which the rent of the average sar is computed is interesting. It involves a *waṣûm*, the masculine form corresponding to the feminine *waṣītum*, here translated the "projecting". The idea seems to be that the square nindan is regarded as a line 1 nindan long and provided with a projecting breadth of

1 nindan. Bisection of the area can hence be made by bisecting this breadth into two "moieties" of 30′ nindan each, and to find the rent by "raising" the specific rents to this number (which, compared to the full "projecting 1", can indeed be regarded as a factor of proportionality).

TMS XVI[110]

The next text is as atypical as it is informative. It presents a problem – a genuine first-degree equation – but does not solve it; what it does is to explain the meaning of the steps by which the equation is transformed, and thus to make explicit what is implicit in most of the material at our disposal. This character could have to do with the origin of the text: it was written in Susa, a peripheral area, toward the very end of the Old Babylonian period, and teachers from a peripheral school may possibly have felt the need for written instructions where those from the core could rely on a more firmly established tradition of oral explanations (we shall encounter other didactic texts from Susa below). But there is no reason to believe that the written explanation of our Susa text deviates from the oral expositions given elsewhere. Though rare in the rest of the corpus, similar traces of the actual teaching do turn up elsewhere – one unpublished text from Eshnunna[111] has first a didactic explanation of the configuration similar to the one given here, and next a solution (the same combination is found in TMS VII and TMS IX, below, pp. 89 and 181, respectively); YBC 8633 inverts the order and prescribes the solution before it discusses the configuration.[112]

1. [The 4th of the width, from] the length and the width to tear out, 45′. You, 45′
   [4-at sağ *i-na*] uš *ù* sağ zi 45 za.e 45
2. [to 4 raise, 3 you] see. 3, what is that? 4 and 1 posit,
   [*a-na* 4 *i-ší* 3 *ta*]-*mar* 3 *mi-nu šu-ma* 4 *ù* 1 ğar
3. [50′ and] 5′, to tear out, ⌈posit⌉. 5′ to 4 raise, 1 width. 20′ to 4 raise,
   [50 *ù*] 5 zi ⌈ğar⌉ 5 *a-na* 4 *i-ší* 1 sağ 20 *a-na* 4 *i-ší*

---

110 Based on the hand copy and transliteration in [TMS, pl. 25, pp. 91*f*], with corrections from [von Soden 1964]. Cf. revised edition of the full tablet in [Høyrup 1990: 299–302].

111 IM 43993, a problem on a rectangle where the width is said to be ⅔ of the length, and the accumulation of surface, length and width 1 [Friberg and al-Rawi 1994a].

112 The interpretation of a passage in TMS IX as evidence of specific Susian methods on the level of elementary "algebra" (underscored in the preface to the volume, and often quoted in the secondary literature) relies on a double misunderstanding – cf. [Høyrup 1990: 326], and below, note 128.

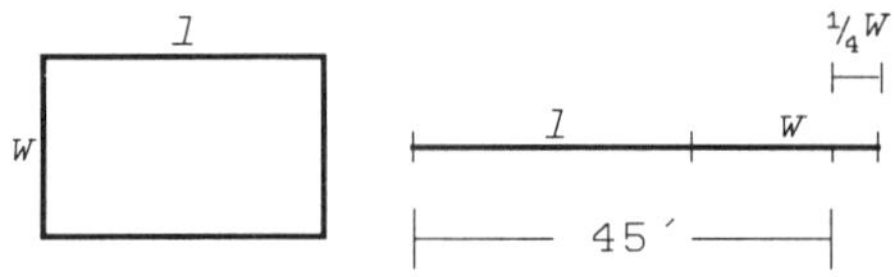

Figure 16. The situation of TMS XVI #1.

4. 1° 20´ you ⟨see⟩, 4 widths. 30´ to 4 raise, 2 you ⟨see⟩, 4 lengths. 20´, 1 width, to tear out,
1,20 *ta-⟨mar⟩* 4 sağ 30 *a-na* 4 *i-ší* 2 *ta-⟨mar⟩* 4 uš 20 1 sağ zi

5. from 1° 20´, 4 widths, tear out, 1 you see. 2, the lengths, and 1, 3 widths, accumulate, 3 you see.
i-na 1,20 4 sağ zi 1 *ta-mar* 2 uš *ù* 1 3 sağ UL.GAR 3 *ta-mar*

6. Igi 4 de[ta]ch, 15´ you see. 15´ to 2, lengths, raise, [3]0´ you ⟨see⟩, 30´ the length.
igi 4 *pu-[tú-ú]r* 15 *ta-mar* 15 *a-na* 2 uš *i-ší* [3]0 *ta-⟨mar⟩* 30 uš

7. 15´ to 1 raise, [1]5´ the contribution of the width. 30´ and 15´ hold.
15 *a-na* 1 *i-ší* [1]5 *ma-na-at* sağ 30 *ù* 15 *ki-il*

8. Since "The 4th of the width, to tear out", it is said to you, from 4, 1 tear out, 3 you see.
aš-šum 4-*at* sağ *na-sà-ḫu qa-bu-ku i-na* 4 1 zi 3 *ta-mar*

9. Igi 4 de⟨tach⟩, 15´ you see, 15´ to 3 raise, 45´ you ⟨see⟩, 45´ as much as (there is) of [widths].
igi 4 *pu-⟨tú-úr⟩* 15 *ta-mar* 15 *a-na* 3 *i-ší* 45 *ta-⟨mar⟩* 45 *ki-ma* [sağ]

10. 1 as much as (there is) of lengths posit. 20, the true width take, 20 to 1´ raise, 20´ you see.
1 *ki-ma* uš ğar 20 gi.na sağ *le-qé* 20 *a-na* 1 *i-ší* 20 *ta-mar*

11. 20´ to 45´ raise, 15´ you see. 15´ from $^{30}{}_{15'}$ [tear out],
20 *a-na* 45 *i-ší* 15 *ta-mar* 15 *i-na* $^{30}{}_{15}$[113] [zi]

12. 30´ you see, 30´ the length.
30 *ta-mar* 30 uš

The equation whose transformation is analyzed deals with the length (l) and the width (w) of a rectangle – see Figure 16; in the actual case, however, this concrete meaning is relatively unimportant. In line 1, we are indeed told (in symbolic translation) that

$$(l+w)-\tfrac{1}{4}w = 45' \ .$$

The addition is expressed merely as uš *ù* sağ, "length and width", which elsewhere appears to be an ellipsis for UL.GAR uš *ù* sağ, "the accumulation of length and width". There is thus no reason to think, for instance, of the length in relation to the rectangle of which it is the length, and the width moved so as

---

[113] Thus, indeed ($^{30}{}_{15}$) instead of the regular writing "45".

to prolong the length – the two are treated as nothing but measurable segments, fully independent. The left-hand part of Figure 16 is thus irrelevant even in the first step – all that is important is in the right-hand part. Here we also see that the fourth of the width, *qua* part of the width, is part of the total aggregation of length and the width – whence it may be "torn out".

In lines 1–2, we are told to "raise" the number 45′ by 4. The outcome is 3, the meaning of which is asked for.

The explanation shows that the values of l (30′) and w (20′) are supposed to be known. At first we are to "posit" 4 and 1 (for the multiplied and the original equation). We may imagine something like Figure 17 (without believing in the exact details of the representation). Next it is explained that 5′ "raised" to 4 yields 20′, one width; that 20′, one width, "raised" to 4 gives 1°20′, four widths; and that 30′ "raised" to 4 gives 4 lengths. "Tearing out" 20′, the one width coming from the "raising" of 5′, from 1°20′, the 4 widths, leaves 1, that is, 3 widths. Together with the 2 that represent the (4) lengths, this gives us the 3 that were to be explained (line 5).

But the explanation does not stop here. Next it multiplies with igi 4, thus reverting to the upper part of Figure 17, and starts this section of the analysis by identifying 30′ and 15′ as the contributions of length and width, which are to be held (in the head). Next comes a determination of the coefficients of the width and (with no need for calculation) the length, referred to with the phrase "as much as (there is) of widths/lengths". For the width, a single false position is used: removing the fourth from 4 would leave 3; in order to see how much is left of 1, we have to reduce 4 to 1, which is done by a multiplication with igi 4; the result is 45′, which is indeed the coefficient of the width.

In lines 10–11, the coefficient is seen to produce the contribution. At first the "true width" is multiplied by 1 in order to give "the width"; as suggested in note 60, the factor 1 may indeed mean 1′, which would serve to reduce an imaginary "real width" of 20 nindan to a "model width" of 20′ nindan. In any case, in the next step this width is "raised" to the coefficient 45′, which gives us the contribution 15′; removing this from the total 45′ (already written as an aggregate of the two contributions that were held in the head) leaves 30′, that is, the length.

All in all, as we see, the text is thus a highly pedagogical exposition, moving back and forth between the various levels so as to create full understanding of their mutual connections; but no attempt is made to achieve anything like a deductive structure.[114]

The mathematical texts from Susa were published by E. M. Bruins and M. Rutten in 1961 as [TMS]. The hand copies of the texts had been made by Rutten in great haste in the late thirties, before the Louvre Museum stored the tablets away because war was supposed to be imminent; they are as good as such copies can be when

114 The tablet contains another didactic exposition, namely, of the equation "the 4th of the width to that which length over width went beyond I have appended: 15′". This problem is transliterated and translated in [Høyrup 1990: 301–302].

secondary checks after preliminary interpretation are not possible.[115] The transliteration proper seems to be the result of collaboration, but the translation of ideograms into Akkadian, the translation, and the interpretive commentary are certainly due to Bruins (even though the preface asserts that the translation is collaborative).

The interpretation was certainly no easy job, and Bruins invested all his legendary stubborniness and creative fantasy in the work – sometimes for better but not rarely for worse (an example is discussed in note 207). The translation should therefore be used with great circumspection, and never without reference to von Soden's polite but careful review [1964]; even the translation of logograms into Akkadian is often trivially mistaken. For many texts (including the present one, whose particular character is overlooked completely), the mathematical commentary is best disregarded.

Whereas my transliterations of texts that were originally published by Neugebauer, Thureau-Dangin, Sachs, and Baqir follow theirs with only marginal exceptions (cf. p. 15), my versions of the Susa texts therefore differ from the original publication on many important points even on this level.

In the present text, the following observation can be made:

Line 3: [TMS] transliterates "[50 *ù*] 5 ZI.A ⟨GAR⟩ ...", and Bruins interprets ZI as a phonetically motivated writing error for SI, which would give the meaning "50´ and 5´ which went beyond posit" (SI.A = dirig); [TMS] renders both sign names and Sumerograms as small caps. However, if the hand copy can be trusted, the supposed A is damaged and clearly separated from the ZI; the traces might as well represent the supposedly missing g̃ar; moreover, it is clearly stated in lines 1, 4, 5, and 8 that 5´ is to be "torn out" (as is indeed the 4·5´ that results from it); it is never referred to as something "going beyond".

Line 7: "Contribution" translates *manātum*, an abstract noun derived from *manûm*, "to count". Etymologically, the meaning would be "the count"/"the counting". The term is found only in this place and in the Susa texts TMS XII and XXIV; in XII it stands for the "counting" of a lenght in terms of a width (i.e., the ratio – [Muroi 2001a: 4]), in XXIV it is isolated by a break. [AHw, 602a] suggests a conjectural identification with Hebrew and Aramaic *m*^{e}*nat*, which in [HAHw, 439a] is exemplified by "Anteil der Priester und Leviten" and "d. Teil (Beitrag) des Königs". The ensuing "share/contribution of the width" fits the present text excellently.

Line 7: *ki-il*, "hold" was proposed conjecturally in [von Soden 1964: 49]. The form and the function has now been confirmed by the text BM 96957 (I 7 and *passim*, in [Robson 1996: 183]). [TMS] has *ḫulum*, Assyrian for "way", and interprets as "method".

Line 8: the zi of the statement is quoted as a syllabic infinitive, not as a finite form; as mentioned on p. 33, this *could* mean that the logograms of certain compact

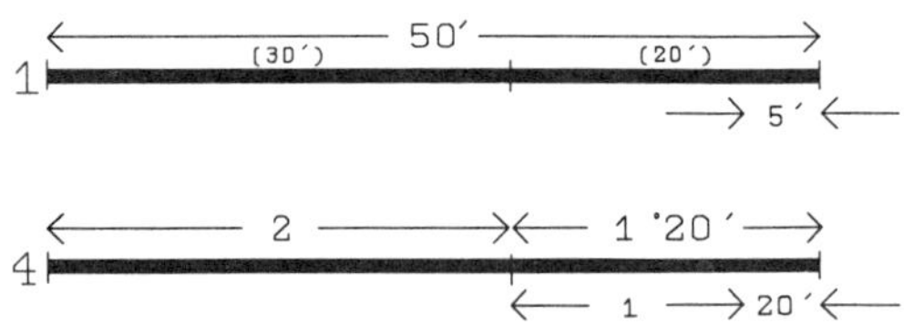

Figure 17. The transformations of TMS XVI #1.

[115] After the War, several tablets (including number XVI) were irretrievable, which is the reason that photographs are lacking in the edition (personal communication from Jim Ritter, who has now located the missing tablets; the disappearance is unacknowledged in the volume).

texts were to be read (at least in Susa) as relatively abstract, de-personalized, and de-temporalized operations.

Line 10: [TMS] claims that an indubitable gi.na, "true", is a mistake for *ki-ma*, "as much as". If this were correct, *kīma* sağ, "as much as of widths", would represent both the coefficient of the width (45′, in line 9) and the value of the width (20, in line 10).

TMS IX[116]

The following text (once again from Susa) belongs to the same didactically explicit genre as the previous one. The present specimen, however, explains some of the basic second-degree techniques in #1 and #2, before applying these to a complex problem in #3. It does so in a way which might have changed the received interpretation of the Old Babylonian "algebraic" genre decades ago, if only the original publication had been adequate.

#1

1. The surface and 1 length accumulated,[117] 4[0′. ¿30, the length,? 20′ the width.]
a.šà *ù* 1 uš UL.GAR 4[0 ?30 uš? 20 sağ][118]

2. As 1 length to 10′ ˡthe surface, has been appended,]
i-nu-ma 1 uš *a-na* 10 ˡa.šà daḫ]

3. or 1 (as) base to 20′, [the width, has been appended,]
ú-ul 1 KI.GUB.GUB[119] *a-na* 20 [sağ daḫ]

---

116 Based on the hand copy and transliteration in [TMS, pl. 17, pp. 63*f*], with corrections from [von Soden 1964]. Cf. revised edition in [Høyrup 1990: 320–323].

117 Since even this text is from Susa and related to TMS XVI in its didactic approach, I have chosen to translate the UL.GAR and other logograms of the statement as infinite forms; as already discussed on p. 33, this is only a possibility and not compulsory.

118 This restitution is mine, as are many of those that follow. From the quotation in line 6 the statement can be seen to have given the value of the width; whether the length was also stated explicitly or just presupposed routinely remains a guess, but the reference to the value of the surface in line 2 shows that it is supposed to be known.

119 "Base" translates the logogram KI.GUB.GUB, which is not known from elsewhere (the Late Babylonian value ki.du.du~*kidudûm* is clearly irrelevant). GUB has two different Sumerian interpretations, du/RÁ etc., "to go" [SLa § 268], and gub, "to stand, to erect" [SLa § 267]; to judge from the logographic occurrences, the reduplication is used to indicate iterative and durative aspects. ki may function as a virtual locative verbal prefix, "on the ground" [SLa §306]. A possible reading of the complex thus seems to be ki.gub.gub, "to stand/that which stands erected

4. or 1°20′ [¿is posited?] to the width which 40′ together ˡwith the length ¿holds?]
ú-ul 1,20 *a-na* sağ *šà* 40 *it-ˡti* uš ¿NIGIN ğar?]

5. or 1°20′ toge⟨ther⟩ with 30′ the length hol[ds], 40′ (is) [its] name.
ú-ul 1,20 *it-*⟨*ti*⟩ 30 uš NIG[IN] 40 *šum-*[*šu*]

6. Since so, to 20′ the width, which is said to you,
aš-šum ki-a-am a-na 20 sağ *šà qa-bu-ku*

7. 1 is appended: 1°20′ you see. Out from here
1 daḫ-*ma* 1,20[120] *ta-mar iš-tu an-ni-ki-a-am*

8. you ask. 40′ the surface, 1°20′ the width, the length what?
ta-šà-al 40 a.šà 1,20 sağ uš *mi-nu*

9. [30′ the length. T]hus the procedure.
[30 uš *k*]*i-a-am ne-pé-šum*

#2

10. [Surface, length, and width accu]mulated, 1. By the Akkadian (method).
[a.šà uš *ù* sağ U]L.GAR 1 *i-na ak-ka-di-i*

11. [1 to the length append.] 1 to the width append. Since 1 to the length is appended,
[1 *a-na* uš daḫ] 1 *a-na* sağ daḫ *aš-šum* 1 *a-na* uš daḫ

12. [1 to the width is app]ended, 1 and 1 make hold, 1 you see.
[1 *a-na* sağ d]aḫ 1 *ù* 1 NIGIN 1 *ta-mar*

13. [1 to the accumulation of length,] width and surface append, 2 you see.
[1 *a-na* UL.GAR uš] sağ *ù* a.šà daḫ 2 *ta-mar*

14. [To 20′ the width, 1 appe]nd, 1°20′. To 30′ the length, 1 append, 1°30′.[121]
[*a-na* 20 sağ 1 da]ḫ† 1,20 *a-na* 30 uš 1 daḫ 1,30

15. [¿Since? a surf]ace, that of 1°20′ the width, that of 1°30′ the length,
[¿*aš-šum*? a.š]à *šà*† 1,20 sağ *šà* 1,30 uš

16. [¿the length together with? the wi]dth, are made hold, what is its name?
[¿uš *it-ti*? sa]ğ† *šu-ta-ku-lu mi-nu šum-šu*

---

permanently on the ground".

The reading "coefficient of the length" proposed by Kazuo Muroi [1994] can be safely disregarded, both because it suggests (without collation of the tablet) the reading to be changed into *ki.gub uš, and because the supposedly corroborative evidence in the text BM 15285 is indeed counter-evidence – cf. [Høyrup 1995b].

120 This follows the hand copy of [TMS], against the transliteration.

121 My restitutions of lines 14–16 are somewhat tentative, even though the mathematical substance is fairly well established by the parallel in lines 28–31.

17. 2 the surface.
2 a.šà

18. Thus the Akkadian (method).
ki-a-am ak-ka-du-ú

#3

19. Surface, length, and width accumulated, 1 the surface. 3 lengths, 4 widths accumulated,
a.šà uš *ù* sağ UL.GAR 1 a.šà 3 uš 4 sağ UL.GAR

20. its [17]th to the width appended, 30′.
[17]-*ti-šu a-na* sağ daḫ 30

21. [You], 30′ to 17 go: 8°30′ [yo]u see.
[za.]e 30 *a-na* 17 *a-li-ik-ma* 8,30 [*t*]*a-mar*

22. [To 17 widths] 4 widths append, 21 you see.
[*a-na* 17 sağ] 4 sağ daḫ-*ma* 21 *ta-mar*

23. [21 as] much as of widths posit. 3, of three lengths,
[21 *ki-*]*ma* sağ ğar 3 *šà-la-aš-ti* uš

24. [3, as] much as lengths posit. 8°30′, what is its name?
[3 *ki*]-*ma* uš ğar 8,30 *mi-nu šum-šu*

25. [3] lengths and 2[1 wid]ths accumulated.
[3] uš *ù* 2[1 sa]ğ UL.GAR

26. 8°30′ you see
8,30 *ta-mar*[122]

27. [3] lengths and 21 widths accumu[lated.]
[3] uš *ù* 21 sağ UL.[GAR]

28. [Since 1 to] the length is appended [and 1 t]o the width is appended, make hold:
[*aš-šum* 1 *a-na*] uš daḫ [*ù* 1 *a*]-*na* sağ daḫ NIGIN-*ma*

29. 1 to the accumulation of surface, length, and width append, 2 you see,
1 *a-na* UL.GAR a.šà uš *ù* sağ daḫ 2 *ta-mar*

30. [2 the sur]face. Since the length and the width of 2 the surface,
[2 a.]šà *aš-šum* uš *ù* sağ *šà* 2 a.šà

31. [1°30′, the length, toge]ther with 1°20′, the width, are made hold,
[1,30 uš *it*†]-*ti* 1,20 sağ *šu-ta-ku-lu*

32. [1 the appe]nded of the length and 1 the appended of the width,
[1 *wu-ṣú-*]*bi*[123] uš *ù* 1 *wu-ṣú-bi* sağ

---

[122] The transliteration of [TMS] supposes that something is missing in the beginning of the line; The hand copy indicates that the line is simply written with indention.

[123] "Zu WA.ZU-*bi* im math. Susatext nr. IX: Ich hatte mich für die Rezension von MDP 34 [= von Soden 1964 – JH] ziemlich gründlich damit beschäftigt und als mögliche Lesung *wu-ṣú-bi* als St. constr. eines sonst nicht bekannten *wuṣubbûm* notiert, diese Lesung aber dann als zu wenig gesichert nicht veröffentlicht" (von Soden, personal communication).

33. [make hold ¿1 you see? 1 and 1,] the various (things),[124] accumulate, 2 you see.
[NIGIN ¿1 *ta-mar*? 1 *ù* 1 ¿...?] ḫi.a UL.GAR 2 *ta-mar*

34. [3 ..., 21 ..., and 8°30′ accumulate,] 32°30′ you see;
[3 ... 21 ... *ù* 8,30 UL.GAR] 32,30 *ta-mar*

35. [s]o you ask.
[*ki-a*]*-am ta-šà-al*

36. [...] of widths, to 21, that accumulation:[125]
[...].TI sağ *a-na* 21 UL.GAR-*ma*

37. [...] to 3, lengths, raise,
[...][126] *a-na* 3 uš *i-ší*

38. [1`3 you see. 1`3 t]o 2, the surface, raise:
[1,3 *ta-mar* 1,3 *a*]*-na* 2 a.šà *i-ši-ma*

39. [2`6 you see, ¿2`6 the surface?.] 32°30′ the accumulation break, 16°15′ you ⟨see⟩.
[2,6 *ta-mar* ¿2,6 a.šà?] 32,30 UL.GAR *ḫe-pé* 16,15 *ta-*⟨*mar*⟩

40. {...}. 16°15′ the counterpart posit, make hold,
{1[6,15 *ta-*]*mar*}[127] 16,15 gaba ğar NIGIN

41. 4`24°3′45″ you see. 2`6 [¿erasure?]
4,[24,]3,45 *ta-mar* 2,6 [¿erasure?]

42. from 4`[2]4°3′45″ tear out, 2`18°3′45″ you see.
i-na 4,[2]4,3,45 zi 2,18,3,45 *ta-mar*

43. What is equalside? 11°45′ is equalside, 11°45′ to 16°15′ append,
mi-na íb.si 11,45 íb.si 11,45 *a-na* 16,15 daḫ

44. 28 you see. From the 2nd tear out, 4°30′ you see.
28 *ta-mar i-na* 2-kam zi 4,30 *ta-mar*

45. Igi 3, of the lengths, detach, 20′ you see. 20′ to 4°[30′]
igi 3-*ti* uš *pu-ṭúr* 20 *ta-mar* 20 *a-na* 4,[30]

46. {...} raise: 1°30′ you see,
{20 *a-na* 4,30} *i-ši-ma* 1,30 *ta-mar*

---

124 "the various (things)" translates ḪI.A. It presupposes, either, that the Sumerian suffix .ḫi.a (which designates a plurality of different entities) is used as a nominal pseudo-Sumerogram, indicating the collection of surface, length, and width; or, alternatively, that .ḫi.a stands as a pseudo-Sumerographic complement to a preceding noun (represented by "the things" in the translation) that characterizes the third number *1* of the line – still indicating it to be a plurality.

125 The translation of this opaque line in [TMS] mistakes length for width and "accumulation" for "appending".

126 [TMS] reads a ḪI.A before *a-na*, the ḪI of which, however, is very different from that of line 33 in the hand copy.

127 Following the hand copy against the transliteration of [TMS].

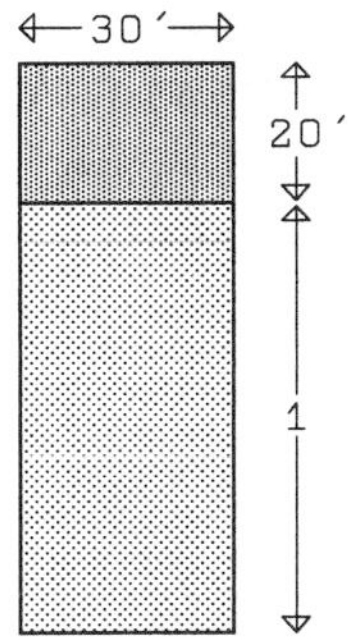

Figure 18. The configuration described in TMS IX #1.

47. 1°30′ the length of 2 the sur[face. What] to 21, the widths, [may I posit]
1,30 uš *šà* 2 a.š[à *mi-na*] *a-na* 21 sağ [*lu-uš-ku-un*]

48. which 28 give[s me? 1°20′ po]sit, 1°20′ the width
šà 28 *i-na-di*[†][*-na* 1,20 ğ]ar 1,20 sağ

49. of 2 the surface. Turn back. 1 from 1°[30′ tear out,]
šà 2 a.šà *tu-úr* 1 *i-na* 1,[30 zi]

50. 30′ you see. 1 from 1°20′ tear [out,]
30 *ta-mar* 1 *i-na* 1,20 z[i]

51. 20′ you see.
20 *ta-mar*

All three parts of the text deal with the same rectangle ⊏⊐(30′,20′). The tablet is damaged, but #1 clearly presupposes in its explanation that these dimensions as well as the resulting area 10′ are known. It discusses what to do when the sum of area and length is given, ⊏⊐$(l,w)+l$ = 40′. It is immediately taken for granted ("As ...") that this means that the width is prolonged by 1 – cf. Figure 18. Then follows a sequence of reformulations ("or ... or ... or ...", much in the vein of modern mathematical parlance). All in all, the total area 40′ is seen to be that of a rectangle held by the length and the width prolonged by the "base" 1 (a term which suggests the orientation of Figure 18). In the end it is explained how, if the width 20′ and the total area 40′ are given, the length can be found to be 30′.

#2 still presupposes the known values of l and w in its explanations but now treats the situation where ⊏⊐$(l,w)+l+w$ = 1, and tells us how to apply "the Akkadian (method)".[128] As shown in Figure 19, this implies that both

---

128 Bruins somehow has not noticed that two different equations are treated, and therefore claims that #1 presents the Susian more elegant method, after which #2 explains the "Akkadian method" as a more clumsy alternative. Nor has he seen that #3 makes use of the method presented in #2 and not of the calculator's supposedly "own" method.

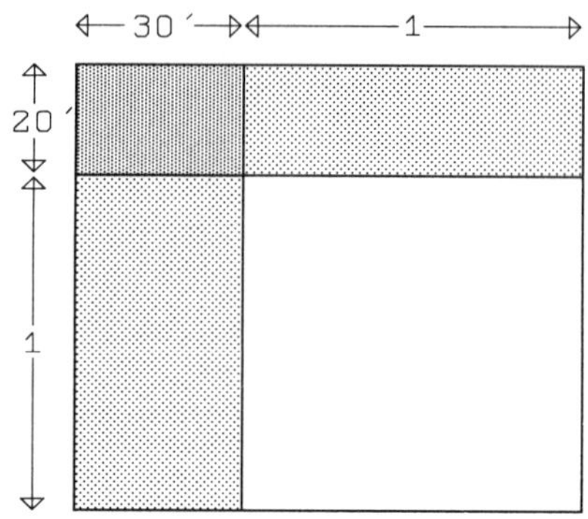

Figure 19. The configuration of TMS IX #2.

length and width are prolonged by 1, and hence also that a completing square □(1,1) be "appended" to the surface. The resulting "surface 2" then has the length 1°30´ and the width 1°20´.

Since the feature which distinguishes the procedure of #2 most clearly from that of #1 is the quadratic completion, we may safely assume that this trick is what carried the name "the Akkadian (method)".

#3 combines the equation of #2 with an abstruse linear condition:

$$⊏⊐(l,w)+l+w = 1\ , \quad \tfrac{1}{17}(3l+4w)+w = 30'\ .$$

At first it transforms the latter equation by means of the techniques taught in TMS XVI: multiplying by 17, and finding the total coefficients of l and w. As summed up in line 26*f*,

$$3l+21w = 8°30'\ .$$

Next it repeats the trick of #2, showing that a "surface 2", with length 1°30´ and width 1°20´, presupposes that 1 is "appended" to both length and width – putting $\lambda = l+1$, $\omega = w+1$, we get $⊏⊐(\lambda,\omega) = 2$. Moreover (the damages to lines 33 and 34 prevents us from knowing the exact formulation), $3\lambda+21\omega = 32°30$. Finally, if $\Lambda = 3\lambda$, $\Omega = 21\omega$ (Λ and Ω, in contrast to λ "the length of 2 the area" and ω "the width of 2 the area", carry no name of their own in the text, whereas their sum is spoken of in line 39 as "the accumulation")

$$\Lambda+\Omega = 32°30\ , \quad ⊏⊐(\Lambda,\Omega) = (3\cdot 21)\cdot 2 = 2`6\ .$$

This standard form of the problem is obtained in line 39, after which a normal cut-and-paste procedure starts (see Figure 20, and cf. Figure 13, which also finds the sides of a rectangle from the area and the sum of the sides): the sum of length and width is bisected and the "counterpart" of the "moiety" positioned so as to "hold" a completed square, whose surface must be □(16°15´) = 4`24°3´45´´. From this is "torn out" the surface of the rectangle, transformed into a gnomon, leaving for the completing square an area 2`18°3´45´´. By this, 11°45´ is "equalside", which is "appended" to the first side of the completed square (horizontal in Figure 13) and "torn out" from its

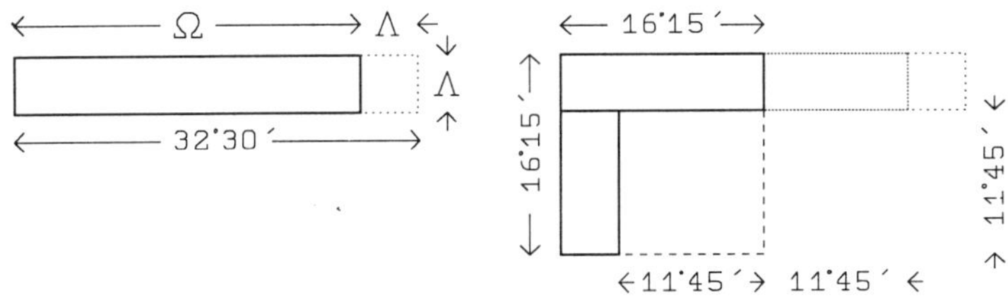

Figure 20. The transformed system of TMS IX #3.

vertical "counterpart". As in all cases where it is not *the same* piece that is "torn out" and "appended", "appending" precedes "tearing-out".

"Appending" and "tearing-out" give the values of Λ (= 4°30´) and Ω (= 28), from which λ = 1°30´ and ω = 1°20´ follow. Finally we "turn back" to the original rectangle by "tearing out" the "appended" 1 from each of these.

Apart from giving another example of the didactic style of Old Babylonian mathematics teaching and also showing the routines by which the Old Babylonian calculators would solve complex second-degree problems if they did not happen to discover a shortcut, the text is of interest because of its underlying distinction between *that which is given* and *that which is merely known*. That which is given may be used in the calculation; the magnitude of that which is merely known may serve as an identifying tag in the description of the procedure (a useful alternative in the absence of that naming by letters which we know from Greek geometry and modern algebra), but it cannot be used in the calculation.[129] Failing to appreciate this distinction, many workers have seen overdetermination, clumsiness, or outright mathematical stupidity in texts which are characterized instead by a very acute awareness of the status of the various kinds of information which the calculator possesses.

129 We remember that the "name" (*šumum*) of an entity might be both the explanation of its composition ("[3] lengths and 2[1 wid]ths accumulated", line 25) and the corresponding number (lines 5, 16).

Chapter IV
Methods

Before we proceed with further presentation and analysis of texts, a brief summary of what we have seen so far regarding the basic methods employed in the Old Babylonian "algebraic" texts may be useful.

"Naive" Cut-and-Paste Geometry

The starting point for the whole development of an "algebraic" technique is constituted by normalized problems about square or rectangular areas and their sides. Those that deal with squares combine linear entities and areas – when adding often by means of accumulation in the statement of the problem, but invariably representing the sides by "broad lines" in the solving procedure; when sides are subtracted from squares, and in certain additive cases, this geometric representation already appears in the statement.

In the basic problems about rectangles – those where the area and the sum of or the difference between the sides are given – we encounter no similar ontological difficulties; here, linear entities are combined with linear entities and not with areas. As shown by TMS IX, however, other rectangular problems might make exactly this combination.

Thanks to the use of broad lines, however, all cases reduce to what in Greek geometry would be seen as "application of an area with square excess or deficiency". Comparison with the way this matter is dealt with in the *Elements* may make the characteristics of the Old Babylonian technique stand out more clearly.

The diagram belonging with *Elements* II.6 – the theorem that corresponds to problems where sides are added to or subtracted from square areas, and to

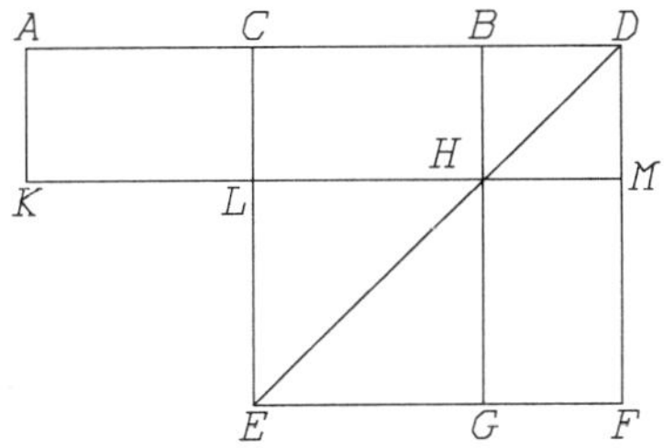

Figure 21. The diagram of *Elements* II.6.

rectangular problems where the difference between the sides is given[130] – is shown in Figure 21. If we base our reasoning on the rectangular problem, *AD* corresponds to the length and *DM* to the width of the rectangle. The Euclidean theorem states that if line *AB* is bisected at *C*, and prolonged by line *BD*, then the rectangle contained by *BD* and *AD* together with the square on *BC* will equal the square on the line *CBD*. The proof starts by constructing the latter square *CEFD* and drawing the diagonal *DE*. Next through *B* the line *BHG* is drawn parallel to *CE* or *DF* (*H* being the point where the line cuts *DE*) and through *H* the line *KM* parallel to *AB* or *EF*. Finally, through *A* the line *AK* is drawn parallel to *CE* or *DF*.

Now the diagram is ready, and with reference to the way the construction was made ⊏⊐*AL* is shown to equal ⊏⊐*HF*. Adding ⊏⊐*CM* to both, the gnomon *CDFGHL* is seen to equal ⊏⊐*AM*. Further addition of □*LG* shows that ⊏⊐*AM* together with □*LG* equals □*CF*, as stated in the theorem. We may add to the theorem (by analogy with VI.29, which treats the problem of applying a given area to a given line with an excess similar to a given parallelogram), that if a given area (becoming in the end the rectangle ⊏⊐*AM*) is to be applied to the line *AB* with square excess, then we add to it the square □*BC* on the bisected line *AB*; the result is laid out as a square along *CB* but exceeding this line by an amount which is designated *BD* – and the application has been successfully performed, that is, we have constructed a rectangle with the given area and whose length exceeds the width by *AD*.

This is exactly what is done, for instance, in YCB 6967 (p. 55) – if we look at the net result. The Old Babylonian text, however, says nothing about parallel lines, and does nothing to demonstrate that one rectangle equals another. It tacitly takes for granted that if the excess length is bisected, then the part of the rectangle that lies along this excess belongs together with it. Instead of proving that another rectangle is equal to the moiety of the excess it moves the piece around, making it "hold" a square area (without arguing

130 In Edward Heath's translation [1926: I, 385]: "If a straight line be bisected and a straight line be added to it in a straight line, the rectangle contained by the whole with the added straight line and the added straight line together with the square on the half is equal to the square on the straight line made up of the half and the added straight line".

explicitly that this area *is* square) together with that part of the excess that remains in place. After adding the quadratic complement (□*LG* in Figure 21) it calculates what the side *CD* = *DF* of □*CF* must be – and by moving the rectangle ⊏⊐*GM* back into its original position it finds *AD* and *DM*.

A first difference is thus that the Old Babylonian areas and lines are measurable, that their magnitude can be calculated in numbers; nothing similar is found in the *Elements*. Another contrast can be formulated with reference to Kant's notion of a *critique*. Kant did not formulate an alternative to Newton's mechanics in his *Critique of Pure Reason*, nor a better set of moral rules in his *Critique of Practical Reason*. What he set out to do was to find out *in which sense and to which extent* Newton was right in formulating his laws in absolute time and absolute space, and to analyze under what conditions a rule of conduct can be regarded as a moral rule. These were questions which Newton and traditional moralists had felt no need to ask, knowing already – *naively* – that time and space were there, and what kind of stuff morality was.

In this sense, the Babylonian method is *naive*. We *see* immediately that the procedure is correct, and we have to make an effort to see what precisely we have presupposed when believing that ⊏⊐*AL* when moved to fit unto *HM* will contain a square. Instead, what the *Elements* present us with is *a critique*. Instead of intuitively cutting pieces off the rectangle and moving them around, the first part of the proof constructs a diagram, in which equality of various parts can be argued with theoretical rigour; what goes on in the second part is then equivalent to the naive procedure, but because of the way the diagram is constructed it is *no longer naive*.

This has a pedagogical cost, which can be expressed in terms of another revered dichotomy. The naive procedure is *analytical*: it presupposes that the rectangle we are looking for is already there, draws it (we shall return to the question about *how* and *how concretely* it draws), and performs all steps in such a way that their correctness is obvious (and even their relevance, once the central trick of the quadratic completion has been grasped). The procedure is, as formulated by Karine Chemla ([1991, 1996], and elsewhere) regarding Chinese mathematics, algorithm and proof in one.

The Euclidean proof, in contrast, is *synthetic*. It starts by constructing a diagram which serves for this particular theorem and nowhere else – and then, suddenly, everything can be fitted into the diagram, and the proof is there. No wonder that the mathematicians of the Early Modern epoch, familiar both with Greek geometry and with algebra, were convinced that the proofs of the Greek mathematicians had been constructed on the basis of a preceding analysis, and reproached Euclid, Archimedes, and Diophantos for concealing this step – "so that his ingenuity and quickness of mind should be more admired", as Viète says about Diophantos.[131]

131 *In artem analyticen isagoge*, ch. V [ed. Hofmann 1970: 10].

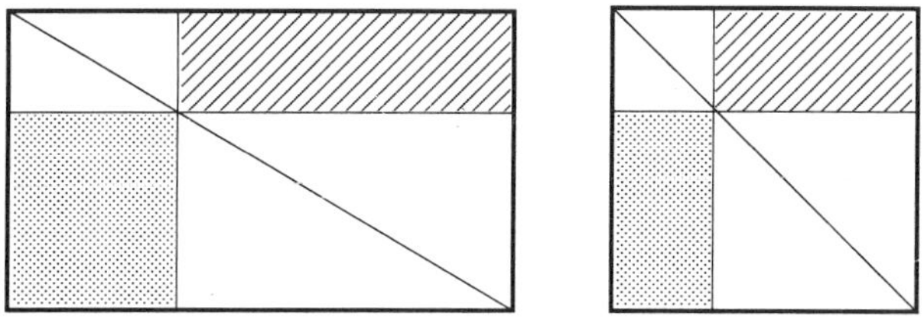

Figure 22. The diagram for *Elements* VI.24, in rectangle and (trivial) square version.

In certain cases, even the Old Babylonian procedures are rather synthetic, in the sense that a diagram seems to be present before it is used in the argument; thus BM 13901 #8–#9 (above, p. 66). Even here, however, an important difference exists: Greek synthesis refers to diagrams that are constructed *ad hoc* for the particular demonstration; in instances like *Elements* II.6, where the diagram repeats a familiar naive-geometric configuration,[132] the theorem has been reformulated so as to veil the link and make the proof appear as a rabbit coming out of the sleeves of the magician. Old Babylonian synthesis, in contrast, seems to refer to only a couple of familiar configurations: the concentric squares, and the subdivided square on the sum of two segments. For the same reason, there is no need to describe the construction of the diagrams – which anyhow is not easily done within the linguistic habits of Old Babylonian mathematics, which allow only explicit operation with entities whose measure can be stated (the operation with magnitudes which are not given but "merely known" seems to be reserved to particular text groups, cf. note 320).

Scaling and Other Changes of Variable

Normalized problems can be solved by cutting and pasting alone. As soon as problems are not normalized, a supplementary technique has to intervene. Within the framework of Greek geometry, this would correspond to the application of an area with rectangular but non-square excess or deficiency (*Elements* VI.28–29, which generalize from rectangle to parallelogram).

The Euclidean solution of the problem builds on the gnomon theorem (*Elements* VI.24) which, when applied to a rectangle as in Figure 22, above, expresses that the shaded and the dotted partial rectangles are equal. The underlying heuristics of the proof considers the rectangle as blown up (e.g.)

---

132 So far it is of course only a gratuitous assumption that the near-coincidence of Figure 2 with Figure 21 corresponds to a real historical link; that it does so will be argued in Chapter XI.

horizontally from the corresponding square, where the equality is intuitively obvious; blowing up the figure will change all areas with the same factor of proportionality, from which the theorem follows. In the technical proof, this heuristics is evidently transformed ("critically") by means of the theory of proportions.

Once again, there are no traces of a similar critique in the Old Babylonian texts, even though BM 13901 #14 as well as a number of other texts make use of the rule $\sqsubset\!\sqsupset(\alpha p,q) = \sqsubset\!\sqsupset(p,\alpha q)$.[133] Their procedure was shown in Figures 4 an 15, and consists in a "raising" by the coefficient α of the square as a linear factor, that is, in a change of linear scale in the vertical direction, in agreement with the heuristics underlying *Elements* VI.24; thereby the problem is transformed into a normalized problem about a "confrontation" $\sigma = \alpha s$ and its area $\square(\sigma)$, and there is no need for steps corresponding to those of *Elements* VI.29. What happens after the "scaling" may be seen as a simple "change of variable".

Other problems have presented us with more complex uses of this principle. Several clear examples are found in TMS IX #3; in the first of these – when both length and width are augmented by *1* – the new variables $\lambda = l+1$ and $\omega = 1°20'$ are spoken of explicitly with names of their own, as "the length of 2 the surface" and "the width of 2 the surface"; those introduced in the second step ($\Lambda = 3\lambda$, $\Omega = 21\omega$) are not referred to explicitly but only become visible through the operations, and may have possessed a less certain status as independent variables than λ and ω. In the same sense, the exact status of the variable $\sigma = \alpha s$ that is used during the treatment of non-normalized problems is of course uncertain – I remember no single text where it gets a designation of its own.

In cases of representation, exemplified by BM 13901 #12, it seems inappropriate to speak about a change of variable; the two original "confrontations" are certainly unknown and asked for, but they do not play the role of *variables*, that is, of entities that occur and serve within an analytical argument but to which no specific numerical value is as yet ascribed – they only enter the argument by being represented.

Accounting, Coefficients, Contributions

Scaling and change of variable are procedures that allow the reduction of complex equations to simple standard cases. The "accounting technique"

133 As we shall see (p. 382), the Old Babylonian school undertook a different kind of critique. Critique is always relative and never definitive; it never finds the firm bedrock on which everything can be built for eternity, it hammers its piles through enough strata of mud and clay for the building to stand firm for another while.

allows us to formulate problems as equations, determining their coefficients and thus permitting reduction by the former procedures.

We first encountered it, in a simple version, in BM 13901 #10, in the computation of the total number of small squares as 7×7+6×6; BM 13901 #14 presented us with a more sophisticated but not fundamentally dissimilar case. In both texts, the numbers resulting from the accounting are presented but not provided with any descriptive epithet which might permit us to know exactly to which extent the Babylonian calculator would possess a general concept of the technique and the outcome.

Here the didactic Susa text TMS XVI is helpful. When identifying the coefficient by the phrase "as much as (there is) of widths" it shows the existence of a formal concept, and by its determination of "the contribution of the width" it expresses an idea corresponding to the numerical value of the member αw of a polynomial. That such ideas were present can certainly come as no surprise – if the labour costs of a building project were to be determined from the amount of bitumen and earth and the number of bricks, and the cost of each of these to be determined from technical coefficients for production and carrying, the total would be composed exactly of such "contributions", and each of these would be found by a multiplying the relevant technical constant by a coefficient.

What TMS XVI taken alone cannot tell us is whether its phrases constituted a technical terminology or described the idea in non-technical language, and thus in terms that in other cases might be used differently. Other Susa texts shows the latter to be the case. In TMS VIII (below, p. 188) we shall encounter the phrase 4 *kīma* sag̃ in the sense "4, as much as the width (is, namely, in terms of a unit equal to its fourth)"; this can certainly also be regarded as a coefficient, but not as the coefficient of the width. In the text TMS XVII, line 11, the expression is used in still another way.

Determination of coefficients and contributions was thus a standard procedure that was familiar from ordinary applied mathematics; but it was not supported by a standard terminology – or if it was, then by a terminology that was not sufficiently standardized to be transferred to the discussion of equations as we find them in the didactic Susa texts.

In the series text YBC 4713 (below, p. 203) we shall see how coefficients with respect to the operation of *repetition* might be expressed.

Single (and Other) False Positions – and Bundling

The "single false position" or "rule of false" is an analytical but non-algebraic technique that until the sixteenth century was often used by practical reckoners to solve homogeneous problems. An example that turns up in numerous late medieval manuscripts is the following:[134]

> The head of a fish weighs $\frac{1}{3}$ of the whole fish, and the tail of the fish $\frac{1}{4}$ of the whole fish. And the body in the middle weighs 8 ounces. Tell me, how much is the weight of the head, how much that of the tail, and how much that of the whole fish. Do like this and say, $\frac{1}{3}$ and $\frac{1}{4}$ can be found in 12. And take the $\frac{1}{3}$ and the $\frac{1}{4}$ of 12, which together make 7, and say, from 7 to 12 there are 5, and this is the divisor. Now, because the body of the fish weighs 8 ounces, multiply 12 by 8; it makes 96. Divide by 5, and 19 and $\frac{1}{5}$ results. And so much is the weight of the whole fish, that is, 19 ounces and $\frac{1}{5}$. And if you want to verify it, and want to know what each one weighs by itself, take away the $\frac{1}{3}$ of 19 and $\frac{1}{5}$, from which results 6 and $\frac{2}{5}$. And so much is the weight of the head. And then take away the $\frac{1}{4}$ of 19 and $\frac{1}{5}$, which is 4 and $\frac{4}{5}$. And so much is the weight of the tail. Now join together 6 and $\frac{2}{5}$ and 4 and $\frac{4}{5}$, which make 11 and $\frac{1}{5}$. Detract it from 19 and $\frac{1}{5}$. Remains 8 ounces, and 8 ounces is the weight of the part in middle. And it is correct, and thus all are made.

We thus posit, falsely but conveniently in view of the presence of the fractions $\frac{1}{3}$ and $\frac{1}{4}$, that the total weight be 12. *If* that had been the case, the remainder would have been 5; since it is 8, the initial position must be increased by an adequate factor of proportionality, namely, $\frac{8}{5}$.

In Renaissance texts, this was often coupled to the rule of three,[135] for instance, with the scheme

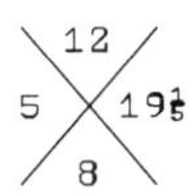

No similar schemes are found in the cuneiform material, but if we ask only for the underlying idea of the false position – the choice of a convenient reference quantity of which the requested fraction can conveniently be taken, and ensuing proportional scaling – then that idea is certainly present though not coupled to the scheme of the rule of three.

TMS XVI, lines 8–9, is an indubitable example. In order to see what remains of *1* when its fourth is taken away, *4* is chosen as reference quantity. After removal of the fourth, 3 remains; but since *4* should be scaled to *1*, which is obtained if we multiply by igi 4 = 15′, 3 is scaled to 3·15′ = 45′.

Similar examples turn up elsewhere in the corpus. But there are also examples where it is not certain whether the idea behind the numerical steps is really a "false position" in the sense that a reference quantity is taken as *model* for the quantity that is to be determined, or this quantity is *itself* subdivided in a convenient way. One such example was discussed above (BM 13901 #10, p. 58).

---

134 I translate from Jacopo da Firenze's *Tractatus algorismi*, ms. Vat. Lat. 4826, fol 24r.

135 Though no scheme appears, the rule of three is certainly also in Jacopo's mind, as he shows by referring in definite form to "the divisor" and multiplying before he divides (which deprives the intermediate result of concrete interpretation): it corresponds to his presentation of the rule of three on fol. 17r.

We also find other types of falses assumptions in use – for instance, the assumption in VAT 8389 #1 (above, p. 77) that the two plots be equal. We also find the epithet "false", which seems to imply that the Babylonians saw "false assumptions" where we do not expect this way of thinking – for instance, in VAT 8389 and VAT 8391, where the rent per sar is spoken of as the "false grain".

The double false position (see above, p. 82) seems never to have been used by thc Old Babylonian calculators. The first appearances seem to be in the Chinese *Nine Chapters on Arithmetical Procedures* and in Hero's *Metrica* III.20 (same epoch), where the author uses it to find (an approximate value of) $\sqrt[3]{100}$.[136]

The alternative to the interpretation of BM 13901 #10 as a false position is one of the two methods to which the texts give a name (the other being the "Akkadian method" of TMS IX): the *makṣarum* or "bundling" – see p. 66. We shall return repeatedly to this method below.

Drawings? Manifest or Mental Geometry?

A geometrical representation seems to presuppose space and drawings. How does this fit the textual evidence? Are the transformations as shown in Figure 2, Figure 3, Figure 4, etc., to be found on the tablets? And if not, how is that to be explained?

They are *not* found on the tablets. All that occurs are diagrams that illustrate the statements of problems. With one atypical and partial exception (YBC 8633, below, p. 254), the texts never contain drawings that illustrate what goes on in the procedure. The rare drawings we find only serve to elucidate the meaning of the statement (we shall encounter some examples below).

Before we go on with the quest for an explanation we should have a look at a text that gives us a glimpse of Old Babylonian geometrical concepts: the field plan MIO 1107 – see Figure 23. The plan dates from the 21st century BCE, but we may safely assume that the conclusions we may draw from it would hold true a few centuries later.

The first observation to make is that the proportions of the plan are obviously fitted to the round tablet on which it is drawn and not to those of the terrain it describes. The next is that right angles are rendered as right

---

136 See, respectively, [Chemla 1997] and [Ma 1993: 10–12]. Both texts are from the 1st c. CE; Hero, one should notice, demonstrates his failure to understand the formula he is employing by leaving out a factor *1*.

On the absence of the double false position from the Babylonian material, cf. [Vogel 1960].

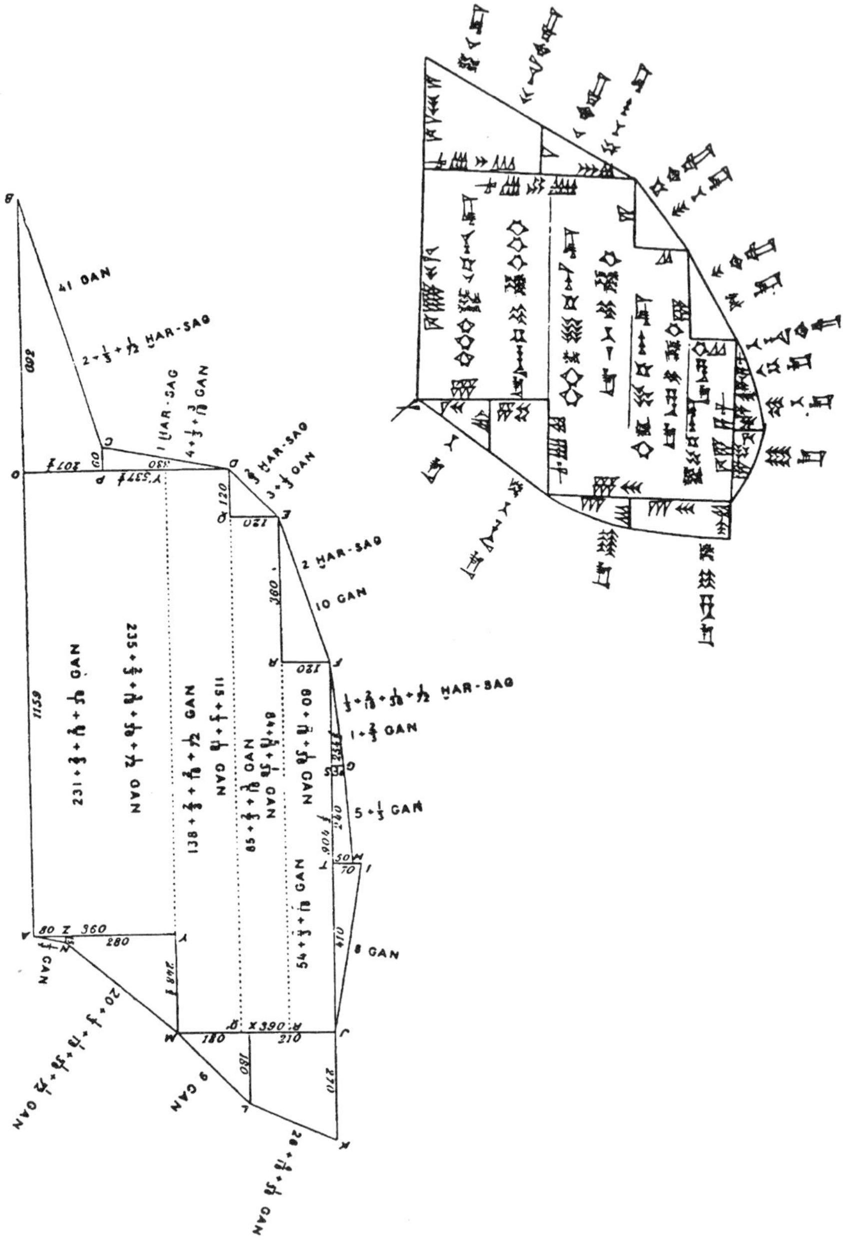

Figure 23. Plan of the fields belonging to the district Šulgi-sib-kalama as rendered on the tablet MIO 1107 (left) and redrawn in true proportions (right). From [Thureau-Dangin 1897: 13, 15].

angles; these are the corners between lengths and widths that will be multiplied in order to give the areas of the fields. The third and (in this context) final observation is that no attempt is made to render other angles correctly (as we see, an angle of c. 120° at bottom appears as 180°), nor even to draw lines that point in the same direction in proportion to their stated length – as exemplified to the left, where *the same* drawn line has lengths 50 and 70.

What we find on the tablet is thus no real map in our sense, no metrically faithful rendering of the geometry of the terrain; it is a *structure diagram* whose main function is to identify and summarize the role of measured segments in the area calculation.

This kind of diagram is hence what an Old Babylonian surveyor would need as drawn support for his geometrical thought. That it was also sufficient for the user of the mathematical texts is clear from some of the drawings that illustrate statements – the most glaring example probably being the initial diagram of YBC 4675 (see p. 244).

This does not mean that the Babylonians were not able to distinguish an oblong rectangle from a square or an angle of 20° from one of 65°: though made without measurement and by impression of the ruler in the clay, the drawings of the geometrical text BM 15285 and the regular heptagon and hexagon of TMS II are fairly precise – even on the convex reverse of BM 15285, the supposedly right angles never err by more than c. 5°; the circle in TMS I is compass-drawn, and the inscribed isosceles triangle is impeccable. What it implies is exactly what was stated: the Old Babylonian calculator did not *need* this degree of metrical fidelity for thinking about areas and their mutual relations. He will have had no difficulty in seeing the upper configuration of Figure 4 as if it were the lower configuration, and thus have had no need for redrawing; at most, he would have to "posit" a new number to the vertical side.

This lessens a bit the problem which the absence of diagrams illustrating the progress of calculations seems to present; but it does not eliminate it.

It turns out that "procedure diagrams" are absent not only from the "algebraic texts" but also from properly geometric texts where we may be quite certain that the calculations correspond to transformations of a geometric configuration (see below, VAT 8512, p. 234). Even a return to the arithmetical interpretation of the "algebra" texts would thus not free us from the problem: If geometrical transformations were performed but are not drawn in the clay tablets, *where* were they made?

A further reduction of the problem (which still does not resolve it) is obtained by the observation that much may have been made as "mental geometry": after some amount of work with the cut-and-paste methods and appurtenant scalings one is able to imagine the underlying configurations well enough to be able to perform the calculational steps without making an actual drawing, or with the support of only a most rudimentary sketch. Calculators with a minimum of training will probably have had no need to perform the actual cuttings and displacements of Figure 1, etc. – a mere drawing of the rectangle itself will have been sufficient.

But mental arithmetic with two-digit numbers (that is, numbers that require *calculation* because they fall outside the tables that we have learned by heart) is a secondary acquisition and presupposes a stage of material arithmetic (in our world made on paper, but an abacus would do). Similarly, mental geometry will have been a secondary ability, which could only be learned by working on manifest geometry. And the problem remains: If this was not made on the clay tablets, *where then*?

In principle, clay and drawings are not the only possibilities; pebbles, in the style of Greek figurate arithmetic, are one possibility; sticks representing "broad lines" or rectangles are another. But both have drawbacks, and proper drawings or structure diagrams remain the most adequate medium.

Two decades ago, excavations of an Old Babylonian school at Tell ed-Dēr made M. Tanret [1982: 49] propose that the initial training of cuneiform writing was based on models drawn "in the sand of the school yard".[137] Eleanor Robson (personal communication) objects that the courtyard of the building in question was paved with baked brick. However, sand or some similar material might be strewn upon a pavement or upon part of it, either for facilitating cleaning[138] or as a medium for drawing.

Our incessant use of cheap paper has made us forget, but for purposes where numbers or geometrical figures have to be cancelled or modified during the process, sand (spread on an even surface) or a dustboard are often more convenient than paper or clay. For reckoning with Arabic numerals, the dustboard was used well into the second millennium CE; when paper took over, a whole set of new algorithms had to be developed which did not ask for constant cancellation and rewriting – see [Van Egmond 1986: 59*f*].

The anecdote of Archimedes tracing geometrical figures in sand is familiar. It is true that the story may have resulted from a misunderstanding, as argued by Dijksterhuis [1956: 31*f*], and that the usual medium of Greek geometrical argument was the dustboard; but the very possibility of the misunderstanding demonstrates that geometrical diagrams were sometimes made in sand. The anecdote about the shipwrecked philosopher Aristippos finding geometrical figures traced in the sand of the Rhodian shore leads to the same conclusion (Vitruvius, *De architectura* VI.i [ed. Granger 1931: II, 2]). Aristotle, finally, refers quite explicitly to the drawings underlying geometric

---

137 He found, indeed, that the exercise tablets belonging to the higher levels of writing contained in parallel the instructor's model and the student's attempt to imitate it. In contrast, the tablets belonging to the elementary level (stylus exercises and the first training of signs) contain no instructor's model. If not due to an excavation accident, however, this may represent a particular characteristic of the Tell ed-Dēr-school – in Nippur the situation seems to be different (Eleanor Robson, personal communication).

138 In Denmark (and in Northern Europe in general, I would guess), sand was strewn and carefully distributed on wooden floors with this purpose until the advent of varnishing in the late nineteenth century.

proofs (including "premisses") as being made "on the ground" (ἐν τῇ γῇ – *Metaphysica* 1078ᵃ19 [ed. Tredennick 1933: II, 190]).

That Greek geometers would sometimes trace their diagrams in the sand or on the ground only shows that it could be done. Their dustboard, however, turns out to be imported from the Near East. The name used by the Greeks, indeed, is ἄβαξ, "abacus", the same as was used to designate the board where computation by means of counters were performed. The word, as first observed by Nesselmann [1842: 107 n.5], is derived from the Semitic root ʾbq, "to fly away", "light dust" – which implies, firstly, that the dust board precedes the counter abacus, and secondly, that the device was borrowed from Semitic speakers, most likely in the Syro-Phoenician area (as even the wax tablet was borrowed). Given the close cultural connections between the Syrian region and Mesopotamia (connections which comprise all we know about Syro-Phoenician practical mathematics), it is likely that an instrument that was used in Syria or Phoenicia was also known in Mesopotamia in the mid-first millennium BCE. It does not follow that it was known already in the Old Babylonian epoch (the wax tablet seems to belong exclusively to the first millennium); but the very possibility shows that Tanret's sand of the school yard was not the only possible medium for geometrical operation. Still other possibilities suggested by Eleanor Robson are "rough tablets, lumps of clay; potsherds; charcoal on plastered mud-brick walls ...".

If I am to point out the most likely medium, I shall opt for sand spread over an even surface, perhaps in the school yard. Archimedes's demonstrations may have been too complex for the use of common sand, as claimed by Dijksterhuis; but as far as its geometry is concerned, no Old Babylonian "algebraic" or related text goes beyond what could easily be traced as structure diagrams on the ground. Charcoal drawings on a wall are well suited for representing the fixed configuration dealt with by a whole sequence of problems (e.g., the standard 20×30-rectangle), and may explain the predilection for suc̣h sequences; but they are not the ideal medium for procedures involving cancelling and redrawing.

Chapter V
Further "Algebraic" Texts

In this chapter we shall look at a larger range of "algebraic" texts – in part occupying entire tablets, in part single problems drawn from longer compositions, but in all cases texts which permit us to see new facets of Old Babylonian "algebraic" thought.

BM 13901 #18[139]

The following problem comes from the theme text about squares from which we have already examined #1–3, #8–10, #12, and #14. Apart from dealing with three instead of two squares, it may be regarded as a simpler version of #14 – or, better, #14 may be seen as a combination of the statement of the differences between squares in absolute terms which we find in the present problem, and the definition in relative terms found in #10 (and several others). Therefore, the problem is first of all interesting because of its deviations from and similarities to #14.

Rev. I

39. The surface of my t[h]ree confrontations I have accumulated: 23′ 20″.
a.šà *š[a-]la-aš mi-it-ḫa-ra-ti-ia ak-mur-ma* 23,20

40. The confrontation over the confrontation, 10′ it went beyond.
mi-it-ḫar-tum ugu *mi-it-ḫar-tim* 10 *i-te-er*

139 Based on the transliteration in [MKT III, 4].

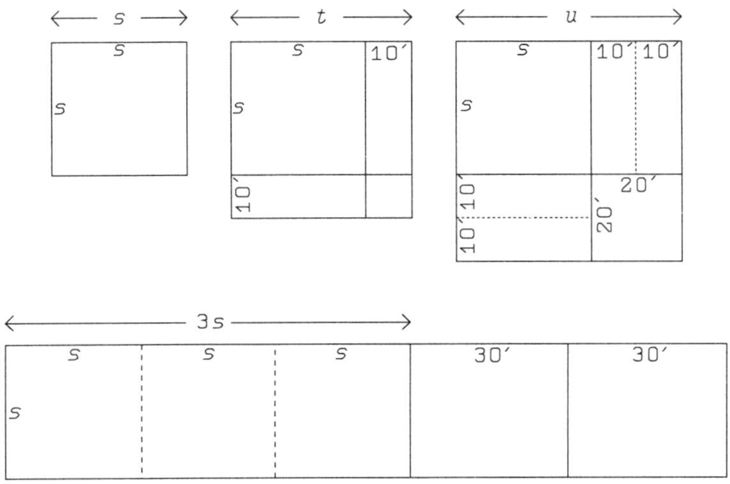

Figure 24. The three squares of BM 13901 #18.

41. 10´ which it went beyond to 1 you raise, 10´; to 2 you raise, 20´ and 20´
10 *ša i-te-ru a-na* 1 *ta-na-ši* 10 *a-na* 2 *ta-na-ši* 20 *ù* 20

42. ⟨you make hold⟩, 6´40˝ is it. 10´ and 10´ you make hold, 1´40˝ to 6´40˝ you append, 8´20˝
⟨*tu-uš-ta-kal*⟩ 6,40.e 10 *ù* 10 *tu-uš-ta-kal* 1,40 *a-na* 6,40 *tu-ṣa-ab* 8,20

43. in the inside of 23´20˝ you tear out: 15´ to 3, the confronta[tion]s,
lìb-ba 23,20 *ta-na-sà-aḫ-ma* 15 *a-na* 3 *mi-it-ḫa-ra*[*-ti*]

44. you raise, 45´ you inscribe. 10´ and 20˝ you accumulate:
ta-na-ši 45 *ta-la-pa-at* 10 ù 20 *ta-ka-mar-ma*

45. 30´ and 30´ you make hold, 15´ to 45´ you append:
30 *ù* 30 *tu-uš-ta-kal* 15 *a-na* 45 *tu-ṣa-ab-ma*

46. by 1, 1 is equalside. 30´ which you have made hold you tear out, 30´ you inscribe.
1.e 1 íb.ṣi$_8$ 30 *ša tu-uš-ta-ki-lu ta-na-sà-aḫ-ma* 30 *ta-la-pa-at*

47. The igi of 3, of the confrontations, 20´ to 30´ you raise, 10´ the confrontation.
igi 3 *mi-it-ḫa-ra-ti* 20 *a-na* 30 *ta-na-ši* 10 *mi-it-ḫar-tum*

48. 10´ to 10´ you append: 20´ confrontation nº 2. 10´ to 20´
10 *a-na* 10 *tu-ṣa-ab-ma* 20 *mi-it-ḫar-tum* ki.2 10 *a-na* 20

49. you append: 30´ confrontation nº 3.
tu-ṣa-ab-ma 30 *mi-it-ḫar-tum* ki.3

For simplicity we may label the three sides s, $t = s+1\cdot10´ = s+10´$, and $u = t+10´ = s+2\cdot10´ = s+20´$ (Figure 24, upper part). These pedantic computations correspond to line 41; the multiplications involved are "raisings", as they should be. In line 42, the areas of the lower right corners of $\square(t)$ and $\square(u)$ are found ("being held", again quite regularly) and accumulated, and the result is "torn out" from the total area 23′ 20″, leaving 15′ as the value of the three squares □(*s*) and the "wings" of □(*t*) and □(*u*) (Figure 24, lower part). The presence of three squares □(*s*) calls for a scaling with the factor 3, giving a total area of the scaled figure equal to 45′. Only after this is found, and when the width of the wings must be "made hold", is it found as 10′+20′ = 30′ (by "accumulation", i.e., in an accounting procedure, not through "appending" materially the wings of □(*u*) to those of □(*t*)). Lines 45 to 47 follow the familiar pattern, until "the confrontation" is found by inverse scaling. Lines 48 and 49 then find *t* and *u* by successive additions of 10.

In the discussion of #14 we saw that a new unit *s* was made use of (perhaps equal to *u*, perhaps to 1′·*u*; see p. 77). This is obviously not the case here. The initial accounting process dissects □(*t*) and □(*u*) into areas and sides of the [first] square. This is reflected in the absence of a multiplication of 10 by 1 in line 47 (whereas the lack of an epithet "*first* confrontation" is in agreement with general stylistic patterns and thus not significant). More important is, however, the absence of "making 1 and 1 hold" before line 41: The square areas are already present, and regarded as identical with the square whose area and sides is the object of accounting.

An important observation can be made from the precedence of the scaling over the computation of the width of the "wings". This order agrees with a general norm which was also present in #14; it recurs in #24, which is a three-square variant of #14; and in other texts as well. The arrangement implies that bringing about the fundamental (normalized) one-square situation is the important point in the solution of problems of this kind, while the computation of the precise dimensions of the figure is a secondary matter. This would be anomalous if the geometric technique was based on drawings made more or less to scale. In mere structure diagrams, on the other hand, delayed computation of the width of the wings would be possible, even if still a bit strange. In the case that the geometry appealed to was mental and not material, however, early computation of this width would only burden the calculator's memory with a number to be remembered, or (rather) force him to make a note. Computation of the precise proportions of the figure would better be delayed until the moment where they were to be used.

We cannot conclude from the order of the operations that every actual solution of our problem was made mentally; after all, the text is prescriptive, not a description of actual solutions. An immutable order also suggests that we are dealing with a convention. Even a convention, however, will have been derived through generalization from typical usage. The order in question thus intimates that at least the basic naive-geometric operations were performed mentally by trained calculators.

YBC 4714[140]

This catalogue text belongs to the genre of series texts (cf. p. 9); according to the final colophon, it is the fourth tablet of the series to which it belongs. It deals with squares, and its subject-matter is closely related to that of BM 13901 even if dealing on the whole with more sophisticated variants of the problems. We may imagine with Neugebauer [MKT III, 10] that the problems of the third tablet were at the level of BM 13901.

The language of the tablet is highly elliptic. Almost everything is written logographically, grammatical complements are rare, and the style is utterly laconic – at times demonstrably ambiguous (cf. note 152).[141] I have translated the ideographically written verbs as infinite forms in order to reproduce the abrupt impression conveyed by the text.

A noteworthy feature of the language is the use of íb.si$_8$ as a logogram for *mitḫartum*. Even more remarkable is the use of the suffix .e on numerals that count the number of squares (both cardinally and ordinally). It has evidently lost whatever connection to the Sumerian ergative suffix it may still have in BM 13901 and YBC 6295 (see pp. 52 and 66, respectively). A number of other Sumerian suffixes, on the other hand, are used correctly, or at least as correctly as in Old Babylonian Sumerian in general.

In contrast to BM 13901, but like other series texts, the present tablet does not explain the procedures to be used. But it does inform us about the solutions, which at least helps us interpret certain ambiguous problem statements.

The tablet also differs from BM 13901 by stating given and resulting lengths not as seemingly pure numbers but with explicit units (mostly nindan, but also UŠ – 1 UŠ = 1` nindan). Areas, on the other hand, are given in an implicit unit sar (with the single exception of #29). As in the other series texts, the order of magnitude is integer nindan, not minutes (sixtieths).

---

140 Based on the transliteration in [MKT I, 487–492].

141 At an early moment, we remember, Neugebauer maintained that the ideograms of the mathematical texts functioned "precisely as mathematical symbols" (see note 56). The ambiguities resulting from the condensed style of the present tablet shows that this is not so: compact ideographic writing may have alleviated writing, and was certainly needed if the present text was to be fitted into a tablet of 11 cm by 8½ cm – but just as certainly it did not facilitate intelligibility or provide transparency. *A fortiori* it could never lead to that *operation directly at the level of symbols* which distinguishes *symbolic* from *rhetorical* and *syncopated algebra*.

Obv. I

#1

******

[...]

1. accumulated⌈: ...]
ğar.ğar-*m*⌈*a* ...]

2. The 3 (mistake for "2") confrontations, [what?]
íb.si$_8$ 3.e[142] [en.nam]

3. ¿30? nindan, the 1st.
¿30? nindan 1.e

4. 20 nindan, the 2nd.
20 nindan 2.e

#2

5. The surface of 4 confrontations
a.šà íb.si$_8$ 4.e

6. accumulated: 1`` 30`.
ğar.ğar-*ma* 1,30

7. The 4 confrontations
íb.si$_8$ 4.e

8. accumulated: 2` 20.
ğar.ğar-*ma* 2,20

9. The 4 co[nfront]ations, what?
í[b.s]i$_8$ 4.e en.nam

10. 50 [nindan], the 1st.
50 [nindan] 1.e

11. 40 nindan, the 2nd.
40 nindan 2.e

12. 30 nindan, the 3rd.
30 nindan 3.e

13. 20 nindan, the 4th.
20 nindan 4.e

#3

14. The surface of 6 confrontations
a.šà íb.si$_8$ 6.e

15. accumulated: 1`` 52` 55.
ğar.ğar-*ma* 1,52,55

16. The 6 confrontations
íb.si$_8$ 6.e

---

142 Certainly a writing error for 2.e, both because this is the number of "confrontations" that is actually found and because the problem is part of a sequence that deals with an even number of "confrontations".

17. accumulated: 3`15.
ĝar.ĝar-*ma* 3,15

18. The 6 confrontations, what?
íb.si$_8$ 6.e en.nam

19. 45 nindan, the 1st.
45 nindan 1.e

20. 40 nindan, the 2nd.
40 nindan 2.e

21. 35 nindan, the 3rd
35 nindan 3.e

22. 30 nindan, the 4th.
30 nindan 4.e

23. 25 nindan, the 5th.
25 nindan 5.e

24. 20 nindan, the 6th.
20 nindan 6.e

#4

25. The surface of 3 confrontations
a.šà íb.si$_8$ 3.e

26. accumulated: 30`50.
ĝar.ĝar-*ma* 30,50

27. The 7th part of the 1st
igi 7 ĝál 1.e

28. [and 1]5 nindan, the 2nd.
[*ù* 1]5 nindan 2.e

29. Half of the 2nd
šu.ri.a 2.e

30. and 5 nindan, the 3rd.
ù 5 nindan 3.e

31. The 3 confrontations, what?
íb.si$_8$ 3.e en.nam

32. 35 nindan, the 1st.
35 nindan 1.e

33. 20 nindan, the 2nd.
20 nindan 2.e

34. [15 ni]ndan, the 3rd.
[15 ni]ndan 3.e

Obv. II.

**

#5

1. [... ¿1?]st [...]
[...].e [...]

2. [...] 2nd [...]
[...] 2.e [...]

3. The 3 confrontations, [w]hat?
íb.si$_8$ 3[.e e]n.nam

4. ⌈5⌉5 nindan, the 1st.
⌈5⌉5 nindan 1.e

5. 24 nindan, the 2nd.
24 nindan 2.e

6. 22 nindan, the 3rd.
22 nindan 3.e

#6

7. The surface of 3 confrontations
a.šà íb.si$_8$ 3.e

8. and the ⌊3] confrontations
ù íb.si$_8$ ⌊3].e

9. accumulated: 27`⌈50⌉.
ĝar.ĝar-*ma* 27,⌈50⌉

10. Confrontation to confrontation,
íb.si$_8$ íb.si$_8$.ra

11. The 17th part was smaller,
igi 17 ĝál ba.lal

12. ½ nindan (being) appended to the 1st,
½ nindan 1.e daḫ

13. 3 nindan appended to the 2nd,
3 nindan 2.e daḫ

14. 2 nindan appended to the 3rd.[143]
2 nindan 3.e daḫ

15. The 3 confrontations, what?
íb.si$_8$ 3.e en.nam

16. 25 nindan, the 1st.
25 nindan 1.e

17. 24 nindan, the 2nd.
24 nindan 2.e

18. 20 nindan, the 3rd.
20 nindan 3.e

#7

19. The surface of 3 confrontations
a.šà íb.si$_8$ 3.e

---

143 The formulation is either corrupt or too elliptic (even according to the norm used in other problems of the tablet) to allow proper interpretation. The meaning might be that $(x+\frac{1}{2})-y = \frac{1}{17}(x+\frac{1}{2})$, $y-(z+2) = (y+3)-x$, as suggested by Neugebauer [MKT I, 499], and will plausibly have to be something of that kind.

20. accumulated: 1``17`30.
ĝar.ĝar-*ma* 1,17,30

21. The 11th part of the 1st
igi 11 ĝál 1.e

22. and 30 nindan, the 2nd.
ù 30 nindan 2.e

23. The 7th part of the 2nd
igi 7 ĝál 2.e

24. and 15 nindan, the 3rd.
ù 15 nindan 3.e

25. The 3 confrontations, what?
íb.si$_8$ 3.e en.nam

26. 55 nindan, the 1st.
55 nindan 1.e

27. 35 nindan, the 2nd.
35 nindan 2.e

28. 20 nindan, the 3rd.
20 nindan 3.e

#8

29. The surface of ⸤4⸥ confrontations
a.šà íb.si$_8$ ⸤4⸥.e

30. accumulated: 2`` 2[3]` 20.
ĝar.ĝar-*ma* 2,2[3],20

31. The 3rd part of the small confrontation,
igi 3 ĝál íb.si$_8$ tur.ra

32. confrontation over confrontation went beyond.
íb.si$_8$ ugu íb.si$_8$ dirig

33. The 4 confrontations, what?
íb.si$_8$ 4.e en.nam

34. 1 UŠ the 1st.
1 UŠ 1.e

35. 50 nindan, the 2nd.
50 nindan 2.e

36. 40 nindan, the 3rd.
40 nindan 3.e

37. 30 nindan, the 4th.
30 nindan 4.e

#9

38. The surface of 4 co[nfront]ations
a.šà í[b.s]i$_8$ 4.e

39. and thc 4 confrontations
ù íb.si$_8$ 4.e

Obv. III

******

1. 30 [nindan, the 4th].
30 [nindan 4.e]

#10

2. The surface of ⌈4 confrontations]
a.šà í[b.si$_8$] ⌈4.e]

3. accumulated: 1``15`[50.]
g̃ar.g̃ar-*ma* 1,15,[50]

4. The 1st confrontation
íb.si$_8$ 1.e

5. over the 2nd, 10 nindan went beyond.
ugu 2.e 10 nindan dirig

6. The seventh part of the 2nd
igi 7 g̃ál 2.e

7. and 25 nindan, the 3rd.
ù 25 nindan 3.e

8. $\frac{2}{3}$ of the 3rd, the 4th.
$\frac{2}{3}$ 3.e 4.e

9. The 4 confrontations, what?
íb.si$_8$ 4.e en.nam

10. 45 nindan, the 1st.
45 nindan 1.e

11. 35 nindan, the 2nd.
35 nindan 2.e

12. 30 nindan, the 3rd.
30 nindan 3.e

13. [20 nindan, the 4th.]
[20 nindan 4.e]

#11

14. [The surface of 4 confrontations]
[a.šà íb.si$_8$ 4.e]

****** [c. 4 lines are destroyed]

15. [The 4] con[frontations, what?]
í[b.si$_8$ 4.e en.nam]

16. 50 [nindan, the 1st.]
50 [nindan 1.e]

17. 50 nindan, [the 2nd.]
50 nindan [2.e]

18. 45 nindan, [the 3rd].
45 nindan [3.e]

18. 40 nindan, the 4th.
40 nindan 4.e

#12

20. The surface of 4 confrontations
a.šà íb.si$_8$ 4.e

21. accumulated: 1``36`15.
. ğar.ğar-*ma* 1,36,15

22. the 4th part of (that which) the 11th part of the great,
igi 4 ğál igi 11 ğál gal

23. over the 7th part of the 2nd went beyond,
ugu igi 7 ğál 2.e dirig

24. confrontation over confrontation went beyond.[144]
íb.si$_8$ ugu íb.si$_8$ dirig

25. The 4 confrontations, what?
íb.si$_8$ 4.e en.nam

26. 55 nindan the 1st.
55 nindan 1.e

27. 35 nindan, the 2nd.
35 nindan 2.e

28. 30 nindan, the 3rd.
30 nindan 3.e

29. 25 nindan, the 4th.
25 nindan 4.e

#13

30. The surface of 3 confrontations
a.šà íb.si$_8$ 3.e

31. accumulated: 2``47`5.
. ğar.ğar-*ma* 2,47,5

32. 1`20 nindan the 1st confrontation.
1,20 nindan íb.si$_8$ 1.e

33. The 7th part of (that which) the 1st
igi 7 ğál 1.e

34. over the 2nd went beyond,
ugu 2.e dirig

35. the 3rd (compared) to the 2nd was smaller.
3.e 2.e.ra ba.lal

36. The 2 confrontations what?
íb.si$_8$ 2.e en.nam

37. 45 nindan the 2nd, 40 nindan the 3rd.
45 nindan 2.e 40 nindan 3.e

---

144 The meaning of lines 22–24 appears to be $\frac{1}{4} \cdot (x_1 - x_2) = \frac{1}{11} \cdot x_1 = \frac{1}{7} \cdot x_2 = x_2 - x_3 = x_3 - x_4$.

Obv. IV

#14[145]

1. [The surface of 3 confrontations]
[a.šà íb.si$_8$ 3.e]

2. [accumulated: 2``47`5.]
[ĝar.ĝar-*ma* 2,47,5]

3. [The 7th part of (that which) the 1st]
[igi 7 ĝál 1.e]

4. [over the 2nd went beyond,]
[ugu 2.e dirig]

5. [the 3rd (compared to) the 2nd was smaller.]
[3.e 2.e tur.ra]

6. [45 nindan the 2nd confrontation.]
[45 nindan íb.si$_8$ 2.e]

7. [The 2 confrontations, what?]
[íb.si$_8$ 2.e en.nam]

8. [1`20 nindan, the 1st].
[1,20 nindan 1.e]

9. [40 nindan, the 3rd.]
[40 nindan 3.e]

#15

10. [The surface of 3 confrontations]
[a.šà íb.si$_8$ 3.e]

11. [accumulated: 2``47`5.]
[ĝar.ĝar-*ma* 2,47,5]

12. [The 7th part of (that which) the 1st]
[igi 7 ĝál 1.e]

13. ov[er the 2nd went beyond,]
ug[u 2.e dirig]

14. the 3rd [(compared to) the 2nd was smaller.]
3.e [2.e tur.ra]

15. 40 [nindan the] 3rd [confrontation.]
40 [nindan íb.si$_8$] 3.e

16. The [2 con]frontations, what?
íb.s[i$_8$ 2.e] en.nam

17. 1`20 [nindan, the $1_{|}$st.
1,20 [nindan $1_{|}$.e

[145] Even though #14–16 are almost completely destroyed, the main lines of the reconstructions can probably be relied upon. Details, however, are uncertain – we cannot possibly know, e.g., whether #17 line 34 and not #13 line 35 should be the general model for the fifth line of the problems.

18. 45 [nindan, the 2nd.]
45 [nindan 2.e]

#16

19. The surface [of 3 confrontations]
a.šà [íb.si$_{8}$ 3.e]

20. [accumulated: 2``47`5.]
[g̃ar.g̃ar-*ma* 2,47,5]

21. [The 7th part of (that which) the 1st]
[igi 7 g̃ál 1.e]

22. [over the 2nd went beyond,]
[ugu 2.e dirig]

23. [the 3rd (compared to) the 2nd was smaller].
[3.e 2.e tur.ra]

24. [The ... part ...]
[igi ... g̃ál ...]

25. [and ... went beyond ...]
[*ù* ... dirig ...]

26. [The 3 confrontations, what?]
[íb.si$_{8}$ 3.e en.nam]

27. [1`20 nindan the 1st.]
[1,20 nindan 1.e]

28. [45 nindan the 2nd.]
[45 nindan 2.e]

29. [40 nindan the 3rd.]
[40 nindan 3.e]

#17

30. The surface [of 3 confrontations]
a.šà [íb.si$_{8}$ 3.e]

31. accumulated: 2``[47`5.]
g̃ar.g̃ar-*ma* 2,[47,5]

32. The 7th part [of (that which) the 1st]
igi 7 g̃ál [1.e]

33. [ov]er the 2nd went beyond,
[ug]u 2.e dirig

34. the 3rd (compared to) the 2nd was smaller.
3-[e] 2-[e] tur.ra

35. The 3rd part of the 1st
igi 3 g̃ál 1.e

36. and 13°20′, the 3rd.
ù 13,20 3.e

37. The 3 confrontations, what?
íb.si$_{8}$ 3.e en.nam

38. 1`20 nindan, the 1st.
1,20 nindan 1.e

39. 45 nindan, the 2nd.
45 nindan 2.e

40. 40 nindan, the 3rd.
40 nindan 3.e

#18

41. The surface of 3 confrontations
a.šà íb.si$_8$ 3.e

42. accumulated: 2``47`5.
ĝar.ĝar-*ma* 2,47,5

43. The 7th part of (that which) the 1st
igi 7 ĝál 1.e

44. over the 2nd went beyond,
ugu 2.e dirig

45. the 3rd (compared to) the 2nd is becomes smaller.
3.e 2.e tur.ra

46. The 4th part of the 1st
igi 4 ĝál 1.e

47. and 20 nindan, the 3rd.
ù 20 nindan 3.e

48. The 3 confrontations, what?
íb.si$_8$ 3.e en.nam

49. 1`20 nindan, the 1st.
1,20 nindan 1.e

50. 45 nindan, the 2nd.
45 nindan 2.e

51. 40 nindan, the 3rd.
40 nindan 3.e

#19

52. [The surf]ace of 3 con[fronta]tions
[a.š]à í[b.s]i$_8$ 3.e

53. accumulated: 2``47`5.
ĝar.ĝar-*ma* 2,47,5

Obv. V

****** [...]

#20

****** [...]

1. accu[mulated: ...]
ĝar.[ĝar-*ma* ...]

2. [...] part [...]
igi [...]

3. ov[er ...]
ug[u ...]

4. 3rd [...]
3.e [...]

5. Ha[lf ...]
šu[.ri.a ...]

6. [...]

****** C. 4 lines destroyed

#21[146]

****** [The surface of 4 confrontations]
[a.šà íb.si$_8$ 4.e]

****** [accumulated: 52`30.]
[ĝar.ĝar-*ma* 52,30]

****** [The 7th part of the 1st,]
[igi 7 ĝál 1.e]

****** [confrontation over confrontation went beyond.]
[íb.si$_8$ ugu íb.si$_8$ dirig]

7. [The] 4 [confrontations]
[íb.si$_8$] 4-[e]

8. [w]hat?
[e]n.nam

9. 35 nindan the 1st.
35 nindan 1.e

10. 30 nindan the 2nd.
30 nindan 2.e

11. 25 nindan the 3rd.
25 nindan 3.e

12. 20 nindan the 4th.
20 nindan 4.e

#22

13. The surface of 4 confrontations
a.šà íb.si$_8$ 4.e

14. and the 4 confrontations
ù íb.si$_8$ 4-[e]

15. accumulated: 54`20.
ĝar.ĝar-*ma* 54,20

16. The 7th part of the 1st,
igi 7 ĝál 1.e

17. confrontation over confrontation went be[yond].
íb.si$_8$ ugu íb.si$_8$ di[⌜rig$^?$]

---

146 The unnumbered lines are my reconstruction, based on #22–28.

18. The 4 confrontations,
íb.si$_8$ 4.e

19. what?
en.nam

20. 35 nindan, the 1st.
35 nindan 1.e

21. 30 nindan, the 2nd.
30 nindan 2.e

22. 25 nindan, the 3rd.
25 nindan 3.e

23. 20 nindan, the 4th.
20 nindan 4.e

Rev. I

#23

1. [The surf]ace of 4 confrontations
[a.š]à íb.si$_8$ 4.e

2. [acc]umulated: [5]2`30.
[ĝar].ĝar-*ma* [5]2,30

3. The 4th part of the smallest confrontation,
igi 4 ĝál íb.si$_8$ tur.ra

4. confrontation over confrontation went beyond.
íb.si$_8$ ugu íb.si$_8$ dirig

5. The 4 confrontations, what?
íb.si$_8$ 4.e en.nam

6. 35 nindan, the 1st.
35 nindan 1.e

7. 30 nindan, the 2nd.
30 nindan 2.e

8. 25 nindan, the 3rd.
25 nindan 3.e

9. 2[0 nindan], the 4th.
2[0 nindan] 4.e

#24

10. The su[rface] of 4 [con]frontations
a[.šà íb.]si$_8$ 4.e

11. and the 4 confrontations
ù íb.si$_8$ 4.e

12. accumulated: 54`20.
ĝar.ĝar-*ma* 54,20

13. The 4th part of the smallest,
igi 4 ĝál tur.ra

14. confrontation over confrontation went beyond.
íb.si$_8$ ugu íb.si$_8$ dirig

15. 35 nindan, the 1st.
35 nindan 1.e

16. 30 nindan, the 2nd.
30 nindan 2.e

17. 25 nindan, the 3rd.
25 nindan 3.e

18. 20 nindan, the 4th.
20 nindan 4.e

#25

19. The surface of 4 confrontations
a.šà íb.si$_8$ 4.e

20. accumulated: 52`30.
g̃ar.g̃ar-*ma* 52,30

21. The 5th part of the 3rd,
igi 5 g̃ál 3.e

22. confrontation over confrontation went beyond.
íb.si$_8$ ugu íb.si$_8$ dirig

23. The 4 confrontations, what?
íb.si$_8$ 4.e en.nam

24. 35 nindan, the 1st.
35 nindan 1.e

25. 30 nindan, the 2nd.
30 nindan 2.e

26. 25 nindan, the 3rd.
25 nindan 3.e

27. 20 nindan, the 4th.
20 nindan 4.e

#26

28. The surface of 4 confrontations
a.šà íb.si$_8$ 4.e

29. and the [4] confrontations
ù íb.si$_8$ [4].e

30. accumulated: [54`20.]
g̃ar.g̃ar-*ma* [54,20]

31. [The 5t]h part [of the 3rd,]
igi [5 g̃á]l [3.e]

32. co[nfrontation over confrontation went beyond.]
í[b.si$_8$ ugu íb.si$_8$ dirig]

33. [The 4] co[nfrontations, what?]
í[b.si$_8$ 4.e en.nam]

34. 3[5 nindan, the 1st.]
3[5 nindan 1.e]

35. [30 nindan, the 2nd.]
[30 nindan 2.e]

36. [25 nindan, the 3rd.]
[25 nindan 3.e]

37. [20 nindan, the 4th.]
[20 nindan 4.e]

#27

38. [The surface of 4 confrontations]
[a.šà íb.si$_8$ 4.e]

39. [accumulated: 52`30.]
[g̃ar.g̃ar-*ma* 52,30]

40. [Half of the 3rd part]
[šu.ri.a igi 3 g̃ál]

41. [of the 2nd confrontation,]
[íb.si$_8$ 2.e]

42. [confrontation over confrontation went beyond.]
[íb.si$_8$ ugu íb.si$_8$ dirig]

Rev. II

1. The 4 confrontations,
íb.si$_8$ 4.e

2. what?
en.nam

3. 35 nindan, the 1st.
35 nindan 1.e

4. 30 nindan, the 2nd.
30 nindan 2.e

5. 25 nindan, the 3rd.
25 nindan 3.e

6. 20 nindan, the 4th.
20 nindan 4.e

#28

7. The surface of 4 confrontations
a.šà íb.si$_8$ 4.e

8. and the 4 confrontations
ù íb.si$_8$ 4.e

9. accumulated: 54`20.
g̃ar.g̃ar-*ma* 54,20

10. Half of the 3rd part
šu.ri.a igi 3 g̃ál

11. of the 2nd confrontation,
íb.si$_8$ 2.e

12. confrontation over confrontation went beyond.
íb.si$_8$ ugu íb.si$_8$ dirig

13. The 4 confrontations, what?
íb.si$_8$ 4.e en.nam

14. 35 nindan, the 1st.
35 nindan 1.e

15. 30 nindan, the 2nd.
30 nindan 2.e

16. 2[5 nindan], the 3rd.
2[5 nindan] 3.e

17. 20 [nindan], the 4th.
20 [nindan] 4.e

#29

18. The surface of 2 confrontations
a.šà íb.si$_8$ 2.e

19. accumulated: 48`45.
ĝar.ĝar-*ma* 48,45

20. A surface having been built,
a.šà ŠU.BA.AN.TU[147]

21. 2 ešè 1 ½ iku.
2(ešè) 1(iku) 1(½iku)^{iku}[148]

22. The 2 confrontations
íb.si$_8$ 2.e

23. what
en.nam

24. [4]5 nindan, the 1st.
[4]5 nindan 1.e

25. 30 nindan, the 2nd.
30 nindan 2.e

#30

26. ⅔ of the 1st confrontation,
⅔ íb.si$_8$ 1.e

27. the 2nd confrontation.
íb.si$_8$ 2.e

28. The 2nd confrontation
íb.si$_8$ 2.e

29. together with 2!5!, the 2nd width,
ki 2!5! saĝ 2.e[149]

---

147 For lack of rival explanations I have understood this word tentatively as a (strangely Sumerianized) form of *banûm*. The word is found in other series texts in the same meaning and context and nowhere else (it can have nothing to do with the šu.ba.an.ti of AO 6770, which is a familiar logogram for *leqûm* used in agreement with parallel passages). So much seems certain from parallel passages (AO 8862, VAT 8390, and YBC 4608, *passim*) that a form of *banûm* or some equivalent is meant by the present ŠU.BA.AN.TU, in syllabic or logographic meaning.

148 1 (½iku) designates ½ iku, written with the corresponding special sign. The total area in the line is 22`30 sar.

149 Since this saĝ/"width" is "second", the "confrontation" itself must also be a

30. being made hold:
ì.gu$_7$-*m*[*a*]

31. (That which) the 1st surface
a.šà 1.e

32. over the 2nd surface goes bey[ond].
ugu a.šà 2.e dir[iğ]

33. The 2 confrontations, what?
íb.si$_8$ 2.e en.nam

34. 30 nindan, the 1st.
30 nindan 1.e

35. 20 [nind]an, the 2nd.
20 [nind]an 2.e

#31

36. Half of the 1st
šu.ri.a 1-[e]

37. and 5 nindan, the 2nd.
ù 5 nindan 2[.e]

38. 25 nindan, the [2nd w]idth,
25 nindan [s]ağ [2.e]

39. together with the 2nd confrontation,
ki íb.si$_8$ 2[.e]

40. being made hold: (that which) the 1st sur[face]
ì.gu$_7$-*ma* a.[šà 1.e]

41. over the 2nd surface
ugu a.šà 2.e

42. {a line from the edge erroneously counted as part of the text}

Rev. III

1. The 2 confrontations,
íb.si$_8$ 2.e

2. what?
en.nam

3. 30 nindan, the 1st.
30 nindan 1.e

4. 20 nindan, the 2nd.
20 nindan 2.e

#32

5. 25 nindan, the alternate width,
25 nindan sağ kúr.ra[150]

"width". This agrees with "the four fronts" of BM 13901 #23 and UET V, 864 (below, pp. 224 and 250), whereas certain catalogue texts from Nippur, Susa, and Eshnunna identify uš and *mitḫartum* (in TMS V–VI written LAGAB or NIGIN).

150 I choose Thureau-Dangin's reading from [TMB], but translate it in agreement with

6. together with the 2nd confrontation,
ki íb.si$_8$ 2.e

7. being made hold:
ì.gu$_7$-*ma*

8. As much as the 1st surface
a.šà 1.e

9. (over) the 2nd surface.
a.šà 2.e gin$_7$.nam[151]

10. The 3rd part of the 1st confrontation
igi 3 ĝál íb.si$_8$ 1.e

11. and 10 nindan, the 2nd.
ù 10 nindan 2.e

12. The 2 confrontations,
íb.si$_8$ 2.e

13. what?
en.nam

14. [30 nindan], the 1st.
[30 nindan] 1.e

15. 20 [nindan], the 2nd.
20 [nindan] 2.e

#33

16. The 3rd part of (that which) the 1st
igi 3 ĝál 1.e

17. over the 2nd [went beyond,]
ugu 2.e [dirig]

18. (To) the 1st confrontation
íb.si$_8$ 1-[e]

19. appended: ⌈33° 20′⌉.
daḫ-*ma* ⌈33,20⌉

20. The 2nd confrontation
íb.si$_8$ 2.e

21. together with 25, the 2nd width,
ki 25 saĝ 2.e

---

rev. II 29, III 21, etc.

151 This reading (which is taken over from Thureau-Dangin) fits Neugebauer's hand copy better than Neugebauer's own reading ba.zi. Similarly in following occurrences.

The translation "as much as" is given in the beginning of the previous line, where it would belong in grammatical Akkadian (since the Sumerian expression is an equative suffix, there is nothing wrong with its actual location). Similarly below.

22. being made hold:
ì.gu$_7$-*ma*

23. As [much as] the 1st surface (over) the 2nd.
a.šà 1.e 2.e g[in$_7$.nam]

24. The 2 confrontations,
íb.si$_8$ 2.e

25. what?
en.nam

26. 30 nindan, the 1st.
30 nindan 1.e

27. ⌊20⌋ nindan, the 2nd.
⌊20⌋ nindan 2.e

#34

28. The 3rd p[art] of (that which) the 1st
i[gi] 3 ğál 1.e

29. over the 2nd went beyond,
ugu 2.e dirig

30. (from) the 1st confrontation
íb.si$_8$ 1.e

31. torn o[ut]: 26°40´.
ba.z[i]-*ma* 26,40

32. 2[5 nindan,] the 2nd [wi]dth,
2[5 nindan s]ağ 2.e

33. together with the 2[nd confrontation,$^{!}$
ki [íb.si$_8$ 2$^{!}$.e

34. being made hold:
ì.gu$_7$-*ma*

35. As much as the 1st
a.šà 1.e

36. (over) the 2nd surface.
a.šà 2.e gin$_7$

37. The 2 confrontations,
íb.si$_8$ 2.e

38. what?
en.nam

Rev. IV

#35

1. The 3rd par[t of (that which) the 1]st
igi 3 ğá[l 1-]e

2. over the 2nd went beyond,
ugu 2.e dirig

3. to the 2nd confrontation
íb.si$_8$ 2.e.ra

4. appended: 23° 20´.
daḫ-*ma* 23,20

5. The 2nd confrontation,
íb.si$_8$ 2.e

6. together with 25 nindan, the 2nd width,
ki 25 nindan sağ 2.e

7. being made hold:
ì.gu$_7$-*ma*

8. As much as the 1st
a.šà 1.e

9. surface (over) the 2nd.
a.šà 2.e gin$_7$

10. The 2 confrontations,
íb.si$_8$ 2.e

11. what?
en.nam

12. 30 nindan, the 1st.
30 nindan 1.e

13. 20 nindan, the 2nd.
20 nindan 2.e

#36

14. The 3rd part of (that which) the 1st
igi 3 ğál 1.e

15. over the 2nd went beyond,
ugu 2.e dirig

16. (from) the 2nd confrontation
íb.si$_8$ 2.e

17. torn out: 16°40´.
ba.zi-*ma* 16,40

18. The 2nd confrontation,
íb.si$_8$ 2.e

19. together with 25 nindan, the 2nd width,
ki 25 nindan sağ 2.e

20. being made hold:
ì.gu$_7$-*ma*

21. A[s much as] the 1st surface
a.šà 1.e

22. (over) the 2nd surface.
a.šà 2.e g[in$_7$.nam]

23. The 2 confron[tations],
íb.[si$_8$] 2.e

24. w[ha]t?
e[n.na]m

25. [30 nindan, the 1]st.
[30 nindan 1].e

#37

26. [20 nindan, the 2nd.]
[20 nindan 2.e]

27. [The 3rd part of] (that which) the 1st
[igi 3 ĝál] 1.e

28. ⌈over the 2nd⌉ went beyond
⌈ugu 2.e⌉ dirig

29. (to) the 2 confrontations[152]
íb.si$_8$ 2.e

30. appended: 53° 20′.
daḫ-*ma* 53,20

31. 25 nindan, the 2nd width,
25 nindan saĝ 2.e

32. together with [the 2nd] confrontation,
ki íb.si$_8$ [2.e]

33. being made hold:
ì.gu$_7$-*ma*

34. As much ⌊as] the 1st
a.šà 1.e

35. surface (over) the 2nd surface.
a.šà 2.e gin$_7$.n⌊am]

36. The 2 confrontations,
íb.si$_8$ 2.e

37. what?
en.nam

Rev. V

#38

1. The 3rd [pa]rt of (that which) the 1st
[ig]i 3 ĝál 1.e

2. [ov]er the 2nd went beyond,
[ug]u 2.e dirig

3. (from) the 2 confrontations
íb.si$_8$ 2.e

---

152 Comparison with rev. IV 3–4 shows that no difference is made in writing between "appending" to "the second confrontation" and "the two confrontations". The dative suffix .ra, indeed, corresponding to an Akkadian *ana*, can hardly have been meant to distinguish one interpretation from the other, as also confirmed by the completely identical formulations, in rev. IV 16–17 and rev. V 3–4, respectively, of "second" and "two". The written text was a support for memory, and for the original user of the tablet no less than for the modern interpreter, the numbers it contains had to determine the reading of the single problem.

4. torn out: 46°40′.
ba.zi-*ma* 46,40

5. 25 nindan, the 2nd width,
25 nindan sağ 2.e

6. together with the 2nd confrontation,
ki íb.si$_8$ 2.e

7. being made hold:
ì.gu$_7$-*ma*

8. As much as the 1st surface
a.šà 1.e

9. (over) the 2nd surface.
a.šà 2.e gin$_7$

10. The 2 confrontations,
íb.si$_8$ 2.e

11. what?
en.nam

12. 30 nind[an], the 1st.
30 nind[an] 1.e

13. 20 [nindan], the 2nd.
20 [nindan] 2.e

#39

14. The [3]d part of (that which) the 1st
igi [3] ğál 1.e

15. over the 2nd went beyond,
ugu 2.e dirig

16. (to) the 1st appended: 33°20′.
1.e daḫ-*ma* 33,20

17. 25 nindan, the ⸢2nd width,⸣
25 nindan ⸢sağ 2.e⸣

18. together with the [2nd] confrontation,
ki íb.si$_8$ [2.e]

19. being made hold: As muc⸢h as⸣ ⟨the 1st surface⟩
ì.gu$_7$-*ma* ⟨a.šà 1.e⟩

20. (over) the 2nd surface.
a.šà 2.e g⸢in$_7$⸣

21. The 2 confrontations,
íb.si$_8$ 2.e

22. what?
en.nam

23. 30 nindan the 1st.
30 nindan 1.e

24. 20 nindan, the 2nd.
20 nindan 2.e

⌈4⌉3 ⌈exer⌉ci⌈ses⌉[153]
⌈4⌉3 ⌈I⌉M.⌈ŠU⌉

This is the 4th tablet.
dub 4-kam-*ma*

#1–#3

These three problems all deal with an even number of "confrontations". In contrast to what happens further on, nothing is said about the nature of the differences; given the custom of assuming ("by default", cf. note 58) the situation to be as simple as allowed by the information that is given, this will have been understood as an implicit message that all differences were alike. It can be observed that it might be difficult to formulate this explicitly within the standard terminology (which is all the very compact logographic writing can refer to) without stating the actual value of the difference.

If constant differences are taken for granted, all three problems can be solved via an accounting procedure similar to the one used in BM 13901 #18. If c designates the side of the smallest square, z the constant difference, and if $P_n = 1+2+...+(n-1)$, $Q_n = 1^2+2^2+...+(n-1)^2$, the general case reduces to

$$Q_n\cdot\Box(z)+2P_n\sqsubset\!\sqsupset(c,z)+n\Box(c) = A \qquad\qquad P_n\cdot z+nc = B$$

which can be solved by familiar methods.[154] General problems of this kind might even have roused interest in finding Q_n, knowledge of which is attested to in the Seleucid tablet AO 6484, obv. 3–5.

However, two things should be noticed that point in a different direction. Firstly, #1 appears to coincide with BM 13901 #8; secondly, the number of squares involved is always even: 2, 4, or 6, which is likely to reflect the intended procedure. Both #8 and #9 of that tablet, as we remember, made use of the "average confrontation" and the semi-difference. If the same trick is applied here, and the squares are coupled two by two, finding the square on the semi-difference requires nothing but some patient counting, since the average side follows from the sum of the sides (cf. Figure 25, which shows the situation for four squares).

---

153 Even though one problem counts as two IM.ŠU if divided between two columns, as observed by Neugebauer [MKT I, 492], Eleanor Robson [1999: 176] has shown beyond doubt that an IM.ŠU is a specific task or exercise.

154 This exchange of the respective roles of unknowns and parameters was certainly not beyond the horizon of the calculators that produced the mathematical series texts: in the first 12 problems of the parallel texts YBC 4668 and YBC 4713, not only the length and width of a rectangle but also their coefficients (*šá* [e]tab, "that which repeated") are unknown and involved in relations of the first or the second degree – see below, p. 203.

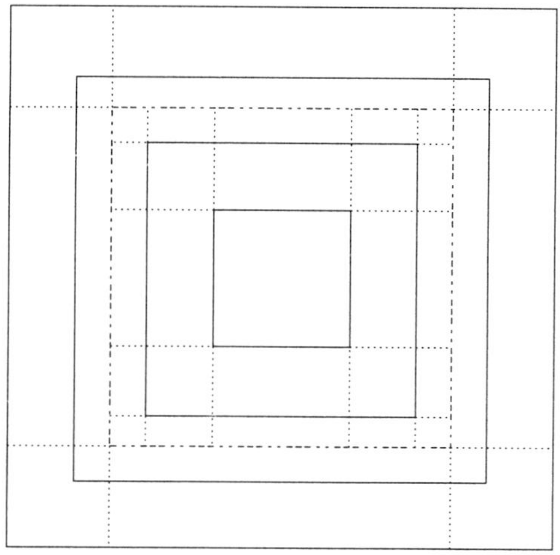

Figure 25. The four squares of YBC 4714 #4, with average square.

#4–7, 10–12

With a slight proviso for #5, #6, and #11 (which are corrupt or damaged), all these problems could be solved by the method of BM 13901 #14. The labour involved would of course be more arduous, since the coefficients are quite complex; but no difficulties of principle would arise.

#8–9

#8 could be solved precisely as BM 13901 #10–11. Assuming that #9 is the companion piece to be expected (cf. #21–28), stating the sum of sides and areas to be 2`` 26` 20 and the sides to differ again by $^1/_3$ of the smallest side, a single false position ($x = 3u$, $y = 4u$, $z = 5u$) would lead to a mixed, non-normalized equation with integer coefficient analogous to BM 13901 #7, in which the accumulation of 7 times the "confrontation" and 1 times the area is said to be 6°15´.

#13–20

#13–15 could be solved by a variant of the method of BM 13901 #14, inverting the role of the decrease (which is unknown here) and the side of a "basic" square (which is given in the present case). As far as the naive-geometric and accounting procedures are concerned, all principles would be the same.

#17–18 (and apparently also #16) would require reductions of the first-degree conditions before the same method could be applied. These might be of the kind taught in TMS XVI, and would thus offer no problems of principle.

Since the sum of the three areas is the same in #19 as in #13–18, it can be safely assumed to belong to the same group. With less certainty this may also be presumed of #20, since the traces that are left differentiate that problem clearly from the following group.

Comparison between #13 and #17–18 permits us to make a minor philological observation: the fifth line of all three problems has exactly the same function and meaning, yet #13 says that 3.e 2.e.ra ba.lal, while the others have 3.e 2.e tur.ra. There is thus no doubt that the two subtractive terms lal and tur are used synonymously.

#21–28

This sequence consists of problem pairs, each of which is analogous to what seems to be the pair #8–9. There is nothing of significance to be added to what was said on possible solving procedures, but an interesting observation may be made on the arrangement of the problems. All problems concern the same squares on 35, 30, 25, and 20. The constant difference can thus be expressed as $\frac{1}{7}$, $\frac{1}{6}$, $\frac{1}{5}$, or $\frac{1}{4}$ of the respective sides. $\frac{1}{6}$, however, is expressed as $\frac{1}{2}$ of $\frac{1}{3}$ – and this variation of the problem is put in the end, maybe because it is regarded as the most sophisticated.

Like #2–3 and #8–9, these problems deal with squares whose sides are in arithmetical progression. It is likely that these squares were thought of in concentric position (cf. above, note 94). It is noteworthy that all problems of this type on the tablet deal with an even number of squares.

#29

#29 is of the same type as BM 13901 #12 (above, p. 71), and we may assume that it was intended to be solved in the same manner. The use of the verb *banûm*, "to build", or some equivalent term is noteworthy. Repeatedly, the phrase "a surface I have built" (probably meaning "I have marked out a field") occurs in the Old Babylonian mathematical texts as added explanation when a length and a width have been "made hold" (thus in VAT 8390, above, p. 61). In the present text, the only occurrence of the phrase is followed by the only use of practical area units. We may make the conjectural inference that the phrase is thus a reference to a surveyors' idiom – which is in fact corroborated by the occurrence of *banûm*, e.g., in the tablet AO 8862 (below, p. 162), where the author claims to have walked around the field which has just been "built".

#30–39

This last group brings us beyond the range of problems known from or related to problems from BM 13901. All deal with two squares, the sides of which are eventually found to be $u = 30$ and $v = 20$. Furthermore, a "second" or "alternate" width $w = 25$ is given.[155]

In all cases we are told that the rectangle held by v together with w equals the excess of □(u) over □(v). The other condition is linear, ranging from the simple ($v = \frac{2}{3}u$ – the only homogeneous case) to the fairly complex (the value of $t \pm \frac{1}{3}(u-v)$ is given, where t is u, v, or $u+v$).

If we leave aside the simple case $v = \frac{2}{3}u$, no tricks seem to be at hand which would permit easy reduction. The intended procedure appears to have been a reduction of the linear condition followed by a transformation of the other equation in agreement with the pattern which we know from BM 13901 #14. The most adequate procedure will have been to express u in terms of v, $u = av+b$, which would yield the excess area by precisely the means employed in BM 13901 #14: the rectangle ⊏⊐(w,v), we observe, already has the ontological status needed for the geometrical procedure – no numerical multiple of the length but a rectangular surface $w = 25$ long and v broad. The introduction of the "alternate width" may thus be another way to avoid the implicit notion of the "broad line", an alternative to the "projection".

Considerations of the possible procedure cannot bring us any further. It may be interesting, however, to look at the configuration on which all statements are built (see Figure 26). The area condition states that the shaded excess area equals the cross-hatched rectangle. This is visually evident, since the width of the excess gnomon is half of v, and because $w = (u+v)/2$. Together, the two partial rectangles of which the gnomon is composed thus have a width v, while their average length is precisely w.

A related geometrical manoeuvre seems to have been what led the calculator astray in YBC 6504 #4 (below, p. 174). There is thus some reason to believe that the cut-and-paste idea which is illustrated in Figure 26 inspired the construction of the problems 30–39 of the present tablet.

---

155 #39 is an exact repetition of #33 as far as the mathematical substance and most of the formulations are concerned.

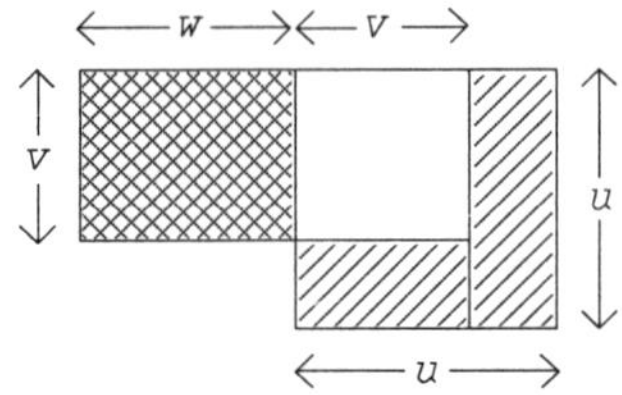

Figure 26. The configuration of YBC 4714 #30–39.

General Commentary

If we take a global look at the tablet, it turns out to be fairly well-ordered, both as far as the organization of each group is concerned and in the progression from group to group. As in BM 13901, problems dealing with fewer squares precede problems dealing with more. Except for the sequence 1–3, this principle overrides organization built on shared mathematical principles: hence, #8–9 are inserted between #4–7 and #10–12. At the same time, problems dealing with the same number of squares tend to be arranged according to a fixed progression between types: hence, #13–20 precede #21–28, just as #4–7 precede #8–9. Within this latter group (as within the corresponding pairs of the group #21–28), the simple problem precedes the complex version.

However, the principle that simplicity precedes complexity does not seem to be too important – #4–7 are as difficult and complex as #9. We are rather faced with a result of an attempt at three-dimensional systematic organization of the text. The system is not administered with full consequence, and elsewhere we encounter a much more perfect four-dimensional organization (see below, p. 203, in a sequence from YBC 4668); yet the attempt is unmistakable, and a far cry from BM 13901. Even the organization of Str 363, a tablet that contains three variations of the pattern of BM 13901 #14, only corresponds to variation along a single dimension where YBC 4714 involves three.

When it comes to mathematical substance, the distance between the three texts is much more modest. YBC 4714, as BM 13901 and Str 363, is engaged in the sub-discipline "(algebraic) study of square and squares". Solution of the problems of YBC 4714 will have asked for more labour and more patience, but no new mathematical principles apart from the reduction of first-degree conditions (again laborious but well-known) seem to be involved. Not only will the techniques of BM 13901 carry through, it is also difficult to imagine alternative methods that would be more effective.

It is striking that a similar observation would be made if we compare the methods of BM 13901 with modern techniques for solving second-degree equations. Apart from the shift from a geometric to an arithmetical conceptualization and the ensuing streamlining of the panoply of distinct operations, we would indeed use much the same methods today if confronted with the problems one by one. With us, as with the authors who produced the "series text" version of the discipline, change is to be found in the higher-level organization within which problems are imbedded, and in the metatheoretical framework through which we understand this organization. Our organization and our framework, it is true, are totally different from those displayed in YBC 4714 – demonstrating, if it should be needed, that Babylonian mathematics was fairly similar to ours (at the level which it reached), but that the way it was looked upon and understood by those who were engaged in it was quite different.

BM 85200 + VAT 6599[156]

The tablet whose contents is given in full in the present section was broken, and one part was bought by the British Museum while another ended up in the Berlin collection of Vorderasiatische Texte – whence the composite name. Line numbers from the Berlin fragment are labelled by an asterisk*, whereas unlabelled numbers refer to the BM fragment.

All 30 problems of the text deal with a túl.sag̃, i.e., a rectangular prismatic excavation, as made clear by the mathematical operations and by the fact that no parameters beyond length, width, and depth are needed to describe it. Some problems have the mathematical structure of second-degree equations, and are in fact solved by means of the characteristic second-degree cut-and-paste techniques; others are effectively of the third degree, and correspondingly they are solved by other means (factorization and recourse to a table, as we shall see). It is thus obvious that the Babylonian calculators knew the practical difference between the two algebraic degrees. It is equally obvious, however, that the characteristic feature shared by all problems of the tablet is the geometric configuration that is dealt with. As we shall see, this primacy of geometric constitution over algebraic structure applies even on lower levels, which shall provide us with clues to the technique of didactic exposition.

The volume of the "excavation" is represented by the amount of saḫar, "earth", that is dug out, while the area of its base is spoken of as the KI*ri* (logogram for *qaqqarum*), "ground". As always, length and width are supposed to be measured in nindan, depth in kùš, and volume as well as area in sar

156 Based on the transliteration in [MKT I, 194–200], cf. corrections in [MKT III, 54*f*], [TMB], and [Thureau-Dangin 1937]. Partial discussions of interest are [Vogel 1934], [Gandz 1937], and [Thureau-Dangin 1940] (where further bibliographic information is found on p. 1). Vogel's treatment of the cubic problems makes a geometric approach; the others are all based on the customary arithmetical interpretation.

(nindan²·kùš and nindan², respectively). When the depth is stated to be equal to (e.g.) the width, this is meant to concern “real” or “physical” extension, not measuring numbers. This holds true even when the depth is equal to a width defined as *igibûm* (#19). Accordingly, length and width spoken of as *igûm* and *igibûm*, and hence apparently as a pair of numbers from the table of reciprocals, are still to be understood as palpable extensions that happen to fulfil a specified numerical condition concerning the area they contain, not as mere numbers.

Obv. I

#5

14*. [An excavation. So much as the length, that is the depth. The earth I have torn out. My ground and the earth I have accumulated, 1°10´]
[túl.sağ *ma-la* uš GAM-*ma* saḫar.ḫi.a ba.zi KI? *ù* saḫar.ḫi.a UL.GAR 1,10]

15*. [...]

16*. [...][157]

1. [... length and wid]th, what?
[... uš sa]ğ en.nam

2. ⌈... 3 you see⌉. ½ of 3 break. 1°30´ you see,
⌈... 3 *ta-mar*⌉ ½ 3 *ḫe-pé* 1,30 *ta-mar*

3. [... igi 1°30´ detach], 40´ you see, the conversion of the width. Igi 12, the conversion of the depth, detach;
[... igi 1,30 du₈.a] 40 *ta-mar* bal sağ igi 12 bal GAM du₈.a

4. [5´ you see. To 1⌉ raise, 5´ you see. To 40´ raise, 3´20˝ you see.
[5 *ta-mar a-na* 1⌉ *i-ši* 5 *ta-mar a-na* 40 *i-ši* 3,20 *ta-mar*

5. ⌈3´20˝⌉ to 5´ raise, 16˝40‴ you see. Igi 16˝40‴ detach, 3`36 you see. 3`36
⌈3,20⌉ *a-na* 5 *i-ši* 16,40 *ta-mar* igi 16,40 du₈.a 3,36 *ta-mar* 3,36

6. to 1°10´ raise, 4`12 you see, 6 is equalside. 6 to 5´ raise, 30´ you see. 6 to 3´20˝ raise,
a-na 1,10 *i-ši* 4,12 *ta-mar* 6 íb.si₈ 6 *a-na* 5 *i-ši* 30 *ta⟨-mar⟩* 6 *a-na* 3,20 *i⟨-ši⟩*

7. 20´, the width. 6 to 1 raise, 6 you see, the depth. Thus
20 sağ 6 *a-na* 1 *i-ši* 6 *ta-mar* GAM *ki-a-am*

8. the procedure.
ne-pé-šum

---

157 It is not quite clear whether #5 begins in line 14*, as suggested by Neugebauer, or only in line 15* or even 16*, as suggested by Thureau-Dangin [TMB, 11]. Traces suggesting the end of the term [*ne-pé-šu*]*m* in line 13* support Neugebauer's assumption; no other statements, on the other hand, extend over more than two lines, which supports Thureau-Dangin.

#6

9. An excavation. So much as the length,[158] that is the depth. 1 the earth I have torn out.[159] My ground and the earth I have accumulated, 1°10′. Length and width, 50′. Length, width, what?
túl.sağ *ma-la* uš GAM-*ma* 1 saḫar.ḫi.a ba.zi KI^{ri} *ù* saḫar.ḫi.a UL.GAR 1,10 uš *ù* sağ 50 uš sağ en⟨.nam⟩

10. You, 50′ to 1, the conversion, raise, 50′ you see. 50′ to 12 raise, 10 you see.
za.e 50 *a-na* 1 bal *i-ši* 50 *ta-mar* 50 *a-na* 12 *i-ši* 10 *ta-mar*

11. Make 50′ confront itself, 41′40″ you see; to 10 raise, 6°56′40″ you see. Its igi detach, 8′38″24‴ you see;
50 *šu-tam⟨-ḫir⟩* 41,40 *ta-mar a-na* 10 *i-ši* 6,56,40 *ta-mar* igi-*šu du*$_8$.*a* 8,38,24 *ta⟨-mar⟩*

12. to 1°10′ raise, 10′4″48‴ you see, 36′, 24′, 42′ are equalsides.
a-na 1,10 *i-ši* 10,4,48[160] *ta-mar* 36 24 42 íb.si$_8$

13. 36′ to 50′ raise, 30′, the length. 24′ to 50′ raise, 20, the width; 36′ to 10 raise, 6, the depth.
36 *a-na* 50 *i-ši* 30 uš 24 *a-na* 50 *i-ši* 20 sağ 36 *a-na* 10 6 GAM

14. [The pr]ocedure.
[*n*]*e-pé-šum*

#7

15. An excavation. So much as the length, that is the depth. 1 the earth I have torn out. My [gr]ound and the earth I have accumulated, 1°10′. Length over width, 10′ went beyond.
túl.sağ *ma-la* uš GAM-*ma* 1 saḫar.ḫi.a ba.zi [K]I^{ri} *ù* saḫar.ḫi.a UL⟨.GAR⟩ 1,10 uš ugu sağ 10 dirig

16. You, 1 and 12, the [conv]ersions, posit. 10′ the ⌈going-beyond t⌉o 1 raise, 10′ you see; to 12 raise, 2 you see.
za.e 1 *ù* 12 [b]al ğar.ra 10 ⌈dirig *a-n*⌉*a* 1 *i-ši* 10 *ta-mar a-na* 12 *i-ši* 2 *ta-mar*

17. 10′ make confront itself, 1′40″ you see; to 2 raise, 3′2[0″ y]ou see. Igi 3′20″ detach, 18 you see;
10 *šu-tam⟨-ḫir⟩* 1,40 *ta-mar a-na* 2 *i-ši* 3,2[0 *t*]*a-mar* igi 3,20 du$_8$.a 18 *ta-mar*

---

158 I read MA as the Akkadian enclitic particle *-ma*, which, when following after a noun, serves emphasis of identification. It is, however, possible, that the sign is simply a phonetic complement indicating that the preceding GAM is to be read gam, not gúr. GAM-*ma* is then to be replaced throughout the tablet by gam.ma, and the translation "that is the depth" by "the depth".

I prefer the first reading because the affix is found invariably when GAM closes an expression beginning with *ma-la*, and never in the final section of the problem when its numerical value is stated, nor in questions for this value. Such consistency is not found in other cases where a Sumerian phonetic indicator is used – compare the use of daḫ.ḫa in obv. II 6* with that of daḫ in obv. II 13*.

159 A more idiomatic translation would be "removed" or "dug out". It is, however, worthwhile keeping in mind that the text uses the same term for digging out earth and for mathematical "subtraction".

160 Written with a conspicuous space between 10 and 4 to distinguish 10,4 from 14.

18. to 1°10′ raise, 21 you see, 3, 2, 21 (error for 3°30′) are equalsides. [10′ to 3 r]aise, 30′, the length.
a-na 1,10 *i-ši* 21 *ta-mar* 3 2 21$^{(sic)}$ íb.si$_8$ [10 *a-na* 3 *i*]*-ši* 30 uš

19. 10′ to 2 raise, 20′, the width. 3 to 2 raise, ⸢6⸣ you see, [6,] the depth.
10 *a-na* 2 *i-ši* 20 sağ 3 *a-na* 2 *i-ši* ⸢6⸣ *ta-mar* [6] GAM

20. The procedure.
ne-pé-šum

#8

21. An excavation. So much as the length, that is the depth. The eart[h] I have torn out. My ground and the earth I have accumulated: 1°10′. 30′, the length. The width, w[hat]?
túl.sağ *ma-la* uš GAM*-ma* saḫar.[ḫi].a ba.zi KI^{ri} *ù* saḫar.ḫi.a UL.GAR*-ma* 1,10 30 uš sağ e[n.nam]

22. You, 30′, the length, to 12 raise, 6 you see, the depth. 1 to 6 append, 7 you see.
za.e 30 uš *a-na* 12 *i-ši* 6 *ta-mar* GAM 1 *a-na* 6 daḫ.ḫa 7 *ta-mar*

23. Igi 7 is not detached. What to 7 may I posit which 1°10′ gives me? 10′ posit. Igi 30′ detach,
igi 7 nu du$_8$.a en.nam *a-na* 7 ğar.ra *ša* 1,10 sum.mu 10 ğar.ra igi 30 uš du$_8$.a

24. 2 you see. 10′ to 2 raise, 20′, the width, you see.
2 *ta-mar* 10 *a-na* 2 *i-ši* 20 sağ *ta-mar*

25. The procedure.
ne-pé-šum

#9

26. An excavation. So much as the length, that is the depth. The earth I have torn out. My ground and the earth I have accumulated: 1°10′. 20′, the width. The length, ⟨what⟩?
túl.sağ *ma-la* uš GAM*-ma* saḫar.ḫi.a ba.zi KI^{ri} *ù* saḫar.ḫi.a UL.GAR*-ma* 1,10 20 sağ uš ⟨en.nam⟩

27. You, 20′ to 12 raise, 4 you see. 4 to 1°10′ raise, 4°40′ you see.
za.e 20 *a-na* 12 *i-ši* 4 *ta-mar* 4 *a-na* 1,10 *i-ši* 4,40 *ta-mar*

28. ½ of 20′, the width, break, 10′ you see. 10′ make confront itself, 1′40″ you see; to 4°40′ append,
½ 20 sağ *ḫe-pé* 10 *ta-mar* 10 *šu-tam-ḫir* 1,40 *ta-mar a-na* 4,40 daḫ.ḫa

29. 4°41′40″ you see, 2°10′ is equalside. 10′ which you have made hold tear out:
4,41,40 *ta-mar* 2,10 íb.si$_8$ 10 *ša* ì.gu$_7$.gu$_7$ ba.zi*-ma*

30. 2 you see. Igi 4 detach, 15′ you see; to 2 raise,
2 *ta-mar* igi 4 du$_8$.a 15 *ta-mar a-na* 2 *i-ši*

31. 30′ you see, the length.
30 *ta-mar* {erasure} uš

32. The procedure.
ne-pé-šum

Obv. II

#12

5*. An excavation. So much as the length, that is the depth. The earth I have torn out. My ground and the earth I have ac[cumulated,]
túl.sağ *ma-la* uš GAM-*ma* saḫar.ḫi.a ba.zi KI^{ri} *ù* saḫar.ḫi.a U[L.GAR]

6*. the 7th part I have taken, to the ground I have appended: 20′ you see. 30′ [the length].
igi 7 ğál *él-qé a-na* KI^{ri} daḫ.ḫa-*ma* 20 *ta*⟨-*mar*⟩ 30 [uš]

7*. You, 30′ to 12 raise, 6 you see, the depth. 1 to 6 [append,]
za.e 30 *a-na* 12 *i-ši* 6 *ta-mar* GAM 1 *a-na* 6 [daḫ.ḫa]

8*. 7 you see. The 7th part take, 1 you see. 1 and 1 ac[cumulate],
7 *ta-mar* igi 7 ğál *le-qé* 1 *ta-mar* 1 *ù* 1 U[L.GAR]

9*. 2 you see. Igi 2 detach, 30′ you see, 30′ to 20′ the accumulation r⌈aise],
2 *ta-mar* igi 2 du$_8$.a 30 *ta-mar* 30 *a-na* 20 UL.GAR ⌈*i*$^{?}$-*ši*]

10*. 10′ you see. Igi 30′, the length, detach, 2 you see. 2 to 10′ rai[se, 20′ the width].
10 *ta-mar* igi 30 uš du$_8$.a 2 *ta-mar* 2 *a-na* 10 *i-š*[*i* 20 sağ]

11*. The procedure.
ne-pé-šum

#13

12*. An excavation. So much as the length, that is the depth. The earth I have torn out. My ground and the earth I have accumulated,
túl.sağ *ma-la* uš GAM-*ma* saḫar.ḫi.a ba.zi *qá-qá-ri ù* saḫar.ḫi.a UL.[GAR]

13*. 1°10′. Its 7th part I have taken, to [my] ground I have appended, 20′. 20′, the width.
1,10 igi 7 ğál-*šu él-qé a-na* KI[^{ri}] daḫ 20 20 sağ

14*. You, 20′ to 7 raise, 2°20′ you see. 20′, the width, to 12 raise,
za.e 20 *a-na* 7 *i-ši* 2,20 *ta-mar* 20 sağ *a-na* 12 *i-ši*

15*. 4 you see. 4 to 2°20′ raise, 9°20′ you see. To 7, 1 appen[d],
4 *ta-mar* 4 *a-na* 2,20 *i-ši* 9,20 *ta-mar a-na* 7 1 daḫ-ḫ[a]

16*. 8 you see. 20′ to 8 raise, 2°40′ you see. ½ of 2°40′ break, [make confront itself],
8 *ta-mar* 20 *a-na* 8 *i-ši* 2,40 *ta-mar* ½ 2,40 *ḫe-pé* [*šu-tam*⟨-*ḫir*⟩]

17*. 1°46′40″ you see, to 9°20′ append, 11°6′40 yo[u see,]
1,46,40 *ta-mar a-na* 9,20 daḫ.ḫa 11,6,40 *t*[*a-mar*]

18*. 3°20′ is equalside. 1°20′ which you have made hold tear out, 2 you [see.]
3,20 íb.si$_8$ 1,20 *ša* ì.gu$_7$.gu$_7$ ba.zi 2 *ta*[-*mar*]

19*. Igi 4 detach, 15′ you see. 15′ to 2 raise, 30 [the length.]
igi 4 du$_8$.a 15 *ta-mar* 15 *a-na* 2 *i-ši* 30 [uš]

20*. The proced[ure].
ne-pé-š[*um*]

Obv. II

#14

1. An excavation. So much as the *igûm*, the length. So much as the *igibûm*, the width. So much as the *igûm* {over the *igibûm* went beyond},
túl.sağ *ma-la* igi uš *ma-la* igi.bi sağ *ma-la* igi {ugu igi.bi dirig}[161]

2. That is the depth. 1⸤6 of earth I have torn] out. Length, width, and depth, what?
GAM-*ma* 1⸤6 saḫar.ḫi.a ba.z]i uš sağ *ù* GAM en.nam

3. You, igi 12 detach, [5´ you] see. 5´ [to 16] rai[se, 1°2]0´ you see,
za.e igi 12 du$_8$.a [5 *ta-*]*mar* 5 [*a-na* 16] *i-š*[*i* 1,2]0 *ta*⟨*-mar*⟩

4. 1°20´ the igi. Igi 1°2[0´ detach, 45´ you s]ee, 4⟨5⟩´ the *igibûm*. [16] the depth.
1,20 igi igi 1,2[0 du$_8$.a 45 *ta-m*]*ar* 4⟨5⟩ igi.bi [16] GAM

5. The pro-[ce]-dure.
ne-[*pé*]*-šum*

#15

6. An excavation. So much as the *igûm*, the length. So muc[h as the *igibûm*, the wi]dth. So much as that which the *igûm* over the *igibûm* went beyond, ⟨that is the depth⟩.
túl.sağ *ma-la* igi uš *ma-*[*la* igi.bi sa]ğ *ma-la ša* igi ugu igi.bi dirig ⟨GAM-*ma*⟩

7. 36 of earth I have torn out[: *igûm*, *igibûm*, and depth], what?
36 saḫar.ḫi.a ba.zi-*m*[*a* igi igi.bi *ù* GAM] en.nam

8. You, igi 12 detach, [5´ you see. 36⸣ to 5´ {...†} raise,
za.e igi 12 du$_8$.a [5 *ta-mar* 36⸣ *a-na* 5 {⸤*a-na* ...][162]} *i*⟨*-ši*⟩

9. 3 you see. ½ of 3 b[reak, 1°30´ you see,]1°30´ the *igûm*. [40´ the *igibûm*, 36⸣ the depth.
3 *ta-mar* ½ 3 *ḫ*[*e-pé* 1,30 *ta-mar*] 1,30 igi [40 igi.bi 36⸣ GAM

10. The pr[o]ced[ure]
ne-[*p*]*é-š*[*um*]

#16

11. An excavation. So much as the *igûm*, the length. So much as [the *igibûm*, the width.] So much as the total of *igûm* and *igibûm*, that is the depth.
túl.sağ *ma-la* igi uš ma-la [igi.bi sağ] *ma-la* niğin[163] igi *ù* igi.bi GAM-*ma*

12. 26 of earth I have torn out. *igûm*, *igibûm*, and depth, what?
26 saḫar.ḫi.a ba.zi igi igi.bi *ù* GAM en.nam

---

161 With this emendation, the following calculation (as reconstructed by Neugebauer) is correct. The wrong formulation (which is not solvable in rational numbers, and from which the *ša* of problems 15 and 17 is absent) seems to be a contamination from the following problem.

162 Thus according to Eleanor Robson (personal communication).

163 According to its use an abbreviation for šu.niğin, the "total" or "summa summarum" of accounts.

13. You, igi 12 detach, 5´ you see; 5´ to 26 raise,
za.e igi 12 du$_8$.a 5 *ta-mar* 5 *a-na* 26 *i-ši*

14. 2°10´ you see. ½ of 2°10´ break, make confront itself, 1°10´25″ you se[e]. ⟨1 from 1°10´25″ tear out, 10´25″ you see,⟩
2,10 *ta-mar* ½ 2,10 *ḫe-pé šu-tam-⟨ḫir⟩* 1,10,25 *ta-ma[r]* ⟨1 *i-na* 1,10,25 ba.zi 10,25 *ta-mar*⟩[164]

15. 25´ is equalside, to ⟨1°⟩5´ append and tear out. 1°30´ and 40´ yo[u see;]
25 íb.si$_8$ *a-na* ⟨1,⟩5 daḫ.ḫa *ù* ba.zi 1,30 *ù*[165] 40 *t[a-mar]*

16. 1°30´ the *igûm*; 40´ the *igibûm*; 26 the depth.
1,30 igi 40 igi.bi 26 GAM

17. The procedure.
ne-pé-šum

#17

18. An excavation. So much as the *igûm*, the length. So much as the *igibûm*, the width. So much as that which the *igûm* over the *igibûm* we[nt beyond]
túl.sağ *ma-la* igi uš *ma-la* igi.bi sağ *ma-la ša* igi ugu igi.bi d[irig]

19. from the *igûm* I have torn out, that is the depth. 6 of earth I have torn out. *igûm* and *igib[ûm*, what?]
i-na igi ba.zi GAM-*ma* 6 saḫar.ḫi.a ba.zi igi *ù* igi.b[i en.nam]

20. You, igi 12 detach, 5´ you see; to 6 raise, 30´ you see.
za.e igi 12 du$_8$.a 5 *ta-mar a-na* 6 *i-ši* 30 *ta-mar*

21. [I]gi 3[0´ de]tach, 2 you see. 2, the *igûm*, 30´, the *igibûm*. 6, the depth.
[i]gi 3[0 d]u$_8$.a 2 *ta-mar* 2 igi 30 igi.bi 6 GAM

22. The procedure.
ne-pé-šum

#18

23. An excavation. So much as the *igûm*, the length. So much as the *igibûm*, the width. So much as the total, *igûm*, *igib[ûm*, that is the depth]. 30 of ea[rth I have torn out].
túl.sağ *ma-la* igi uš *ma-la* igi.bi sağ *ma-la* niğin igi igi.b[i GAM-*m]a* 30 s[aḫar.ḫi.a ba.zi]

24. You, igi 12 detach, 5´ y[ou s]ee. 5´ to 30, the earth, raise,
za.e igi 12 du$_8$.a 5 *t[a-ma]r* 5 *a-na* 30 saḫar.ḫi.a *i-ši*

---

[164] The omission of this passage is one of several indications that the tablet is copied from another tablet, and is neither an original nor the direct reproduction of an oral presentation. Cf. the corresponding omission in rev. II, 4, equally called forth by the presence of two identical sequences of signs close to each other, and the apparent dittography in obv. II, 8.

[165] I follow Thureau-Dangin's reading of the sign as the first part of an *ù*, which agrees with obv. III 27 and a number of other passages. Neugebauer's reading as a full igi makes no sense.

25. 2°30′ you see. ½ of 2°30′ break, make [confront] itself, [1°33′4]5″ you [see].
2,30 *ta-mar* ½ 2,30 *ḫe-pé šu*[*-tam-ḫir* 1,33,4]5 *ta*[*-mar*]

26. 1 from 1°33′45″ tear out, 3[3′4]5″ you see, ⌞4⌟5′ is equalside.
1 *i-na* 1,33,45 ba.zi 3[3,4]5 *ta-mar* ⌞4⌟5 íb.si₈

27. To 1°15′ append and tear out, 2 and 30′ you s[ee].
a-na 1,15 daḫ.ḫa *ù* ba.zi 2 *ù* 30 *ta-m*[*ar*]

28. The procedure.
ne-pé-šum

#19

29. An excavation. So much as the *igûm*, the length. So much as the *igibûm*, the width. So much as the *igibûm*, that is the depth.
túl.sağ *ma-la* igi uš *ma-la* igi.bi sağ *ma-la* {erasure} igi.bi GAM-*ma*

30. 20 of earth I have torn out. *igûm*, *igibûm*, and depth, what?
20 saḫar.ḫi.a ba⟨.zi⟩ igi igi.bi *ù* GAM en.nam

31. You, igi 12 detach, to 20 raise, 1°40′ you see.
za.e igi 12 du₈.a *a-na* 20 *i-ši* 1,40 *ta-*{erasure}*mar*

32. 1°40′, the *igûm*. 36′, the *igibûm*. 20, the depth.
1,40 igi 36 igi.bi 20 GAM

33. The procedure.
ne-pé-šum

Rev. I

#20

1. An excavation. So much as I have made confront itself, and 7 kùš, that is the depth. 3′20″ of earth I have torn out.
túl.sağ *ma-la uš-tam-*⟨*ḫir*⟩ *ù* 7 kùš GAM-*ma* 3,20 saḫar.ḫi.a ba.zi

2. Length, width, and depth, what?
uš sağ *ù* GAM en.nam

3. You, the 7th part of 7 take, 1 you see. Igi 12 detach, 5′ you see.
za.e igi 7 ğál 7 *le-qé* 1 *ta-mar* igi 12 du₈.a 5 *ta-mar*

4. 5′ to 1 raise, 5′ you see. 5′ to 12 raise, 1 you see.
5 *a-na* 1 *i-ši* 5 *ta-mar* 5 *a-na* 12 *i-ši* 1 *ta-mar*

5. 5′ make confront itself, 25″ to 1 raise, 25″ you see. Igi 25″ detach, 2`24
5 *šu-tam*⟨*-ḫir*⟩ 25 *a-na* 1 *i-ši* 25 *ta-mar* igi 25 du₈.a 2,24

6. you see. 2`24 to 3′20″, the earth, raise, 8 you see. What are equalsides?
ta-mar 2,24 *a-na* 3,20 saḫar.ḫi.a *i-ši* 8 *ta-mar* en.nam íb⟨.si₈⟩

7. 1, 1, 8, are equalsides. 5′ to 1 raise, 5′ you see, 5′, a kùš, the length.
1 1 8 íb.si₈ 5 *a-na* 1 *i-ši* 5 *ta-mar* 5 kùš uš

8. 8 to 1 raise, 8 kùš the depth.
8 *a-na* 1 *i-ši* 8 kùš {erasure} GAM

9. The pro[c]edu[re]
ne-[*p*]*é-š*[*um*]

#21[166]

10. An excavation. So much as I have made confront itself, and 7 [kù]š, that is the depth. ⟨[3°15´]⟩ of earth I have torn out.
túl.sağ *ma-la uš-tam-ḫir ù*[167] 7 [kù]š GAM-*ma* 13 ⟨[3,15]⟩ saḫar ba.zi

11. Length, width, and depth, what?
uš sağ *ù* GAM en.nam

12. You, [a]s in the very counterpart, proceed[168] [7`48], what are equalsides?
za.e [*ki-m*]*a maḫ-ri-ma e-pu-uš* 4,48 ⟨[7,48]⟩ en.nam íb.si$_8$

13. ⌈6⌉ 6 13 are equalsides. 6 confronts itself, 13 the depth.
⌈6⌉ 6 13 íb.si$_8$ 6 *im*⟨*-ta-ḫar*⟩ 13 GAM

14. The procedure.
ne-pé-šum

#22

15. An excavation. So much as I have made confront itself, the depth. 1°30´ of earth I have torn out. Length, width, [and] depth, [what?]
túl.sağ *ma-la uš-tam-ḫir* GAM-*ma* 1,30 saḫar.ḫi.a ba.zi uš sağ [*ù*] GAM ⟨en.nam⟩

16. You, igi 12 detach, 5´ you see. 5´ to 1°30´ raise, 7⌊´30″ you see]
za.e igi 12 du$_8$.a 5 *ta-mar* 5 *a-na* 1,30 *i-ši* ⌊7,30 *ta-mar*]

17. 30´ is equalside. 30´ to 1 raise, 30´ confronts itself. 30´ t[o 12⌉ raise, 6 the depth.
30 íb.si$_8$ 30 *a-na* 1 *i-ši* 30 *im-ta-ḫar* 30 *a*[*-na* 12⌉ *i*⟨*-ši*⟩ 6 GAM

18. The procedure.
ne-pé-šum

#23

19. An excavation. So much as I have made confront itself, and 1 kùš, going beyond, that is the depth. 1°45´ of earth [I have] torn out.
túl.sağ *ma-la uš-tam-ḫir ù* 1 kùš dirig GAM-*ma* 1,45 saḫar.ḫi.a [ba].zi

20. You, 5´, going beyond, to 1, the conversion, raise, 5´ you see; to 12 rai[se, 1⌋ you see.
za.e 5 dirig *a-na* 1 bal *i-ši* 5 *ta-mar a-na* 12 *i-š*[*i* 1⌋ *ta-mar*

21. 5´ make confront itself, 25″ you see. 25″ to 1 raise, 25″ you see. Igi ⌊25 detach,]
5 *šu-tam*⟨*-ḫir*⟩ 25 *ta-mar* 25 *a-na* 1 *i-ši* 25 *ta-mar* igi ⌊25 du$_8$.a]

22. 2`24 you see. 2`24 to 1°45´ raise, 4`12 [you see.]
2,24 *ta-mar* 2,24 *a-na* 1,45 *i-ši* 4,12 [*ta-mar*]

---

166 The text as it stands is corrupt. In ⟨[...]⟩ I give Thureau-Dangin's corrections as proposed in his [1936: 181], from where the reading of line 12 is also taken.

167 In this place, a GAM seems to have been written first. Afterwards, the scribe discovered the mistake and covered it by the *ù*.

168 That is, "proceed precisely as in the corresponding (immediately preceding) case".

23. from[169]"equalside, 1 appended", 6 ¿1? is/are equal[side(s).] ˻6 to 5′˼ ra[ise, 30′] you see, confronts itself. 6 (error for 7) the depth.
i-na íb.si₈ 1 daḫ.ḫa 6 ¿1?[170] íb.s[i₈] ˻6 *a-na 5*˼ *i-*[*ši* 30] *ta*⟨*-mar*⟩ *im*⟨*-ta-ḫar*⟩ 6[sic] GAM

24. The proced[ure].
ne-pé-š[um]

#24

25. An excavation. 3° 20′, that is the depth. 27°46′40 of earth I have torn out. The length over the width, 50′ we[nt beyond.]
túl.sağ 3,20 GAM-*ma* 27,46,40 saḫar.ḫi.a ba.zi uš ugu sağ 50 d[irig]

26. You, igi 3° 20′, the depth, detach, 18′ you see; to 27°46′40″, the earth, you raise,
za.e igi 3,20 GAM du₈.a 18 *ta-mar a-na* 27,46,40 saḫar.ḫi⟨.a⟩ *i-ši*

27. ˻8°˼20′ you see. ½ of 50′ break, make confront itself, 10′25″ you see;
˻8˼,20 *ta-mar* ½ 50 *ḫe-pé šu-tam*⟨*-ḫir*⟩ 10,25 *ta-mar*

28+1*. to 8° 20′ append, [8° 3˼0′25″ you see,
a-na 8,20 daḫ.ḫa [8,3˼0,25 *ta-mar*

29+2*. 2°55′ is equalside; until [2 posit], to 1 append, from 1 tear out.
2,55 íb.si₈ *a-di* [2 ğar.ra] *a-na* 1 daḫ.ḫa *i-na* 1 ba.zi

30+3*. 3° 20′ the length, 2° 30′ the width you see.
3,20 uš 2,30 sağ *ta-mar*

31+4*. The proced[u]re.
ne-p[é-]šum

#25

5*. An excavation. 3° 20′, that is the depth. 27°46′4[0″ of earth I have torn out. Length and width I have accumu]lated, 5°5˻0′.˼
túl.sağ 3,20 GAM-*ma* 27,46,4[0 saḫar.ḫi.a ba.zi uš *ù* sağ UL.GA]R 5,˻50˼

6*. You, igi 3° 20′, the depth, detach, 18′ you see; ˻to 27°46′40 raise,]
za.e igi 3,20 GAM du₈.a 18 *ta-mar* ˻*a-na* 27,46,40 *i-ši*]

7*. 8° 20′ you see. ½ of 5°50′ break, make confront itself, ˹8° 30′25″ you see.]
8,20 *ta-mar* ½ 5,50 *ḫe-pé šu-tam*⟨*-ḫir*⟩ ˹8,30,25 *ta-mar*]

---

169 All the way through the text, íb.si₈ is a verb: the question is en.nam íb.si₈, while nouns are asked for with the phrase *X* en.nam; resulting values corresponding to nouns, moreover, are "seen", whereas *tammar* never occurs together with íb.si₈. Not being a noun, the word íb.si₈ cannot be what is governed by the preposition *ina*/"from, in, by means of". The alternative is that it is the whole phrase íb.si₈ 1 daḫ.ḫa that is governed by the preposition; the passage will then have to be understood "by means of (the table) 'equalside, 1 appended'". A similar expression ba.si 1-lal, "equilateral, 1 diminished", is used about the side *n* of a prism $n \times n \times (n-1)$ in VAT 8521, rev. 12.

170 As possible alternative readings, Neugebauer suggests "6 1 1" and "6 nindan", none of which make sense. "6 7" seems to be ruled out by the hand copy. [TMB] appears to regard the traces following "6" as an erasure, neglecting them entirely.

8*. 8° 20′ from the inside tear out, 10′ 2[5″ you see, 25′ is equalside;]
8,20 *i-na lìb-ba* ba.zi 10,2[5 *ta-mar* 25 íb.si$_8$]

9*. to 2°55′ append and tear out, 3° 20′ [the length, 2° 30′ the width.]
a-na 2,55 daḫ.ḫa *ù* ba.zi 3,20 [uš 2,30 sağ]

10*. The proce[dure].
ne-pé[-šum]

#26

11*. An excavation. 3° 20′ the depth. 27°46′ 40″ of earth [I have torn out. That which the width over the depth went beyond, ⅔ of the length.]
túl.sağ 3,20 GAM-*ma* 27,46,40 saḫar.ḫi.a [ba.zi *ša* sağ ugu GAM dirig ⅔ uš]

12*. You, igi 3° 20′ detach, 18′ you see; to 2[7°46′ 40″ raise,]
za.e igi 3,20 du$_8$.a 18 *ta-mar a-na* 2[7,46,40 *i-ši*]

13*. 8° 20′ you see. 8° 20′ to 40′ raise, 5° 33′[20″ you see. 3° 20′, the depth, to 5′ raise, 16′ 40″.]
8,20 *ta-mar* 8,20 *a-na* 40 *i-ši* 5,33,[20 *ta*⟨*-mar*⟩ 3,20 GAM *a-na* 5 *i-ši* 16,40]

14*. Turn back. ½ of 16′ 40″ break, 8′ 20″ you see, make confront itself, 1′[9″ 26‴ 40⁗ to 5° 33′ 20″ append.]
niğín.na ½ 16,40 *ḫe-pé* 8,20 *ta-mar šu-tam*⟨*-ḫir*⟩ 1,[9,26,40 *a-na* 5,33,20 daḫ.ḫa]

15*. What is equalside? 2° 31′ 40″ (error for 2° 21′ 40″) until 2 posit; 8′[20″] app[end and tear out]
en.nam íb.si$_8$ 2,31,40[sic] *a-di* 2 ğar.ra 8⟨,20⟩ da[ḫ.ḫa *ù* ba.zi]

16*. 2° 30′, the width; (and) 2°13′ 20″ you see. Igi 40′ detach, 1° 30′ you see. [To 2°13′ 20″ raise,]
2,30 sağ 2,13,20 *ta-mar* igi 40 du$_8$.a 1,30 *ta-mar* [*a-na* 2,13,20 *i-ši*]

17*. 3° 20′ the length you s[ee.]
3,20 uš *ta-m[ar]*

18*. The procedure.
ne-pé-šum

#27

19*. An excavation. 1°40′ the length. The 7th part of that which length over width went beyond, that is the depth. 1°40′ of eart[h I have torn out;]
túl.sağ 1,40 uš igi 7 ğál *ša* uš ugu sağ dirig GAM-*ma* 1,40 saḫar.ḫ[i.a ba.zi]

20*. Length,[171] width, and depth, wh[at?]
uš sağ *ù* GAM en[.nam]

21*. You, 1°40′, the length, to 12, the conversion of depth, raise, 20 you [see].
za.e 1,40 uš *a-na* 12 bal GAM *i-ši* 20 *ta[-mar]*

22*. Igi 20 detach, 3′ you see. 3′ to 1°40′, the e[arth, raise, 5′ you see.]
igi 20 du$_8$.a 3 *ta-mar* 3 *a-na* 1,40 s[aḫar.ḫi.a[172] *i-ši* 5 *ta-mar*]

---

171 Already given.

172 I prefer this reconstruction (proposed in [Thureau-Dangin 1937: 11] and [TMB]) to Neugebauer's, both because its fits the hand copy best, and because of the parallel to the procedure in #29, rev. II 3-4. It also happens to make much better sense of

23*. 7 to 5´ raise, 35´ you s[ee. ½ of 1°40´ break, make confront itself, 41´40˝].
7 a-na 5 i-ši 35 ta-m[ar ½ 1,40 *ḫe-pé šu-tam⟨-ḫir⟩* 41,40]

24*. 35´ ⌞from ins⌟i[de tear out, 6´40˝ you see, 20´ is equalside.]
35 ⌞*i-na lìb*⌟-[*bi* ba.zi 6,40 *ta-mar* 20 íb.si$_8$]

25*. T[o 50´ append and tear out, 1°10´ and 30´, the width].
a-n[*a* 50 daḫ.ḫa *ù* ba.zi 1,10 *ù* 30 sağ ...]

26* [The 7th part of 1°10´ take, 10´, the depth.]
[igi 7 ğál 1,10 *le-qé* 10 GAM][173]

27* [The procedure.]
[*ne-pé-šum*]

Rev. II

#29

1. An excavation. 1°40´ the length. The 7th part of that which the length over the width goes beyond, and 2 kùš, that is the depth. 3°20´ of [ear]th ⟨I have torn out.⟩
túl.sağ 1,40 uš igi 7 *ša* uš ugu sağ dirig *ù* 2 kùš GAM-*ma* 3,20 [saḫ]ar.ḫi⟨.a⟩ ⟨ba.zi⟩

2. Width and depth, what?
sağ *ù* GAM en.nam

3. You, 1°40´, the length, to 12, the conversion of depth, raise, 20 you see. Igi 20 detach, 3´ you see;
za.e 1,40 uš *a-na* 12 bal GAM *i-ši* 20 *ta-mar* igi 20 du$_8$.a 3 *ta-mar*

4. 3´ to 3°20´ raise, 10´ you see. ⟨10´ to 7 raise, 1°10´ you see⟩. 10´ going beyond[174] to 7 raise, 1°[10´ yo]u see.
3 *a-na* 3,20 *i-ši* 10 *ta-mar* ⟨10 *a-na* 7 *i-ši* 1,10 *ta-mar*⟩ 10 dirig *a-na* 7 *i-ši* 1[,10 *t*]*a-mar*

5. 1°40´, the length, to 1°10´ append, 2°50´ you see. ½ of 2°⌈50´ break, make con⌈front itself.
1,40 uš *a-na* 1,10 daḫ.ḫa 2,50 *ta-mar* ½ 2,⌈50 *ḫe-pé šu-tam*⌈*-ḫir*

6. 2°25˝ you see. From 2°25˝ 1°10´ tear out, 50´25˝ you see,
2,25 *ta-mar i-na* 2,25 1,10 ba.zi 50,25 *ta-mar*

7. 55´ is equalside; to 1°25´ append and tear out:
55 íb.si$_8$ *a-na* 1,25 daḫ.ḫa *ù* ba.zi-*ma*

8. 2°20´ and 30´, the width, you see. The 7th part of 2°20´ t[ak]e, 20´, the depth.
2,20 *ù* 30 sağ *ta-mar* igi 7 ğál 2,20 *l*[*e-q*]*é* 20 GAM

9. The procedure.
ne-pé-šum

---

the procedure.

173 Reconstruction suggested by rev. II 8.

174 I.e., the 2 kùš of line 1.

#30

10. An excavation. 1°40´ the length. The 7th part of that which the length over the width goes beyond, and 1 kùš diminishing, that is the depth.
túl.sağ 1,40 uš igi 7 ğál *ša* uš ugu sağ dirig *ù* 1 kùš ba.l[al] GAM-*ma*

11. 50´ of earth I have torn out. The width and the depth, what?
50 saḫar.ḫi.a ba.zi sağ *ù* GAM en.nam

12. You, 1°40´, the length, to 12, the conversion of depth, raise, 20 you see. Igi 20 detach, 3´ you see;
za.e 1,40 uš *a-na* 12 bal GAM *i-ši* 20 *ta-mar* igi 20 du$_8$.a 3 *ta*⟨*-mar*⟩

13. 3´ to 50´ raise, 2´30″ you see. 2´30″ to 7 raise, 17´30″ yo[u see].
3 *a-na* 50 *i-ši* 2,30 *ta-mar* 2,30 *a-na* 7 *i-ši* 17,30 *t*[*a-mar*]

14. 7 to 5´, 1 kùš, raise, 35´ you see. 35´ from 1°40´, the length, tear out,
7 *a-na* 5 1 kùš *i-ši* 35 *ta-mar* 35 *i-na* 1,40 uš ba.zi

15. 1°5´ you see. ½ of 1°5´ break, 32´30″ make confront itself, 17´36″15‴ you see,
1,5 *ta-mar* ½ 1,5 *he-pé* 32,30 *šu-tam*⟨*-ḫir*⟩ 17,36,15 *ta*⟨*-mar*⟩

16. from the inside 17´30″ tear out, 6″15‴ you see; 2´30″ is equalside
i-na lìb-bi 17,30 ba.zi 6,15 *ta-mar* 2,30 íb.si$_8$

17. to 32´30″ append and tear out, 35´ and 30´, the width, you see. (The) 7(th of) 35´, 5´ the depth.
a-na 32,30 daḫ.ḫa *ù* ba.zi 35 *ù* 30 sağ *ta-mar* 7 35 5 GAM[175]

18. The procedure.
ne-pé-šum

In the symbolic translations of the structure of the problems in the following I shall use l and w for the length and the width, respectively. For the depth measured in nindan I shall use d, whereas G stands for the depth measured in kùš; if "the depth is as much as the length", we thus have $d = l$, $G = 12d$.

The ordering of problems in the tablet is not derived from principles of mathematical structure, and there is thus no reason to follow it in the discussion. Instead, I shall group problems together which make use of the same characteristic technique; it is evidently no coincidence that this will also be a grouping according to algebraic degree.

The Third-Degree Problems

The present tablet is the only known Old Babylonian procedure text that treats of inhomogeneous problems of the third degree. The key to the procedure is to be found in the homogeneous cubic problem YBC 6295 (above, p. 65) and its solution by the *maksarum* technique.

---

175 This sequence of numbers could be filled out as ⟨igi⟩ 7 ⟨ğál⟩ 35 ⟨*le-qé*⟩ 5 GAM, but is remarkable enough to stand in its original elliptic formulation.

Let us first look at #6. We are told that the length equals the depth ($d = l$), that the accumulation of earth and ground equals 1°10´ ($l{\times}w{\cdot}G{+}l{\times}w$ = 1°10´), and that length and width equal 50´ ($l{+}w$ = 50´).[176] Even the volume of the earth is stated, but this is to be understood not as a given quantity but as an entity which is "merely known" for identification purposes (the same happens in #7; cf. discussion below, p. 161).

According to the normal conceptualization of second-degree problems of the type "surface+sides", we must expect the sum of "earth and ground" to be imagined as the volume of the excavation prolonged downwards by an extra kùš (cf. Figure 27). This *a priori* expectation is confirmed by #8, which "appends" an extra kùš to the depth (obv. I 22), and by an error in #23 (rev. I 23) which comes from identification of the two formulations – cf. note 181. The presence of two different volumes will be the reason that the original volume, though not given, has to possess a (numerical) identification tag.

The first step in the procedure is the computation of the volume of a cube. That a volume and no mere product is involved is made clear by the distinction between multiplications: length and width are "confronted" as sides of a square, which is then "raised" to the height. The side of the cube is chosen as the sum of the length and the width of the excavation.

The treatment of the three dimensions is remarkably symmetric: all are found by a multiplication by the appropriate conversion factor (bal): 1 for length and width (both thus 50´ [nindan]), and 12 for the depth (thus 10 kùš).

Next, the volume of the extended excavation is found by means of a customary "igi-division" to be N = 10´4˝48‴ times the reference volume. This ordering of the computational steps is another indication that a concrete

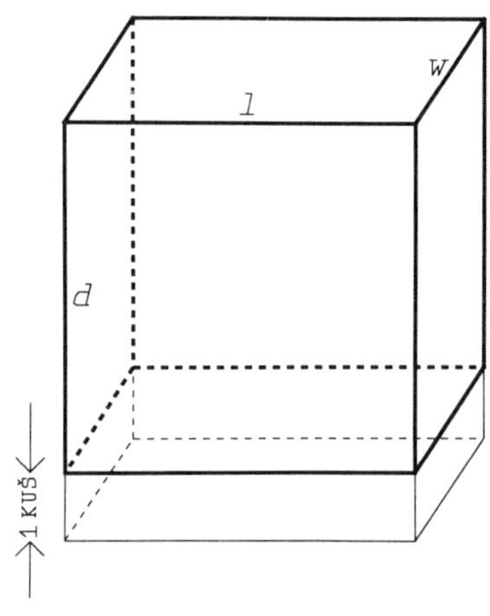

Figure 27. The situation of BM 85200 + VAT 6599 #6.

[176] We observe that the addition is once again performed by a mere "and" – cf. TMS XVI l. 1, see p. 86.

reference entity is involved; in the case of a mere normalization,[177] the volume of the excavation would (according to the habits which we know from other texts) have been divided by 50′, 50′, and 10 one after the other, not once and for all by their product. The "equalsides" of the "quotient volume" N (actually sides which are *not* equal) are given without explanation to be 36′, 24′, and 42′. What has to be looked for is, indeed, a factorization $N = p \cdot q \cdot r$ where $p+q = 1$, $r = p+6'$ (6′ represents the extra kùš "appended" to the depth as measured by 10 kùš, i.e., by the depth of the reference volume).[178]

The length, finally, is found as 36′ times the length of the reference cube, i.e., as 36′·50′ [nindan] = 30′ [nindan]; the width is found to be 24′·50′ [nindan] = 20′ [nindan], and the depth as 36′·10 [kùš] = 6 kùš (while the extended depth would have been 42′·10 kùš = 7 kùš).

#7 is a close parallel; this time, however, the excess of length over width is given (and equal to 10′ [nindan]). The reference volume is a cube with sides equal to this excess. It is constructed and found to be 3′20″ [sar], yielding a quotient volume equal to 21; this is said without explanation to have "equalsides" 3, 2, and 21 (mistaken for 3°30′).[179] Since the side of the reference volume is $l-w$ = 2 kùš, it is indeed required that $p-q = 1$, $r-p = 30'$.

The two factorizations into sets of "equalsides" may have been found by systematic search – even though the number of possible factorizations is infinite (Babylonian sexagesimals make no distinction between integers and non-integers), start from the simplest possibilities combined with a bit of mathematical reflection would soon lead forward.[180] However, the complete absence of calculation (e.g., of the 6′ and 30′ representing $r-p$ in the two problems) and justification – as compared, e.g., to the careful multiplication

---

177 This was proposed by Thureau-Dangin [1940: 3], in an interpretation which otherwise seems to come close to the one presented here, apart from its arithmetical dress (the formulation given in [TMB, xxxv*ff*] makes the difference stand out more clearly). Even MKT [I, 211] speaks of a "normal form".

178 In his geometrical interpretation of this and the following problem, Vogel [1934: 91–93] does not build on the actual sequence of operations but rather on mathematical feasibility. It is thus not astonishing that his explanation differs from the one given here while being closely related.

The relation between original volume V, reference volume v, and quotient volume **N** may be clearer to the modern reader if made explicit in symbols. In the present case, V represents the prolonged excavation, $V = l \times w \cdot \delta$, $\delta = d+1$ kùš = $d+5'$ nindan; $v = a \times b \cdot c$ = 50′×50′·50′ nindan^3 = 50′×50′·10 nindan^2·kùš; $N \doteq (^{l}/_{a}) \times (^{w}/_{b}) \cdot (^{\delta}/_{c}) = (p \times q) \cdot r$. Thus, since $l+w = a = b$, $p+q = {}^{l+w}/_{a} = 1$; and since $d = l$ and $a = c$, $r = {}^{\delta}/_{c}$ = (l+1 kùš)/c = $p+$ $^{(1 \text{ kùš})}/_{(10 \text{ kùš})}$ $= p+6'$.

179 While other copyist's mistakes in the tablet (jumps from one occurrence of a sequence of signs to another) could have been made by a scribe who copied word for word without thinking about what goes on in the text, this one intimates that the copyist was aware of its mathematical content, and inserted by mistake a 21 which was still on his mind – or that the author did so, and the copyist followed him. (The same cause seems to have produced the "13" of #21, rev. I 10).

180 Cf. Vogel's tabulations [1934: 92*f*].

with a factor *1* in obv. I 10 and 16 – suggests that they are drawn from the sleeves. Since the problems have been constructed backwards from known dimensions, this will have been quite feasible. On the other hand, the fact that even the factor r for the extended depth is listed – though of no use further on – demonstrates that what may plausibly have been drawn from the sleeves is still meant as a solution by factorization.

#23 is of a similar though simpler structure. It is stated that the depth exceeds "as much as I have made confront itself" by 1 kùš, which means that length and width confront each other as sides of a square; thus $l = w$, $d = l+1$ kùš. Furthermore, the volume is $(l{\times}w)\cdot G = (l{\times}w)\cdot 12d = 1°45'$; the same structure would have come about if we had added the base and a cubic volume.

This time, the reference volume is chosen as a cube with side equal to the excess of depth over length, i.e., to 1 kùš = 5′ [nindan]. Its volume is found to be $5'{\times}5'\cdot 1$ [nindan^2·kùš] = 25″ [sar], the quotient volume being therefore equal to 4`12. This must be factorized as $p{\times}p\cdot(p+1)$, and if the habits from #6 and #7 had been followed, the listing of three "equalsides" 6, 6, and 7 would have been expected. Instead we are told "from equalside, 1 appended, 6 ¿1? is/are equalside(s)", which seems to mean, firstly, that one side should be obtained by adding 1 to the others (which are equal); and, secondly, that the resulting equalside is 6.[181] A tabulation of the type "equilateral, 1 appended" (but without this heading), that is, of $n^2\cdot(n+1)$, is actually known (VAT 8492, [MKT I, 76]), which identifies only one number (n) as the "equilateral"; furthermore, the only other problem of the present tablet which might be solved by means of such a table (#5) also lists only one "equalside", while all others that make use of a quotient volume indicate three. It is thus quite plausible that the phrase "equalside, 1 appended" was the habitual designation of such a table (or possibly to its contents), and that a table has indeed been used for the solution of these (and only these) two problems.[182]

We should now be ready to tackle #5. The beginning is lost, but it is clear from the following that the accumulation of earth and ground will have been given as 1°10′, and that depth equals length. A supplementary condition leads in lines 2–3 to the conclusion that the length is equal to 1°30′ widths, and the width hence equal to 40′ times the length;[183] thus, the "conversion of the

---

181 The dubious ¿1? might be another result of the copyist's thinking about the procedure while writing (cf. note 179) and perhaps attempting to stamp out a number *1* written by mistake.

The error of line 23, "6 the depth", is likely to have been induced by an interference with the equivalent configuration "earth+ground = 1°45′", where the depth would indeed be 6 kùš.

182 Neugebauer, who already proposed in [MKT I, 210*f*] that #5 and #23 were solved by means of the table $n^2\cdot(n+1)$, also presumed #6–7 to have made use of tables – confessing at the same time, however, that he could not imagine their make-up.

183 The wording of the original condition is not obvious at all, but it is at least clear

width" – the factor converting the measuring number for the length into that of the width, if we are to believe the parallel to the "conversion of the depth" – will indeed be 40´.

Both this number and the corresponding number 1 (which by analogy will have been the "conversion of the length") are then "igi-divided" by 12, and "raising" one of the resulting numbers to the other gives the reference volume. The most likely explanation of this surprising procedure is that #5 closes a family of increasingly complex problems (the tablet contains several series of that kind); such a family might have started with a cubic excavation where the sum of "the ground" and "the earth" was given (a "ground+earth"-variant of #23), and have taken as its reference volume the cube with side 1 kùš, which could be found by shrinking a (volume) sar by a factor 5´ in each horizontal direction; in #5, the corresponding shrinking factors will instead have to be 5´ and 3´20˝. However, the absence of concrete information on the preceding problems and on the beginning of #5 prevents us from knowing.

Once the reference volume is found, everything goes as usual, and as in #23 the quotient between the volumes is found as 4`12, which is said to have the (single) "equalside" 6, corresponding to a factorization $p{\times}p\cdot(p+1)$.

Other variations on #23 might have been produced where the excess of depth over length was a regular number. Arithmetically speaking, the system

$$l{\times}l\cdot(12l+\mathrm{a}) = b$$

may be reduced to

$$(\tfrac{12}{a}\cdot l)^2\cdot(\tfrac{12}{a}\cdot l+1) = (\tfrac{12}{a})^2\cdot(\tfrac{b}{a}) .$$

Such problems, however, are not to be found in the preserved parts of the tablet. Instead, #20 and (presumably) #21 demonstrate how to proceed if a is irregular (and its third power does not divide b).

In #20 it is first observed that the 7th part of 7 is one, i.e., that a reference cube 1 kùš high divides the excess height 7 times. Next the reference volume is constructed and computed in painstaking detail: its height, 1 kùš and thus 5´ nindan, is reconverted into 1 kùš. The quotient volume is found to be 8, which has to be factorized as $p\cdot p\cdot(p+7)$, and which is indeed stated to have the (three) "equalsides" 1, 1, and 8.

#21 as it stands is corrupt, but so much sense remains that Thureau-Dangin's emendations can probably be relied upon. It is then a close parallel to #20, jumping with the (very unusual) instruction to proceed as in the corresponding (preceding) case directly to the value of the quotient volume, and factorizing it into the "equalsides" 6, 6, and 13. At this point it stops, having shown the essential step and omitting the conversions of the length and

---

that an intermediate step finds 3 widths to be equal to a double length, since 3 follows from a computation and is then "broken", the operation resulting in a "natural" half. One possibility (cf. #26) might be that the two lengths are said to exceed the two widths by one width.

width from 6 lengths/widths of the reference volume (that is, 6 kùš) into 30′ nindan.

The final third-degree problem is #22, which is homogeneous and quite simple. All three dimensions of the excavation are said to be equal, and the method seems to be a simple conversion of the volume 1°30 [sar] into 7′30″ [nindan^3]. 7′30″ is found in the standard table of cube roots (for which reason its served as reference volume in YBC 6295, we remember), and its (single, and genuine) "equalside" is said in agreement with this table to be 30′. "Raising" 30′ to 1, the "conversion" of horizontal extension, yields 30′ [nindan] as side of the square base of the excavation; "raising" it to 12, the "conversion of depth", gives 6 [kùš] as the depth.

Apart from #22, all third-degree problems of our text (which, as observed, is the only evidence for Old Babylonian work with inhomogeneous third-degree problems) make use of a reference volume and thus of a generalized version of the *makṣarum* or "bundling" technique. This method only works because the resulting quotient volumes allow simple factorizations; it could never provide a general solution of inhomogeneous problems of the third degree. As formulated by Thureau-Dangin [TMB, xxxviii], the present solutions amount to a confession that the Old Babylonian calculators did not possess a (general) method to solve problems of the third degree. It is the duty of the modern historian to confess, on the other hand, that the solutions offered by our text are much more than ingenious artifices: they are the very best that could be done by means of the mathematical techniques at hand, unless one should stumble upon the trick described by Cardano in the *Ars magna*.[184] We may add to this that they are more closely related to latter-day theory of numbers than to algebra. *If* we want to measure them with a contemporary gauge, it is therefore of scarce relevance whether they are generalizable *as algebra*; judged as *what they are* they are as sophisticated as Old Babylonian second-degree algebra at its best.

The second degree: Length–Width, Depth–Width, and Length–Depth

The tablet contains several groups of second-degree problems, which can be grouped conveniently according to their dress: different dress was indeed used by the author of our tablet as a way to differentiate the approach. Of greatest interest are the two sequences 24–25–26 and 27–[28?]–29–30.

---

184 As a matter of fact, all elements of this trick are present in Old Babylonian texts, and Cardano well may have derived his insight from recognizing in Tartaglia's formula the familiar solution of the rectangle problem in which the area and the difference between the sides are given; but this is a different story, and in the Old Babylonian texts the elements are never used together.

In #24, the volume of the excavation (27°46′40″), the depth (3°20′) and the excess of length over width (50′) are given. Elimination of the depth leaves us with a problem that can be translated

$$\sqsubset\!\sqsupset(l,w) = 8°20′ \qquad\qquad l = w+50′ ,$$

and which is solved by the customary cut-and-paste methods (cf. Figure 2, etc.):, transforming the rectangle into a gnomon of the same area, completing it as a square, finding the "equalside" of this square and "positing" it twice. Only the concise formulation is unusual, particularly the end: in rev. I 27, we are not told the result of the "breaking", only that the outcome shall "confront"; in 29+2*, instead of stating, in agreement with the "norm of concreteness" (p. 58), that the bisected excess which was "made hold" shall be "torn out" from one and "appended" to the other, the text just prescribes "to 1 append, from 1 tear out", without specifying *what* is appended and torn out.[185]

#25 is the usual companion-piece, giving the sum instead of the difference between length and width, and presents no noteworthy features apart from an even more concise formulation – a recurrent feature of the tablet when minor variations on already known patterns are presented.

#26, on the other hand, though from the viewpoint of mathematical structure nothing but a slightly more complex variant, provides important information.

The depth is still 3°20′ [kùš] (transformed in line 13* into 16′40″ [nindan]), and the volume is given again as 27°46′40″ (all three problems have the same solution). We are informed, finally, that the excess of the width over the depth equals $\frac{2}{3}$ of the length. Division by the depth thus transforms the problem into one which in symbols can be expressed

$$\sqsubset\!\sqsupset(l,w) = A \qquad\qquad w = \tfrac{2}{3}l+d$$

(A = 8°20′, d = 16′40″), or as

$$\tfrac{2}{3}\square(l)+d\cdot l = A\ .$$

A is multiplied by 40′, corresponding either to

$$\sqsubset\!\sqsupset(\tfrac{2}{3}l,w) = 40′\cdot A \qquad\qquad w = \tfrac{2}{3}l+d$$

or to

$$\square(\tfrac{2}{3}l)+d\cdot(\tfrac{2}{3}l) = 40′\cdot A.$$

Both versions are standard problems, as we have seen: the former is similar to #24 (rectangle with given area and excess of length over width); the second is

185 The same abbreviated formula is used in IM 53965, IM 54559, and Db_2-146, all from Eshnunna; and in YBC 4662–63 (see pp. 320, 324, and 346). I have not observed the formula in a variant where removal precedes "appending", and thus not in any text that obeys the "norm of concreteness".

an instance of the problem "sides added to square area" which we have encountered repeatedly in BM 13901; both, as we know, follow the same cut-and-paste procedure. If the latter interpretation of the procedure was correct, however, we should expect the solution to give only the side ($\frac{2}{3}l$) of the square, and to find from there first l and next w. Instead, the text "appends and tears out" precisely as in #24, and presents immediately the larger resulting number (2°30′) as the width, finding the length as igi 40′ times the smaller resulting number. It has thus been kept in mind throughout that the longer side of the rectangle with area $40'\cdot A$ coincides with the original width, while the shorter side is 40′ times the original length. The details of the procedure hence leave no doubt that the transformed problem was thought of in terms of a "rectangle with given excess length" and not as a square with "appended" sides.[186]

The observation is interesting, not because it has general value for Old Babylonian mathematics but rather because it shows that the opposite observation (which one might be tempted to make after similar close reading of other late Old Babylonian texts[187]) cannot be generalized (cf. also below on #9 and #13). Depending on expediency or personal preference, Babylonian calculators might conceptualize problems of this type one way or the other.

#24–26 can be characterized as *length-width*-problems. Correspondingly, #27–30 (with a proviso concerning the missing #28) can be seen as *depth-width*-problems. Their particular interest lies in their relation to the previous group.

In #27, the length is given to be 1°40′, the volume equally 1°40′, and the depth equal to $\frac{1}{7}$ of the excess of length over width.

The first step in the procedure is to tip the excavation around mentally, putting the length in vertical position: The length is "raised" to 12, identified as "conversion of depth", and thus converted into 20 kùš.[188] It is then eliminated, and the rectangle held by the width w and the depth d is seen to be 5′ [nindan^2].

Since $d = \frac{1}{7}(1°40'-w)$, the next step is to find the area $7\cdot5' = 35'$ of another rectangle with sides $7d$ and w. In this rectangle, the sum of length and width is indeed known, and we are thus brought back to the situation known from #25. The procedure is the same in the part of the text which is preserved – and indeed all the way through, if we can trust the parallel

---

186 Neugebauer's interpretation of the procedure [MKT I, 217] refers to a square area and sides (that is, to a quadratic equation in one variable), while Thureau-Dangin supports the two-variable option. None of them give arguments for their choice.

187 Namely, IM 52301 #2 and BM 85194, rev. II, 7–21 (below, pp. 213 and 217, respectively).

188 Once again, that this is what goes on is demonstrated not only by the identification of the factor 12 but also by the exact ordering of steps. A mere elimination of l and a conversion of the resulting area from nindan·kùš into nindan^2 would indeed, according to Babylonian customs, have been performed through successive "divisions" by 1°40′ and 12, not through a single "division" by their product.

passages in #29 and #30. According to the parallels, one resulting side is identified immediately as the width, while the other side is divided by 7, and the outcome 10´ [nindan] stated to be the depth without being converted into kùš, in agreement with the reconceptualization of the depth as a horizontal dimension.

#29 is strictly similar, containing the slight complication that $d = \frac{1}{7}\cdot(1°40'-w)+10'$ [nindan], and thus $7d = 1°40'-w+7\cdot10' = 2°50'-w$. Apart from that everything is analogous. The same applies to #30, where the complication is a subtraction of 1 kùš. Together the three problems (and, we may suspect, #28) appear to present an attempt at systematic training of the mutual conversion between horizontal and vertical dimensions.

Comparison with another group of closely related problems (#9 and #13, *length-depth*-problems) shows that a particular and not the normal procedure is thought of in the sequence #27–30. In #9, the accumulation of ground and earth is given as 1°10´, the width is 20´, and the depth equals the length. In symbolic translation,

$$l\times w\cdot G+l\times w = 1°10' \qquad d = l \qquad w = 20' \ .$$

The first step in the procedure is to multiply 20´ with the number 12, which is *not* presented as the conversion of depth or in any similar way. A simple arithmetical recalculation of $l\times w\cdot G$ as a certain number of squares with side l (*viz.*, 4 such squares) appears to be the best interpretation,

$$l\times w\cdot G+l\times w = l\times w\cdot 12d+l\times w = l\times 20'\cdot 12l+20'\times l =$$
$$12\cdot 20'\cdot\square(l)+20'l = 4\square(l)+20'l= 1°10' \ .$$

In the next step, this is transformed into a genuine square-area-and-sides problem with the side equal to $4l$,

$$\square(4l)+20'\cdot(4l) = 4\cdot 1°10' = 4°40' \ ,$$

which is solved by the usual cut-and-paste technique, giving $4l = 2$, and hence $l = 15'\cdot 2 = 30'$ (the depth is not spoken about).

#13 is similar but more sophisticated. In symbolic translation

$$1/7\cdot(l\times w\cdot G+l\times w)+l\times w = 20' \qquad d = l \qquad w = 20' \ .$$

(The sum of the earth and the ground is also given, but as a magnitude that is "merely known"). Once again, the initial steps may be explained in symbols (remembering, as always, that the "products" are areas and volumes, and not mere numbers):

$$(l\times w\cdot G+l\times w)+7\cdot l\times w = 7\cdot 20' = 2°20' \ ,$$

whence

$$12\cdot 20'\cdot\square(l)+l\times w+7\cdot l\times w = 4l^2+l\times w+7\cdot l\times w = 2°20'$$

and thus

$$\square(4l)+(4l)\times w+7\cdot(4l)\times w = 4\cdot 2°20' = 9°20' \ .$$

It is only at this point, when the problem has been transformed into one concerning $4l$, that the total number of sides to be added to the square area $\square(4l)$ is found, as 1+7 "raised" to w = 20′, i.e., as 8·20′ = 2°40′ – a recurrent feature, the implications of which were discussed in some depth on p. 110. Then finally everything can go by cut-and-paste geometry, and $4l$ and eventually l be found. Once again, the depth goes unmentioned.

The depth is not only absent from the answer but also from the question of both problems (while, e.g., #22 asks for and gives the depth, even though it is said to equal the length). Inspection of the steps of the procedure show, furthermore, that they obliterate the very possibility of linking one side of the rectangle which is cut and pasted to the depth (while the other is easily identified as 4 lengths). We can thus be fairly confident, firstly, that even these two problems should be understood as training *a specific* technique; and secondly, that this technique was the use of the "square area and sides" model, as presupposed in my symbolic translation.

Second-Degree igûm-igibûm-Problems

A final cluster of second-degree problems (#15, 16, and 18) determine the length and the width as *igûm* and *igibûm*, "the igi and its igi". In all cases, the volume is also given, implying that the depth follows trivially ($V = l{\times}w{\cdot}G = l{\times}l^{-1}{\cdot}G = G = 12d$). #16 and 18 furthermore identify the depth with the total of *igûm* and *igibûm*, which leads to a problem of the same type as #25: a rectangle with given area, ⊏⊐(l,w) = 1, and given sum of length and width, $l+w = d$. Their only specific interest lies in their use of the elliptic formula "append and tear out", which is shared by #25, 26, 27, 29, and 30 and appears nowhere else in the tablet. Since #16, 18, 25, and 26 are indubitable "rectangle-" and not "square-problems", this provides us with corroborative evidence that #27, 29, and 30 should be understood in the same way.

Rectangle problems with given sum of length and width normally go together with problems where the excess of length over width is given. So also here: #15 states the depth to be equal to the excess of *igûm* over *igibûm*, while the volume is 36 and d thus 3 nindan. The interesting feature is that this problem has no rational (and hence no *Babylonian*) solution. Nonetheless the text proceeds in a way which demonstrates that 36 is no writing error, and proceeds as done in all similar problems until the point where the excess is bisected. Then suddenly it breaks off and states the result of the bisection to be the *igûm*, which is impossible whatever the area of the rectangle, as long as this area exceeds 0.

Evidently, either the text or the procedure of the problem is somehow corrupt. On the other hand the presence of a companion piece to #16 and #18 with given excess instead of total is next to compulsory. A textual mix-up which could produce as much sense as actually present is not very likely; it seems rather as if somebody (not necessarily the author of the first version of the text) has inserted a problem which for once was *not* constructed backwards from given results, and has then broken off and cheated at the point where the insolubility became evident: when 1°30´ follows from the bisection, even a moderately trained calculator would immediately know its square (2°15´) as well as the result of the quadratic completion (3°15´), and hence that this latter number does not appear in the table of square roots.

In #16 and #18, the singular use of the accounting term nig̃in, "total", for the sum may give us a glimpse of a grid of fine terminological distinctions, not as much according to mathematical meaning as depending on problem dress and thus perhaps historical connections. This could of course be an instance of floating terminological boundaries – but since similar couplings between particular dresses and particular terminological choices are found elsewhere in the material,[189] this is not likely to be the case. The fact that the term is found precisely in connection with *igûm-igibûm* problems might thus be no accident. Through their appurtenance with the table of reciprocals, *igûm* and *igibûm* refer to the same sphere of social activity as does nig̃in: scribal accounting and planning rather than surveying. According to the principle that recreational problems are to be considered as a "supra-utilitarian" superstructure on mathematical practice,[190] this might point to an origin of *igûm-igibûm* problems within this specific orbit,[191] and be another witness of a striking tendency to conserve a characteristic vocabulary.

First-Degree Problems

The tablet contains two groups of first-degree problems, #8+12 and #14+ 17+19, respectively. Both are quite simple as far as mathematical substance is concerned.

In #8 and #12, the length is given as 30´, and the depth is said to equal the length. In #8, the "accumulation" of earth and ground is further said to be 1°10´, while #12 states that $\frac{1}{7}$ of this accumulation "appended" to the ground gives 20´. The procedures are quite similar, and we shall only follow that of #12 in the geometrical diagram (Figure 28) that is suggested by the "appending" in obv. II 6.

189 Thus, the interrogative phrase *kiyā*, "how much each", has an outspoken propensity to go together with partition between brothers/šeš (thus in VAT 8522 #2, VAT 6597, and YBC 4608); in part this is of course due to the general semantics of the word, but other questions for several values are mostly formulated differently. *inanna*, "now", has a penchant for brick carrying and other problems that involve technical constants. Both tendencies recur in geographically widely separated text groups.

190 The notion of the "supra-utilitarian" is discussed in more detail below, p. 366.

191 Though certainly *not* to an independent focus for the creation of second-degree "algebra" – as we have seen in YBC 6967 (above, p. 55), the unknown numbers of *igûm-igibûm*-problems were represented by the geometrical magnitudes of normal "surveying" cut-and-paste geometry.

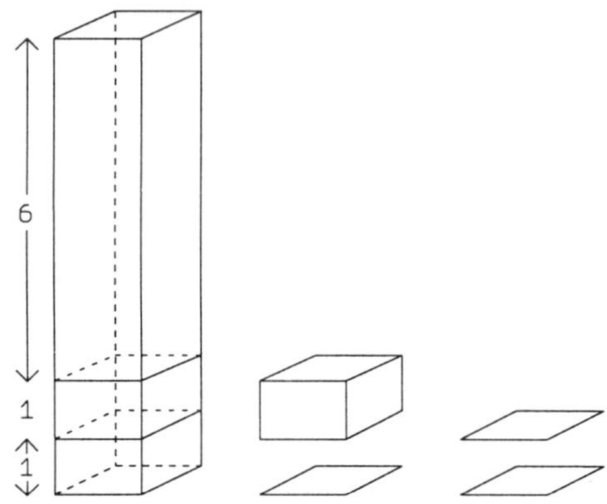

Figure 28. A geometric interpretation of the steps of BM 85200 + VAT 6599 #12.

As a first step, the length is multiplied by 12, resulting in "6 [kùš], the depth". To this 1 [kùš] is "*appended*", giving 7 [kùš] – the depth of the figure representing "earth plus ground". Its 7th is found as 1 [kùš] – implying that the corresponding volume coincides with the ground. That this observation is in fact tacitly made is suggested by the next step: 1 and 1 (representing the ground that is "appended" in the statement) are *"accumulated"*, i.e., the number 20′ is understood as two times the ground (which is probably the reason that it is regarded as an "accumulation", in spite of its origin in an "appending" process[192]), not as a volume 2 kùš high and with base equal to the ground. 20′ is thus multiplied by the igi of 2, resulting in 10′ [the ground]. Division by "30′ the length" yields 20′, the width.

The shift between the two "additions" thus reveals something about the pattern of thought involved: submitting the "accumulation" of "earth" and "ground" to a further operation (taking its seventh) automatically produces a geometric interpretation, so that the result can now be "appended" to the "ground". On the other hand, a height equal to 1 kùš calls forth an immediate identification of basis and volume (in perfect agreement, of course, with the coinciding metrologies and the coinciding values of the two in sar).

The other group of first-degree problems determine length and width as *igûm* and *igibûm* (whence G = volume). Furthermore, the volume is given. In #19, the depth is said to be identical with the *igibûm* (even though the result makes it coincide with the *igûm* instead); in #17, the depth results if the excess of *igûm* over *igibûm* is "torn out" from the *igûm* – a trivially complicated way

192 Neugebauer [MKT I, 196] as well as Thureau-Dangin [TMB, 13] read the logogram UL.GAR in 9* in its function as a verb, "accumulate", take it to be an error for *i-ši*, "raise", and read traces of the ensuing sign as the beginning of a *-ma*. According to the hand copy (in particular the way *i* is written elsewhere), however, the reading *i-ši* is just as plausible as *-ma* while avoiding the (always unpleasant) hypothesis of a scribal blunder.

to say that it equals the *igibûm*; in #14, finally, the procedure and the solution forces us to believe that the depth should have been stated to coincide with the *igûm*, even though the statement contains some extra words which might make us expect another – grammatically clumsy – companion piece to #16 and 18. In all cases, the solution follows from a simple division of the volume (and thus of G) by 12, which yields either *igûm* or *igibûm*. In #17, no word is wasted upon the identification of $l-(l-w)$ with w; the problem looks more like a challenge or a puzzle than as a step in a didactic sequence.

Clues to Teaching Methods

Two features of text allow us to draw conclusions about the way mathematics teaching was organized. One of them is the appearance of "merely known" magnitudes in #6, #7, and #13. Since everything else in these problems is perfectly clear and points to the goal, we may safely assume that the "superfluous" information has a purpose, which is clearly not to serve as data.

Instead, the presence of these numbers can be understood if we think of the purpose and use of the text as a tool for actual teaching. We should imagine the teacher explaining beforehand the total situation: the excavation, its dimensions, the earth, and the ground, giving also their numerical values inasfar as these may be useful as identifying labels. Afterwards, he shows how to extricate the dimensions from a specific set of data; in the oral exposition of the procedure he will have had the possibility to identify, say, the original volume as "1 the earth", in contrast to the extended volume – just as a modern exposition will distinguish V from V'.

In the present case, the written text only conserves traces of this oral exposition technique – but enough to allow us to establish the link to the didactic Susa texts TMS XVI and IX. Even in these, we remember, "merely known" quantities were present. What a modern mathematical reading tends to see as a manifestation of incompetence or deficient understanding is thus a rudiment of an oral technique achieving by other means what *we* are accustomed to achieve in writing by algebraic symbols.

The other informative feature is the seemingly chaotic organization of the tablet, with its bewildering going back and forth between linear, quadratic, and cubic problems. This impression of chaos, however, is an artefact, produced by the imposition of the modern analytical perspective on the text. In only shows that mathematical structure and techniques in our sense do *not* constitute the primary ordering principle of the tablet, and their absence does not prove the general absence of such a principle.

In order to find the ordering principle, we may look at the statements. Firstly, of course, the uniting principle of the tablet as a whole is the excavation, and not the investigation of a specific mathematical structure or training of a particular technique. This corresponds to the character of the "square text" BM 13901. But there is more to it. #5–9 all state the accumulation of earth and ground to be 1°10′. Whatever the mathematical character of the problem, be it of the first, the second, or the third degree, it will thus have to be discussed with reference to an excavation prolonged 1 kùš downwards. #10 and 11 are missing. #12, which as far as mathematical substance is concerned is nothing but a slight variation on #8, starts from a corresponding variation of the configuration, as does #13, which regarding mathematical substance has the same relation to #9. Instead of exhausting first the possibilities of the *method* of #8, which would make #12 follow it immediately, all possibilities *of the configuration* shown in Figure 27 are exploited before further training of the various methods is undertaken.

#14–19 are then *igûm-igibûm* problems; #20–23 deal with excavations with a square ground; #24–26 all have the same volume and depth given and a rectangular base; and #27–30 (with a proviso for the missing #28) all have the length given as 1°40′ and make use of the entity $\frac{1}{7}(l-w)$.

Whereas a categorization according to mathematical structure and techniques only suggests fragments of local order within a generally chaotic structure, the categorization according to configuration thus uncovers a genuine global order and explains the most striking examples of seeming disorder. There is no reasonable doubt that the global order of the tablet is determined by the way didactic exposition was organized, and that this organization was centred on the configuration (maybe, as suggested by Eleanor Robson, with charcoal on the wall) – the *object* of the problem – and not on mathematical structure as reflected in methods.

Below the level of global order, and subordinated to its principles, we find an ordering of shorter sequences according to mathematical principles and progression. Recognition of the important role of didactic exposition should not overshadow the fact that understanding of mathematical principles is *also* demonstrated by the tablet. There is certainly no reason to dismiss it as "merely didactic opportunism and *hence* no testimony of real mathematical thought".

AO 8862 #1–4[193]

For palaeographic reasons, Neugebauer and Thureau-Dangin supposed this tablet to be one of the oldest mathematical texts. As we shall see (pp. 338*ff*), it seems indeed to be a witness of the experimental phase when the scribal school first took over a set of surveyors' riddles and made it the starting point for the creation of a discipline. For the moment we shall prepare this argument

193 Based on the transliteration in [MKT I, 108–111] and the hand copy in [MKT II, pl. 36], cf. corrections in [MKT III, 53] and discussion of grammatical details in [Høyrup 1990: 313].

by noticing a number of deviations from the usual ways of the Old Babylonian texts.

In total, the text contains seven problems. #1–4 – those which are included here – deal with rectangular fields and their sides. #5–7 deal with brick-carrying, and refer to a technical constant that fixes the work norm. #5–6 are first-degree problems about proportional sharing between workers; the third is of the second degree, adding men, days, and bricks (which are proportional to the number of man-days), and thus of a kind related in structure to #1–4 (but even more closely to TMS IX, #2–3).

The following particular features of the text can be pointed out in advance:

- The outcome of an accumulation is constantly spoken of as *kimrātum*, a plural of the verbal substantive *kimirtum* corresponding to *kamārum*, "to accumulate". In one case, as we shall see, the *kimrātum* is the spatially located aggregate of the constituents; in others, the term suggests that the accumulation is still thought of concretely in terms of these constituents. In order to render both the grammatical form of the term and the inherent connotation, I translate it "the things accumulated".
- Linear quantities are "appended" directly to the surface, without any use of a "projection" (as in BM 13901) or "alternate width" (as in YBC 4714); they are evidently treated as "broad lines".
- The text makes use of both *nasāḫum*, "to tear out" and *ḫarāṣum*, "to cut off"; "cutting off" is used exclusively for removal from linear entities or when an extreme part is detached, in nice agreement with the everyday connotations of the term; "tearing-out" is used where a part is removed from a surface, but also at times when linear extensions are involved; the choice seems not to depend on any genuine conceptual difference but rather to be a choice between synonyms, governed by non-technical connotations.
- When taking a "moiety", the text uses the phrase *ba-a-šu ša X*, "its moiety, that of *X*", implying that the entity *X* has some kind of independent existence and can be pointed at. The "accidental half" in I 32 is spoken of as *mišlum*.
- When the unknown length and the width of the field have been "made hold", the outcome of the process is explained as the "building" of a surface (but not when their equally unknown sum and differences are the sides in III 3 – the phrase is thus no simple stand-in for that statement of the resulting area which ensues when the sides are known).
- Quite exceptionally in procedure texts, the term a.rá, "steps of", turns up recurrently. In some instances it stands alone, but often it is part of an indubitable double construction, preceded either by a statement that two lines are "made hold" (I 25, II 14) or that they are "inscribed twice" (II 22). Whereas the bulk of the corpus always takes the numerical computation of the area that results from a rectangularization as being implicit in this process, the present text thus distinguishes the two. With this in mind we may assume that even the occurrences in I 13 and II 14 are meant as double constructions, and that the preceding "breaking" is meant not only

as a bisection but to imply also (in good agreement with the general sense of the word) the formation of a corner with the two "moieties" as legs.

I

edge Nisaba[194]
^{d}Nisaba

#1

1. Length, width. Length and width I have made hold:
uš saĝ uš *ù* saĝ *uš-ta-ki-il$_5$-ma*

2. A surface have I built.
a.šà^{lam} *ab-ni-i*

3. I turned around (it). As much as length over width
as-sà-ḫi-ir ma-la uš *e-li* saĝ

4. went beyond,
i-te-ru-ú

5. to inside the surface I have appended:
a-na li-ib-bi a.šà^{lim} *u-ṣi-ib-ma*

6. 3`3. I turned back. Length and width
3.3 *a-tu-úr* uš *ù* saĝ

7. I have accumulated: 27. Length, width, and surface w[h]at?

27	3`3	the things accumulated
15	the length	3` the surface
12	the width	

ĝar.ĝar-*ma* 27 uš saĝ *ù* a.šà *mi-[nu$^?$]-um*

27	3.3	*ki-im-ra-tu-ú*
15	uš	3 a.šà
12	saĝ	

8. You, by your proceeding,
at-ta i-na e-pe-ši-ka

9. 27, the things accumulated, length and width,
27 *ki-im-ra-at* uš *ù* saĝ

10. to inside [3`3] append:
a-na li-bi [3.3] *ṣi-ib-ma*

11. 3`30. 2 to 27 append:
3.30 2 *a-na* 27 *ṣi-ib-ma*

12. 29. Its moiety, that of 29, you break:
29 *ba-a-šu ša* 29 *te-ḫe-ep-pe-e-ma*

13. 14°30′ steps of 14°30′, 3`30°15′.
14.30 a.rá 14.30 3.30.15

194 *Inter alia* goddess of the scribal art. This invocation is the closest any Old Babylonian mathematial text comes to connecting its topic with religious or other esoteric matters.

14. From inside 3`30°15′
i-na li-bi 3.30.15

15. 3`30 you tear out:
3.30 *ta-na-sà-aḫ-ma*

16. 15′ the remainder. By 15′, 30′ is equal[side.]
15 *ša-pi-il*$_5$*-tum* 15.e 30 íb.[si$_8$]

17. 30′ to one 14°30′
30 *a-na* 14.30 *iš-te-en*

18. append: 15 the length.
ṣi-ib-ma 15 uš

19. 30′ [fr]om the second 14°30′
30 [*i*]*-na* 14.30 *ša-ni-i*

20. you cut off: 14 the width.
ta-ḫa-ra-aṣ-ma 14 sağ

21. 2 which to 27 you have appended,
2 *ša a-na* 27 *tu-uṣ*$_4$*-bu*

22. from 14, the width, you tear out:
i-na 14 sağ *ta-na-sà-aḫ-ma*

23. 12 the true width.
12 sağ gi.na

24. 15, the length, and 12, the width, I have made hold:
15 uš 12 sağ *uš-ta-ki-il*$_5$*-ma*

25. 15 steps of 12, 3` the surface.
15 a.rá 12 3 a.šà

26. 15, the length, over 12, the width,
15 uš *e-li* 12 sağ

27. what goes beyond?
mi-na wa-ta-ar

28. 3 it goes beyond. 3 to inside 3` the surface append,
3 *i-te-er* 3 *a-na li-bi* 3 a.šà *ṣi-ib*

29. 3`3 the surface.
3.3 a.šà

#2

30. Length, width. Length and width
uš sağ uš *ù* sağ

31. I have made hold: A surface I have built.
uš-ta-ki-il$_5$*-ma* a.šà^{lam} *ab-ni*

32. I turned around (it). The half of the length
a-sà-ḫi-ir mi-ši-il$_5$ uš

33. and the third of the width
ù ša-lu-uš-ti sağ

34. to the inside of my surface
a-na li-bi a.šà*-ia*

35. [I have] appended: 15.
[*ú-*]-*ṣi-ib-ma* 15

36. [I tu]rned back. Length and width
[*a-t*]*u-úr* uš ù sağ

37. [I have ac]cumulated: 7.
[*ak-*]*mu-ur-ma* 7

II

1. Length and width what?
uš ù sağ *mi-nu-um*

2. You, by your proceeding,
at-ta i-na e-pe-ši-i-ka

3. [2 (as) in]scr[i]ption of the half
[2 *n*]*a-al-p*[*a*]*-at-ti mi-iš-li-im*

4. [and] 3 (as) inscription
[*ù*] 3 *na-al-pa-ti*

5. of the [th]ird you ins[cr]ibe:
[*ša-*]*lu-uš-ti ta-l*[*a*]*-pa-at-ma*

6. Igi 2, 30´, you detach:
igi 2-bi 30 *ta-pa-ṭar-ma*

7. 30´ steps of 7, 3°30´; to 7,
30 a.rá 7 3,30 *a-na* 7

8. the things accumulated, length and width,
ki-im-ra-tim uš ù sağ

9. I bring:
ub-ba-al-ma

10. 3°30´ from 15, my things accu[mu]lated,
3,30 *i-na* 15 *ki-i*[*m*]*-ra-ti-i-a*

11. cut off:
ḫu-ru-uṣ$_{?}$*-ma*

12. 11°30´ the remainder.
11,30 *ša-pi-il*$_5$*-tum*

13. Do n[ot] go beyond. 2 and 3 make hold:
l[*a*] *wa-t*[*ar*] 2 ù 3 *uš-ta-kal-ma*

14. 3 steps of 2, 6.
3 a.rá 2 6

15. Igi 6, 10´ it gives you.
igi 6 ğál 10 *i-na-di-kum*

16. 10´ from 7, your things accumulated,
10 *i-na* 7 *ki-im-ra-ti-i-ka*

17. length and width, I tear out:
uš ù sağ *a-na-sà-aḫ-ma*

18. 6°50′ the remainder.
6,50 *ša-pi-il₅-tum*

19. It[s] moiety, that of 6°50′, I break:
ba-a-š[u] ša 6,50 e-ḫe-pe-e-ma

20. 3°25′ it gives you.
3,25 *i-na-di-ku*

21. 3°25′ until twice
3,25 *a-di ši-ni-šu*

22. you inscribe; 3°25′ steps of 3°25′,
ta-la-pa-at-ma 3,25 a.rá 3,25

23. 11°40′[25″]; from the inside
11,40,[25] *i-na li-bi*

24. 11°30′ I tear out:
11,30 *a-na-sà-aḫ-ma*

25. 10′25″ the remainder. ⟨By 10′25″, 25′ is equalside⟩.
10,25 *ša-pi-il₅-tum* ⟨10,25.e 25 íb.si₈⟩

26. To the first 3°25′
a-na 3,25 *iš-te-en*

27. 25′ you append: 3°50′,
25 *tu-ṣa-am-ma* 3,50

28. and (that) which from the things accumulated of
ù ša i-na ki-im-ra-at

29. length and width I ha[ve] torn out
uš *ù* sag̃ *a[s]-sà-aḫ-ma*

30. to 3°50′ you append:
a-na 3,50 *tu-ṣa-am-ma*

31. 4 the length. From the second 3°25′
4 uš *i-na* 3,25 *ša-ni-im*

32. 25′ I tear out: 3 the width.
25 *a-na-sà-aḫ-ma* 3 sag̃

32a. 7 the things accumulated.
7 *ki-im-ra-tu-ú*

32b. 4, the length
3, the width — 12, the surface
4 uš
3 sag̃ — 12 a.šà

#3

33. Length, width. Length and width
uš sag̃ uš *ù* sag̃

34. I have made hold:
uš-ta-ki-il₅-ma

35. A surface I have built.
a.šà*lam* *ab-ni*

III

1. I turned around (it). So much as length over width
a-sà-ḫi-ir ma-la uš *e-li* sağ

2. goes beyon[d], together with the things accumulated,
i-te-ru-[*ú*] *it-ti ki-im-ra-at*

3. my length and [width], I have made hold:
uš *ù* [sağ]*-ia uš-ta-ki-il*$_5$*-ma*

4. To the in[si]de of my surface
a-na l[*i-b*]*i* a.šà*-ia*

5. I have app[end]ed:
u-ṣi-i[*b*]*-ma*

6. 1``13`20. I [tu]rned back. Length and width
1,13,20 *a-*[*tu*]*-úr* uš *ù* sağ

7. I have accumul[ated:] 1`40.

1`40	1``13`20 the things accumulated
1` the length 40 the width	40` the surface

ak-mu-u[*r-ma*] 1,40

1,40	1,13,20 *ki-im-ra-tu-ú*
1 uš 40 sağ	40 a.šà

8. You, by your proceeding,
at-ta i-na e-pe-ši-i-ka

9. 1,40 the things accumulated, length and width,
1,40 *ki-im-ra-at* uš *ù* sağ

10. 1`40 steps of 1`40, 2``46`40.
1,40 a.rá 1,40 2,46,40

11. From 2`46`40, 1``13`20 the surface
i-na 2,46,40 1,13,20 a.šà

12. you tear out: 1``33`20.
ta-na-sà-aḫ-ma 1,33,20

13. Do not go beyond. The moiety, that of 1`40,
la wa-tar ba-a-šu ša 1,40

14. you break: 50 steps of 50,
te-ḫe-pe-e-ma 50 a.rá 50

15. 41`40 to 1``33`20 you append.
41,40 *a-na* 1,33,20 *tu-ṣa-am-ma*

16. By 2``15`, 1`30 is equalside.
2,15.e 1,30 íb.si$_8$

17. 1`40 over 1`30 what goes beyond?
1,40 ugu 1,30 *mi-na i-te-er*

18. 10 it goes beyond. 10 to 50 append,
10 *wa-tar* 10 *a-na* 50 *ṣi-ib*

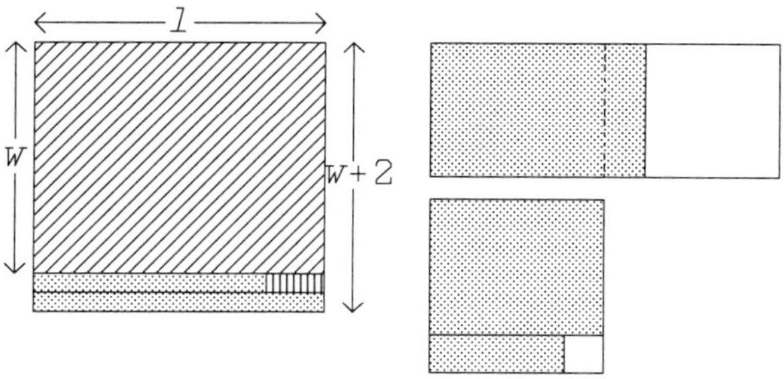

Figure 29. The situation and procedure of AO 8862 #1.

19. 1` the length. 10 from 50 cut off,
1 uš 10 *i-na* 50 *ḫu-ru-uṣ*$_4$*-ma*

20. 40 the width.
40 sağ

#4

21. Length, width. Length and width
uš sağ uš *ù* sağ

22. {...} I have made hold: A surface I have built.
{uš *ù* sağ} *uš-ta-ki-il*$_5$*-ma* a.šà *ab-ni*

23. I turned back. Length and width I have accumulated:
a-tu-úr uš *ù* sağ *ak-mu-ur-ma*

24. together with the surface, they confront each other (as equals).
it-ti a.šà *mi-it-ḫa-ar*

25. Length, width, and surface I have accumulated:
uš sağ *ù* a.šà *ak-mu-ur-ma*

26. 9. Length, width, and surface what?
9 uš sağ *ù* a.šà *mi-nu-um*

#1

In I 23, this problem distinguishes a "true width" from the width that has just been found, and thus shows how the elegant change of variable that it employs was spoken about.

The text starts by stating that a rectangular surface or field (obliquely hatched in the upper diagram of Figure 29) is built, that is, marked out; after pacing off its dimensions, the speaker "appends" the excess of the length over the width (regarded as a broad line – vertically hatched) to it; the outcome is 3`3. Even this is done quite concretely in the terrain. Then he "turns back" and reports the accumulation of the length and the width to be 27.[195] The

195 The reading of *saḫārum* "to turn around", and *târum*, "to turn back" as referring to

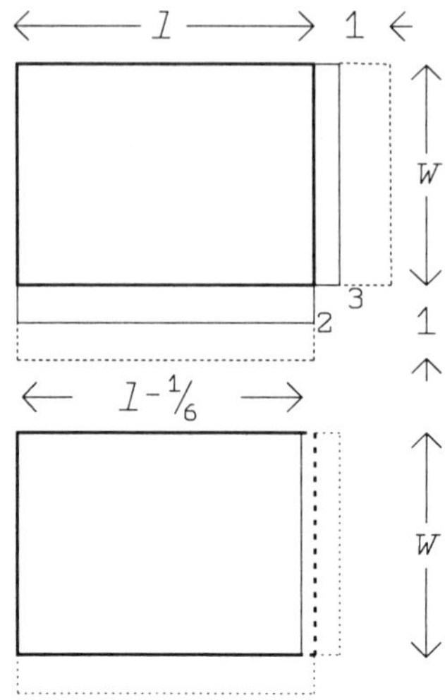

Figure 30. The configuration of AO 8862 #2.

procedure starts by "appending" these latter "things accumulated" (dotted) to the hatched surface, from which we get a new rectangle with the same length and a width that has been augmented by 2. The surface is 3`3+27 = 3`30, and the sum of the length and the new width is obviously 27+2 = 29. This standard problem is solved by means of the usual procedure, as shown in distorted proportions in the lower part of Figure 29. The length turns out to be 15, and the width 14; the original or "true" width is therefore 12.

#2

As we shall see, the preceding problem is likely to be one of those from which the Old Babylonian school started the creation of its algebra; #2 shows how it began experimenting and to construct variants by using fractional coefficients.

The anecdote is similar. This time, only ½ of the length and ⅓ of the width are "appended", which gives a total surface 15 – see the upper diagram in Figure 30. Accumulation of length and width yields 7.

The numbers 2 and 3 are "inscribed" as representations of ½ and ⅓; perhaps we should imagine this to be done as in Figure 30. In the next step (II 6–7) it is found out what the "things accumulated, length and width", would amount to if the breadth was only 30´ instead of 1 – namely, 3°30´.

imagined "real" movements in the terrain (as originally suggested to me by Aage Westenholz) and not as mere demarcations of sections is supported by the use of a completely different phrase when sections are demarcated in the prescriptions (II 13, III 13) – *lā watar*, the same verb as in the phrase "*a* over *b*, *d* goes beyond", meaning something like "no further!".

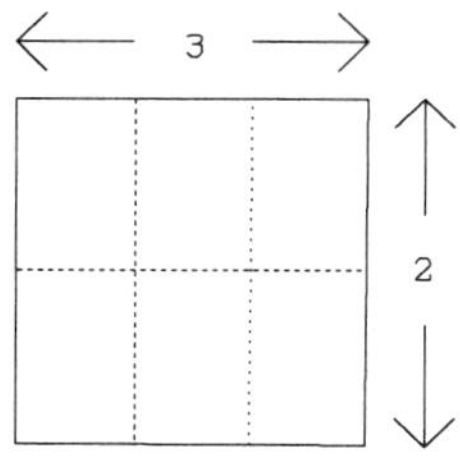

Figure 31. The geometrical determination of $^1/_2$–$^1/_3$.

This entity is "brought" "to 7, the things accumulated, length and width", and then "cut off" from 15; "bringing" is thus not a process with numerical implications, but it allows that an entity *A* which is not originally part of another entity *B* be "cut off" from it – that is, we may suppose that "bringing to the place of *B*" entails a spatial superposition. In the present case, 3°30′ is brought to the edges of the field where the aggregation of length and width are located.

When the length and width provided with a breadth 30′ are "cut off" (leaving 11°30′), the "appended" half of the length is eliminated; but more than the third of the width is removed.

In order to know how much more than the third of the width is removed, a queer trick is used, introduced by the phrase "Do not go beyond!" that informs us that the course of the procedure is interrupted. The two numbers that represented the fractions are "made hold", probably as in the corner of the upper diagram of Figure 30; Figure 31 shows the situation in detail. The argument is not filled out, but the gist must be that $\frac{1}{2}$ corresponds to 3 small squares, each of value $\frac{1}{6}$, and $\frac{1}{3}$ to only 2 small squares; therefore the difference is $\frac{1}{6}$ = 10′.

The purpose of the procedure must go beyond the mere determination of this result: in the table that gives igi 6 as 10′ we also find that $\frac{1}{2}$ is 30′ and $\frac{1}{3}$ is 20′. Perhaps the passage is meant to show how such calculations were performed outside that scribal environment which made routinely use of the igi table – perhaps didactic heuristics was aimed at.

In any case the length is seen to decrease by 10′, for which reason the sum of length and width of the resulting rectangle is 7–10′ = 6°50′; its surface, as we remember, is 11°30′.

The ensuing resolution of the rectangle problem runs as before, though without explicit reference to the new length as a length, and hence with no need to distinguish a "true length".

#3

#1 of the present tablet makes use of the trick that is explained in TMS IX #1 (in which adding one length to the area was seen to be equivalent to increasing

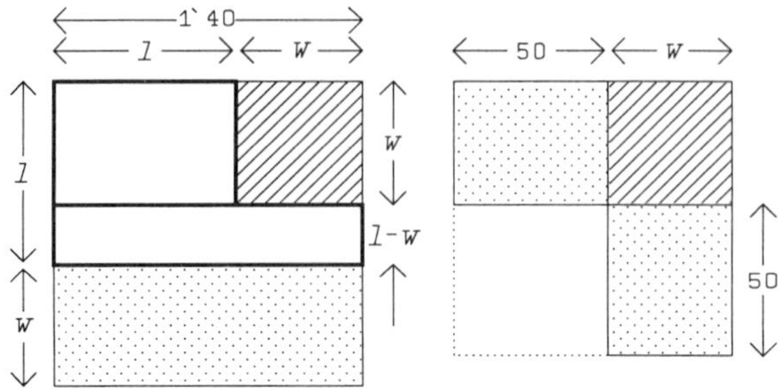

Figure 32. The configuration and procedure of AO 8862 #3.

the width by *1*). #2 *might* have used the trick of TMS IX #2, adding $\frac{1}{2}\times\frac{1}{3}$ in the corner and thus completing the rectangle; instead, even this time the model of TMS IX #1 is followed. The reason may be aesthetic, namely, that the adoption of the other model would have entailed use of the completion technique twice within the same problem (as happens in TMS IX #3).

In the present problem we are told that ⊏⊐(*l*,*w*)+⊏⊐(*l*+*w*,*l*–*w*) = 1``13`20, and that *l*+*w* = 1`40.[196] The former condition might have been transformed into ⊏⊐(*l*,*w*)+100·(*l*–*w*), which (with addition of 100·(*l*+*w*), in analogy with #1) would once more have allowed us to return to the model of TMS #1. What goes on, however, avoids this arithmetical artifice and is related instead to the standard diagram that was shown in Figure 12, the diagram on which the solution of BM 13901 #8 is likely to have been based – see Figure 32: the square □(*l*+*w*) on the given aggregate of length and width is constructed, and the given (non-shaded) area ⊏⊐(*l*,*w*)+⊏⊐(*l*+*w*,*l*–*w*) is "torn out" ("cutting off" would indeed be a most unfortunate metaphor, we see). This leaves 1``33`20, which can be seen to be composed of a square □(*w*) and a rectangle ⊏⊐(*w*,100). This rectangle is broken, which allows rearrangement of the area 1``33`40 as a gnomon. Completion and finding of the "equalside" shows that $w+\frac{l-w}{2}$ = 1`30.

Average and Deviation

The following steps are awkward from our point of view: since $\frac{l-w}{2}$ is known to be 50, *w* could easily be found as $(w+\frac{l-w}{2})-\frac{l-w}{2}$ = 1`30–50 = 40, whence *l* = (*l*+*w*)–*w* = 1`40–40 = 1`.

The way the text proceeds, however, is no symptom of clumsiness: it

---

196 ⊏⊐(*l*,*w*)+⊏⊐(*l*+*w*,*l*–*w*) = 1°13′20″, *l*+*w* = 1°40′ is also possible.

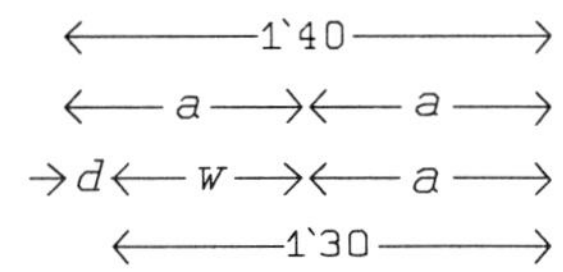

Figure 33. The relation between 1`40, 1`30, width, and deviation.

follows from, and betrays, a fundamental conceptual substructure:[197] when confronted with two quantities (say, p and q) that form a genuine pair, the Old Babylonian calculator would perceive them symmetrically, as composed from *average* (a) and *deviation* (d), $p = a+d$, $q = a–d$. Technically, the average is of course the semi-sum, $a = \frac{p-q}{2}$, while the deviation is the semi-difference, $d = \frac{p-q}{2}$; but these characterizations are misleading inasmuch as they transform fundamental entities into derived quantities.

In III 19–20, indeed, we see that l and w are found by "appending" to and "cutting off" from the average a = 50. The quantity that is "appended" and "cut off" is d = 10, found as the excess of 1`40 over 1`30. In order to see how the deviation may be immediately understood as the excess of 1`40 (the "double average") over 1`30 (width and average), we may look at Figure 33.

Once the role of average and deviation as fundamental entities has been understood from this text, new light is cast on a number of other texts which at the first approach did not let this structure stand out so clearly:

- The elliptic "append and tear out" of BM 85200 + VAT 6599 turns out to express the relation between the twin magnitudes p and q and the appurtenant average and deviation; this relation to a very familiar structure with appurtenant procedures explains that the ellipsis could develop and be coupled precisely to rectangle problems (see p. 158).
- The solution of VAT 8389 #1 can be seen to have been very close at hand (see p. 77): given the accumulation of the two fields, the average is found by simple "breaking"; what remains is then to find the deviation.
- All two-square and rectangle problems make use of the average and deviation, whereas square-and-sides problems use a reduced version. As demonstrated by Seleucid and later ancient texts, this choice is by no means compulsory. Given the pervasive use of the average-deviation structure this suggests that rectangle problems should be seen as at least conceptually, perhaps also historically primary, and square problems as secondary (as also intimated by the statistical preponderance of rectangle problems).

---

197 "Substructure" or "conceptual habit", not simply "concept" or "conceptual structure": in spite of the recurrent use of average and deviationin many connections no general technical term appears to have corresponded to them, they are always spoken of in words referring to the actual situation. This, however, remained the case in Greek mathematics (cf. pp. 402, 405).

– Finally, recognition of the role of average and deviation allows us to formulate "the norm of concreteness", the rule determining in which cases "tearing-out" should precede "appending" if this norm is respected (p. 58): If the deviation is given, and the average derived in two copies (by "breaking" or as an "equalside" and its "counterpart"), then the two quantities are to be found by transferring the deviation from one to the other – which means that "tearing-out" ought to come first. If the deviation is derived as the side of a completing square (and present in two copies), addition may safely be performed first

#4

Since no procedure is indicated, all we can say is that the obvious strategy is to find the surface and the sum of the sides separately as 4°30′ each; this gives us the same standard form as #1–2, and sides 3 and 1°30′.

In itself, the problem is thus not very interesting. As we shall see, however, its presence in precisely this text turns out to be highly significant, not least in connection with the seemingly insignificant order of the sequence "length, width, and surface" in III 25.

YBC 6504[198]

This tablet contains four problems, all dealing with the same geometrical configuration – a rectangle □(*l*,*w*) from which the square □(*l*–*w*) on the excess of the length over the width has been "torn out", and whose area is 8′20″. The other datum is *l*–*w* (#1), *l*+*w* (#2), *l* (#3) and *w* (#4). The text thus presents us with another example of that didactic primacy of the configuration which we encountered in the excavation text BM 85200 + VAT 6599 (see p. 162).

The methods used are very flexible – so much so, indeed, that #4 is solved by an illegitimate shortcut.

198 Based on the transliteration im [MKT III, 22*f*]. Since most verbal Sumerograms are provided with conjugation prefixes, they are translated as finite forms (chosen in agreement with the habits of purely syllabic texts, since the syllabically written verbs follow the normal pattern).

Obv.

#1

1. [So much] as length over width goes beyond, I have ma[de confront itself, from the inside of the surface]
[ma-l]a uš ugu sağ SI íb.s[i$_8$ *i-na lìb-ba* a.šà]

2. [I have torn it] out: 8´20˝. Length over width [10´ goes beyond.]
[ba.z]i-*ma* 8,20 uš ugu sağ [10 SI]

3. ⌈By y⌉our proceeding, 10´ you mak[e hold:]
⌈*i-na*⌉ *e-pe-ši-k[a]* 10 *tu-uš-t[a-kal-ma]*

4. 1´⌈4⌉0˝ to 8´20˝ you appen[d: 10´⌉ you [pos]it.
1,⌈40⌉ *a-na* 8,20 bí.d[aḫ-*ma* 10⌉ i[n.ğa]r

5. Half of 10´ you brea[k:] 5´ you posit.
šu.ri.a 10 *te-ḫe-ep-p[e-m]a* 5 in.ğar

6. 5´ you make hold: 25˝ you posit.
5 *tu-uš-ta-kal*[199]*-ma* 25 in.ğar

7. 25˝, the surface, to 10´ you append: 10´25˝ you posit.
25 a.šà *a-na* 10 bí.daḫ-*ma* 10,25 in.ğar

8. By 10´25˝, 25´ is equalside. 5´ to 25´ you a[pp]end:
10,25.e 25 íb.si$_8$ 5 *a-na* 25 b[í.d]aḫ-*ma*

9. 30´, the length, you posit. 5´ from 25´ your tear out:
30 uš in.ğar 5 *i-na* 25 ba.zi-*ma*

10. 20´, the width, you posit.
20 sağ in.ğar

#2

11. So much as length over width goes beyond, I have made confront itself, from the inside of the surface I have torn it out:
ma-la uš ugu sağ SI íb.si$_8$ *i-na lìb-ba* a.šà ba.zi-*ma*

12. 8´20˝. Length and width accumulated: 50´. By [your] proceeding,
8,20 uš *ù* sağ ğar.ğar-*ma* 50 *i-na e-pe-ši[-ka]*

13. 50´ you make hold: 41´40˝ you posit.
50 *tu-uš-ta-kal-ma* 41,40 in.ğar

14. ⌊41´40˝ to] 8´20˝ you append: 50´ you posit.
⌊41.40 *a-na*] 8,20 bí.daḫ-*ma* 50 in.ğ[ar]

---

199 For two reasons, I follow Thureau-Dangin's reading of the sign, against Neugebauer's ḫir. Firstly, the hand copy seems to suit this reading best; secondly, the syllabic value *ḫir* for the sign in question (EZEN) is problematic in an early or middle Old Babylonian text. According to MEA, #152, it is only testified from the Middle Babylonian period onwards. This is not quite true, since the value is found in BM 85200+VAT 6599 (obv. I, 28; rev. I, 10, 15, 19); still, that tablet is late Old Babylonian, whereas the present one is likely to belong to the early phase – cf. pp. 337*ff*.

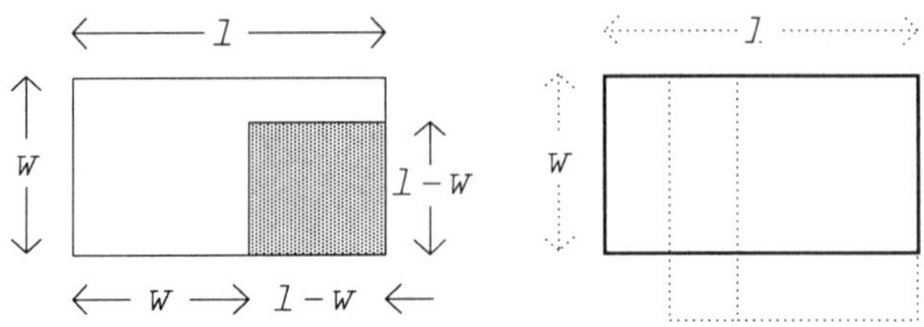

Figure 34. The procedure of YBC 6504 #1.

15. Igi !5] you detach:[1]2′ you p[osit].
igi !5 ğá]l *ta-pa-ṭar-m*[*a* 1]2 in[.ğar]

16. 12′ to 50′ you raise[: 1]0′ you [posit]
12 *a-na* 50 *ta-na-aš-ši*[*-ma* 1]0 in.[ğar]

17. [Hal!f of 50′ you break: [2]5′ you [posit]
[šu.ri!.a 50 *te-ḫe-ep-pe-ma* [2]5 in.[ğar]

18. 25′ you make hold[: 10′2!5″ you posit
25 *tu-uš-ta-kal*[*-ma* 10,2!5 in.ğar

19. 10′ from 10′2[5″ you tear out:! 25″ you [posit].
10 *i-na* 10,2[5 ba.zi-*m*]*a* 25 in.[ğar]

20. By 25″, 5′ i[s equalside]. 5′ to 25″ you [append:]
25.e 5 í[b.si$_8$] 5 *a-na* 25 bí[.daḫ-*ma*]

21. 30′, the length, you posit.
30 uš in.ğar

22. 5′ from 25′ you tear out:
5 *i-na* 25 ba.zi-*ma*

23. 20′, the width, you posit.
20 sağ in.ğar

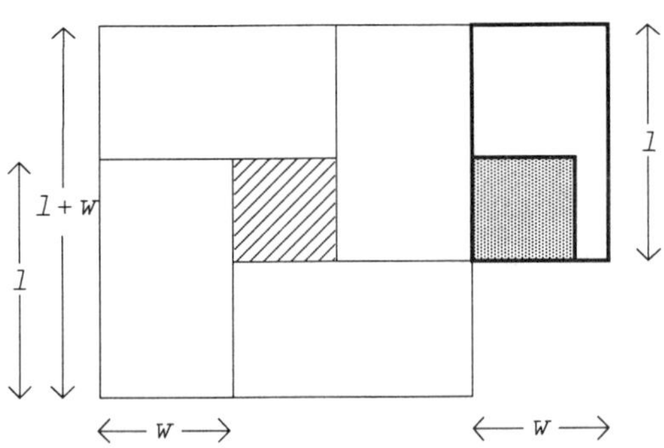

Figure 35. The procedure of YBC 6504 #2.

Rev.

#3

1. ⌈So] much as length over ⟨width⟩ goes beyond, made encounter,[200] from inside the surface I have torn out,
⌈*ma-*]*la* uš ugu ⟨sağ⟩ SI du$_7$.du$_7$ *i-na lìb-ba* a.šà ba.zi

2. 8´20˝. 30´ the length, its width what?[201]
8,20 30 uš sağ.bi en.nam

3. 30´ made encounter: 15´ you posit.
30 du$_7$.du$_7$*-ma* 15 in.ğar

4. 8´20˝ from inside 15´ you tear out, 6´40˝ you posit.
8,20 *i-na lìb-ba* 15 ba.zi*-ma* 6,40 in.ğar

5. Half of 30´ you break:
šu.ri.a 30 *te-ḫe-ep-pe-ma*

6. 15´ made encounter: 3´45˝ you posit.
15 du$_7$.du$_7$*-ma* 3,45 in.ğar

7. 3´45˝ to 6´40˝ you append: 10´25˝ you [posit.]
3,45 *a-na* 6,40 bí-daḫ*-ma* 10,25 in[.ğar]

8. By 10´25˝, 25´ is equalside. 15´ from 25´ you tear out[:]
10,25.e 25 íb.si$_8$ 15 *i-na* 25 ba.zi-[*ma*]

9. 10´ you posit. 10´ from 30´ you tear out:
10 in.ğar 10 *i-na* 30 ba.zi*-ma*

10. 2[0´, the wid]th, you posit.
2[0 sa]ğ in.ğar

#4

11. So much as length o[ve]r width goes beyond, made encounter, from inside the surface I have torn ou[t:]
ma-la uš u[g]ù sağ SI du$_7$.du$_7$ *i-na* a.šà ba.z[i$^?$*-ma*$^?$]

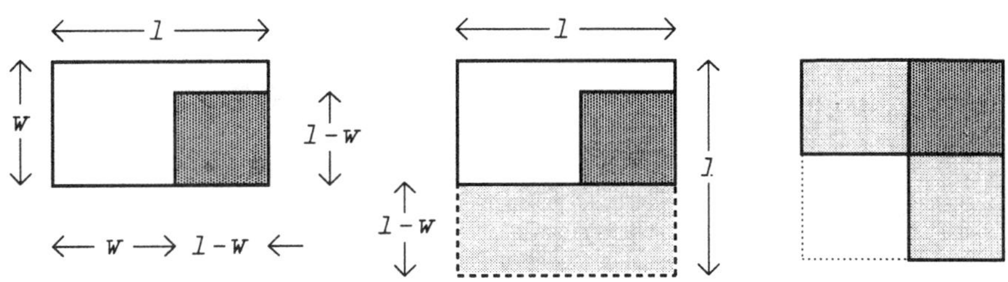

Figure 36. The procedure of YBC 6504 #3.

200 Contrary to other logograms used in the text, du$_7$.du$_7$ is provided with neither Sumerian nor Akkadian grammatical elements. So, translation to an infinite form seems appropriate.

201 The abbreviated character of the two last problems is reflected not only in the use of du$_7$.du$_7$ instead of íb.si$_8$ but also in the absence of a reference to "your proceeding" corresponding to that of obv. 3 and 12.

12. 8′ 20″. 20′ the width, its length what?
8,20 20 sağ uš.bi en.nam

13. 20′ made encounter: 6′ 40″ you posit.
20 du$_7$.du$_7$-*ma* 6,40 in.ĝar

14. 6,4$_{|}$0′ t$^{|}$o 8′ 20″ you append: 15′ you posit.
6,$_{|}$40 *a*$^{|}$-*na* 8,20 bí.daḫ-*ma* 15 in.ĝar

15. By 15′, 30′ is equalside. 30′, the length, you posit.
15.e 30 íb.si$_8$ 30 uš in.ĝar

In #1, *l–w* is given, which leads immediately to □(*l–w*). Adding this to the truncated rectangle restores the full rectangle ⊏⊐(*l*,*w*), and thus gives us the standard problem where a rectangular area and the excess of length over width is known. The solution is found accordingly (but without respecting the "norm of concreteness").

Even #2 is reduced to a rectangular standard problem, but with a quite ingenious method – see Figure 35: Since *l*+*w* is given, the square on this quantity is formed. This square consists of 4 rectangles equal to ⊏⊐(*l*,*w*) plus that square □(*l–w*) which was "torn out" from the original rectangle. Together, the truncated rectangle and the square □(*l*+*w*) thus equal 5 times the original rectangle. Multiplying with igi 5 gives the area of the rectangle; at the same time, the sum of the sides is known.

In #3, the length *l* is given as 30′. Removal of the truncated rectangle from □(*l*) leaves the square □(*l–w*) and a rectangle ⊏⊐(*l*,*l–w*) = ⊏⊐(30′,*l–w*) – see Figure 36. This situation is similar to what we encountered in AO 8862 #3, and the solution is obtained by a similar procedure.

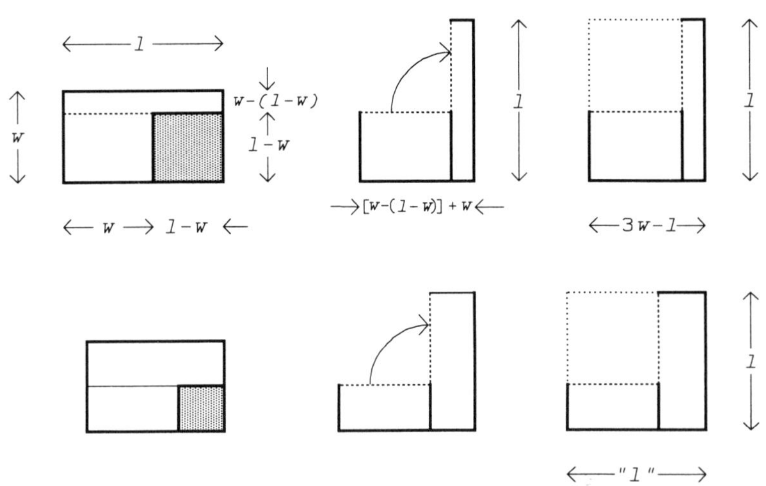

Figure 37. The procedure of YBC 6504 #4, above in distorted, below in correct proportions.

In #4, an error occurs which seems to result from the geometrical method in combination with the habit to use numbers as identifiers – see Figure 37. For us, a similar effect may follow from a drawing in correct proportions, since the error is a consequence of the fact that $w-(l-w) = l-w$ (in order to avoid the resulting fallacy, the diagrams corresponding to #1–3 were drawn in distorted proportions).

The width w is given to be 20′. The square □(w) on this entity is "appended" to the truncated rectangle, and the result is supposed to be □(l). In the lower part of Figure 37 we see how this idea may be engendered; in the upper part we see that the actual outcome of the process is ⊏⊐(l,3w–l).

From the terminological point of view, the present text is unique. It is alone in "positing" all results, intermediate as well as final; it is one of very few texts that employ íb.si$_8$ logographically for *šutamḫurum*, "to make confront itself"; it is alone in using SI as an abbreviation for dirig (= SI.A), "to go beyond"; and it is also alone in using šu.ri.a (normally a Sumerogram for *mišlum*, the "accidental half") logographically for *bāmtum*, the "moiety".

AO 6770 #1[202]

Like the two preceding texts, this one deviates from standard terminology (as we shall see, orthography suggests that all three come from the same environment – below, p. 337); it is also unusual by attempting to formulate a rule in general terms and not through a paradigmatic example. In combination, these two features make the text opaque, and they have given rise to several fanciful interpretations.[203]

Obv.

1. Length and width, so much as the surface let it confront (as equal).
uš *ù* sağ *ma-la* aša$_5$ *li-im-ta-ḫa*[*r*][204]

202 Based on the transliteration in [MKT II, 37], cf. [TMB, 71].

203 Thus, Thureau-Dangin [TMB, 71], among other bold guesses, inserts a full extra line. Brentjes and Müller [1982] misread the glossary of [MKT], which refers to the use of *ù*, "and", in the expression "make *p and q* hold" to the effect that *ù* itself means multiplication, and assert (again misinterpreting the glossary) that "positing to twice" may mean "multiply by 2".

204 [Brentjes and Müller 1982: 21] propose a reading as a plural *li-im-ta-a*[*ḫ-ru*]. Whether the accumulation of length and width will have been thought of as a singular or as a plural (as "the things accumulated" of AO 8862) cannot be decided. If the Brentjes-Müller reading is correct, the translation becomes "let them confront the surface". However, according to Jeremy Black, who collated the tablet for Eleanor Robson, "there's barely room for ⌈*har*⌉ – certainly not for ⌈*ah*⌉-[*ru*]" (Robson, personal communication).

2. You, by your proceeding
at-ta i-na e-pe-ši-ka

3. The step to twice you posit,
a-ra-am a-na ši-ni-šu ta-ša-ka-an

4. from the inside 1 you tear out,
i-na li-ib-bi-im 1 *ta-na-as-sà-aḫ*

5. the igi you detach,
i-[g]i$_4$*-a-am ta-pa-aṭ-ṭa-a[r]*

6. together with the step which you have posited
it-ti a-re-e-em ša ta-aš-ku-nu

7. you make hold,
tu-uš-ka-al-ma

8. the width it gives you.
sağ *i-na-ad-di-ik-kum*

One of the stumbling-blocks for the interpretation of the preceding few lines is the "step" (*arûm*), an Akkadian derivation from the multiplication a.rà, "steps of". For the moment we shall leave it open whether it refers to the outcome (that is, the product) or the single step (that is, a multiplicand or factor); both meanings are found elsewhere in the corpus. What is certain is that this quantity *Q* is "posited" twice, that *1* is "torn out" from one copy, that igi *Q*–1 is "detached", and that this igi is "made hold" together with the other copy of *Q*, from which the width is said to result.

In line 1, GÁN has mostly been interpreted as the metrological unit iku instead of aša$_5$, and interpreted as meaning 1 iku. This turns out either to require that violence be done to the rest of the text, or to result in mathematical errors. Therefore, an interpretation in agreement with the first condition of AO 8862 #4 (as originally proposed by Solomon Gandz [1948: 39]) is worth trying – if only for the reason that both texts state the equality as a confrontation of equals instead of using the normal *kīma*, "as much as".

The situation is shown to the left in Figure 38. The sides are interpreted as broad lines, as they have to be if the statement be meaningful, and in agreement with the model AO 8862. What we are told in obv. 1 is that the hatched rectangle ⊏⊐(*l*,1) and the dotted rectangle ⊏⊐(*w*,1) taken together are as much as the original rectangle ⊏⊐(*l*,*w*). Hence, as we see in the mid-part of the diagram – where a part equal to ⊏⊐(*w*,1) has been removed from

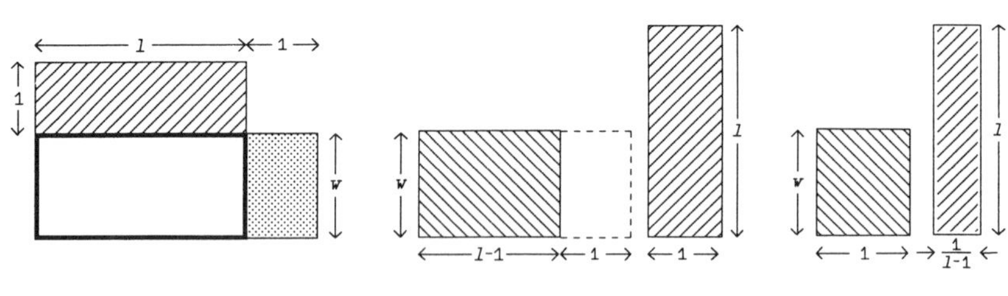

Figure 38. The procedure of AO 6770 #1.

$\sqsubset\!\sqsupset(l,w)$ – the two hatched rectangles $\sqsubset\!\sqsupset(l,1)$ and $\sqsubset\!\sqsupset(l–1,w)$ are equal. If the latter is to be reduced to a rectangle $\sqsubset\!\sqsupset(1,w)$ – and hence be equal to the width itself – it must be reduced in horizontal direction by a scaling factor equal to igi(l–1); submitting the rectangle $\sqsubset\!\sqsupset(l,1)$ to the same scaling factor amounts to constructing a rectangle still with length l but with breadth igi(l–1) instead of 1 – as in the lower right of the figure.

All we need in order to make this correspond to the text is to identify the "step" with l, in agreement with a computation of the area (once it has been "built") as "w steps of l", that is, in accordance with AO 8862 #1–4 – and in agreement with the usage of the series text YBC 4715, obv. I 3, 8, and 13, cf. [TMB, 190*f*], and TMS VII (see presently). One apparent anomaly remains, namely, that an igi is "made hold" and not involved in a "raising" multiplication. This, however, will remain anomalous irrespective of interpretation, since it happens in no other text; in the present reading it is at least meaningful.

This is one of very few Old Babylonian attempts to formulate a general rule instead of letting the rule be implicit in one or more paradigmatic examples; it is also atypical by dealing with an indeterminate problem – the kindred AO 8862 #4, we remember, was made determinate by the extra condition $l+w+\sqsubset\!\sqsupset(l,w) = 9$. It is of course equivalent to the symbolic-algebraic transformations

$$lw = l+w \;\Rightarrow\; w\cdot(l–1) = l \;\Rightarrow\; w = \frac{l}{l–1}\;.$$

The rule is evidently not meant for practical use; since the construction of indeterminate rules was not a normal routine of the Old Babylonian school, we may ask for the purpose of this particular specimen. No definitive answer is at hand. A possibility is that is was a tool for constructing problems similar to AO 8862 #4: for any choice of $l>1$, it allows the determination of the corresponding w (provided l–1 be regular), whence also of the corresponding sum of sides and surface. If more problems of the kind had been known, this would have been less of a gratuitous conjecture.

TMS VII[205]

This text belongs to the same didactic genre as TMS XVI and TMS IX; it teaches how to deal with indeterminate equations of the first degree (which was totally overlooked in the commentary of the original edition).

The text presents us with some interesting extensions of the terminology. Firstly, it makes use of nominal derivatives of gerundive type of the verbs

205 Based on the hand copy and the transliteration, [TMS, pl. 14*f*, 52–55], cf. [von Soden 1964: 48] and [Høyrup 1993].

waṣābum, "to append", and *nasāḫum*, "to tear out". The meaning of *wāṣbum* is approximately "that which shall/should be appended" (in German, "das Hinzuzufügende"); I shall translate it inadequately but concisely as "the appending". Correspondingly, *nāsḫum*, "that which shall/should be torn out" ("das Hinauszureißende") I shall translate "the tearing-out".[206]

Secondly, *alākum*, "to go", is used in a way that corresponds to the multiplication by a.rá, "going steps". In the text TMS VIII (below, p. 188) we shall see, however, that it ranges further, and may also be used about repeated "appending" of the same magnitude. In the present text, the "step" is written logographically, as a.rá. In AO 6770 #1, we remember, syllabic writings of *arûm* were found; the term had hence been taken over in Akkadian as a loanword, on a par with *igûm* (< igi), *igibûm* (< igi.bi) and *basûm* (< ba.si$_8$).

Like TMS XVI, both problems of the present text treat of an uš, "length", and a sag̃, "width", even though they are of the first degree. It will turn out, however, that whereas the role of the two quantities as sides of a rectangle was wholly immaterial in TMS XVI, it may have suggested the heuristics of the solution in the present case.

#1

1. The 4th of the width to the length I have appended, its 7⟨th⟩ to 10 [I have gone,]
4$^{\textit{-at}}$ sag̃ *a-na* uš daḫ 7$^{\text{-}\langle \textit{ti}\rangle}$-*šu* [207] a-na 10 [*al-li-ik*]

2. as much as the accumulation of length and ⟨width⟩. You, 4 posit; 7 [posit;]
ki†*-ma* UL.GAR uš *ù* ⟨sag̃⟩ za.e 4 g̃ar 7 [g̃ar]

3. 10 posit; 5′ to 7 raise, 35′ you see.
10 g̃ar 5 *a-na* 7 *i-ší* 35 *ta-mar*

4. 30′ and 5′ single out. 5′, the step, to 10 raise,
30 *ù* 5 *be-e-er* 5 a.rá† *a-na* 10 *i-ší*[208]

---

206 The construct states *waṣib* and *nasiḫ* demonstrate that we really have to do with nouns; that they are of gerundive type follows from their role in the calculation.

207 In Bruins's interpretation, this *-šu* becomes a Sumerian šu, "hand", which gives rise to a fanciful invention of a heuristic "method of the hand". In later years, this method has made its way into the general literature – thus [Gericke 1984: 25-32], where it constitutes the very introduction to Babylonian algebra. That a *-ti-* has simply been omitted and a 7th is thus meant can be seen from a number of parallel passages (thus line 17 of the present tablet; TMS IX, line 20; VAT 8520, obv. 1, rev. 4, in [MKT I, 346*f*]). The reading is confirmed by the consistency of the text which is obtained.

208 The superscript dagger† still indicates that a reading is changed with respect to the publication of the text which I refer to; cf. note 106.

5. 50′ you see. 30′ and 20′, posit. 5′, the step, to 4, of the fourth of the width,
50 *ta-mar* 30 *ù* 20 ĝar 5 a.rá† *a-na* 4 *re-⟨ba-ti⟩* sağ

6. raise: 20′ you see, 20′, the width. 30′ to 4, of the fourth,
i-ší-ma 20 *ta-mar* 20 sağ 30 *a-na* 4 *re-ba-⟨ti⟩*

7. raise, 2 you see. 2 posit, lengths. 20′ from 20′ tear out,
i-ší 2 *ta-mar* 2 ĝar uš 20 *i-na* 20 zi

8. and from 2, 30′ tear out, 1°30′ you see.
ù i-na 2 30[209] zi 1,30 *ta-mar*

9. From 4, of the fourth, 1 tear out, 3 {...} you see.
i-na 4 *re-ba-ti* 1 zi 3{,20}[210] *ta-mar*

10. Igi 3 detach, 20′ you see. 20′ to 1°30′ raise:
igi 3 *pu-ṭú-⟨úr⟩* 20 *ta-mar* 20 *a-na* 1,30 *i-ší-ma*

11. 30′ you see, 30′ the length. 30′ from 50′ tear out, 20′ you see, 20′ the width.
30 *ta-mar* 30 uš 30 *i-na* 50 zi 20 *ta-mar* 20 sağ

12. Turn back. 7 to 4, of the fourth, raise, 28 you see.
tu-úr 7 *a-na* 4 *re-ba-⟨ti⟩ i-ší* 28 *ta-mar*

13. 10 from 28 tear out, 18 you see. Igi 3 detach,
10 *i-na* 28 zi 18 *ta-mar* igi 3 *pu-⟨ṭú-úr⟩*

14. 20′ you see. 20′ to 18 raise, 6 you see, 6 (for) the length.
20 *ta-⟨mar⟩* 20 *a-na* 18 *i-ší* 6 *ta-mar* 6 uš

15. 6 from 10 tear out, 4 (for) the width. 5′ to 6 [raise,]
6 *i-na* 10 zi 4 sağ 5 *a-na* 6 [*i-š*]*í*

16. 30′ the length. 5′ to 4 raise, 20′ you see, 20′ the ⟨width⟩.
30 uš 5 *a-na* 4 *i-ší* 20 *ta-⟨mar⟩* 20 ⟨sağ⟩

#2

17. The fourth of the width to the length I have appended, [its] seventh
4-*at* sağ *a-na* uš daḫ 7-*ti*[-*šu*]

18. until 11 I have gone, over the [accumulation]
a-di 11 *al-li-ik* ugu† [UL.GAR]

19. of length and width, 5′ it went beyond. You, [4 posit;]
uš *ù* sağ 5 dirig za.e [4 ĝar]

20. 7 posit; 11 posit; and 5′ posit.
7 ĝar 11 ĝar *ù* 5 ĝar

---

209 It appears from the hand copy *either* that the two numbers were at first written together but 30 then deleted and rewritten with distance, *or* that some small wedges marking a separation are written between the two numbers.

210 As observed by Bruins, the scribe has tried to correct this number (which should be 3) but has done so incorrectly. 3°20′ will have been on the scribe's mind as 4·(length+width) – cf. below.

21. 5′ to 7 raise, 3[5′ you see.]
5 *a-na* 7 *i-ší* 3[5 *ta-mar*]

22. 30′ and 5′ posit. 5′ to 1[1 raise, 55′ you see.]
30 *ù* 5 ğar 5 *a-na* 1[1 *i-ší* 55 *ta-mar*]

23. 30′, 20′, and 5′, to tear out, posit. 5′ [t]o 4
30 20 *ù* 5 zi ğar 5 [*a-n*]*a* 4

24. raise, 20′ you see, 20 the width. 30′ to 4 raise:
i-ši 20 *ta-⟨mar⟩* 20 sağ 30 *a-na* 4 *i-ší-ma*

25. 2 you see, 2, lengths. 20′ from 20′ tear out.
2 *ta-mar* 2 uš 20 *i-na* 20 zi

26. 30′ from 2 tear out, 1°30′ posit, and 5′ t⌈o ...]
30 *i-na* 2 zi 1,30 ğar *ù* 5 *a-⌈na* ...]

27. 7 to 4, of the fourth, raise, 28 you see.
7 *a-na* 4 *re-⟨ba-ti⟩ i-ší-ma* 28 *ta-mar*

28. 11, the accumulations, from 28 tear out, 17 you see.
11 UL.GAR *i-na* 28 zi 17 *ta-mar*

29. From 4, of the fourth, 1 tear out, 3 ⌊you] see.
i-na 4 *re-⟨ba-ti⟩* 1 zi 3 ⌊*ta*]-*mar*

30. Igi 3 detach, 20′ you see. 20′ ⌊to] 17 raise,
igi 3 *pu-ṭú-⟨úr⟩* 20 *ta-⟨mar⟩* 20 ⌊*a-na*] 17 *i-⟨ší⟩*

31. 5°40′ you see, 5°40′, (for) the [le]ngth. 20′ to 5′, the going-beyond, raise,
5,40 *ta-⟨mar⟩* 5,40 [u]š 20 *a-na* 5 dirig *i-ší*

32. 1′40″ you see, 1′40″, the appending of the length. 5°40′, (for) the length,
1,40 *ta-⟨mar⟩* 1,40 *wa-ṣí-ib* uš 5,40 uš

33. from 11, accumulations, tear out, 5°20′ you see.
i-na 11 UL.GAR zi 5,20 *ta-mar*

34. 1′40″ to 5′, the going-beyond, append, 6′40″ you see.
1,40 *a-na* 5 dirig daḫ 6,40 *ta-mar*

35. 6′40″, the t[ea]ring-out of the width. 5′, the step,
6,40 *n*[*a*]-*sí-iḫ* sağ 5 a.rá

36. to 5°40′, lengths, raise, 28′20″ you see.
a-na 5,40 uš *i-ší* 28,20 *ta-mar*

37. 1′40″, the appending of the length, to 28′20″ [append⌋,
1,40 *wa-ṣí-ib* uš *a-na* 28,20 [daḫ⌋

38. 30′ you see, 30′ the length. 5′ t[o 5°20′]
30 *ta-mar* 30 uš 5 *a-*[*na* 5,20]

39. raise: 26′40″ yo[u see. 6′40″,]
i-ši-ma 26,40 *t*[*a-mar* 6,40]

40. the tearing-out of the width, from [26′40″ tear out,]
na-sí-iḫ sağ *i-na* [26,40 zi]

41. 20´ you see, 20´ the wid[th.]
20 *ta-mar* 20 sa[g̃]

#1

#1 is the simpler of the two. Lines 1-2 state that a 4th of the width has been "appended" to the length, that the 7th of the outcome has been taken 10 times, and that this amounts in total to the accumulation of length and width. In anachronistic symbols:

$$\tfrac{1}{7}(l+\tfrac{1}{4}w)\cdot 10 = l+w$$

Lines 2–5 now explain the meaning of this equation by extensive "positing", which we may follow in Figure 39 – to which extent this diagram corresponds to the Old Babylonian representation of the transformations must remain an open question.

First the numbers 4, 7, and 10 – divisors and multiplier – are recorded. Then 5´ (the "step" of line 5 – taken at this point to be known) is "raised" to 7, giving 35´, which can be decomposed as 30´+5´ (i.e., $l+\tfrac{1}{4}w$). Next it is "raised" to 10, which gives 50´, decomposable as 30´+20´ (i.e., $l+w$).

So far the text has presented us with a didactic exposition of the numerical foundation of the original equation. Lines 5–11 go on with a similar exposition of the meaning of the first step of the transformation of the equation, which is a multiplication by 4,

$$\tfrac{1}{7}[4l+w]\cdot 10=4\cdot(l+w)\ .$$

We look at the decomposition 35´ = 30´+5´ (*B* in Figure 39): multiplying 5´ by 4 gives 20´, the width; multiplying 30´ yields 3, (4) lengths. Now one width and one length are removed. "Tearing out" a width (20´) from 20´ leaves, literally, *nothing worth speaking about*; "tearing out" a length (30´) from the 4 lengths (2) leaves 1°30´; this is found in line 9 to correspond to 3 (*viz.*, lengths). Multiplying by $\tfrac{1}{3}$=20´ indeed gives 30´, the length; "tearing this out" from 50´ leaves 20´, the width.

The *tu-úr*, "turn back", of line 13 expresses that the explanations are now finished and the procedure is to begin (in the present case, a spatial return

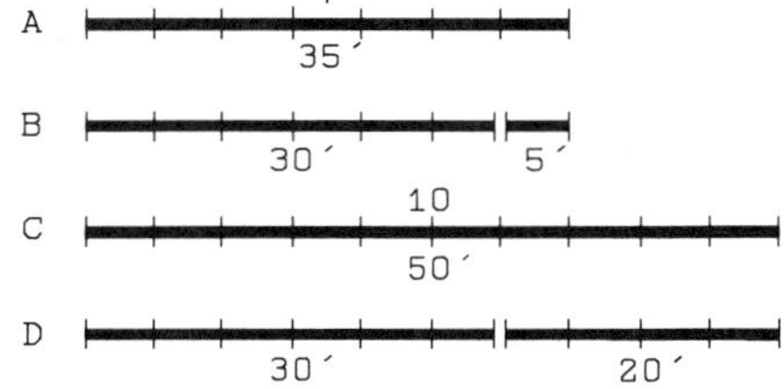

Figure 39. A graphical representation of the steps of TMS VII #1.

appears to be out of the question). After the multiplication by 4 which was already explained, the equation is multiplied by 7, which leads from

$$\tfrac{1}{7}[(4-1)l+(l+w)]\cdot 10 = 4\cdot(l+w)$$

to

$$3l\cdot 10+(l+w)\cdot 10 = 28\cdot(l+w)$$

and hence to

$$3l\cdot 10 = 18\cdot(l+w).$$

Dividing this by 3 ("raising" to igi 3 = 20´) yields

$$l\cdot 10 = 6\cdot(l+w)\ . \qquad (*)$$

The most obvious solution to this indeterminate equation is found if we put the first factors equal, $l = 6$, and the second factors equal, $10 = l+w$, from which w can of course be found as 10–6. This seems in fact to be what happens, either on this arithmetical level or perhaps, as it may be intimated by #2, by imagining the factors as sides of a rectangle (see Figure 40).

Evidently, this is not the only set of solutions to the equation (whence the "(for)" in the translation); in fact, every set $6z$, $4z$ will do. By multiplying by 5´ the text obtains the solution presupposed in the beginning, l = 30´, w = 20´.

The 5´ used here is the "step" from line 4, as indicated in the parallel passage in line 35. There is no compelling reason that precisely this factor should transform a solution obtained from (*) into the set originally thought of – unless the multiplication by igi 3 that leads to (*) is chosen because it gives a preliminary value equal to the number of "steps" going into the intended length. We must therefore assume that this is precisely what was done. This would also explain why the equation is not reduced even further, *viz.*, to $l\cdot 5 = 3\cdot(l+w)$.

#2

This problem too deals with the length and width of a rectangle, and takes the same 7th. This time, however, 11 steps are made, which makes us go 5´ beyond the accumulation of length and width:

$$\tfrac{1}{7}(l+\tfrac{1}{4}w)\cdot 11 = l+w+5´\ .$$

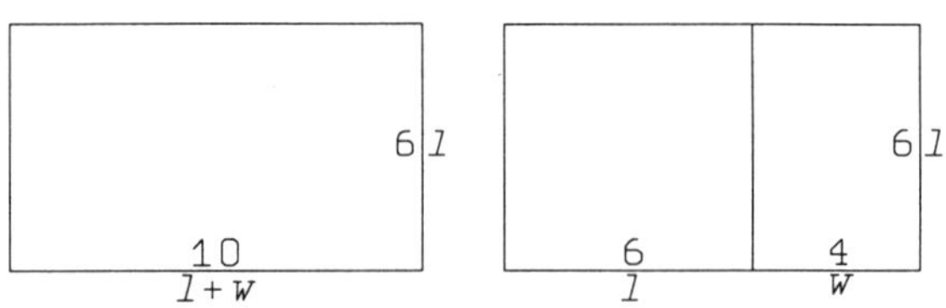

Figure 40. A graphical interpretation of the penultimate step of TMS VII #1.

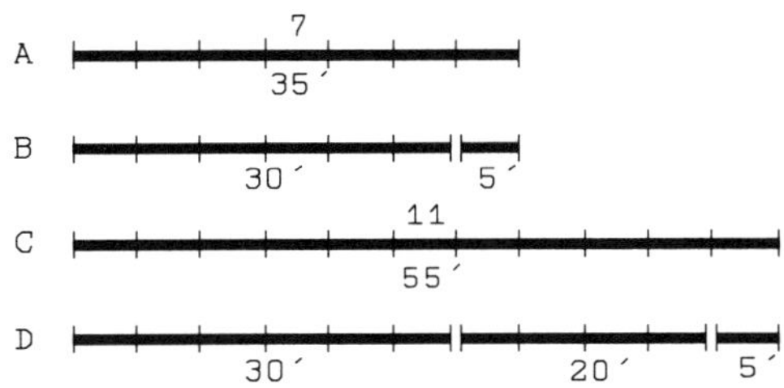

Figure 41. Graphical representation of the equation and the initial transformations of TMS VII #2.

Once again, we may follow what goes on in lines 19 to 23 in a diagram – see Figure 41: lines 23 to 26 explain the multiplication of $(l+\frac{1}{4}w)$ by 4 and the ensuing transformation. The end of line 26 suggests that a further transformation corresponding to

$$\tfrac{1}{7}\,[(4-1)l-5'+(l+w+5')]\cdot 11 \;=\; 4\cdot(l+w+5')$$

is prepared, but damage to the tablet prevents us from knowing precisely how.

In line 27, the procedure starts for good. We may follow it in symbols, from

$$\tfrac{1}{7}\,[(4-1)l-5'+(l+w+5')]\cdot 11 \;=\; 4\cdot(l+w+5')$$

to

$$11\cdot[3(l-\tfrac{1}{3}\cdot 5')]+(l+w+5')\cdot 11 \;=\; 28\cdot(l+w+5')$$

and hence to

$$11\cdot(l-1'\,40'') \;=\; 5°40'\cdot(l+w+5')\ .$$

Already at this stage, 5°40´ is ascribed to the length. The entity to which it corresponds is, however, not the original length l but $\lambda = l-1'\,40''$, where 1´40˝ is explicitly introduced as the "appending of the length".

Concomitantly, 11 will have to correspond to $l+w+5'$. "Tearing out" 5°40´ for λ leaves 5°20´ to correspond to ω ($\lambda+\omega = l+w+5'$), which in line 34 is found to be $(l+w+5)-(l-1'\,40'') = w+(5'+1'\,40'') = w+6'\,40''$, where 6´40˝ is called the "tearing-out of the width". The ease by which this computation is carried out corroborates the conjecture that a geometric or similar intuitively transparent representation is used – cf. Figure 42. In any case, the obvious solutions $\lambda = 5°40'$, $\omega = 5°20'$ are multiplied by the step 5´, giving the

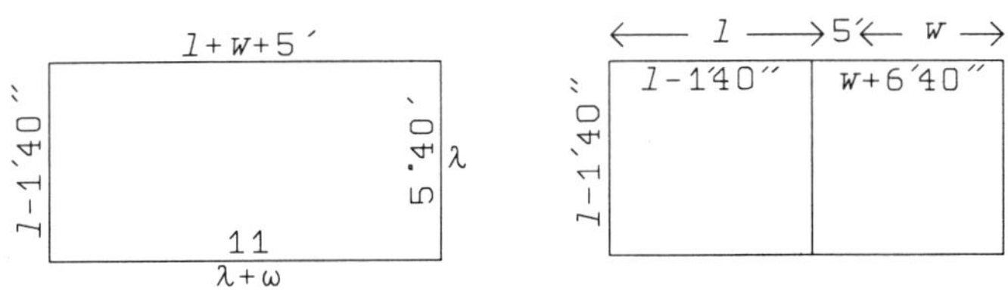

Figure 42. Geometrical interpretation of the solution of TMS VII #2.

preferred solutions $\lambda = 28'20''$, $\omega = 26'40''$. "Appending" and "tearing out" what according to their names *should be* appended and torn out finally gives $l = 30'$, $w = 20'$.

A Concluding General Observation

After its didactic introduction, #1 thus demonstrates how to solve a homogeneous indeterminate first-degree problem. #2, for its part, shows how to reduce an inhomogeneous to a homogeneous problem through a shift of variables made very explicit by the "appending" and "tearing-out".

Obviously, an equation $a \cdot u = b \cdot v + c$ can be brought much more easily to homogeneous form by means of the principles used in our text if either a or b is regular – *viz.*, as $a \cdot u = b \cdot (v + c/b)$ or $a \cdot (u - c/a) = b \cdot v$. The method which is actually followed, which reduces the problem into one in l and $l+w$ (or λ and $\lambda+\omega$), shows that the "accumulation" (UL.GAR, $l+w+5' = \lambda+\omega$) is regarded as an entity of its own; it even seems to betray that this entity is regarded as fundamental, not just introduced *ad hoc* for the construction or solution of certain problems. In TMS VIII we shall find the term dirig, the Sumerian verb corresponding to *watārum*, "to go beyond", treated as a noun and seemingly with a similar proper existence: a noteworthy contrast to the apparent lack of technical terms for average and deviation (see note 197), the respective moieties.

TMS VIII[211]

Even the following text from Susa considers the length and the width of a rectangle. This time, however, the rectangle is indubitably real, since it possesses a surface 10′ (in both #1 and #2; #3 will presumably have dealt with the same rectangle, but too little of it is conserved to prove it). This may make us suspect that the sides, once again, are $l = 30'$ and $w = 20'$. However, the present problems contain no introductory didactic explanation, so we may be supposed not to know.

---

211 Based on the hand copy and the transliteration, [TMS, pl. 16, 58], cf. [von Soden 1964: 48], [Høyrup 1993], [Muroi 1994a]; see [Høyrup 1996b] for my reasons not to believe in Muroi's reconstructions of lines 1, 8, 11, and 17. I use the opportunity to correct a mistake in this review: The sign SAR *is* found in another mathematical Susa text, namely, TMS III, 70; it is written here in normal second-millenium manner, wholly different from the sign read *sar* by Muroi – which makes Muroi's reconstructions of lines 1 and 11 even more suspicious.

The process of "going", we observe, is used in line 1 about repeated "appending".

#1

1. [The surface 10′. The 4th of the width to the width I have appended], to 3 I have gone [... over]
[a.šà 10 4-*at* sağ *a-na* sağ daḫ] *a-na* 3 *a-li-*⌈*ik* ⌈... ... ...⌉? ugu]

2. [the length 5′ went] beyond. You, [4,] of [the fo]urth, as much as width posit. The four[th of 4 take, 1 you see.]
[uš 5 dir]iğ za.e [4 *r*]*e-ba-ti ki-ma* sağ ğar *re-b*[*a-at* 4 *le-qé* 1 *ta-mar*]

3. [1 to 3] go, 3 you see. 4 fourths of the width to 3 app[end, 7 you see.]
[1 *a-na*] 3 *a-li-ik* 3 *ta-mar* 4 *re-ba-at* sağ *a-na* 3 d[aḫ 7 *ta-mar*]

4. ⌈7⌉ as much as length posit. 5′ the going-beyond to the tearing-out of the length posit. 7, of the length, to 4, [⌈of the width?, raise,]
⌈7⌉ *ki-ma* uš ğar 5 dirig *a-na na-sí-iḫ* uš ğar 7 uš *a-na* 4 [⌈sağ? *i-ší*]

5. 28 you see. 28, of the surfaces, to 10′ the surface raise, 4°40′ you see.
28 *ta-mar* 28 a.šà 28 *a-na* 10 a.šà *i-ší* 4,40 *ta-mar*

6. [5′⌋, the tearing-out of the length, to four, of the width, raise, 20′ you see. ½ break, 10′ you see. 10′ make hold,
[5⌋ *na-sí-iḫ* uš *a-na* 4 sağ *i-ší* 20 *ta-mar* ½ *ḫe-pe* 10 *ta-mar* 10 NIGIN

7. [1′40″] you see. 1′40″ to 4°40′ append, 4°41′40″ you see. What is equalside? 2°10′ you se[e].
[1,40] *ta-mar* 1,40 *a-na* 4,40 daḫ 4,41,40 *ta-mar mi-na* íb.si 2,10 *ta-ma*[*r*]

8. [10′ the e]qual (?) to 2°10′ append, 2°20′ you see. What to 28, of the surfaces, may I posit which 2°20′ gi[ves] me?
[10 ⌈s]i$_8$†.si$_8$? *a-na* 2,10 daḫ 2,20 *ta-mar mi-na a-na* 28 a.šà ğar *šà* 2,20 *i-na-*[*di-n*]*a*

9. [5′ posit.] 5′ to 7 raise, 35′ you see. 5′, the tearing-out of the length, from 35′ tear out,
[5 ğar] 5 *a-na* 7 *i-ší* 35 *ta-mar* 5 *na-sí-iḫ* uš *i-na* 35 zi

10. [30′ you] see, 30′ the length. 5′ the length[212] to 4 of the width raise, 20′ you see, 20 the length (mistake for width).
[30 *ta-*]*mar* 30 uš 5 uš *a-na* 4 sağ *i-ší* 20 *ta-mar* 20 {uš} ⟨sağ⟩

#2

11. [The surface 10′.] The 4th of the width to the length (probably erroneously for width[213]) append, to 1 go, (the outcome falls 5′ short of the length[214])

---

212 This could refer to the "length" of the square □(w/4). Both TMS V and TMS VI, indeed, refer to the square side as a "length".

213 Unless the meaning (or the reason for this slip of the stylus) is "append (to the equivalent of the width) along the (direction of the) length".

214 As pointed out by von Soden [1964: 48a], a better interpretation of this (so far nonsensical) passage can only be attained through collation. Since neither the GAR nor the *le-qé* in the next line look as they should on the hand copy, lines 11-12

[a.šà 10] 4-*at* sağ *a-na* uš daḫ *a-na* 1 *a-li-ik a-na* ¿KI/DI [u]š ugu? sağ ¿*i-ší-ma*?

12. [Yo]u, 4, of the fourth, as much as width posit. The fourth of 4 take, 1 you see. To 1 g[o.]
[za].e 4 *re-ba-ti ki-ma* sağ ĝar *re-ba-at* 4 *le-qé* 1 *ta-mar* 1 *a-na* 1 *a-li*[*-ik*]

13. [¿1?]. 4 the counterpart, 4 posit. 1, the going, to 4 append, 5 you see, as much as length posit.
[¿1?] 4 gaba 4 ĝar 1 *ta-lu-ka a-na* 4 daḫ 5 *ta-*⟨*mar*⟩ *ki-ma* uš ĝar

14. [5′ to] the appending of the length posit. 5, of the length, to 4, of the width, raise, 20 you see, 20 surfaces.
[5 *a-na*] *wa-ṣí-ib* uš ĝar 5 uš *a-na* 4 sağ *i-ší* 20 *ta-mar* 20 a.šà

15. [20 to 10′] ⌊raise], 3° 20′ you see. 5′, the appending of ⟨the length⟩, to 4, of the width, raise, 20′ you see.
[20 *a-na* 10] ⌊*i-ší*] 3,20 *ta-mar* 5 *wa-ṣí-ib* ⟨uš⟩ *a-na* 4 sağ *i-ší* 20 *ta-mar*

16. [½ break, 10′ you see.] Make hold, 1′40″ you see. 1′40″ to 3° 20′ append, 3° 21′40″ you see.
[½ *ḫe-pe* 10 *ta-mar*] NIGIN 1,40 *ta-mar* 1,40 *a-na* 3,20 daḫ 3,21,40 *ta-mar*

17. [What is equalside? 1°50′ you see. ~~10′~~] the equal (?), from 1°50′ tear out, 1°40′ you see.
[*mi-na* íb.si 1,50 *ta-mar* 10] ¿si$_8$†.si$_8$†?[215] *i-na* 1,50 zi 1,40 *ta-mar*

18. [Igi 20, of the surfaces, detach , 3′ you see. 3′] to 1°40′ raise, 5′ to 5, of the length,
[igi 20 a.šà *pu-tú-úr* 3 *ta-mar* 3] *a-na* 1,40 *i-ší* 5 *a-na* 5 uš

19. [raise, 25′ you see. 5′, the appending of the length, t]o 25′ append, 30′ you see, 30′ the length.
[*i-ší* 25 *ta-mar* 5 *wa-ṣí-ib* uš *a*]*-na* 25 daḫ 30 *ta-mar* 30 uš

20. 20 , the width
20 sağ

#3

21. [... ... ... as⌋ much as the accumulation of 3 lengths and 4 widths
[... ... ... *ki*⌊*-ma* UL.GAR 3 uš ù 4 sağ

22. [... ... ... yo⌋u see
[... ... ...*t*⌊*a-mar*

---

must presumably be quite damaged (TMS contains no photo of the tablet – cf. note 115).

215 Bruins and Rutten read the first signs in line 8 as *a-di*, and the analogous group here as *le-qé*, none of which make sense. The hand copy agrees acceptably well with the assumption that the same signs are used in both cases. The second sign appears to be a DI; the first is too close to the breaks to be read with certainty, but might be another DI (this possibility is confirmed by Jim Ritter after collation). Though no mathematical standard expression, this makes sense if read si$_8$.si$_8$ or sá.sá~*šāninum*, "that which is equal", corresponding to the conceptualization of the side of a square (concerning the writings sá/si$_8$, see note 45)

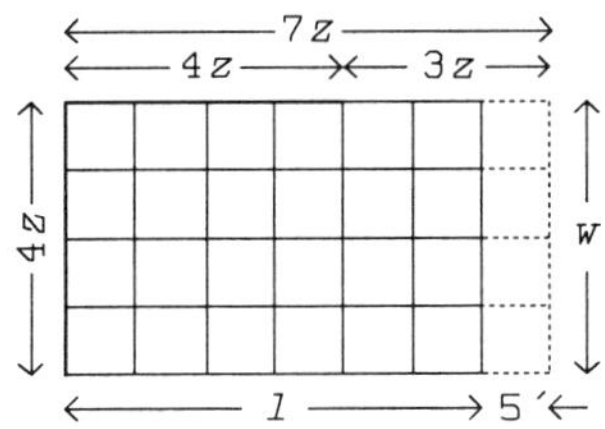

Figure 43. The prolonged and subdivided rectangle of TMS VIII #1.

#1

Here, beyond the surface, we are told that 3 4ths of the width "appended" to the width exceeds the length by 5´ – in symbols

$$w+3\cdot\tfrac{1}{4}w = l+5' \ .$$

In line 2 we are told to "posit" 4 "as much as the width". The "4 fourths" (in grammatical absolute state) in line 3 shows that this means taking $z = \frac{1}{4}w$ as a new unit. Returning to line 2, we see that one 4th (of the width) is then 1 – namely, 1 z. Repeating it 3 times gives 3, for which reason the length *including a "tearing-out" of 5´* will be $7z$ (line 4).

Lines 4-5 calculate the number of surfaces $\square(z)$ contained by $4z$ times $7z$ to be "28 surfaces" – cf. Figure 43. That this, and not "28 the surface", is how the expression is to be read, follows from the use of a "raising" multiplication – if "a surface 28" had been found, the multiplication would have been a *construction*, in the present text presumably represented by NIGIN. "Raising" presupposes that a geometrical configuration (namely, $\square(z)$) is *already there*; only the number of times it is present has to be calculated.

What the text is aiming at is thus a problem of the type "square area minus sides equals number" (minus, because the 28 surfaces correspond to a rectangle which is 5´ longer than the given area). So far, we have obtained

$$28\cdot\square(z)-n\cdot z = 10' \ .$$

This is normalized in the habitual way "28, of the surfaces", being "raised" "to 10´ the surface":

$$\square(28z)-n\cdot(28z) = 28\cdot 10' = 4°40' \ , \qquad (**)$$

As usually, it is only at this point that the text finds the number n of subtracted sides: Since the excess of $28\square(z)$ over the original rectangle has the sides $4z$ and 5´, its area is $4\cdot 5'\cdot z = 20'\cdot z$; therefore, $n = 20'$ (line 6). The inherent transformation can be expressed in the formula $\sqsubset\!\sqsupset(\alpha p,q) = \sqsubset\!\sqsupset(p,\alpha q)$ – a rule which was also used in the corresponding transformation of the "wings" in BM 13901 #14, and which corresponds to the Euclidean "gnomon theorem (see p. 99). Now the procedure of Figure 3 can be used, as shown in Figure 44: 20´ is bisected, "made hold" so as to give as supplementary square with surface $\frac{n}{2}\times\frac{n}{2}$ = 10´^{2} = 1´40˝ (line 7). "Appending" this quadratic surface to the gnomon completes the square on $\sigma-\frac{n}{2}$ as 4°41´40˝ (σ = 28z), by which 2°10´ "is equalside". "Appending" again 10´ (in other texts "that which was made hold", here something like "the equal") where it was broken off gives S = 2°10´+10´ = 2°20´ (line 8).

This is in all respects the standard way to solve a normalized problem "square area minus sides". Next (lines 8-9), z is found as σ/28 (a division which, again, is performed as all divisions by sexagesimally irregular numbers). The length with the extra 5´ is found as 7·z, and the original length l by "tearing out" this 5´, the "tearing-out of the length".

It may be of some interest to compare the path which is actually followed to an alternative which was within reach but not used. For brevity, I describe it in algebraic symbols. The original problem is

$$w\times(\tfrac{7}{4}w-5') = 10'$$

or

$$1°45'\cdot\square(w)-5'\cdot w = 10' .$$

This could be normalized through multiplication by 1°45´ into

$$\square(1°45'\cdot w)-5\cdot(1°45'\cdot w) = 17'30''$$

and then solved with the same procedure as above. In principle this is equivalent to the actual method, since 1°45´ is just as good a number as 28. Furthermore, it bypasses the apparent detour over an intermediate variable z ("the fourth of the width"). That this structurally simpler method is nonetheless avoided demonstrates that the text, though not provided with a separate didactic explanation, is still meant to function at a level where insight in the meaning of what goes on is important (the alternative way might be followed at the level where training of methods *beyond* the intuitively meaningful was possible and aimed at – most of the mathematical series texts belong to this

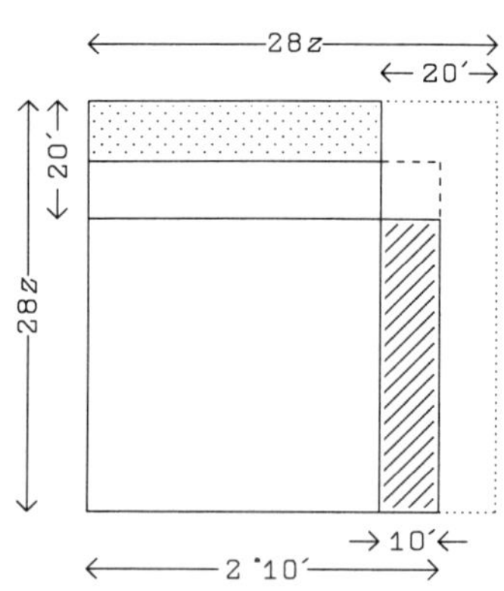

Figure 44. The solution of the normalized equation of TMS VIII #1.

type). A rectangle containing 7 times 4 small squares is, after all, easier to grasp visually than one containing 1,45 times 1 (whether this be understood in the proper order of magnitude, as $1\frac{3}{4}$ times 1, or as 105 times 60).

#2

The appeal to visual insight is made more explicit in #2, which is otherwise a close parallel to #1. This time, we only go once with the 4th of the width, which makes us fall short of the length by 5´[216]. The area is still 10´. In symbols thus

$$w+1\cdot\tfrac{1}{4}w = l-5' \qquad \square(l,w) = 10' \ .$$

Once again, the fourth of the width (z) is taken as the side of an auxiliary square. In spite of some missing signs in the beginning of line 13 the "counterpart 4" seems to tell us that a square of 4 (namely, 4 fourths of the width, as in line 3) times its "counterpart" 4 is drawn (see Figure 45) and "1 the going" then "appended" to the length, giving a rectangle of 5·4 small squares ("raising" again), which fills the original rectangle apart from the dotted strip of 5´ times $4z$.

This time, the problem in z is thus of the type "square areas plus sides equal number". *Mutatis mutandis*, everything runs as before from this point onwards[217].

We have already encountered a similar though homogeneous subdivision into smaller squares in VAT 8390; even there (obv. I 20), the number of these squares was found as a "raising", cf. p. 64); this was the reason, it may be remembered, that I decided to interpret the method of BM 13901 #10 (which makes use of "holding") as a genuine case of false position (cf. p. 60). The present text, in contrast, is close in spirit to the *makṣarum* or "bundling" method and to the use of a reference body in the third-degree problems of BM 85200 + VAT 6599.

---

216 That this must be the meaning of the final part of line 11 is clear from the following. On the whole, the signs to be read on the hand copy make no obvious sense.

217 Since the number of small squares is now regular (*viz.*, 20), we must presume the final division to be performed as a "raising" to the igi. This restitution will also fit into the broken part of line 18, and it is indeed suggested in the hand copy in TMS. Nonetheless, Bruins claims in the commentary (p. 62) that "le scribe se demande de nouveau: par quoi faut-il multiplier 20° pour obtenir 1°40´ et il voit que c'est 5´". Dividing in this manner by a number belonging in the standard table of reciprocals would be totally unprecedented. Comparison with line 8, shows, moreover, that the phrase in question could not be fitted into the lacuna.

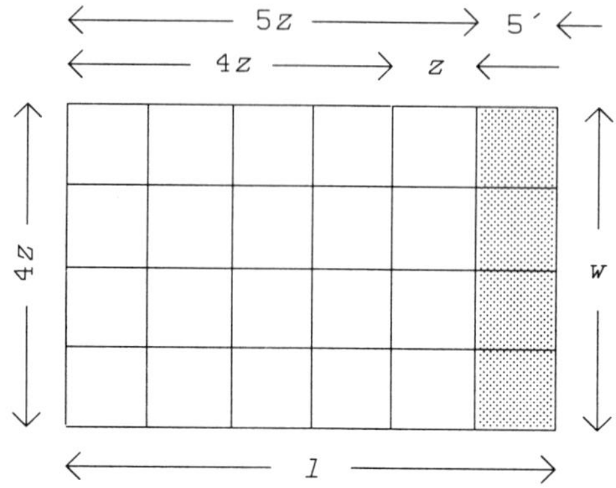

Figure 45. The situation of TMS VIII #2.

TMS XIX[218]

From a modern point of view, this tablet confronts us with an almost paradoxical expression of the primacy of the configuration. It contains two problems, both dealing with a rectangle provided with a diagonal; the first is solved by means of a single false position, the second gives rise to an equation of the eighth degree (solved as a bi-biquadratic). The tablet is exceptionally well preserved.

In both cases, I have chosen to interpret the numbers as adequate for the school yard, not for real fields.

Obv.

#1

1. The width, (compared) to the length, the fourth let it be smaller. 40´ {...} the diagonal.
   saḡ *a-na* uš *ri-ba-ti li-im-ti* 40 {*šà*}[219] BAR.TA
2. Length and width what? You, 1 (for) the length posit, 1 the counterpart posit.
   uš *ù* saḡ *mi-nu* za.e 1 uš ḡar 1 gaba ḡar
3. 15´, of the fourth, from 1 tear out, 45´ you see.
   15 *ri-ba-ti i-na* 1 zi 45 *ta-mar*

---

[218] Based on the hand copy and the transliteration in [TMS, pl. 28*f*, pp. 101–103]. For impenetrable reasons, the two problems are referred to as "C" and "D" in this edition.

[219] Since this unusual *šà* is omitted from the quotation in obv. 7, it should probably be left out. If it is included, the translation is changed into "40´, that of the diagonal".

4. 1 as much as the length posit, 45´ as much as the width posit. 1, the length, make hold, 1 you see.
1 *ki-ma* uš ğar 45 *ki-ma* sağ ğar 1 uš NIGIN 1 *ta-mar*

5. 45´, the width, make hold, 33´45´´ you see. 1 and 33´45´´
45 sağ NIGIN 33,45 *ta-mar* 1 *ù* 33,45

6. accumulate, 1°33´45´´ you see. What is equalside? 1°15´ is equalside.
UL.GAR 1,33,45 *ta-mar mi-na* íb.si 1,15 íb.si

7. Since "40´ the diagonal", it is said to you, igi 1°15`, of the diagonal, detach,
aš-šum 40 BAR.TA *qa-bu-ku* igi 1,15 BAR.TA *p[u]-ṭú-⟨úr⟩*

8. 48´ ⟨you see⟩, 48´ to 40´ the diagonal which is said to you, raise,
48 ⟨*ta-mar*⟩ 48 *a-na* 40 BAR.TA *šà qa-bu-ku i-ší*

9. 32´ you see. 32´ to 1, the length which you have posited, raise,
32 *ta-mar* 32 *a-na* 1 uš *šà* ğar *i-ší*

10. 32´ you see, 32´ the length. 32´ to 45´, the width which you have posited,
32 *ta-mar* 32 uš 32 *a-na* 45 sağ *šà* ğar

11. raise, 24´ you see, 24´ the width.
i-ší 24 *ta-mar* 24 sağ

Rev.

#2

1. A diagonal. 20´ the surface {...}. The length I have made hold, the surface (to) the length I have lifted: together with the diagonal I have made hold:[220]
BAR.TA 20 a.šà {ŠÀ}[221] uš NIGIN a.šà uš *i-la-am-ma* ki BAR.TA NIGIN-*ma*

2. 14´48´´53´´´20´´´´. Length, width, and diagonal what?
14,48,53,20 uš sağ *ù* BAR.TA *mi-nu*

3. You, 20´ the surface ma[ke hold], 6´40´´ you see, may your head hold. Turn back:
za.e 20 a.šà N[IGIN] 6,40 *ta-mar re-eš-ka li-ki-il tu-úr-ma*

4. 14´48´´5[3´´´ 20´´´´ make hold$_{\mathrm{I}}$, 3´39´´[28´´´]44´´´´ 26$^{(5)}$40$^{(6)}$ you see.[222]

---

220 Both the construction with ki~*itti*, "together with", and the *takīltum* of rev. 10 support the reading of NIGIN as a logogram for *šutakūlum*, "to make hold", and not only as an ideogram with the same technical implications.

221 Apparently erased by the scribe, as he discovered to have repeated a sign.

222 This number should have been 3´39´´28´´´43´´´´27$^{(5)}$24$^{(6)}$26$^{(7)}$40$^{(8)}$; it is partially destroyed but can be reconstructed from its reappearance in line 7 (the lacuna is only large enough to contain the number 28). Two errors appear to have been committed: first, 43,27 has become 44,26; next, the repetition of "26" (in both cases preceded by "4") has made the calculator change "44 26 24 26 40" into "44 26 40". Since the number that is produced is used further on, even the latter error must have been committed by the original calculator, which once more shows that the number is found on a separate device and then copied (cf. above, p.

14,48,5[3,20 NIGIN$_1$ 3,39,[28],44,26,40 *ta-mar*

5. ½ of 6´40˝ which your head holds break, 3´20˝ you see.
½ 6,40 *šà re-eš-⟨ka⟩ ú-ki-[lu] ḫe-pe* [3],20 *ta-mar*

6. 3´20˝ make hold, 11˝6‴40⁗ you see. 11˝6‴40⁗
3,20 NIGIN 11,6,40 *ta-mar* 11,6,40

7. to 3´39˝28‴44⁗$26^{(5)}$[$40^{(6)}$][223] append,
a-na 3,39,28,44,26,[40] daḫ

8. 3´50˝36‴43⁗$34^{(5)}26^{(6)}40^{(7)}$[224] you see. What [is equals]ide?
3,50,36,43,34,26,40 *ta-mar mi-na* [íb.s]i

9. 15´11˝6‴40⁗ is equalside.[225] From 15´11˝6‴40⁗
15,11,6,40 íb.si *i-na* 15,11,6,40

10. 3´20˝, the made-hold, tear out, 11´51˝6‴40⁗ you see. What is equalside?
3,20 *ta-ki-il-ta* {2}[226] zi 11,51,6,40 *ta-mar mi-na* íb.si

11. 26´40˝ is equalside. Igi 26´40˝ detach, 2°15 you see.
26,40 íb.si igi 26,40 *pu-ṭú-úr* 2,15 *ta-mar*

12. 2°15´ to 6´40˝, the made-hold which you head holds, raise,
2,15 *a-na* 6,40 *ta-ki-il-ti šà re-eš-ka ú-ki-lu i-ší*

13. 15´ you see. What is equalside? 30´ is equalside, 30´ the width.
15 *ta-mar mi-na* íb.si 30 íb.si 30 sağ

---

73). The first error – the misplacement of a single unit in a wrong order of magnitude – implies that numbers were represented as collections of units in the calculational device, in the manner of calculi placed on a counting board.

223 [TMS, 103] gives the number as "3.39.28.44.26.24.[26.40]"; however, according to the hand copy only "3.39.28.44.26" is legible, and the lacuna has space for only one sexagesimal place – i.e., exactly for the number found in line 4. Bruins has evidently reconstructed with an eye to the correct value, overlooking that the correct value of the square root found in line 9 can be derived from the known end result (and certainly is – try to extract manually the square root of an 8-place sexagesimal!) and not from the intermediate results and therefore is no proof that the tablet is copied from a correct original.

224 In consequence of the errors in line 4 as repeated in line 7, this should have been 3´50˝35‴24⁗$26^{(5)}40^{(6)}$; once again, a unit seems to have been misplaced, namely, in the third instead of the fourth place. Moreover, however, two tens have been added to the fourth place, and an extra "34" has been inserted after "43". It should be observed that the resulting number is not used further on; it is therefore not to be excluded that a copyist's error (or attempts to correct a recognized error) has been superimposed upon an original calculator's error.

225 This number is correct but not the square root of 3´50˝36‴43⁗$34^{(5)}34^{(6)}26^{(7)}40^{(8)}$. Evidently it has been found from the known solution.

226 According to [TMS, 103 n.3], the scribe has tried to erase "*il-ta* 2". The hand copy does not show this, only erasure or damage of *il-ta*, but since this tablet was not mislaid, Bruins may have collated. In any case an erasure of *2* makes sense, unless the text intended to refer to "3,20 the made-hold (number) 2".

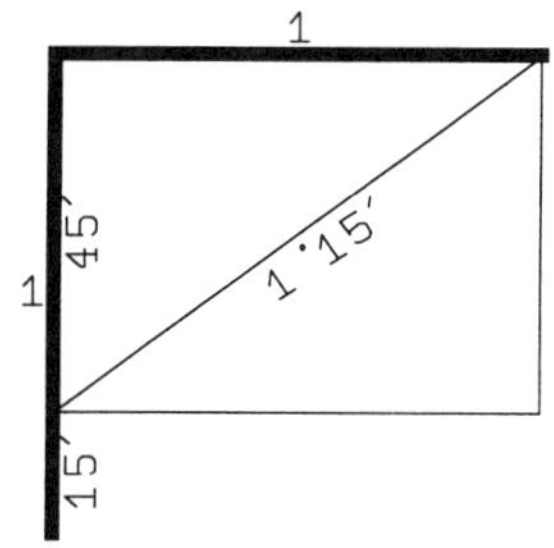

Figure 46. The false position of TMS XIX #1.

#1

After stating that the width falls short of the length by one fourth of the latter, and that the diagonal is 40´, the text prescribes that 1 and its "counterpart" be "posited"; in the next step 15´, representing the fourth, is "torn out" from one of the copies, which makes clear that 1 and the "counterpart" are located as usual, as sides of a right corner – see Figure 46. The length is thus taken to be 1, with the consequence that its fourth is 15´. Presupposing this false position, the width is therefore 1–15´ = 45´, and the diagonal (found via what I shall call the "Pythagorean rule", since the term "theorem" would be a misnomer[227]) 1°15´. We notice that the rule "makes hold" and asks what "is equalside", which implies that the rule is based on geometrical squares, precisely as the Greek theorem – and as we would anyhow expect.

Now the diagonal was actually said to be 40´ and not 1°15´, for which reason all dimensions of the figure have to be reduced – namely, by the factor 32´, obtained by "raising" igi 1°15´ to 40´. "Raising" this factor to the positions 1 and 45´ for length and width, respectively, we find their true values to be 32´ and 24´.

#2

#2 introduces its topic as (something which is characterized by possessing) a diagonal, that is, a rectangle with calculable diagonal. Its surface is said to be 20´. Next, the square □(l) on the length is constructed and then lifted to the length – which means that the volume of the cube ⧉(l) is found. Finally, a

227 Even "Pythagorean" is of course a chronological misnomer, but so blatant that is should cause no problems. The name "diagonal rule" is truer to Babylonian thought, but has the disadvantage of not being immediately meaningful to modern ears.

rectangle is constructed, whose sides are this volume and the diagonal, and whose surface is found to be $14'48''53'''20''''$.

The procedure starts by constructing the square on the area A, which is later treated as the rectangle on the squares on length and width, as in BM 13901 #12:[228]

$$\square(A) = \square\big(\sqsubset\!\sqsupset[l,w]\big) = \sqsubset\!\sqsupset\big(\square[l],\square[w]\big) = 6'40''$$

Next, the square on the rectangle contained by volume and diagonal is constructed, and its surface found to be $Q = 3'39''28'''43''''27^{(5)}24^{(6)}26^{(7)}40^{(8)}$.[229] The steps of lines 5–10 show that they solve a "square-and-sides" problem

$$\square(S)+n\cdot S = Q\ , \qquad (*)$$

where $n = 6'40'' = \square\big(\sqsubset\!\sqsupset[l,w]\big)$; what follows shows S to be $\square\big(\square[l]\big)$. This presupposes the following transformations of Q:

$$Q = \square\big(\sqsubset\!\sqsupset[\text{⧉}(l),d]\big) = \sqsubset\!\sqsupset\big(\square[\text{⧉}(l)],\square(d)\big) = \sqsubset\!\sqsupset\big(\square[\text{⧉}(l)],\square[l]+\square[w]\big)$$
$$= \sqsubset\!\sqsupset\big(\square[\text{⧉}(l)],\square[l]\big)+\sqsubset\!\sqsupset\big(\square[\text{⧉}(l)],\square[w]\big)\ .$$

Further (in this or some similar formulation), the two members are transformed:

$$\sqsubset\!\sqsupset\big(\square[\text{⧉}(l)],\square[l]\big) = \sqsubset\!\sqsupset\big(\square[\square(l)],\square[\square(l)]\big) = \square(S)$$

and

$$\sqsubset\!\sqsupset\big(\square(\text{⧉}(l)),\square(w)\big) = \square\big(\sqsubset\!\sqsupset[\text{⧉}(l),w]\big) = \square\big(\sqsubset\!\sqsupset[\square(l),\sqsubset\!\sqsupset(l,w)]\big)$$
$$= \square\big(\sqsubset\!\sqsupset[\square(l),A]\big) = \sqsubset\!\sqsupset\big(\square[\square(l)],\square[A]\big) = n\cdot S\ ,$$

where $S = \square\big(\square[l]\big)$ and $n = \square(A)$. Thereby we are led to the equation (*). The transformations can be argued from the rules $\sqsubset\!\sqsupset(\alpha p,q) = \sqsubset\!\sqsupset(p,\alpha q)$ (see p. 191), $\square(\alpha p) = \square(\alpha)\cdot\square(p)$, and similar rules where the "raisings" are replaced by rectangularizations or cube formations.

In rev. 11, $26'40''$ is seen to be "equalside" by S – in other words, $\square(l) = 26'40''$; instead of finding the length itself, the square on the surface $\square(A)$ is divided (through igi-division) by $\square(l)$, which gives $\square(w) = 15'$, whence finally $w = 30'$. Though asked for, neither the length ($40'$) nor the diagonal ($50'$) are stated explicitly.

---

228 It would certainly be easier to follow the operations if rectangles and squares were replaced by numerico-algebraic products and powers, and the reader is welcome to do so; this, however, will only show why the solution is correct (apart from computational or copying errors), and not illustrate the kind of conceptual difficulties which any calculator before the advent of symbolic algebra had to circumvent in order to solve a problem like the present one.

229 As we have seen, a wrong figure is actually given here and twice more, whereas the "equalside" found in rev. 9 is correct.

If it had been quite isolated, the present problem would have been nothing but an impressive curiosity. Together with BM 13901 #12, on the other hand, it is evidence of a systematic interest in use of the algebraic technique for "nesting", taking square and rectangular surfaces (and here even a cube) as sides of new rectangles or squares.

If it had been isolated, we might also have been led to think that devising it – that is, discovering that a problem of this type is solvable – was even more impressive than solving it, but we would have had little basis for guessing *how* it was devised. However, the assumption that systematic work on "nested areas" was undertaken allows us to see how the present problem would arise as the ultimate step of a sequence.

At first, of course, the Pythagorean rule states that $\square(d) = \square(l)+\square(w)$. This can be used to see that

$$P = \square(⊏⊐[l,d]) = ⊏⊐(\square[l],\square[d]) = ⊏⊐(\square[l],\square[l]+\square[w])$$
$$= \square(\square[l])+⊏⊐(\square[l],\square[w])$$

If P (or $\sqrt{P} = ⊏⊐(l,d)$) and w are given, this gives rise to a quadratic equation in $\square(l)$, that is, to a biquadratic in l. Further experiments along the same lines might lead to the discovery how a similar problem where A and not w is given might be constructed by further nesting and ingenious transformations – that is, precisely to something like TMS XIX #2.

From the terminological point of view, the use of the verb *elûm*, ("to be/become/make high", standard translation in this function "to lift"[230]) for the formation of the cubic volume from base is interesting. For all other multiplications belonging to this class, the text uses the verb *našûm*, "to raise". The choice of a synonym which has *not* become a technical term (which always entails some loss of concrete connotations) must be meant to emphasize that a real construction of a volume is intended, no mere multiplication. The nesting of area and volume formations must therefore not be interpreted as an indication that the geometrical conceptualizations have been replaced by a purely arithmetical understanding of the problems; a rectangle with a cube as one of its sides is clearly no ordinary rectangle; nor is it, thus we are told by this terminological choice, a mere product of numbers. In their search for striking new problem types, the Babylonian calculators were venturing into abstract geometries which were more related in spirit to the experiments of the nineteenth century, from *Ausdehnungslehre* and quaternions to *Geometrie der Lage*, than to the arithmetization of geometry brought about by analytical geometry (the immense difference of level notwithstanding).

A less momentous terminological observation concerns the term *takīltum*, the "made-hold", as it occurs in rev. 10 and rev. 12. The first instance corresponds to the examples we have seen so far, and is a parallel to the identification by means of the

230 When used to announce a result, I translate it "*n* comes up".

relative clause "which you have made hold". The other concerns a quantity which "you have made your head hold", and is indeed followed by the explanation "which you head holds". This could mean that the term is an open-ended noun that could be used freely about any entity that has been caused to "hold" or to "be held"; alternatively, the second instance is an instance of improper usage or a (voluntary or involuntary) pun, which then calls for the explanatory relative clause. So much is sure that the second occurrence does not mean that a *takīltum* is always an entity that is held by the head, since the large majority of occurrences (including that of rev. 10) do not concern such entities – *pace* von Soden [1964: 49] and others.

YBC 4668, Sequence C, #34, #38–53[231]

What follows is an excerpt from a series text; it is even more compact in style than YBC 4714 (above, p. 111), which is only possible because the format is highly standardized, and because the solution does not vary from one problem to the next. For the same reason, the solution $l = 30$, $w = 20$ is not given for every problem but only at the beginning of sequences.

What we see here is a sub-sequence of 16 problems, together with an earlier problem where the two key parameters are explained somewhat less concisely.

The cores of these key parameters are igi.te.en *šá* sag̃ uš.šè and igi.te.en *šá* uš sag̃.šè. The Sumerian expression igi.te.en (with the corresponding Akkadian loanword *igitēnum*) means "fraction" or "part", as in the phrase *ina* igi.te.en u_4 "during a fraction of the day" (YBC 4996, obv. II 16). In non-spatial and non-temporal contexts, the terminative suffix .še means "to", "as regards", "concerning", and "because of, for the sake of" [SLa § 198]. igi.te.en *šá* sag̃ uš.šè thus means *that fraction which* uš/*the length is of* sag̃/*the width*. In order to render the structure of the phrase I have translated it as "the fraction, of the width, concerning the length". In symbols, it becomes a simple $^l/_w$. Beyond the normal principle of conformality, the purpose of the clumsy translation is to make clear how difficult it will have been to handle this concept in the absence of a technical notion of general fractions or ratios understood as numbers.

The key parameters themselves are more complex, namely, igi.te.en *šá* sag̃ uš.šè uš.šè íl and igi.te.en *šá* uš sag̃.šè sag̃⟨.šè⟩ íl. Here the second terminative suffix .šè is intended spatially, as "to", and the phrases thus mean "the fraction, of the width, concerning the length, to the length raised", and "the fraction, of the length, concerning the width, ⟨to⟩ the width raised" – in symbols $(^l/_w)\cdot l$ and $(^w/_l)\cdot w$, which we may designate for simplicity L and W. This abbreviation is not quite unfaithful to the way the parameters appear after

231 Based on the transliteration in [MKT I, 431–432]. With parallel in YBC 4712, rev. I 19–III 6.

their first introduction, namely, as *šá* uš íl, "that which to the length (is) raised", and *šá* sağ íl, "that which to the width (is) raised" (with variations in #46). L is thus the length increased by a scaling factor $\alpha = l/w$, $L = \alpha l$, whereas W is the width decreased by the same scaling factor, $W = \alpha^{-1}w$.

Rev. III

#34

4. The surface, 1 ešè
a.šà 1(ešè)$^{\lceil \text{iku} \rceil}$

5. The fraction, of the width, [concerning the length]
igi.te.en *šá* sağ [uš.šè]

6. to the length raised, 45.
uš.šè íl [45]

7. The fraction, of the length, [concerning the w]idth,
igi.te.en *šá* uš s[ağ.šè]

8. ⟨to⟩ the width raised: 1⌈3° 20´]
sağ⟨.šè⟩ íl-*ma* 1⌈3.20]

9. its length, width wha[t]?
uš sağ.bi en.n[am]

[...]

#38

19. The 19th p[ar]t of (that which) that which [to the length (is) raised]
igi 19 [ğá]l *šá* [uš íl]

20. over (that which to) the wi[dth (is) raised, goes beyo]nd
[ug]u s[ağ íl diri]g

21. (to that) which to the length (is) raised, appended, 46°40´.
šá uš íl daḫ 46,40

#39

22. (In) steps 2 repeated, appended, 48° 20.
a.rá 2 ^{e}tab daḫ 48.20

#40

23. Torn out: 43° 20´.
ba.zi-*ma* 43,20

#41

24. (In) steps 2 repeated, torn out, 41°40´.
a.rá 2 ^{e}tab zi 41,40

#42

25. (To that) which to the width (is) raised, appended: 15.
šá sağ íl daḫ-*ma* 15

#43

26. (In) steps 2 repeated, appended: 16°40´.
a.rá 2 ^{e}tab daḫ-*ma* 16,40

#44

27. Torn out: 11°40´.
ba.zi-*ma* 11,40

#45

28. (In) steps 2 repeated, torn out, 10.
a.rá 2 ^{e}tab zi 10

#46

29. The surface, 1 ešè
a.šà 1(ešè)^{iku}

30. The 7th part of that which (to) the length, (of the) [wi]dth, (is) raised,
igi 7 ğál *šá* uš [sa]ğ íl

31. (and) that which (to) the width, (of the) length, appended, 53°20´.
šá sağ uš íl daḫ 53,20

#47

32. (In) steps 2 repeated: 1\`1°40´.
a.rá 2 ^{e}tab daḫ-*ma* 1,1,40

#48

33. Torn out: 36°40´.
ba.zi-*ma* 36,40

#49

34. (In) steps 2, torn out, 28°20´.
a.rá 2 ^{e}tab zi 28,20

#50

35. (To that) which to the width (is) raised,
šá sağ íl

36. appended: 21\`40´.
daḫ-*ma* 21,[4]0

#51

37. (In) steps 2 repeated, appended: 30.
a.rá 2 ^{e}tab daḫ-*ma* 30

#52

38. Torn out: 5.
ba.zi-*ma* 5

#53

39. (In) steps 2 repeated:
a.rá 2 ^{e}tab-*ma*

40. 3°20´ it went beyond.
3,20 dirig

#34, as we see, states the surface $\sqsubset\!\sqsupset(l,w)$ together with the values of L and W. We have no direct information about how this was meant to be solved, but the most obvious way – also to a Babylonian calculator – would be to observe that the ratio between L and W is the cube of the scaling factor, $L/W = {}^{l}\!/\!_{w}\cdot\alpha\cdot(\alpha^{-1})^{-1} = \alpha^{3}$. Once α has been found as the cube root of L/W, the rectangle can be transformed into a square – either $\alpha\sqsubset\!\sqsupset(l,w) = \sqsubset\!\sqsupset(l,\alpha w) = \square(l)$ or $\alpha^{-1}\sqsubset\!\sqsupset(l,w) = \square(w)$ – and l or w be found as its "equalside". Alternatively, l can be determined as $\alpha^{-1}\cdot L$, and w correspondingly.

All problems in the sequence #38–53 deal with the same entities L and W. Implicitly given is $\sqsubset\!\sqsupset(l,w) = 1$ ešè $= 10$\` sar, together with a linear equation in L and W – in #38 thus, in symbolic translation,

$$\tfrac{1}{19}\cdot(L-W)+L = 46°40' .$$

From the rule $⊏⊐(\alpha p,q) = ⊏⊐(p,\alpha q)$ it follows trivially that $⊏⊐(L,W) = ⊏⊐(l,w)$. The problem regarded as a problem in L and W is therefore of the same kind as that to which TMS IX #3 is reduced when reformulated in terms of $\lambda = l+1$ and $\omega = w+1$ (see p. 94). Familiar methods would therefore reduce all problems of the sequence to the problem of #34. From that point of view there is no reason to discuss them any further.

It is worth observing, however, that the sequence consists of 2^4 members. This is no accident: the sequence is indeed constructed as a four-dimensional cartesian product, in the following way:

- $\tfrac{1}{19}\cdot(L-W)$ may be replaced by $\tfrac{1}{7}\cdot(L+W)$.
- The second member L may be replaced by W.
- The first member may be subtracted instead of added.
- The first member may be taken twice instead of once.

This systematic spirit was no recent invention in the Old Babylonian age, nor a particularity of mathematical thought. Already the "Profession List", a lexical list from the later fourth millennium – used, we may presume, to teach both the correct writing of occupational titles and the organization of the social hierarchy – is organized in a two-dimensional pattern, one dimension being the variety of occupations, the other being the three-level hierarchy (leader, foremen, common workers) – see [Nissen 1974: 13*f*]. The series and other catalogue texts and the way they are constructed, and in particular the present sequence, can thus be seen as the result of the encounter between the new mathematical genre of "algebra" and the Sumerian school with its age-old fascination with bureaucratic order.

YBC 4713 #1–8[232]

This sequence from yet another series text presents us with a new way of going beyond the too familiar. The problems remain within the standard representation and treat of a rectangle with given area. However, instead of supplying an extra linear condition on the length and the width it states the value of certain multiples of the length and of the width, in which, however, the multipliers or coefficients themselves are unknown. Finally, a linear condition is imposed upon the two unknown coefficients. In modern terms, each problem thus consists of a set of three equations of the second degree and one linear equation, from which four unknowns have to be determined.

The tablet YBC 4713 contains the first 34 problems of YBC 4668, different from these only by containing a few more explanatory lines, and by

---

232 Transliteration [MKT I, 422].

being less damaged. Details in the formulation of YBC 4668 seem to imply that this latter tablet was made from an original series containing YBC 4713, a missing tablet, YBC 4712, and yet another missing text.

Obv. I

#1

1. The surface 1 ešè
a.šà 1(ešè)^{iku}

2. The length repeated: 2`30.
uš ^{e}tab-*ma* 2,30

3. The width repeated: 1`20.
sağ ^{e}tab-*ma* 1,20

4. That which repeated, accumulated: 9.
šá ^{e}tab ğar.ğar-*ma* 9

5. Its length, width, what?
uš sağ.bi en.nam

6. 30 nindan the length, 20 nindan the width
30 nindan uš 20 nindan sağ

#2

7. The surface 1 ešè
a.šà 1(ešè)^{iku}

8. The length repeated: 2`30.
uš ^{e}tab-*ma* 2,30

9. The width repeated: 1`20.
sağ ^{e}tab-*ma* 1,20

10. The repeating of the length over the repeating of the width,
uš ^{e}tab ugu sağ ^{e}tab

11. 1 it goes beyond. Length and width what?
1 dirig uš sağ en.nam

#3

12. The half of the repeating of the length
šu.ri.a uš ^{e}tab

13. and 1°30′, the repeating of the width.
ù 1,30 sağ ^{e}tab

#4

14. $\frac{2}{3}$ of the repeating of the length
$\frac{2}{3}$ uš ^{e}tab

15. and 40′, the repeating of the width.
ù 40 sağ ^{e}tab

#5

16. The third part of (that which) the repeating of the length
igi 3 ğál uš ^{e}tab

17. over the repeating of the width goes beyond,
ugu sağ ^{e}tab dirig

18. (to) the repeating of the length appended: 5°20′.
uš ^{e}tab daḫ-*ma* 5,20

#6

19. (In) steps 2, repeated,
a.rá 2 ^{e}tab

20. (to) the repeating of the length appended, 5°40´.
uš ^{e}tab daḫ-*ma* 5,40

#7

21. (From) the repeating of the length, torn out: 4°40´.
uš ^{e}tab ba.zi-*ma* 4,40

#8

22. (In) steps 2 repeated,
a.rá 2 ^{e}tab

23. (from) the repeating of the width torn out: 4°20´.
uš ^{e}tab ba.zi-*ma* 4,20

The above problem statements all deal with the sides of a rectangle with surface 10`; we are told that the length when "repeated" a certain number of times (say, α) gives 2°30, and the width when "repeated" a certain number of times (say, β) gives 1°20. From this we already guess the solutions, l = 30, w = 20, α = 5, β = 4. This guess should not disturb us, these values will merely be *known* to us, as they were known to the moderately trained Babylonian student; they are clearly not data. But for us as for him, they may help interpret formulations whose conciseness has made them too ambiguous or obscure.

The last datum, all that is needed to solve the problem, is a linear equation in α and β, which varies from problem to problem. The conditions may be expressed as follows in symbolic translation, with implied reduced form:

1. $\alpha+\beta = 9$
2. $\alpha-\beta = 1$
3. $\frac{1}{2}\alpha+1°30' = \beta \; (\Rightarrow 2\beta-\alpha = 3)$
4. $\frac{2}{3}\alpha+40' = \beta \; (\Rightarrow 3\beta-2\alpha = 2)$
5. $\alpha+\frac{1}{3}(\alpha-\beta) = 5°20' \; (\Rightarrow 4\alpha-\beta = 16)$
6. $\alpha+2\cdot\frac{1}{3}(\alpha-\beta) = 5°40' \; (\Rightarrow 5\alpha-2\beta = 17)$
7. $\alpha-\frac{1}{3}(\alpha-\beta) = 4°40' \; (\Rightarrow 2\alpha+\beta = 14)$
8. $\alpha-2\cdot\frac{1}{3}(\alpha-\beta) = 4°20' \; (\Rightarrow \alpha+2\beta = 13)$

The text does not inform about the procedure, but a good guess is that the techniques explained and used in TMS IX, XVI, and XIX #2 were meant to be employed.

At first, we may use the area condition, and form the rectangle contained by the "repeated" length and width; on one hand, of course,

$$⊏⊐(\alpha l,\beta w) = ⊏⊐(2`30,1`20) = 3``20` .$$

On the other, in agreement with the principles used in TMS XIX #2,

$$⊏⊐(\alpha l,\beta w) = ⊏⊐(\alpha,\beta)\cdot⊏⊐(l,w) = 10`\cdot⊏⊐(\alpha,\beta)$$

Therefore,

$$\sqsubset\!\sqsupset(\alpha,\beta) = 3^{``}\,20^{`}/10^{`} = 20\ .$$

This can be combined with the linear condition, reduced in agreement with the principles taught in TMS XVI and used in TMS IX #3. In #1, #7, and #8, this gives us the standard problem of the rectangle with given surface and given sum of the sides; in the others, the standard problem about the rectangle with given surface and given excess of one side over the other.

TMS XVI, we remember, spoke explicitly of the coefficients of the length and width by phrases "as much as (there is of) lengths", etc. (see p. 87). The present text uses a different term, which may be due not solely to the absence of a standardized terminology for the concept (cf. p. 101) but also to its link to a different operation; though the Old Babylonian calculators were fully aware of the homomorphy of the operations, they may well have felt that a coefficient by the operation of "repetition" did not belong to the same species as a coefficient by "raising" (the former, *inter alia*, being by necessity a small integer, the latter any number that could be expressed).

TMS XIII[233]

With this text we turn to a new genre: sham applications of the art of second-degree algebra (real applications did not exist, and none were invented before the technique was used to interpolate in trigonometrical tables in the Middle Ages). The text is from Susa, but the problem recurs in less complete texts from the Babylonian core area[234].

1. 2 gur 2 pi 5 bán of oil I have bought. From the buying of 1 shekel of silver,
   2(gur) 2(pi) 5 bán ì.ğiš šám *i-na* šám 1 gín kù.babbar

2. 4 silà, each (shekel), of oil I have cut away.
   4 silà ta.àm ì.ğiš *ak-ší-iṭ-ma*

3. ⅔ mina {...} of silver as profit I have seen. Corresponding to what
   ⅔ *ma-na* {20 še} kù.babbar *ne-me-la a-mu-úr ki ma-ṣí*

---

233 Based on the hand copy and transliteration in [TMS, pl. 22, 82], cf. corrections and interpretation in [Gundlach and von Soden 1963: 260–263].

234 A strictly parallel problem is YBC 523019 (cf. [Friberg 1982: 57]), which, however, does not explain the procedure. Related is MLC 1842, in which identical quantities of grain are bought at two different rates, and the sum of the rates and the total investment are given. The tablet is heavily damaged but still allows us to see that the same method was used (*mutatis mutandis*) as in the Susa text.

Figure 47. Graphical representation of the data and scaling of TMS XIII.

4. have I bought and corresponding to what have I sold?
a-šà-am ù ki ma-ṣí ap-šu-úr

5. You, 4 silà of oil posit and 40, (of the order of the) mina, the profit posit.
za.e 4 silà ì.ĝiš ĝar *ù* 40 *ma-na ne-me-la* ĝar

6. Igi 40 detach, 1´30˝ you see, 1´30˝ to 4 raise, 6´ you see.
igi 40 *pu-ṭúr* 1,30 *ta-mar* 1,30 *a-na* 4 *i-ší* 6 *ta-mar*

7. 6´ to 12`50, the oil, raise, 1`17 you see.
6 *a-na* 12,50 ì.ĝiš *i-ší-ma* 1,17 *ta-mar*

8. ½ of 4 break, 2 you see, 2 make hold, 4 you see.
½ 4 *ḫi-pi* 2 *ta-mar* 2 NIGIN 4 *ta-mar*

9. 4 to 1`17 append, 1`21 you see. What is equalside? 9 is equalside.
4 *a-na* 1,17 daḫ 1,21 *ta-mar mi-na* íb.si 9 íb.si

10. 9 the counterpart posit. ½ of 4 which you have cut away break, 2 you see.
9 gaba ĝar ½ 4 *šà ta-ak-ší-ṭú ḫi-pi* 2 *ta-mar*

11. 2 to the 1st 9 append, 11 you see; from the 2nd tear out,
2 *a-na* 9 1-kam daḫ 11 *ta-mar i-na* 9 2-kam zi

12. 7 you see. 11 silà each (shekel) you have bought, 7 silà you have sold.
7 *ta-mar* 11 silà ta.àm *ta-šà-am* 7 silà *ta-ap-šu-úr*

13. Silver corresponding to what? What to 11 [¿silà? may I pos]it
kù.babbar *ki ma-ṣí mi-na a-na* 11 [¿silà? *lu-uš-ku*]*-un*

14. which 12`50 of oil gives me? 1,[10 posit, 1 m]ina 10 shekel of s[il-ver.]
šà 12,50 ì.ĝiš *i-na-ad-di-na* 1,[10 ĝar 1 *m*]*a-na* 10 gín k[ù.babbar]

15. By 7 silà each (shekel) which you se[ll of oil,]
i-na 7 silà ta.àm *šà ta-pa-aš-*[*šà-ru* ì.ĝiš]

16. that of 40 of silver corresponding to what? 40 to 7 [raise,]
šà 40 kù.babbar *ki ma-ṣí* 40 *a-na* 7 [*i-ší*]

17. 4`40 you see, 4`40 of oil.
4,40 *ta-mar* 4,40 í.ĝiš

The problem is perplexing already for the reason that it refers to commercial practices which are rather different from ours. A merchant has bought 2 gur 2 pi 5 bán of fine vegetable oil, which later occurs as M = 12`50 [silà],[235] at a rate of (say) p silà per shekel. Selling at the rate of $s = p-4$ silà per shekel he realizes a profit of $\frac{2}{3}$ mina or $\Pi = 40$ [shekel] of silver.

Lines 7–12 find p and s from what must have been the relations

$$p-s = 4\ ,\quad p\cdot s = 1\text{`}17\ ,$$

where 1`17 = $\frac{4}{\Pi}\cdot M$. That $p\cdot s = \frac{4}{\Pi}\cdot M$ follows easily from the equation $\frac{M}{s}-\frac{M}{p} = \Pi$ if we allow ourselves some algebraic manipulation (multiplication by ps). This, however, was hardly the argument from which the Babylonian calculators derived their equation. Firstly, of course, this kind of symbolic manipulation was not available to them; secondly, even if they were able to master it mentally, it would not lead to the order of operations actually found in the text but to the sequence $(M\cdot 4)\cdot\Pi^{-1}$.

Instead, the usual scaling in one dimensions appears to be in play: from line 7 onward, the procedure *is* geometric, and no jump or change of style is visible between line 6 and line 7. Moreover, since the original investment and the profit in oil are calculated in the final section of the text without having been asked for, these entities must be presumed to have played a role. This leads to the following considerations:

The total quantity of oil is the product of the total selling price Σ (original investment plus profit) and the selling rate s (the number of silà sold per shekel). This product we may represent by a rectangle ⊏⊐(Σ,s) as done in the left part of Figure 47, whose total area is 12`50 [silà], and of which the part representing the profit makes up the same fraction as 4 silà of the rate of purchase p – indeed, from what is bought for each shekel (i.e., p), 4 silà is cut away as profit. A scaling operation along the horizontal dimension which reduces the 40 [shekel] to 4 [silà] will hence reduce the selling price to p, thus changing ⊏⊐(Σ,s) into ⊏⊐(p,s). The scaling factor will have to be $4\cdot 40^{-1} = 6'$, as found in line 6, which reduces the area of the rectangle from 12`50 to 1`17 and the original investment to s, and thus the non-shaded rectangle ⊏⊐$(\Sigma-40,s)$ into a square □(s).[236]

We have thus produced the starting point for the habitual transformations of Figure 48, a rectangle with given surface and given excess of the length p over the width s; the rest of the solution can go by the usual cut-and-paste operations. The only deviation from norms (which may have to do with the use of geometry as representation for oil and prices, but may also have other explanations) is the repetition of the "breaking" process in line 10.

---

235 As we remember, 1 silà is the standard unit of capacity metrology, cf. p. 78; the gur is 5` silà, the pi 1` silà, the bán 10 silà.

236 With regard to the rectangle ⊏⊐$(\Sigma-40,s)$, the scaling factor is thus that igi.te.en *šá* uš sag̃.šè, "fraction which the width is of the length", which we encountered in YBC 4668, see p. 200. Beyond its use in the normalization of non-normalized square problems, the scaling that transforms a rectangle into a square (or a right triangle into an isosceles right triangle) will turn up repeatedly in the following, which may explain the existence of a seemingly contorted technical term.

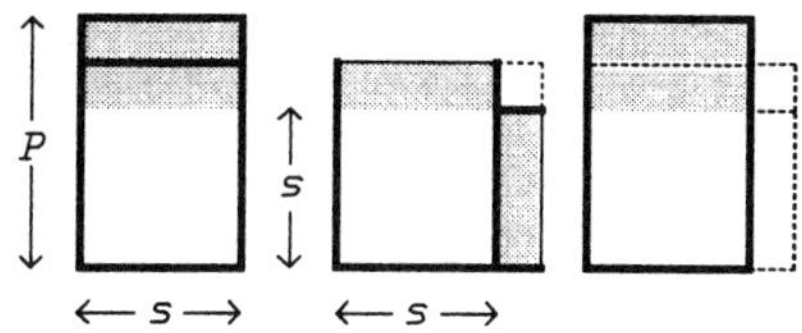

Figure 48. The final cut-and-paste procedure of TMS XIII.

VAT 7532[237]

The problem of the broken reed (gi, Akkadian *qanûm*) seems to have been very popular in the Old Babylonian school; it exists in two main versions, the most common of which is represented in the following; in the other (represented by the problems AO 6770 #5 and Str 362 #5), the reed is shortened stepwise in arithmetical progression. In both versions, the "head of the reed", that is, its initial length, is asked for.

Several features of the text beyond the description of an actual mensuration hint at real-life practice. Firstly, the "reed" equal to 30′ nindan was an actual unit used in practical mensuration; as we see, the initial length of the unknown reed coincides with this unit. Secondly, we have the use of metrological units (the bùr, the nindan). Thirdly, and most interesting, we see how numbers could be expressed in a way that avoids the ambiguities of the place value system – namely, by means of the number word *šūši*, "sixty". The importance of this latter word as well as of the reed is reflected in the fact that both were taken over as loanwords in Greek, as σῶσσος and κάννα/κανών, respectively (the latter, as we see, both in the botanical sense and as a metaphor derived from its metrological role – both still recurring in English, as "cane" and "canon").

That reeds will easily break if used as measuring sticks is also likely to correspond to real-life experience – and the diagram that accompanies the problem statement is as much of a structural diagram as the field plan in Figure 23. Since the graphical progress in the diagram goes from right to left, opposite of the direction of writing (but in the same direction of later Aramaic writing), we may also suspect it to reflect non-school customs. But this is as far as the similarities with real mensurational life go; the problem itself is of the kind Alice might have found behind her mirror.

---

237 Based on the transliteration in [MKT I, 294*f*], cf. [MKT III, 58] and [TMB, 96*f*].

Obv

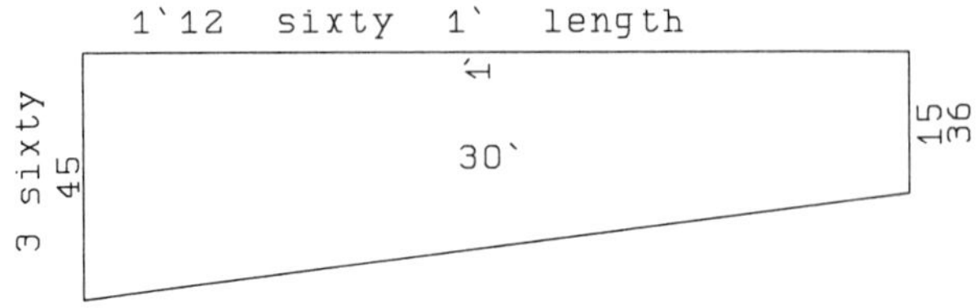

1. A trapezium. I have cut off a reed. I have ta[ken the reed, by] its inte[gr]ity
sağ.ki.gud gi kid gi *e[l-qé-ma i-na š]u-u[l]-m[i]-šu*

2. 1 sixty (along) the length I have gon[e. The 6th par]t
1 *šu-ši uš al-li-i[k* igi 6 ğá]l

3. broke off for me: 1`12 to the leng[th] I have made [fo]llow.
iḫ-ḫa-aṣ-ba-an-ni-ma 1,12 *a-na* u[š] *ú-r[i]-id-di*

4. I turned back. The 3rd part and ⅓ kùš br[oke of]f for me:
a-tu-úr igi 3 ğál *ù* ⅓ kùš *iḫ[-ḫa-aṣ-ba-a]n-ni-ma*

5. 3 sixty (along) the upper width I have go[ne].
3 *šu-ši* sağ an.na *al-li-[ik]*

6. With that which broke off for me I enlarged it[:]
ša iḫ-ḫa-aṣ-ba-an-ni ú-te-er-šum-[m]a

7. 36 (along) the width I went. 1 bùr the surface. The head (initial magnitude) of the reed what?
36 sağ *al-li-ik* 1(bùr)^{iku} a.šà sağ gi en.nam

8. You, by your proceeding, (for) the reed which you do not know,
za.e kìd.da.zu.dè gi *ša la ti-du-ú*

9. 1 may you posit. Its 6th part make break off, 50′ you leave.
1 ḫe.ğar igi 6 ğál-*šu ḫu-ṣú-ub-ma* 50 *te-zi-ib*

10. Igi 50′ detach, 1°12′ to 1 sixty lift:
igi 50 du$_8$-*ma* 1,12 *a-na* 1 *šu-ši* nim-*ma*

11. 1`12 to ⟨1`12⟩ append: 2`24 the false length it gives you.
1,12 *a-na* ⟨1,12⟩ daḫ-*ma* 2,24 uš lul ^{in}sum

12. (For) the reed which you do not know, 1 may you posit. Its 3rd part make break off,
gi *ša la ti-du-ú* 1 ḫe.ğar igi 3 ğál-*šu ḫu-ṣú-ub*

13. 40′ to 3 sixty of the upper width lift:
40 *a-na* 3 *šu-ši ša* sağ an.na nim-*ma*

14. 2` it gives you. 2` and 36 the lower width accumulate,
2 ^{in}sum 2 *ù* 36 sağ ki.ta ğar.ğar

15. 2`36 to 2`24 the false length lift, 6``14`24 the false surface.
2,36 *a-na* 2,24 uš lul nim 6,14,24 a.šà lul

16. The surface to 2 repeat, 1`` to 6``14`24 [l]ift
a.šà *a-na* 2 ^{e}tab 1 *a-na* 6,14,24 [n]im

17. 6```` 14``` 24`` it gives you. And ⅓ kùš which br[ok]e off
6,14,24 ^{in}sum *ù* ⅓ kùš *ša iḫ-h[a-aṣ]-bu*

18. to 3 sixty lift: 5 to 2`24, the false length,
a-na 3 *šu-ši* nim-*ma* 5 *a-na* 2,24 uš lul

19. [l]ift: 12`. ½ of 12` break, 6` make encounter,
[n]im-*ma* 12 ½ 12 gaz 6 du$_7$.du$_7$

Rev.

1. 36`` to 6```` 14``` 24`` append, 6```` 15``` it gives you.
36 *a-na* 6,14,24 daḫ-*ma* 6,15 ^{in}sum

2. By 6```` 15```, 2`` 30′ is equalside. 6` which you have left
6,15.e 2,30 íb.si$_8$ 6 *ša te-zi-bu*

3. to 2`` 30`` append, 2`` 36` it gives you. Igi 6`` 14`24,
a-na 2,30 daḫ 2,36 ^{in}sum igi 6,14,24

4. the false surface, I do not know. What to 6`` 14`24
a.šà lul nu.du$_8$ *mi-nam a-na* 6,14,24

5. may I posit which 2`` 36 gives me? 25′ posit.
ḫe.ĝar *ša* 2,36 ^{in}sum 25 ḫe.ĝar

6. Since the 6th part broke off before,
aš-šum igi 6 ĝál *re-ša-am iḫ-ḫa-aṣ-bu*

7. 6 inscribe: 1 make go away, 5 you leave.
6 *lu-pu-ut-ma* 1 *šu-ut-bi* 5 *te-zi-ib*

8. ⟨Igi 5 detach, 12′ to 25 lift, 5′ it gives you⟩. 5′ to 25′ append: ½ nindan, the head of the reed it gives you.
⟨igi 5 du$_8$-*ma* 12 *a-na* 25 nim 5 ^{in}sum⟩ 5 *a-na* 25 daḫ-*ma* ½ nindan saĝ gi ^{in}sum

Since the language of the statement is somewhat opaque, it may be useful to paraphrase it as follows: The dimensions of a trapezoidal field of surface 1 bùr = 30` sar is measured by means of a reed of unknown length $R$. A first part of the length is measured in 60 steps and is thus equal to $1`\cdot R$.

Then the reed suffers a first loss and is shortened to $r = R-\frac{1}{6}R$. The rest of the length is found to be $1`12\cdot r$. After suffering a further loss of $\frac{1}{3}r+\frac{1}{3}$ kùš, the reed is reduced to a length $z$ (thus, $z = r-\frac{1}{3}r-\frac{1}{3}$ kùš), and the upper width is measured as $3`\cdot z$. Restitution of the piece which was lost in the second instance allows the measurement of the lower width as $36\cdot r$.

The procedure makes use of several false positions; the first is made in obv. 9, and leads to the conclusion that $1`\cdot R = 1`12\cdot r$. The upper length is thus $2`24\cdot r$ – or, as expressed by the text, the "false length" (the length measured with the once shortened reed as unit) is 2`24.

In obv. 12 follows the next explicit false position. For r we are to "posit" *1*, which means that ϱ, r shortened by its ⅓ (but not the ⅓ kùš) will be 40′. 3` steps of $\varrho$ will thus equal $2`\cdot r$. The sum of the two widths measured in the unit r is thus 2`36 – *provided we forget about the* ⅓ kùš. Under the same condition, the "false surface" – the surface of the "repeated" (doubled) trapezium measured in units $\square(r)$ – is $2`24\cdot 2`36 = 6``14`24$ (since trapezoidal surfaces are always found by "raising", we can conclude nothing from the use of this operation).

The true "repeated" surface is 2 bùr = 1` sar`. Therefore,

$$6``14`24\cdot\square(r)-n\cdot r = 1`` ,$$

where the member $n\cdot r$ comes from the neglected ⅓ kùš and corresponds to the shaded area in Figure 49. As usual, the determination of the number n is postponed, and the normalization of the equation by means of a scaling with the factor 6``14`24 is made first. Thereby we get

$$\square(6``14`24\cdot r)-n\cdot(6``14`24\cdot r) = 6````14```24`` .$$

In order to determine the shaded part of the false surface, its height is found in obv. 17–18 to be 3`·⅓ kùš = 5 (nindan) (1 kùš = 5´ nindan). Its length being 2`24·r, the quantity to be subtracted is thus 5·2`24·$r$ = 12`·r.

The resulting square-area-and-sides problem

$$\square(6``14`24\cdot r)-12`\cdot(6``14`24\cdot r) = 6````14```24``$$

is solved in full agreement with the book, and r itself found to be 25´. R is found from rev. 6 onward by yet another false position to be 30´ nindan. The passage is defective in the present problem (the copyist is likely to have jumped from one number 5 to the next), but it can be safely reconstructed from parallel passages in the closely related tablet VAT 7535. Terminologically we observe the use of the word *šutbûm*, "to make go away", for the elimination of the representative of the part that is to be removed (cf. p. 20).

A few words should be added about the numbers that are written into the diagram on the tablet. Most of them belong to the statement, but 45 and 15 do not; nor do they appear in the calculation; instead, they represent the upper and lower width measured in nindan – measures that are never found in the text itself, but may be found once we know that the upper width is 3` times (25´ nindan–⅓ kùš), and the lower width 36 times 25´ nindan. Appearing as they do in the explanatory diagram they may be regarded as "merely known" –

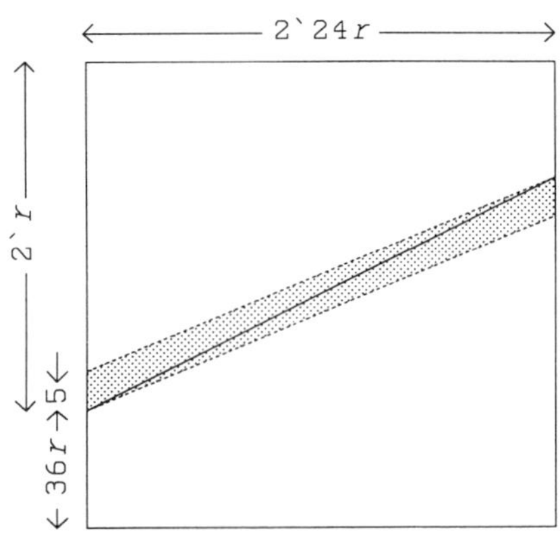

Figure 49. The "false surface" of VAT 7532.

and their absence from the verbal text will reflect an organization of teaching where configurations or situations were explained separately and before the statement of what was considered given in the actual problem, cf. p. 162; cf. also below, p. 243, on Str 367, which belongs to the same text group.

IM 52301 #2[238]

This tablet comes from Eshnunna, and was found together with tablets that are dated c. 1780 BCE. The many erroneous repetitions show that it is a copy made by somebody who did not follow the computations, but the original need not be significantly older than the actual text. #1 of the tablet (obv. 1–15) is similar in character to #2; #3 (rev. 20–24) is a short table of technical constants; #4 (edge 1–4) is an unclear rule formulated in general terms (thus not as a paradigmatic numerical example) dealing somehow with a quadrangle "whose sides are not equal".[239]

Because not all repetitions in #2 have been identified as such by previous workers, they have caused some confusion. Further difficulties for interpretation of the details come from its "unorthographic" Sumerian (cf. [SLa, § 554]) – thus i.gi for igi, ba.se.e for ba.si_8, TUK for dug_4, zú for zu.

Obv.

16. If to two-thirds of the accumulation of upper width
šum-ma a-na ši-ni-ip ku-mu-ri sağ *e-li-tim*

17. and lower, 10, on my hand, I have appended: 20, the length, I have built. The upper width
ù ša-ap-li-tim 10 *a-na qa-ti-ia* daḫ*-ma* 20 uš *ab-ni* sağ

18. over the lower 5 went beyond.
{*e-li*} *e-li-tum e-li ša-ap-li-tim* 5 *i-te-er*

19. The surface 2`30. What my lengths? You, by your saying, 5 which went beyond,
a.šà 2,30 *mi-nu-um* uš*-ia* za.e TUK.zú.dè 5 *ša e-te-ru*

20. 10 which you have appended, 40´ of the two-thirds, my ¿factors of both?, inscribe:
10 *ša tu-iṣ-bu 40 ši-ni-pe-tim a-ra-ma-ni-a-ti-ia lu-pu-ut-ma*

21. Igi 40´, of the two-thirds, detach, 1°30´ you see. 1°30´ {...}
i.gi 40 *ši-ni-pé-tim pu-ṭú-ur-ma* 1,30 *ta-mar* 1,30 {*ḫe-¿pé?-ma*

238 Based on the transliteration in [Baqir 1950a], with revisions in [von Soden 1952: 50], [Gundlach and von Soden 1963: 252*f*], and [Høyrup 1990: 328–330].

239 The best edition of this part of the text together with a possible interpretation is in [Gundlach and von Soden 1963: 253, 259*f*].

22. {...} to 2`30, the surface, raise: 3`45 you see.
4⸢5 *t*⸣*a-mar* 45} *a-na* 2,30 a.šà *i-ši-ma* 3,45 *ta-mar*

23. 3`45 repeat: 7`30 you see. 7`30 your head
3,45 *e-ṣí-ma* 7,30 *ta-mar* 7,30 *re-eš*$_{15}$*-ka*

24. may hold. Turn back. Igi 40′, of the two-thirds, detach,
li-ki-il tu-ur-ma i.gi 40 *ši-ni-pé-tim pu-ṭú-ur*

Rev.

1. 1°30 you see. 1°30 break, 45′ you see, to 10 which you have appended
1,30 *ta-mar* 1,30 *ḫe-pé-ma* 45 *ta-mar a-na* 10 *ša tu-iṣ-bu*

2. raise, 7°30′ you see. {...}
{*i-ši-ma* 7,30 *ta-mar* {7,30 *re-eš*$_{15}$*-ka li-ki-il*}

3. {...}
{*tu-ur-ma* i.gi 40 *pu-tú-ur-ma* 1,30 *ta-mar* 1,40 *ḫe-pé-ma*}

4. {...}
{45 *ta-mar a-na* 10 *ša tu-iṣ-bu i-ši-ma* 7,30 *ta-mar*}

5. 7°30′ the counterpart lay down: Make hold:
7,30 *me-eḫ-*{*ša-*}*ra-am i-di-ma šu-ta-ku-il-ma*

6. 56°15′ you see. 56°15′ to 7`30 which you head
56,15 *ta-mar* 56,15 *a-na* 7,30 *ša re-eš*$_{15}$*-ka*

7. holds, append, 8`26°15` you see. The equilateral
ú-ka-lu ṣi-ib-ma 8,26,15 *ta-mar* ba.se.e

8. of 8`26°15′ make come up: 22°20′ its equilateral. From 22°30,
8,26,15 *šu-li-ma* 22,30 *ba-su-šu i-na* 22,30

9. the equilateral, 7°30′, your made-hold, cut off
ba.se.e 7,30 *ta-ki-il-ta-ka ḫu-ru-us*$_4$

10. 15 the left-over. 15 break, 7°30′ you see, 7°30′ the counterpart lay down:
15 *ši-ta-tum* 15 *ḫe-pé-ma* 7,30 *ta-mar* 7,30 *me-eḫ-ra-am i-di-ma*

11. 5, which width over width went beyond, break:
5 *ša* sağ *e-li* sağ *i-te-ru ḫe-pé-ma*

12. 2°30′ you see. 2°30′ to one 7°30′ append:
2,30 *ta-mar* 2,30 *a-na* 7,30 *iš-ti-in ṣí-im-ma*

13. 10 you see, from the second 7°30′ cut off.
10 *ta-mar i-na* 7,30 *ša-ni-im ḫu-ru-us*$_4$

14. 10 the upper width, 5 the lower width.
10 sağ *e-li-tum* 5 sağ *ša-ap-li-tum*

15. Turn back. 10 and 5 accumulate, 15 you see.
tu-ur-ma 10 *ù* 5 *ku-mu-ur* 15 *ta-mar*

16. Two-thirds of 15 take: 10 you see, and 10 append:
ši-ni-ip-pé-at 15 *le-qé-ma* 10 *ta-mar ù* 10 *ṣí-ib-ma*

17. 20 your upper length. 15 break: 7°30 you see,
20 uš-*ka e-lu-um* 15 *ḫe-pé-ma* 7,30 *ta-mar*

18. 7°30 to 20 raise: 2`30, the surface, you see.
7,30 *a-na* 20 *i-ši-ma* 2,30 a.šà *ta-mar*

19. Thus the procedure.
ki-a-am ne-pé-šum

The problem deals with a trapezium[240] with parallel widths (say, u and v). If these are accumulated and an extra segment 10 which is held at disposition ("on my hand") is "appended" to their two-thirds, we construct ("build") the length; the length is also said to be 20, but this is meant for naming, not as a datum. The surface is given as 2`30.[241]

The procedure can be followed in Figure 51: At first a scaling is made with the igi of $\frac{2}{3}$, the "factors of both",[242] whereby a new length L is obtained that depends directly on u and v, not on their $\frac{2}{3}$ (probably the reason that the original length needs a numerical "name"). Thereby the surface becomes 3`45. This is "repeated", that is, doubled concretely (as the trapezium of VAT 7532 was "repeated"), which gives us, either a rectangle with width $W = u+v$ and length $L = u+v+2\delta = W+2\delta$, or a square $\square(W)$ plus 2δ sides of the same square – possibly understood as in Figure 52 as a square $\square(L)$ from which 2δ sides have been removed. 2δ is of course $10 \div \frac{2}{3} = 15$, but this is never calculated directly.

Both the numerical and the geometrical operations are of course the same in all three interpretations. The explicit "building" of the length (though of l,

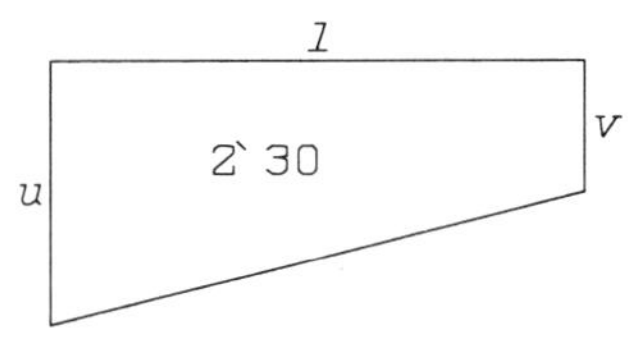

Figure 50. The trapezium of IM 52301 #2.

240 Bruins [1966: 207*ff*] insists that the topic be a triangle which is divided by a parallel transversal; in this view, the "upper" and "lower" width are the segments into which the transversal divides the width. From the numbers alone, both interpretations are possible. However, no problem about a divided triangles is known which is not interested in the partial areas; the "upper length" of rev. 17 rules out the triangle definitively.

241 Since all data are homogeneous, we might have chosen areas two orders and lines one order of magnitude lower. The choice of real instead of school yard magnitudes has the advantage that the three different quantities 7,30 then end up in different orders – otherwise all would be 7´30´´. Since "building" is mostly found as a reference to surveying practice, it is also likely to be the choice the Old Babylonian calculator had in mind.

242 The word is a plural, *aramaniātum*, a loanword probably derived from Sumerian a.rá, "steps of" or "factor", and man, "two" [von Soden 1952: 50].

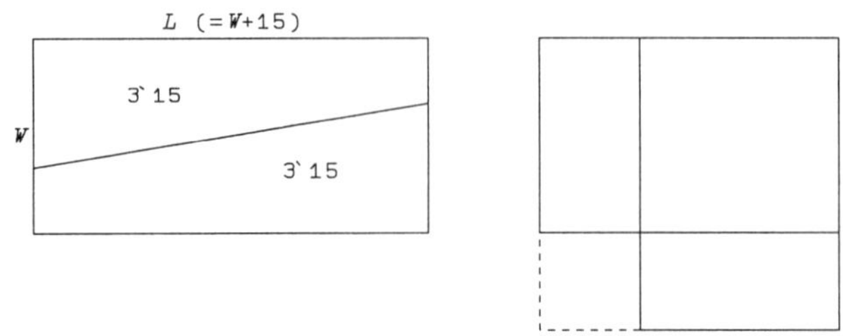

Figure 51. The procedure of IM 52301 #2.

not of L) might speak in favour of a conceptualization based on (L) – but since only W and not L is found, the interpretation in terms of □(W) and its sides seems more plausible.

We have seen in several other cases that the determination of the number of sides was postponed as long as possible; we have also seen in BM 13901 #14 (p. 76) that the total number was never found, since this would have involved a doubling that would have been cancelled by an ensuing "breaking". In the present case the same principle turns up again, even though it is not caused by mutually cancelling steps: instead of finding the excess of L over W as 10 "raised" to igi 40´, a quantity that would then have to be broken, it finds its "moiety" directly by "breaking" the igi before "raising".

Two features of the procedure represent non-standard uses of standard terms or procedures (at least as we know these from the remaining published corpus). Firstly, all other instances of "building" refer to the construction of a (rectangular) surface, never of a linear quantity.[243] Secondly (and similarly), the "laying-down" of the average and its "counterpart" is not uncommon

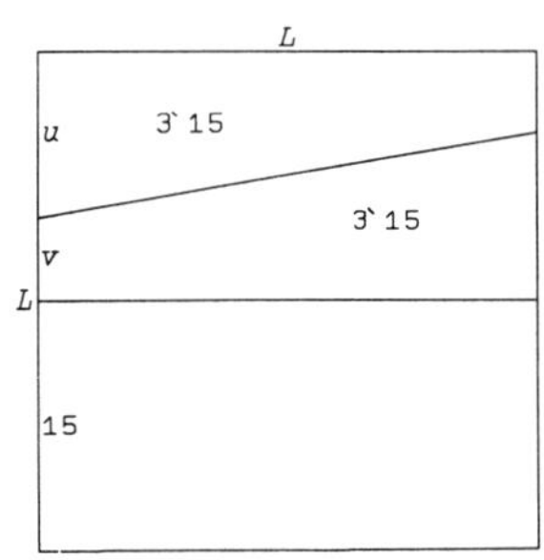

Figure 52. The alternative interpretation of the final part of IM 52301.

---

243 In the following texts, surfaces are built: IM 52685+52304; YBC 4608; VAT 8390; TMS V; TMS XVII, #1; and AO 8862. As we have seen (note 147), the a.šà ŠU.BA.AN.TU of YBC 4714, rev. II, 20 (and other series texts) is somehow related.

íb.si$_8$ and its cognates function as verbs in all of these (except IM 52685+52304 and TMS V, where they do not occur). In the present text, as we observe, *basûm* is a noun.

when two quantities are found from average and deviation – but only if these occur naturally as sides of a quadratic corner, not if they are two moieties resulting from breaking.[244] In the present case, the two copies are likely to be two numbers written down on the tablet for rough work as part of two different calculations.[245]

BM 85194 #25–26[246]

These two problems are extracted from a large anthology text, most of whose problems deal with real or sham practical geometry (as will be obvious, the present specimens belong to the latter category). Both concern a not yet completed siege ramp, whose total content of earth is going to be 1``30` (volume sar) (see Figure 53). So far, a length of l = 32 (nindan) and a height of h = 36 (kùš) has been attained; d = 8 (nindan) in length and δ = 9 (kùš) in height still remain to be completed before the city-wall is attained and the city "inimical to Marduk" (the tutelary deity of Babylon) can be subdued. The width is 6 (nindan). In both problems, the volume and the breadth are given. In #25, the other given quantities are h and d, in #26 they are l and h.

To the left in Figure 53, the situation is drawn in true proportions; however, the text does not bother about the different units, and treats a rectangle with vertical height a kùš and horizontal length a nindan as if it were a square. The right-hand part of the figure therefore presents the situation

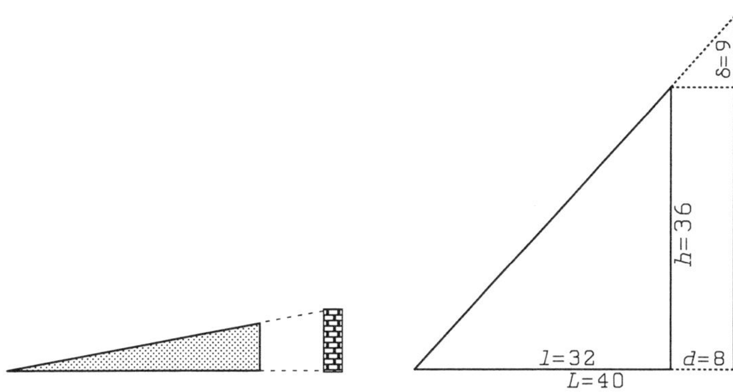

Figure 53. The ramp of BM 85194 #25–26, in true and "arithmetical" proportions.

244 We may be reminded of the "breaking into moieties" as a substitute expression for "making hold" in AO 8862 – see p. 164.

245 The use of *takīltum* in rev. 9, on the other hand, is quite regular. It refers to the 7°30′ of rev. 2, not to the 7`30 of obv. 23 that is to be held in the head. (Misled by the repetition, Gundlach and von Soden [1963: 255] err on this point.)

246 Based on the transliteration in [MKT I, 149]; cf. [TMB, 34*f*].

in what we might term "numerical proportions". In both cases, "raising" to igi 6 (that is, division by the breadth) gives the value 15` for the cross-section *A* in the unorthodox and nameless unit nindan×kùš.

Rev. II

#25

7. Of earth, 1`` 30`. A city inimic[al to Mar]duk I shall seize.
i-na saḫar.ḫi.a 1,30^{iku} uru^{ki} *na-ki-i[r* ^{d}Mar]duk *a-ṣa-ba-at*

8. 6 the (breadth of the) fundament of the earth. ⌞8 should still be made firm bef]ore the city wall is reached.
6 úr saḫar.ḫi.a *ú-ki-in* ⌞8 *a-n]a* bàd *la sà-na-qâm*

9. 36 the pea[k (so far attained) of the ea]rth. Corresponding to what the length
36 *zi-iq-[pu-um ša* saḫ]ar.ḫi.a *ki ma-ṣí* uš

10. I must st[amp] in order to seize [the cit]y? And the length behind
lu-uk-b[u-ús uru^{k}]^{i} *lu-uṣ-ba-at ù* uš egir

11. the *ḫurḫ[urum* (the vertical back front reached so far?) what? You, igi] 6, the fundament of the earth, detach, 10′ you see. 10′ to
ḫur-ḫ[u-ri en.nam za.e igi] 6 úr saḫar.ḫi.a du$_{8}$.a 10 *ta-mar* 10 *a-na*

12. [1`` 30` the earth, raise, 15`] you see. Igi 8 detach, 7′30″ you see.
[1,30 saḫar.ḫi.a *i-ši* 15] *ta-mar* igi 8 du$_{8}$.a 7,30 *ta-mar*

13. [7′30″ to 15` rais]e, 1`52°30′ you see. 1`52°30′ repeat,
[7,30 *a-na* 15 *i-š]i* 1,52,30 *ta-mar* 1,52,30 tab.ba

14. [3`45 you see.] 3`45 to 36 raise, 2`` 15` you see. 1`52°⟨30′⟩
[3,45 *ta-mar*] 3,44⟨+1⟩ *a-na* 36 *i-ši* 2,15 *ta-mar* 1,52,20⟨+10⟩

15. [make hold, 3`` 30`]56°15′ you see. 2`` 15` from 3`` 30`56°15′
[NIGIN 3,30,]56,15 *ta-mar* 2,15 *i-na* 3,30,56,15

16. [tear out, 1`` 1]5`56°15′. What is equalside? 1`7°30′ you see.
[ba.zi 1,1]5,56,14⟨+1⟩ en.nam íb.si$_{8}$ 1,7,30 *ta-mar*

17. [1`7°30′ from] ⌞1`5⌞2°30′ tear out, 45 you see, the height of the city wall.
[1,7,30 *i-na*] ⌞1,5⌞2,30 ba.zi 45 *ta-mar* sukud bàd

18. ½ of 45 break, 22°30′ you [se]e. Igi 22°30′ detach, 2′40″.
[½ 45 *ḫe-pé* 2]2,[30] *ta-[ma]r* igi 22,30 du$_{8}$.a 2,40

19. [15` to] 2′40″ raise, 40, the length. Turn back, 1`` 30`, the earth, see. 22°30′,
[15 *a-na*] 2,40 *i-ši* 40 uš niĝin.na 1,30 saḫar.ḫi.a *a-mur* 22,30

20. [½ of the heig]ht, to 40, the length, raise, 15` you see. 15` to 6 raise,
[½ suku]d *a-na* 40 uš *i-ši* 15 *ta-mar* 15 *a-na* 6 *i-ši*

21. 1`` 30` you see, 1`` 30` is the earth. The procedure.
1,30 *ta-mar* 1,30 saḫar.ḫi.a *ne-pé-šum*

#26

22. Of earth, 1`` 30`. A city inimical to Marduk I shall seize.
i-na saḫar.ḫi.a 1,30 iku uru^{ki} *na-ki-ir* ^{d}Marduk *a-sa-ba-at*

23. Away from the fundament of the earth a length of 32 in front of me I have gone.
iš-tu úr saḫar.ḫi.a 32 uš *a-na pa-ni-ia al-lik*

24. 36 the height of the earth. The length what must I stamp, so that the city
36 sukud saḫar.ḫi.a uš en.nam *lu-uk-bu-ús* uru^{ki}

25. I may seize? The length in front of the *ḫurḫurum* (the vertical back front reached so far) what? You,
lu-us$_4$-ba-at uš *ša pa-ni ḫur-ḫu-ru* en.nam za.e

26. igi 32 detach, 1´52˝30‴ you see. 1´52˝30‴ to 36, the height,
igi 32 du$_8$.a 1,52,30 *ta-mar* 1,52,30 *a-na* 36 sukud

27. raise, 1°7´30`` you see. Igi 6, the fundament of the earth, detach, 10´ you see.
i-ši 1,7,30 *ta-mar* igi 6 úr saḫar.ḫi.a du$_8$.a 10 *ta-mar*

28. 1`` 30`, the earth, to 10´ raise, 15` you see. 15 to ⟨2⟩ repeat,
1,30 saḫar.ḫi.a *a-na* 10 *i-ši* 15 *ta-mar* 15 *a-na* ⟨2⟩ tab.ba

29. 30` you see. 30` to 1°7´30˝ raise, 33`45 you see.
30 *ta-mar* 30 *a-na* 1,7,30 *i-ši* 33,45 *ta-mar*

30. 33`4⟨5⟩, what is equalside? 45 is equalside, 45 the height of the city wall.
33,42⟨+3⟩ en.nam íb.si$_8$ 45 íb.si$_8$ 45 sukud bàd

31. 4⟨5⟩, the height of the city wall, over 36, the height of the earth, what goes beyond? 9 it goes beyond.
44⟨+1⟩ sukud bàd ugu 36 sukud saḫar.ḫi.a en.nam dirig 9 dirig

32. Igi 1°7´30 detach, 53´20˝ you see, 53´20˝ to 9 raise,
igi 1,7,30 du$_8$.a 53,20 *ta-mar* 53,20 *a-na* 9 *i-ši*

33. 8 you see, 8 the length in front of you you stamp.
8 *ta-mar* 8 uš *a-na pa-ni-kà ta-ka-ba-as*

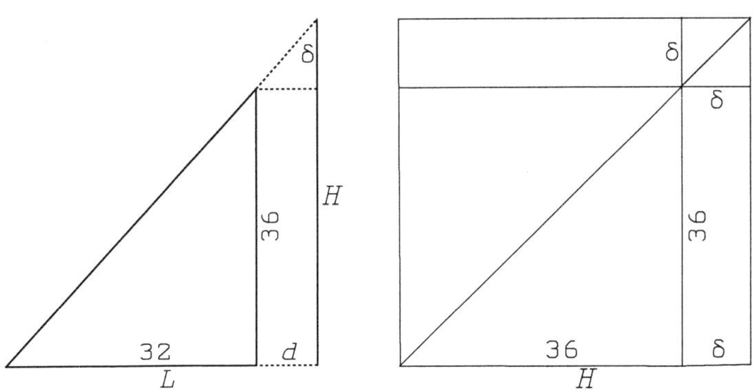

Figure 54. The data and transformation of BM 85194 #26.

#26

#26 is the simpler of the two problems, and will provide us with the key by which we may interpret the procedure of #25 (see Figure 54). The text starts by finding the scaling factor which will make the ramp as long as it is high (in numbers, not in physical terms), as the ratio between h and l. This factor is 1°7′30″. Only as the next step does it calculate the cross-section. This is multiplied by the scaling factor, which corresponds to a transformation into an isosceles right triangle. This triangle is "repeated", that is, doubled into a square with "surface" 33`45, along which 45 is seen to "be equalside". Therefore H is 45 (kùš), and $\delta = H{-}h = 45{-}36 = 9$ (kùš). By reverse scaling, d is finally found to be 8 (nindan).

#25

In this problem, the data are the unpleasant combination of $h = 36$ and $d = 8$ – see Figure 55. The procedure starts by the only step which can be performed without deeper reflection, namely, by finding the cross-section. Its next step is a scaling by a factor igi 8 = 7′30″, followed by a "repetition". This gives us the central configuration of the figure, a rectangle where the part corresponding to the length that is still to be constructed is reduced to 1. The "surface" of the full rectangle $\sqsubset\!\sqsupset(^{L}/_{8},H)$ is 3`45 (line 14).

The aim, however, is the same quadratic configuration as in Figure 54; this is shown as the right-hand configuration in Figure 55. This requires a further scaling with the factor $\delta = H{-}36$, which would change the surface 3`45 into 3`45·(H–36) – a refined instance of inversion of the roles of unknown and given magnitudes! The outcome will be a square $\square(H)$. That is,

$$3`45\cdot H - 3`45\cdot 36 = \square(H) .$$

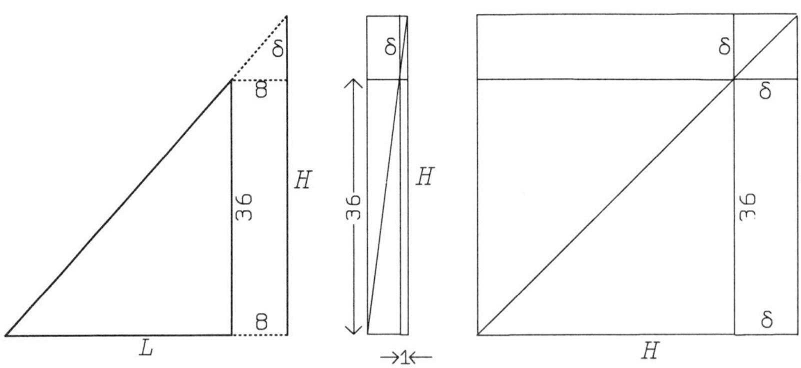

Figure 55. The data and transformations of BM 85194 #25.

In this form, it is true, the equation does not correspond to anything that can be found in a Babylonian text. However, the formulation

$$3`45H-\square(H) = 3`45\cdot 36$$

corresponds precisely to a sequence of problems in the catalogue text TMS V, rev. II 18–20, "*n* confrontations over the surface, *d* goes beyond". It may also be formulated as a rectangular problem, with given surface and given sum of the sides,

$$\sqsubset\!\sqsupset(H,3`45-H) = 3`45\cdot 36\ .$$

In either form, the equation corresponds to the steps that follow, and which we have seen (in rectangle version) in BM 13901 #12, TMS IX #3, AO 8862 #1–2 and YBC 6504 #2: at first 3`45·36 is found in line 14 to be 2``15`. The "breaking" of the coefficient of *H* (respectively, of the sum of the sides *H* and 3`45–*H*) is omitted, since 3`45 is remembered to have resulted from a "repetition", the inverse operation of the "breaking".

The equation is of the type that possesses two (positive) solutions – two possible values for the side if we choose the square problem, a length and a width in the rectangle version. The text uses $H = 40$, the other possibility is $H =$ 3``; the two correspond, respectively, to the left-hand and right-hand configurations in Figure 56. However, precisely because of the geometric procedure, the ambiguity does not present itself to the calculator: already when the figure is laid out it has to be decided whether *H* has to be larger or smaller than 3`45/2 – and this decision did not present itself as one in the Babylonian context, since all problems were constructed backwards from known solutions. In the present problem, the slope corresponding to $H =$ 3`` would of course produce an impossibly steep ramp.

We should hence not wonder that the Babylonians were not aware of the double solution of this kind of equation. They may well have been,[247] but

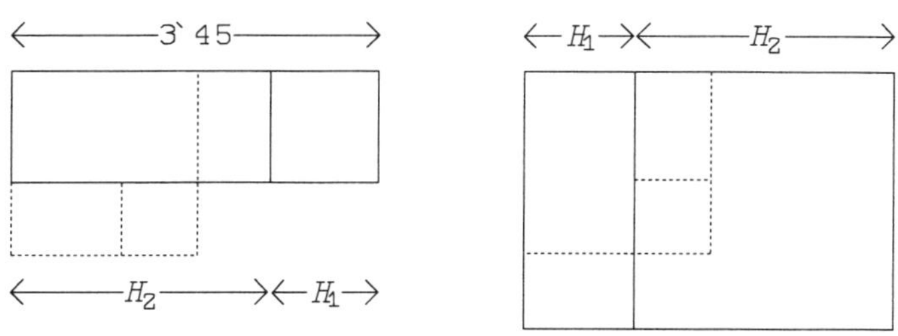

Figure 56. The two diagrams for the double solution of *as*–□(*s*) = *b*.

247 TMS IX #3, by choosing for the modified length Λ the smaller number (4°30′) and for the modified width Ω the larger (28), may demonstrate awareness of the arbitrariness of the assignment of values to unknowns (Ω = 4°30′ would indeed give rise to an irregular ω and, if that were accepted, to a negative *w*). Since even this problem with its "17th part" was evidently constructed backwards from the known solution, we cannot be sure.

their methods and habits would always lead them unambiguously to one of the two; therefore, if they knew about the ambiguity, their texts will not show it.

Once H has been found, the cross-section is divided (via igi-multiplication) by the "moiety" of H, in agreement with the way triangular areas were calculated; the result is that $L = 40$.

In the end the correctness of the result is checked; the fact that #25 is provided with a proof and #26 is not may, if no accident, correspond to the less obvious character of the computation.

BM 13901 #23[248]

This brief problem comes from the tablet BM 13901, quite a few problems from which were already analyzed above. In general, there is a progression from simple to complex problems, in particular from problems involving only one to problems about several squares.

This penultimate problem from the text is a striking exception to this rule; coming just before a three-square version of #14, it treats of a single square; it is normalized; and the coefficient of the sides is a simple integer. This location, however, is only one of several bizarre features.

Rev. II

11. About a surface, the f[o]u[r fronts and the surf]ace I have accumulated, 41′40″.
a.šà[lam] *p*[*a*]-*a*[-*at er-bé-et-tam ù* a.š]à[lam] *ak-mur-ma* 41,40

12. 4, the f[ou]r fronts, yo[u inscr]ibe. Igi 4 is 15′.
4 *pa-a-at er*[-*bé-e*]*t-tam t*[*a-la-p*]*a-at* igi 4 ĝál.bi 15

13. 15′ to 41′40″ [you r]aise: 10′25″ you inscribe.
15 *a-na* 41,40 [*ta-n*]*a-ši-ma* 10,25 *ta-la-pa-at*

14. 1, the projection, you append: by 1°10′25″, 1°5′ is equalside.
1 *wa-ṣi-tam tu-ṣa-ab-ma* 1,10,25.e 1,5 íb.si$_8$

15. 1, the projection, which you have appended, you tear out: 5′ to two
1 *wa-ṣi-tam ša tu-iṣ-bu ta-na-sà-aḫ-ma* 5 *a-na ši-na*

16. you repeat: 10′, nindan, confronts itself.
te-ṣi-ip-ma 10 nindan *im-ta-ḫa-ar*

What already struck Neugebauer just as much as the location of the problem was the method by which it is solved. The text starts by de-normalizing the problem; expressed in symbols, the steps look as follows:

---

248 Based on the transliteration in [MKT III, 4*f*].

$$4s+\square(s) = 41'40''$$
$$\Rightarrow \quad s+\tfrac{1}{4}\square(s) = 10'25''$$
$$\Rightarrow \quad 1+s+\tfrac{1}{4}\square(s) = 1^\circ10'25''$$
$$\Rightarrow \quad 1+\tfrac{1}{2}s = \sqrt{1^\circ10'25''} = 1^\circ5'$$
$$\Rightarrow \quad \tfrac{1}{2}s = 1^\circ5'-1 = 5'$$
$$\Rightarrow \quad s = 2\cdot5' = 10'$$

(Figure 57 shows the geometrical interpretation of the problem and the procedure.)

This deviation from the usual ways of the text, together with a general uneasiness with the language and style, made Neugebauer suggest that the problem in its actual form might be the outcome of a scribal misunderstanding of a problem

$$\sum_1^4 x_i^2 + \sum_1^4 x_i = 51`40\ , \quad x_{i+1}-x_i = 10\ ,$$

similar to YBC 4714 #9. He did not explain how the anomalous procedure should be a consequence of this mistake, nor how the oddities of the formulation (apart from the explicitation of the nindan, familiar from the series texts), should have resulted.

At the time nobody was aware that the problem of the four fronts and the surface turns up time and again in the record during the following three millennia, which is already a reason to reject Neugebauer's way to get rid of the puzzle. Since the many anomalies taken together form a coherent picture, we shall go through them in detail.

The very first word is a presentation of the object, a "surface". Similar presentations occur elsewhere in the corpus, though not in the present tablet. Initial statements of the object mostly use a logogram deprived of phonetic complements. When syllabic writing occurs (e.g., in some of the problems of BM 85196), the form is always the lexical form, i.e., a nominative. The present "surface", however, is provided with a phonetic complement which shows it explicitly to be an accusative; elsewhere the text only uses such complements on its (rather few) logograms when they are needed for

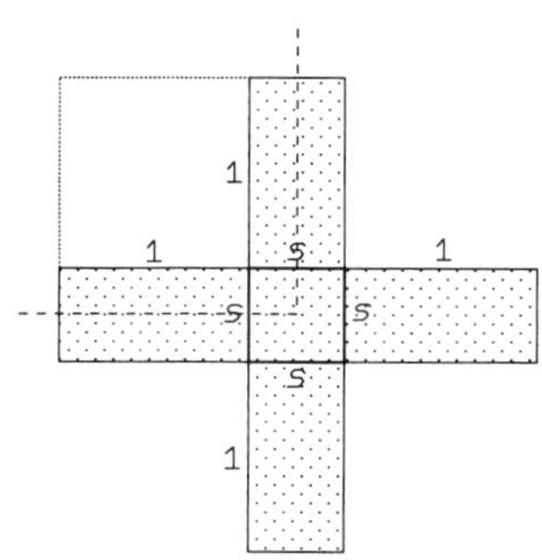

Figure 57. The configuration of "the four fronts and the surface" (BM 13901 #23).

comprehension. The writer must have had a specific intention when making explicit the intended case of the word.

The only context where accusatives occur in the statement of the object is when it is embedded in a full phrase, in the present case something like *šumma eqlam išâluka umma šūma*, "if, about a field, somebody asks you thus".[249] From within the horizon of the Old Babylonian mathematical corpus, the only plausible explanation is thus that a.šà*lam*, to be read *eqlam*, is meant as an ellipsis for this complete phrase.

Further on in the line we see that this surface possesses "four fronts". The grammatical construction (postposited numeral in status rectus) excludes a reading "four times the front"; it is used when the number is an invariable epithet, that is, when *n* items belong inevitably together, as in the Babylonian "the seven mountains" and our "the four cardinal points" (GAG, § 139i). The construction is unique in the whole mathematical corpus, and the references to the specific sides of a square (elsewhere made by way of the expression ta.àm, "each" side) is rare.

Almost unique is the idea that a square possess *pāt*, "fronts", plural of Akkadian *pūtum*, with Sumerogram sag̃ (which I translate "width"). Elsewhere in the present tablet, as square configuration is constantly seen as determined by (and hence identified with) a "confrontation", which only turns up indirectly in line 16, through the verb *imtaḫḫar*, "confronts itself" (<*maḫarum*). In catalogue texts from Nippur, Susa, and Eshnunna the side of a square may be regarded as a "length"; in YBC 4714 #30 a "second width" suggests that the "confrontation" may itself be seen as a width – but it does not say so explicitly).

A newly published text strongly damaged from Nippur (CBS 19761, obv. I 10–11) seems to demand that "your fronts" (*pa-a-ti-ka*, plural again) be detached from a quadratic figure, logographically termed íb.si$_8$ – see note 47 and p. 354. Apart from this, the only explicit reference to the "front" of a square I know of is found in the text UET V, 864, translated and analyzed below (p. 250). This text determines the border between concentric squares; it refers to 4 sag̃, and makes clear by using ta.àm, "each", that this is a reference to *the* four sides. The text is from nineteenth-century Ur; below we shall return to the terminological similarities between this and other texts from early Old Babylonian Ur and from the periphery (see below, pp. 352*ff*).

UET V, 864 uses the Sumerogram sag̃. Whether they refer to rectangles or to squares, both length and width are almost always written with the Sumerograms (deprived of phonetic complements). The only places where the Akkadian terms *šiddum* and *pūtum* are found in syllabic writing in abstract "algebraic" or geometric problems are in a few texts from Eshnunna and in the text from Nippur that was just mentioned; elsewhere, they are only used when the dimensions are thought of concretely – the distance bricks are to be

---

249 This precise turn of the phrase is calqued after Db2-146 (below, p. 257); other texts from Eshnunna have slightly different variants.

carried, the length of a wall, the width of a canal. We must hence presume that the present text chooses the Akkadian term in syllabic writing in order to emphasize that it deals with a *real* field, not with one of the abstract problems that were treated so far.

The next anomaly in the statement is the order of the members: All other problems of the tablet mention the surface before the "confrontations", and it is indeed the consistent habit of all algebraic texts, when sides are "appended" to or "accumulated" with a surface, to speak of the latter before the sides. The only exception I have observed is AO 8862 #4, with its "Length, width, and surface I have accumulated: 9" (above, p. 169). The way the latter problem states the equality of accumulated sides and surface betrays the same explicit will to speak first of the sides: after relating that a surface has been built, the plain way would have been to say that it was "as much as" the accumulated sides; the related problem AO 6770 #1 (p. 179), which formulates the equality in the same way, as a "confrontation" of equals, has the same order.

Obviously, the order of members in BM 13901 #23 is an intentional violation of prevailing school usage; we must presume that it is meant to suggest the same connotations as the syllabic *pāt*.

The last anomaly of the statement is masked but no less real, namely, the implied value of the fronts. No other conserved problem on the tablet that deals with a single square has the side 10′ – nor has any other single-square problem from the whole corpus as far as I know, apart from a virtual exception in the catalogue text BM 80209, obv. 4.[250]. The normal side of a square in the Old Babylonian texts is 30′. This is a simple number in Sumerian and in the sexagesimal system used in the school (namely, $\frac{1}{2}$, and also the length unit gi, the "reed"). 10, instead, is a round number in Akkadian as in other Semitic languages. If the formulations are meant to point away from the school environment, this feature suggests that the environment *toward* which they point was Akkadian-speaking. For the moment this is nothing but at vague suggestion. As we shall see, however, inclusion of later texts in the argument will make it much stronger.

The procedure is particularly adjusted to the presence of precisely 4 sides, but could of course be adapted to other coefficients. As a matter of fact, however, this never happens. No other Old Babylonian text we know of proceeds in a similar way.

It is likely that #22 of the tablet, of which almost nothing remains, stated its result by means of the verbal form *imtaḫḫar*, and possible that it also stated the unit. Insofar as the conserved problems are conserved, #23 is alone in both regards. Elsewhere, this use of *imtaḫḫar*, references to "each" of the sides, and explicitation of the unit nindan tend to go together – often in the same text

250 The exception is "virtual" in the sense that the problem asks for the construction of a quadratic border around the corresponding mid-square □(s) and with the same area; this interpretation of the problem follows from analysis of the procedure text UET V, 864 – see below, p. 250.

and phrase, and always in texts that belong to the same group.

These characteristics of the problem will turn out to be very informative, and we shall return to them. Before leaving the text, we shall comment briefly on the way the procedure is explained. Here, there are no striking deviations from the norms of the tablet as a whole, nor from what we know about the operations from other texts. Whatever references to a non-school context the author includes in the wording of the statement thus does not prevent him from formulating the (singular) procedure in accordance with his school canon. The sides are transformed into broad lines by means of a "projection", the quadratic complement is "appended", after the finding of the equalside the side of the complement is "torn out", and the resulting half of the side is "repeated" concretely to two.

The one feature that may astonish is that the "projection" itself is "appended" in line 14. We remember the care with which 1 was "made hold" in #14 in order to express $\square(u)$ in terms of $\square(s)$ (see p. 75). The probable explanation is that the square held be the "projection" is *already there* (as in the various cases where rectangular or square areas are found by "raising"), and that this square configuration is *parametrized by* and hence potentially *identified with* its side, in agreement with the general view of the square configuration in terms of the "confrontation", the side which confronts itself. Afterwards (in line 15), the "projection" is "torn out" again, but this time in its role as a line.

Chapter VI
Quasi-algebraic Geometry

Introductory Considerations

The texts that were presented in Chapters III and V should give a comprehensive view of that part of Old Babylonian mathematics that can somehow be regarded as "algebraic". However, if the underlying understanding, the "standard representation" and most of the techniques were geometric; if, furthermore, the authors of the theme texts show by their organization of the material to consider geometric configuration primary with respect to type and technique: then inspection of texts that are concerned with geometry but whose methods cannot be characterized as algebraic in the strict sense is likely to throw new light even on the more properly algebraic techniques and concepts.

We shall therefore look at a selection of geometric texts that serve this purpose, and which may hence be characterized as "quasi-algebraic". At first, however, a few preliminary observations on various aspects of Old Babylonian geometric thought may be worthwhile.

Angles and Similarity

The notion of the angle was already touched in passing (p. 105). It has been a recurrent claim that the Babylonians did not possess the concept of the angle – see, for instance, [TMB, xvii], [Gandz 1939: 415–418], and [TMS, 4]. Of these three examples, only Gandz explains precisely which of many versions of this concept is intended when *the* concept is spoken of – namely, the "angle as a measurable quantity in the modern or Greek sense of the word" (p. 416). Thureau-Dangin's tacit understanding may have been similar, but Bruins, when quoting it, takes it to imply that, *a fortiori*, the notion of triangles having the same angles was unknown to the Babylonians – neglecting that similarity ("having the same shape", corresponding to the Euclidean notion of being "given in shape") may be a primitive and not a derived notion (as it has become in *Data*, def. 3[251]).

As Gandz formulates himself, all evidence indicates that he is right. If we return to the field plan in Figure 23, only right angles are rendered as right angles, the others, being immaterial with respect to the area calculation, are distorted in a way that either requires a very conscious decision to distort or, more likely, absence of the idea that the angular magnitude of a corner should be its very essence. What is essential is whether the corner is between segments that contain *and determine* an area, or not. Summarized in a pun, the Old Babylonian *right* angle was understood in its opposition to a *wrong angle*.[252]

Exactly *how* it was understood we do not know – the texts speak too little about it to allow us to understand – but it is beyond doubt that a (probably intuitive) concept of similarity or "same shape" was at hand. For rectangular figures this is already evident from the use of the single false position and the use of the *makṣarum* or "bundling" technique. But we need not restrict the assertion to the rectangular case: the tables of technical igi.gub constants also give constants for the area of many other shapes. They indicate the value of various parameters of such figures in the case where one linear parameter is 1. If instead this parameter is p, coefficients for linear magnitudes are "raised" to p and areas to $\square(p)$ (or p and $\square(p)$, respectively, "raised" to the coefficient).[253]

Perpendicularity and Orientation

Beyond the non-wrongness of the right angle, we may still ask how this particular angle was understood. On this question, Thureau-Dangin [TMB, xviii], once again echoed with approval by Bruins [TMS, 4], asserts that the Babylonians "only understood the perpendicular, which they called the 'descendant', in the etymological sense of the term (the direction of the plumb

---

251 "Rectilinear figures are said to be given in shape if the angles are given one by one and the ratios of the sides to each other are given" [ed. Menge 1896: 1].

252 In Babylonian astronomical texts from the first millennium we do find something which modern mathematical thinking might see as quantified angles (being blind to its own conflation of concepts which only merge when a particular theoretical substructure is presupposed). However, by measuring *distances in heaven* in length units in discordant ways, these texts demonstrate that such distances are not seen in terms of the angle between sighting lines but in the likeness of linear distances or corresponding travel times; see [Powell 1990: §§1.2e, 1.2k-l].

253 Detailed documentation – including both the tables of coefficients and their use in procedure texts – and discussion will be found in [Robson 1999: 34–56].

line)". This is clearly wrong, and an example of the extent to which our own mathematical metaphors have penetrated our thinking and prevents us from seeing that Babylonian metaphors which coincide with ours *are* metaphors, transferred senses. Whether the *muttarittum*, "descendant", is a side that is practically perpendicular to the base of a triangle or it is a genuine height, so much is certain that it is a horizontal and no vertical dimension. The Babylonians would speak of the end of a rectangular field that approaches the irrigation canal as the "upper" and of the other as the "lower" front or width, showing thus to share our verticality metaphor for one of the horizontal dimensions (and one of the abstract dimensions of a tablet which, like the paper, could be turned in any direction but would rarely be precisely vertical). But they might also speak at the same time about the "upper" and "lower length" of a trapezium, thus showing that the metaphorical plumb line might point in several directions.

Normally, the direction of a tablet that is considered "upwards" points toward the left – at least as the tablet is rendered in the modern hand copy, but also more or less as the tablet would be held by the scribe. This directional metaphor is less natural than ours, which makes the metaphorical eye movement "upwards" on the paper coincide with the real upward movement. The more awkward Babylonian metaphor has historical origins – cf. [Edzard 1980: 546*f*]. When writing was invented in the fourth millennium, tablets were small, and the writing direction was "vertical", that is, toward the writer. As tablets grew larger, the scribes were forced to hold them differently, turning them somewhat less than 90° anti-clockwise. Thereby, what was originally "up" was turned more or less toward the left, and writing was made approximately from left to right.

But the actual rotation of tablets (which may perhaps have taken place toward the mid-third millennium) did not affect what was considered "up" and "down" with respect to the direction of writing – which should not astonish readers who understand perfectly well that "above" in the book they are reading is probably in one of the pages that have been turned left- and downwards. It was well understood in the Old Babylonian period that the writing direction on tablets was ideally meant to be vertical, as made clear by documents that are cut in stone, for instance, the Codex Hammurapi: they are still oriented in the old way. It is also fairly evident from the variations of sign forms from the outgoing third millennium that the underlying picture was still taught and thought about – which means that the pictorial upward direction was seen directly in the signs. Although the line of thought seems enticing, it is therefore not possible to argue from the terms "upper" and "lower width" (etc.) that the mathematical problems in which they occur go back to the first half of the third millennium, as once done by Kurt Vogel [1958; 1959: 14*f*].

Rectangles, Triangles, Trapezia, and "Surveyors' Formula"

The whole Old Babylonian second-degree algebra builds on the determination of square and rectangular surfaces as products of length and width; this is also the foundation for the linking of the length and area metrologies (based since long on the nindan and the sar or square nindan, respectively). Triangular areas, as mentioned, were determined by "raising the length to the moiety of the width" (or vice versa); that these dimensions would be thought of as mutually perpendicular at least in practical mensuration seems obvious from field plans like the one of Figure 23. Some of the text examples that follow will also show how the perpendicular side might be distinguished verbally from the non-perpendicular side.

As we have seen repeatedly, the areas of trapezoidal fields were determined as the product of the length and the average width (still by "raising"); similarly, the frequent "repetitions" of trapezia so as to obtain rectangles show that the underlying idea is that the length and the widths are mutually perpendicular; IM 52301, by identifying *the* length of the statement with an "upper length" in the proof also expresses awareness that the two lengths are different but only one of them relevant (and relevant of course only in the case that the trapezium is practically right).

Actual mensuration takes place in the landscape and not in an abstract Euclidean plane, and it is made by means of reeds or other physical tool. The same applies when fields are marked off. In this situation, angles can only be *practically* right; even if rectangles are aimed at, opposing sides will turn out to be only approximately equal; small hills and depressions will give rise to further deviations of the true area from what the product of length and width predicts.

For approximately rectangular fields, the Babylonian calculators would make use of the "surveyors' formula" (which, as this conventional name suggests, was not their exclusive property), according to which the area is determined as the product of average length and average width.[254] The formula is evidently but trivially correct for rectangles, and yields too high results in all other cases.[255] For near-rectangular areas, however, the outcome is acceptable, and in real mensuration the formula would very rarely be used when inadequate – a calculation from the 24th century whose result is some 10% above the true value (and which Allotte de la Fuÿe characterizes as "completely unacceptable") is indeed quite exceptional, and other known cases are much more satisfactory. The surveyor-scribes obviously knew that fields which did not come very close to a rectangle or a rectangular trapezium (whose area appears to have been found as the product of the "good" length and the average width) had to be dissected into adequate pieces[256].

---

254 A terminological *caveat* may be needed. According to our habits, mathematical "formulae" are stated in algebraic letter formalism; in this sense, no formulae are found in pre-Cartesian mathematics. But if we see the essence of the formula in the fixed prescription of mathematical operations (though not necessarily in the details of their order, as in an algorithm), then "average width times average length" and "$^1/_{12}$ of the square on the circumference" are formulae in full right.

255 See, e.g., [Allotte de la Fuÿe 1915: 141*f*].

256 Of the 222 fields inventoried in the "round tablets" from the province of Lagash that are analyzed by Mario Liverani [1990: 160–166 (= Figures 8–14)], 172 are defined by length (mir, "north") and width (kur, "east") alone and thus ideally to be thought of as rectangles, while all of the remaining 50 appear to have been

In Old Babylonian teaching, the formula was sometimes used as a pretext for the formulation of complex mathematical problems without any care for its plausibility. Below, we shall see an example (YBC 4675, p. 244), in which the area that follows from the formula is almost twice as large as the maximal area the sides can contain. But actual surveying documents demonstrate that this lack of interest in the precision of the formula was a consequence of its function in the specific context, not of ignorance.

IM 55357[257]

Like IM 52301, this tablet comes from Eshnunna; it is slightly older, however, probably to be dated c. 1800 BCE. It may therefore be the earliest extant Old Babylonian text.

The diagram in Figure 58 reproduces the features of a drawing on the tablet as faithfully as possible, including the location and orientation of numbers (the letters are evidently added). The numbers show that ΔABC is right, and that the others are cut off by successive heights. All triangles are

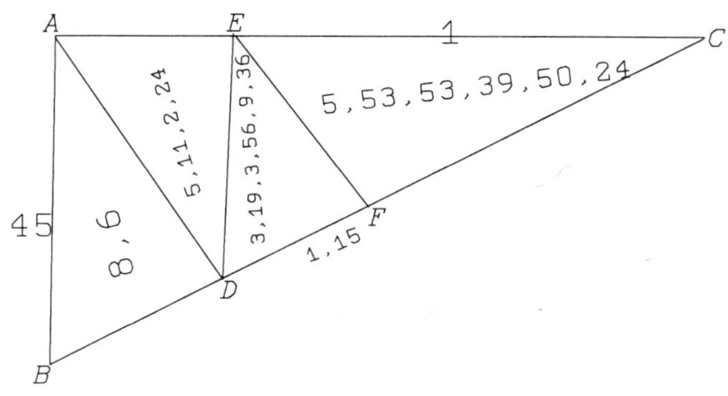

Figure 58. The diagram of IM 55357, with added letters for identification.

thought of as practically rectangular trapezia. The alternative interpretation – that they are practically isosceles, and the one length/width which is stated therefore the identical length of two opposing sides – can be safely dismissed. If such shapes occurred, non-isosceles trapezia defined by four different sides would certainly also turn up.

In a single example from the 21st century BCE, in which the calculated area is at least 51% above what could be the true area if the indicated measures are what they pretend to be, the scribe is likely to have created his numbers backwards after the actual division of a terrain which was confined by a curved stream – for instance, by using a measured height as his length. Alternatively, he used his monopoly on mathematical knowledge to pull the leg of his corrupt superior – cf. [Høyrup 1995].

257 Based on the transliteration, photograph, and hand copy in [Baqir 1950].

thus right, and all are similar to the 3-4-5 triangle which was known to be right (in the Old Babylonian sense discussed above). We observe that the text distinguishes the length *simpliciter*, the length that would serve the determination of the surface, from the "long length", that is, the hypotenuse.

1. A triangle. 1` the length, 1`15 the long length, 45 the upper width.
saĝ.dù 1 uš 1,15 uš gíd 45 saĝ.ki an.ta

2. 22`30 the complete surface. In 22`30 the complete surface, 8`6 the upper surface.
22,30 a.šà til *i-na* 22,30 a.šà til 8,6 a.šà an.ta

3. 5`11°2′24″ the next surface, 3`19°3′56″9‴36⁗ the 3rd surface.
5,11,2,24 a.šà ta 3,19,3,56,9,36 a.šà 3-kam

4. 5`53°53′39″50‴24⁗ the lower surface.
5,53,53,39,50,24 a.šà ki.ta

5. The upper length, the shoulder length, the lower length and the descendant what?
uš an.ta uš.múrgu uš.ki.ta *ù mu-tar-ri-it-tum mi-nu-um*

6. You, to know the proceeding, igi 1`, the length detach, to 45 raise,
za.e ak.ta.zu.un.dè igi 1 uš du$_8$.a *a-na* 45 íl

7. 45′ you see. 45′ to 2 raise, 1°30′ you see, to 8`6 the upper surface
45 igi.dù 45 nam 2 íl 1,30 igi.dù 1,30 nam 8,6 a.šà an.ta

8. raise, 12`˹9˺ you see. (By) 12`9, what is equalside? 27 is equalside.
íl 12,˹9˺ igi.dù 12,9 a.na.àm íb.si$_8$ 27 íb.si$_8$

9. 27 the width. 27 break, 13°30′ you see. Igi 13°30′ detach,
27 saĝ {¿erasure?} 27 *ḫe-pé* 13,30 igi.dù igi 13,30 du$_8$.a

10. to 8`6 the upper [surf]ace raise, 36 you see, the length (which is) the counterpart of the length 45, the width.
nam 8,6 [a.š]a an.ta íl 36 igi.dù uš gaba uš 45 saĝ.ki

11. Turn around. The length 27, of the upper triangle, from 1`15 tear out,
na-ás-ḫi-ir uš 27 saĝ.dù an.ta *i-na* 1,15 ba.zi

12. 48 leave. Igi 48 detach, 1′15″ you see, 1′15″ to 36 raise,
48 íb.tag$_4$.a igi 48 du$_8$.a 1,15 igi.dù 1,15 nam 36 íl

13. 45′ you see. 45′ to 2 raise, 1°30′ you see, to 5`11°2′24″ raise,
45 igi.dù 45 nam 2 íl 1,30 igi.dù 1,30 nam 5,11,2,24 íl

14. 7`46°33′36″ you see. (By) 7`46°33′36″, what is equalside?
7,46,33,36 igi.dù 7,46,33,36 a.na.àm íb.si$_8$

15. 21°36′ is equalside, 21°36′ the width of the 2nd ⟨tria⟩ngle.
21,36 íb.si$_8$ 21,36 saĝ.ki ⟨saĝ⟩.dù 2-kám

16. The moiety of 21°36′ break, 10°48′ you see. Igi 10°48′ detach,
BA 21,36 ⟨*hi*⟩*-pí* 10,48 igi.dù igi 10,48 du$_8$.a

17. to ⟨...⟩
nam ⟨...⟩

Beyond what can be seen in the drawing and deduced from the numbers, the text does not specify how the partial surfaces are defined. From the computations is follows, however, that Δ*ABD* is understood to be similar to Δ*ABC*, that Δ*ADE* is similar to the remainder Δ*ADC*, etc. We may suppose that the calculator who constructed the problem used this definition of the partial triangles to determine their surfaces – Δ*ABD* as □($\frac{3}{5}$) times the "complete surface" Δ*ABC*, Δ*ADE* similarly as □($\frac{3}{5}$) times the remaining surface of Δ*ADC* (which could be found by subtraction), or possibly as □($\frac{4}{5}$) times Δ*ABD*; rather unlikely is a direct computation of the successive lengths *AD*, *DE*, and *EF* and the corresponding bases and ensuing determination of the surfaces.

Even the very identification of the "upper length", the "shoulder length", the "lower length" and the "descendant" is in principle uncertain, since the lines that are computed afterwards are termed differently – but it would be strange if the lengths were not the lengths (*simpliciter*) of the corresponding partial triangles (*AD*, *DE*, and *EF*), and the "descendant" one of the segments *EC* and *FC*. The "upper length" will thus be identical with "the length (which is) counterpart of the length 45, the width", whereas the "shoulder length" is likely also to have been spoken of as "the length of the 2nd triangle".

The solution makes use of that doubling of a right triangle into a rectangle which we encountered *inter alia* in connection with the siege ramp of BM 85194 #25–26, and of the scaling in one dimension which transforms a rectangle into a square. At first, indeed, the text calculates the ratio between the length and the width of Δ*ABC*, and presupposes it to apply to Δ*ABD*. This is what shows that Δ*ABD* is understood to be similar to Δ*ABC*. Multiplying the area 8`6 by twice this ratio gives us the area of the square on the width *BD*,[258] from which *BD* itself is found in line 8 as the corresponding "equalside". In this operation it is thus used that Δ*ABD* is right, as is Δ*ABC*. The height *AD* is found (line 10) from the area and the "moiety" of *BD*.

Line 11 finds *DC* as 1`15–*BD* = 48, and uses this to find the scaling ratio which transforms the doubled Δ*ADC* (and, it is obviously assumed, the doubled Δ*ADE*) into a square on its width – Δ*ADE* being thus understood to be similar to Δ*ADC*. That it is also similar to Δ*ABD* and Δ*ABC* plays no role; by looking at the numbers the calculator will certainly have recognized that by cutting off the triangle Δ*ABD* from Δ*ABC* he obtains a third triangle Δ*ADC* with the same ratio between the sides, and thus of the same shape; but this knowledge is seen to be intentionally avoided in the argument. Continuing as before, *AE* is found in line 15 to be 21°36´. The text breaks off before *DE* is found, but it is evident how the computation would have to go on.

258 Since it is the ratio that is multiplied and not the area, "raising" is used instead of "repetition". The latter operation could only come in play if the surface were multiplied *first* by 45´, thus producing an isosceles triangle that could be combined with its own mirror image as a square.

The use of the term gaba, "counterpart", does not coincide exactly with that of the preceding texts – a length 36 is certainly not identical with a length 45. This aberrant usage shows, however, that *being legs of the same corner* is at least as fundamental an aspect of the mathematical concept as numerical equality, in good agreement with what happened in the algebraic texts.

The scaling operation and transformation of a right triangle into a square coincides with what we saw in the ramp problems (paradoxically, the process is less strictly geometric in this properly geometric text than in the algebraic ramp problem). Similar to what we have seen is also the purist will to distinguish data from that which is merely known.

VAT 8512[259]

Like those of VAT 8389 and 8391 (above, pp. 77*ff*), this problem deals with a field that is subdivided into two "plots". At this point, however, the similarity stops, and the present text is in fact an ingenious piece of pure geometry.

The field is triangular, and for convenience we may assume it to be right, in which case the "descendants" are sections of the length (*simpliciter*)[260]. A *pirkum*, "bar" (a parallel transversal; from *parākum*, "to put oneself across, to bar"), separates the two plots from each other, and we are told the width, the difference between the partial areas, and the difference between the two partial lengths.

Obv.

1. [A triangle. 30 the width. In the inside two p]lots,
[ᶦsağ.dú 30 sağ *i-na li-ib-bi ši-it-ta*ˀ *t*]*a-wi-ra-tum*

2. [the upper surface over] the lower [surface], 7` went beyond.
[ᶦ...ˀ a.šà an.ta ugu a.šà] ki.ta 7 *i-tir*

3. [The lower] de[scendant over the] upper [descendant], 20 went beyond.
m[*u-tar-ri-tum* ki.ta ugu *mu-tar-ri-tim*] an.ta 20 *i-tir*

4. The descendaˡnts and the b]ar wha[t]?
mu-tar-ri-d[*a*]-ˡ*tum ù pi-i-i*]*r-kum mi-nu-*[*u*]*m*

---

259 Based on the transliteration [MKT I, 341*f*]; cf. [TMB, 101–103] and [von Soden 1939: 148].

260 In principle, any triangular shape would do, if only the "descendants" were sections of the height. But since empty generalization was no Babylonian vice we may safely assume that a "right" and no "wrong" triangle was meant.

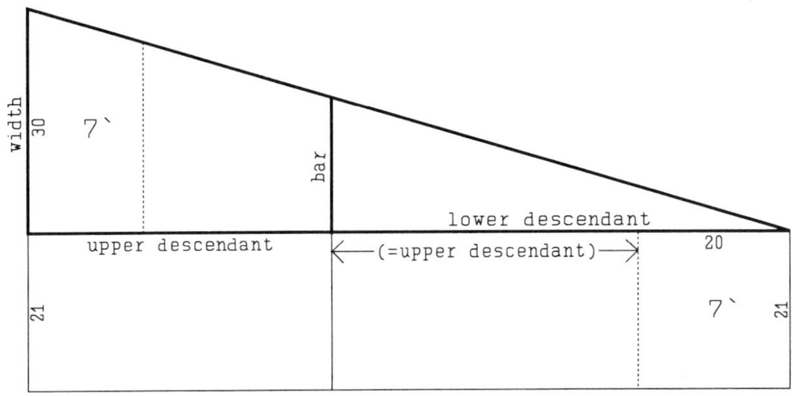

Figure 59. The divided triangle of VAT 8512, with completing rectangle.

5. And the surfa[ces] of the tw[o pl]ots wha[t?]
ù a.š[a] *ši-it*[*-ta ta-wi*]*-ra-tum mi-nu-u*[*m*]

6. You, 30 the width posit, 7` which the upper surface over the lower surface went beyond posit,
at-ta 30 sağ ğar.ra 7 *ša* a.šà an.ta ugu a.šà ki.ta *i-te-ru* ğar.ra

7. and 20 which the low[er d]escendant over the upper descendant went beyond po[si]t.
ù 20 *ša mu-tar-ri-t*[*um* k]i.ta ugu *mu-tar-ri-tim* an.ta *i-te-ru* ğ[ar.r]a

8. Igi 20 which the lower descendant over the upper descendant went beyond
igi 20 *ša mu-tar-ri-tum* ki.ta ugu *mu-tar-ri-tim* an.ta *i-te-ru*

9. detach: 3´ to 7` which the upper surface over the lower surface went beyond
pu-ṭur-ma 3 *a-na* 7 *ša* a.šà an.ta ugu a.šà ki.ta *i-te-ru*

10. raise, 21 may your head hold!
íl 21 *re-eš-ka li-ki-il*

11. 21 to 30 the width append: 51
21 *a-na* 30 sağ *ṣi-ib-ma* 51

12. together with 51 make hold: 43`21
it-ti 51 *šu-ta-ki-il-ma* 43,21

13. 21 which your head holds together with 21
21 *ša re-eš-ka ú-ka-lu it-ti* 21

14. make hold: 7`21 to 43`21 append: 50`42.
šu-ta-ki-il-ma 7,21 *a-na* 43,21 *ṣi-ib-ma* 50,42

15. 50`42 to two break: 25`21.
50,42 *a-na ši-na ḫe-pé-ma* 25,21

16. The equalside of 25`21 what? 39.
íb.si$_8$ 25,21 *mi-nu-um* 39

17. From 39, 21 the made-hold tear out, 18.
i-na 39 21 *ta-ki-il-tam ú-sú-uḫ-ma* 18

18. 18 which you have left is the bar.
18 *ša te-zi-bu pi-ir-kum*

19. Well, if 18 is the bar,
ma šum-ma 18 *pi-ir-kum*

20. the descendants and the surfaces of the tw[o plots what?]
mu-tar-ri-da-tum ù a.šà *ši-i[t-ta ta-wi-ra-tim mi-nu-um]*

21. You, 21 which together with it[self you have made hold, from 51]
at-ta 21 *ša a-na r[a-ma-ni-šu tu-uš-ta-ki-lu i-na* 51]

22. tear out: 30 you le[ave. 30 which you have left]
ú-sú-uḫ-ma 30 *te-z[i-ib* 30 *ša te-zi-bu]*

23. to two break, 1[5 to 30 which you have left raise,]
a-na ši-na ḫe-pé-ma 1[5 *a-na* 30 *ša te-zi-bu* íl]

24. 7`30 may [your] head [hold]!
7,30 *re-eš[-ka li-ki-il]*

Edge

1. 18 the ba[r together with 18 make hold:]
18 *pi-i[r-kam it-ti* 18 *šu-ta-ki-il-ma]*

2. 5`24 [from 7`30 which your head holds]
5,24 [*i-na* 7,30 *ša re-eš-ka ú-ka-lu*]

3. tear [ou]t: 2`6 you l[eave].
ú-sú-[u]ḫ-ma 2,6 te-[zi-ib]

Rev.

1. What to 2`6 may I po[sit]
mi-nam a-na 2,6 *lu-uš[-ku-un]*

2. which 7` which the [upper] surface [over] the lower surface went [beyond] gives me?
ša 7 *ša* a.šà [an.ta ugu] a.šà ki.ta *i-[te-ru] i-na-di-nam*

3. 3° 20′ posit. 3° 20′ to 2`6 raise, 7` it gives you.
3,20 ĝar.ra 3,20 *a-na* 2,6 íl 7 *it-ta-di-kum*

4. 30 the width over 18 the bar what goes beyond? 12 goes beyond.
30 saĝ ugu 18 *pi-ir-ki mi-nam i-tir* 12 *i-tir*

5. 12 to 3° 20′ which you have posited raise, 40.
12 *a-na* 3,20 *ša ta-aš-ku-nu i-ši* 40

6. 40 the upper descendant.
40 *mu-tar-ri-tum* an.ta

7. Well, if 40 is the upper descendant,
ma šum-ma 40 *mu-tar-ri-tum* an.ta

8. the upper surface is what? You, 30 the width,
a.šà an.ta *mi-nu-um at-ta* 30 saĝ

9. 18 the bar accumulate: 48 to two break: 24.
18 *pi-ir-kam ku-mur-ma* 48 *a-na ši-na ḫe-pé-ma* 24

10. 24 to 40 the upper descendant raise, 16`.
24 *a-na* 40 *mu-tar-ri-tim* an.ta íl 16

11. 16` the upper surface. Well, if 16` the upper surface,
16 a.šà an.ta *ma šum-ma* 16 a.šà an.ta

12. the lower descendant and the lower surface what?
mu-tar-ri-tum ki.ta *mi-nu-um ù* a.šà ki.ta *mi-nu-um*

13. You, 40 the upper descendant to 20 which the lower descendant over the upper descendant goes beyond
at-ta 40 *mu-tar-ri-tam* an.ta *a-na* 20 *ša mu-tar-ri-tum* ki.ta ugu *mu-tar-ri-tim* an.ta *i-te-ru*

14. append, 1` the lower descendant.
ṣi-ib-ma 1 *mu-tar-ri-tum* ki.ta

15. 1[8] the bar to two break: 9
1[8] *pi-ir-kam a-na ši-na ḫe-pé-ma* 9

16. to 1` the lower descendant raise, 9`.
a-na 1 *mu-tar-ri-tim* ki.ta íl 9

17. 9` the lower surface.
9 a.šà ki.ta

The computation of the "bar" makes use of an ingenious trick (first unravelled by Solomon Gandz [1948: 36*f*], more clearly explained by Peter Huber [1955]), belonging to the same genre as the quadratic completion. Just as the quadratic completion allows us to replace a rectangle by a square, the present completion reduces the unequal partition of a triangle to the bisection of a trapezium; that problem was solved by Mesopotamian surveyors at least since the 23rd millennium BCE – see [Friberg 1990: 541] concerning the tablet IM 58045. As shown in Figure 59, a rectangle is joined to the triangle, and its width (21) determined in obv. 8–10 in such a way that the rectangular area which is adjacent to the excess of the lower over the upper descendant (20) equals the excess of the upper over the lower plot (7`); it is found (obv. 10) that this requires the rectangle to have the width 21. The prolonged bar will

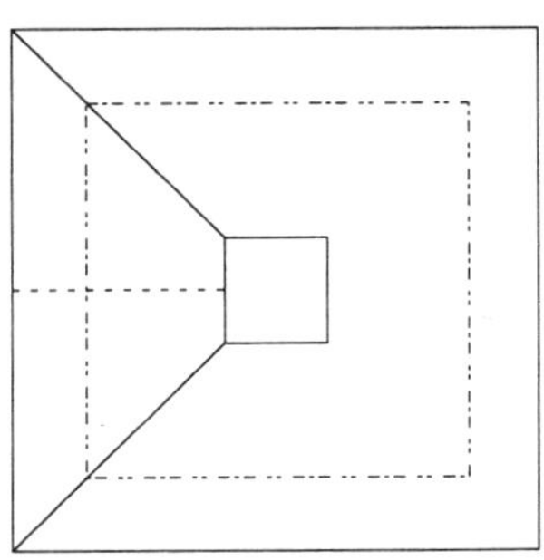

Figure 60. The bisection of a trapezium.

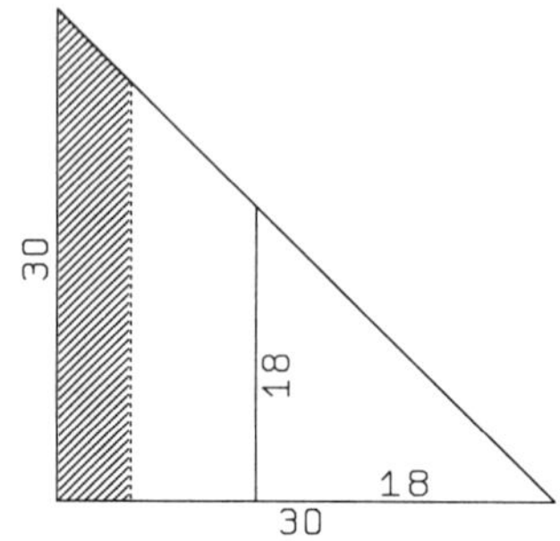

Figure 61. The scaled triangle.

thus bisect the trapezium that results when the rectangle is joined to the triangle.

In obv. 11–15, the areas of the squares on the parallel sides of the trapezium are found and their average (the "moiety" of their sum) is computed. This average is indeed the square on the bisecting transversal, as can be easily argued from Figure 60 (whether one looks at the isosceles or one of the right trapezia – our usual scaling operation in one dimension will have to be applied if the angle at the base is not 45°).[261] Since this transversal turns out to be 39, the original "bar" must be 39–21 = 18 (obv. 18).

What follows next is an elimination of the added rectangle, from which we recalculate the width of the triangle, and then in obv. *22f* the area of the isosceles right triangle on this side (7`30). This means that the triangle has been submitted to a scaling operation which transforms it into a semi-square – cf. Figure 61. In the next step, the square on the bar is found (5`24), i.e., twice the area into which the lower plot is scaled or, indeed, the lower plot and as much of the upper plot as equals the lower plot (the same trick as was used in VAT 8391 #3). Subtraction of 5`24 from 7`30 thus leaves that which results from the scaling of the excess of the upper over the lower plot, i.e., the shaded area of Figure 61 – namely, 2`6.

This allows us to find the inverse scaling factor (3°20′, rev. 3); "raising" the difference (12) between the width and the bar (which equals their distance in Figure 61) to 3°20′ gives their distance in the original triangle (40, rev. *5f*), whence the upper surface can be computed (16`, rev. 11). The lower descendant is found from the upper descendant and from the difference between the two (1`, rev. 14), and the lower surface finally by the usual formula for the triangular area.

261 As we shall see when analyzing the text YBC 4675 (p. 247), the bisection problem is indeed solved by means of scaling and by taking the difference between squares.

Str 367[262]

This text presents us with another variant of the subdivided figure – this time a trapezoidal field divided by a dal, the logogram for *tallum*, a "crossbeam" etc.,[263]. We are told the partial surfaces together with the ratio between the upper and the lower length – all nicely shown on the tablet in the diagram which introduces the statement. Finally, we are told that the difference between the widths amounts to 36 – but it is identified as the accumulation of two components, namely, the excess of the upper width over the "crossbeam" and of the "crossbeam" over the lower width.

Obv.

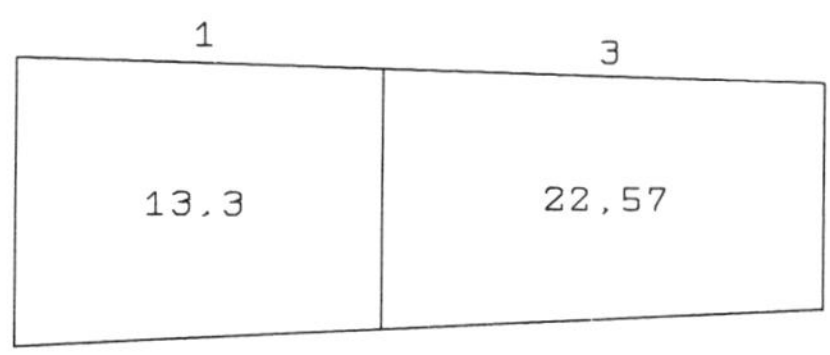

1.

2. A trapezium. In the inside two rivers[264]. 13`3 the upper surface,
sağ.ki.gud *i-na* šà 2 íd.meš 13,3 a.šà an

3. 22`57 surface 2. The 3rd part of the lower length by
22,57 a.šà 2 i[gi] 3 ğál uš ki *i-n*[*a*]

4. the upper length. That which the upper width over the crossbeam went beyond
uš an.na *ša* sağ an.na ugu dal dirig

5. and (that which) the crossbeam over the lower width went beyond, I have accumulated, ⌊36⌋.
ù dal ugu sağ ki.ta dirig ğar.ğar ⌊36⌋

262 Based on the transliteration in [MKT I, 259*f*], cf. [TMB, 90].

263 It is often supposed that dal is a logogram for *pirkum* in the mathematical texts, but inspection of the supposed evidence shows that the conclusion does not follow. Instead, the text YBC 4675 (below, p. 244) uses dal and *tallum* about the same entity. Moreover, as Jöran Friberg reminds me (personal communication), dal and *pirkum* refer to different entities in the igi.gub table TMS III.

264 For unexplained reasons, this and certain other mathematical texts designate partial fields as íd/*nārum*. Thureau-Dangin [1940: 4] proposes that íd may serve in this context as a logogram for *tawirtum* (translated "plot" above), but the text IM 52916 rev. 15*f* [Goetze 1951: 139] seems to show that the Akkadian value even in a mathematical context is the usual *nārum*, "river" (unless, since the function is not precisely the same, already different mathematical contexts correspond to different readings).

6. Their lengths, the widths, and the crossbeam what?
uš.ne.ne sağ.meš *ù* dal en.nam

7. You, by your proceeding, 1 and 3 may you posit.
za.e kìd.da.zu.dè 1 *ù* 3 ḫé.ğar

8. 1 and 3 accumulate, 4. Igi 4 detach, 15′.
1 *ù* 3 ğar.ğar 4 igi 4 du$_8$-*ma* 15

9. 15′ to 36 lift, 9 it gives you. 9 to
15 *a-na* 36 nim 9 ^{in}sum 9 *a-na*

10. 1 lift, 9 it gives you. 9 to 3 lift, 27.
1 nim 9 ^{in}sum 9 *a-na* 3 nim 27

11. 9 that which the upper width over the crossbeam went beyond.
9 *ša* sağ an.na ugu dal dirig

12. 27 that which the crossbeam over the lower width went beyond.
27 *ša* dal ugu sağ ki.ta dirig

13. Igi 1 detach, 1 to 13`3 lift,
igi 1 du$_8$ 1 *a-na* 13,3 nim

14. 13`3 it gives you. Igi 3 detach, 20′ to
13,3 ^{in}sum igi 3 du$_8$ 20 *a-na*

15. 22`57 lift, 7`39 it gives you.
22,57 nim 7,39 ^{in}sum

Rev.

1. 13`3 over 7,39, what goes beyond?
13,3 ugu 7,39 en.nam dirig

2. 5`24 it goes beyond. 1 and 3 accumulate, 4.
5,24 dirig 1 *ù* 3 ğar.ğar 4

3. ½ of 4 break, 2. Igi 2 detach, 30′ to 5,24 ⟨lift,⟩
½ 4 gaz 2 igi 2 du$_8$ 30 *a-na* 5,24 ⟨nim⟩

4. 2`42 it ⟨gives you⟩, the false counting. ⟨Igi⟩ 2`42 is not detached.
2,42 ^{in}⟨sum⟩ *ma-nu* lul[265] ⟨igi⟩ 2,42 nu.du$_8$

5. What to 2`42 may I posit which 9 gives me?
en.nam *a-na* 2,42 ḫé.ğar *ša* 9 ^{in}sum

6. 3′20″ posit. Igi 3′20″ detach, 18 it gives you.
3,20 ḫé.ğar igi 3,20 du$_8$ 18 ^{in}sum

---

265 Neugebauer suggests that the sign MA may be a writing error for the missing sum, which leaves a NU which is rather hard to explain (neither the logogram for *ṣalmum*, "statue, picture", nor a nominalization of the Sumerian prefix nu., "person (taking care) of", seems very adequate). Thureau-Dangin reads MA NU as Akkadian *ma-nu*, which he finds inexplicable, but which should come from *manûm*, "to count", and thus mean "the counting", "the counted" or "the reckoning".

7. 18 to 1 lift, 18, the upper length. 18
18 *a-na* 1 nim 18 uš an.na 18

8. to 3 lift, 54 the lower {...} length.
a-na 3 nim 54 uš {ki uš} ki.ta

9. ½ of 36 break, this 1⟨8⟩ to 1`12 lift,
½ 36 gaz ne 17⟨+1⟩ *a-na* 1,12 nim

10. 21`36 from 36`, the surface, detach, 14`24.
21,36 *i-na* 36 a.šà du₈ 14,24

11. Igi 1`12, of the length, detach, 50´´ to 14`24 lift,
igi 1,12 uš du₈ 50 *a-na* 14,24 nim

12. 12 it gives you. 12 to 36 append: 48.
12 *in*sum 12 *a-na* 36 daḫ-*ma* 48

13. 48 the upper width. 12 to 27 append,
48 sağ an.na 12 *a-na* 27 daḫ

14. 39 the crossbeam. 12 the lower width it gives you.
39 dal 12 sağ ki.ta *in*sum

In rough outline, the procedure may be followed in Figure 62. At first, the total excess of the upper over the lower width is divided proportionally as the respective lengths, from which results that the excess of the upper length over the "crossbeam" is 9, whereas that of the "crossbeam" over the lower width is 27 (obv. 8–12).

From this point (obv. 13) until rev. 8 I follow the gross lines of an interpretation proposed by Jöran Friberg (personal communication), who justly criticized my first explanation (evidently, I remain responsible for the actual shaping of his idea). The parallel multiplications by igi 1 and igi 3 in lines 13–15 suggest that the "positing" of 1 and 3 in line 7 are intended as "false positions" for the two partial lengths.[266] If majuscules indicate true and minuscules false values, we thus have

$$\frac{W_u+D}{2}\cdot L_u = \frac{w_u+d}{2}\cdot l_u = 13`3\ ,\quad \frac{D+W_l}{2}\cdot L_l = \frac{d+w_l}{2}\cdot l_l = 22`57\ ,$$

$$l_u = 1\ ,\ \ l_l = 3\ .$$

The multiplications by igi 1 and igi 3 then tell us that

---

266 Friberg proposes 1` and 3`, in parallel with what I suggested tentatively in my analysis of BM 13901 and TMS XVI (pp. 77 and 87, respectively, cf. note 60). Because of the use of the same numbers as proportionals for the division of the excess in the following lines I find this implausible though not excluded.The disagreement has no influence on the rest of the interpretation apart from what is pointed out in note 267.

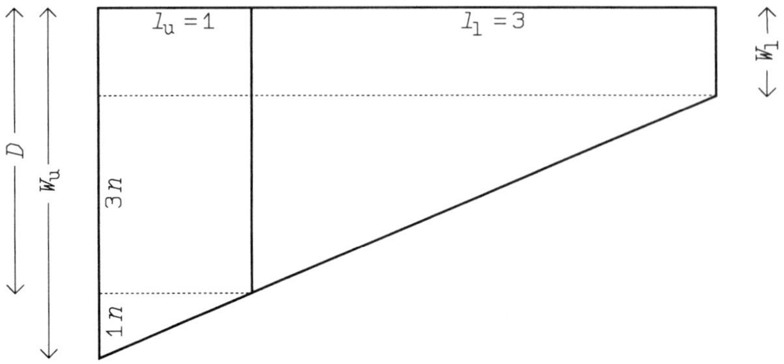

Figure 62. The field of Str 367.

$$\frac{w_u+d}{2} = 13`3\div1 = 13`3, \quad \frac{d+w_l}{2} = 22`57\div3 = 7`39$$

or possibly (but less likely, given that we also divide by 1, the false upper length[267]) in terms of "broad lines",

$$\sqsubset\!\sqsupset(\frac{w_u+d}{2},1) = 13`3\div1 = 13`3, \quad \sqsubset\!\sqsupset(\frac{d+w_l}{2},1) = 22`57\div3 = 7`39 .$$

Rev. 1–2 thus find that

$$\frac{w_u+d}{2} - \frac{d+w_l}{2} = \frac{w_u-w_l}{2} = 5`24 .$$

In rev. 2–4 we see that this magnitude, divided by $\frac{1-3}{2}$, yields a "false counting" (n). This obviously presupposes that

$$\frac{w_u-w_l}{2} = \frac{1+3}{2}\cdot n ,$$

that is, that n is the false value of that step which $w_l–w_l$ contained 1+3 times, and whose true value is known from obv. 9 to be 9.

Rev. 5 now asks for the factor ϕ that will transform the false value 2`42 into the true value 9, and finds it to be 3´20˝. Conversely, the false values of the partial lengths will be transformed into the true values when we multiply by ϕ^{-1} or igi 3´20˝, that is, by 18. This gives us $L_u = 18\cdot1 = 18$, $L_l = 18\cdot3 = 54$, as found in rev. 7–8.

The final part of the procedure can be followed directly in the diagram, where the trapezium is seen to consist of a triangle and an adjacent rectangle.

267 Taking as false values 1` and 3` (and dividing thus by 1` and 3`) would agree better with a "broad-line" interpretation.

In rev. 9 the base 36 of the triangle is broken and then lifted to 1`12, that is, to its length. The resulting surface is "detached" from the total surface 36`, leaving the rectangle. It is noteworthy that neither the total length 18+54 = 1`12 nor the total surface 13`3+22`57 = 36` have been found explicitly; most likely the reason is that the total configuration was already familiar, as we have seen on other occasions.[268] It is no less noteworthy that the same term is used when the triangle is removed from the trapezium as when an igi is found; this leaves little doubt that even the "detaching" of igi *n* is thought of as taking out 1 part from a bundle of *n* – cf. note 47.

The surface of the rectangle that is left when the triangle is "detached" turns out to be 14`24. This is lifted in rev. 11 to the igi of the length, whereby its width is found to be 12. Starting from the left of the diagram we therefore see that the upper width of the trapezium is the width of the triangle with the 12 "appended" (thus 48); the "crossbeam" is found by "appending" 12 to 27; finally, the lower width is seen to be identical with the width 12 of the rectangle. The sequence of operations shows clearly that 12 when first found is not intended as the lower width of the trapezium but only as that of the rectangle. As in VAT 8512, the rectangle is palpably there; at this final point of the calculation, even [MKT] finds it necessary to give a geometrical interpretation.

One of the reasons that the arithmetical interpretation of Babylonian algebra could remain unchallenged for fifty years is that the actual sequence of numerical steps of the Babylonian calculations and algorithms would rarely be different from what was to be expected within the arithmetical framework. Unexpected organizations of algorithms are therefore (like errors) privileged occasions for discriminating between contrasting interpretations – they are, so to speak, the "crucial experiments" of the history of mathematics.[269] Therefore, the sequence rev. 2–7 of the present text is an important clue.

---

268 It is also possible that the repetition in rev. 8 replaces a statement of the total length; it is indeed striking that the number 1`12 is identified in rev. 11 as the length while being unidentified at its first appearance.

36` *is* referred to as the surface in rev. 10, which almost certainly presupposes either that this value was supposed to be known before the problem starts, or that the author or copyist has forgotten the calculation. Since the calculation has no natural location within the preceding text, the former possibility is probably to be preferred. Cf. also p. 212 on the numbers written into the diagram of VAT 7532, which belongs to the same group as the present text.

269 As such, they suffer of course from the same weakness as other crucial discriminators, experimental or otherwise: they are able to discriminate between the possibilities that we already have in mind – but they cannot exclude explanations of the facts which we never thought of. This is where van der Waerden and Neugebauer are misled by the seemingly irrefutable crucial argument for the numerical and against the geometrical interpretation (the inhomogeneous additions of lines and surfaces, and of volumes and areas): after Pedro Nuñez's *Libro de algebra* from 1567, no mathematician seems to have imagined the possibility to think in terms of "broad lines".

Neugebauer characterizes the computation as "roundabout" (*umwegig*, [MKT I, 263]) and says that the meaning of the detour is unclear to him (showing thereby to be aware that it *has* a meaning[270]). The only situation where the presence of two mutually reciprocal factors of proportionality makes sense is indeed when they are used as scaling factors along the two dimensions of a surface, thereby conserving the area;[271] this is exactly what we encountered in YBC 4668, sequence C, #34, #38–53, cf. p. 200. But this means geometrical considerations are really involved, and that the explanation in purely numerico-algebraic terms may show that the Babylonian solution is correct but leaves it as a black box whose functioning we know nothing about.

YBC 4675[272]

This texts confronts us with yet another subdivision, namely, of an irregular quadrangle, defined neither as a rectangle (uš sağ) nor as a trapezium (sağ.ki.gud, literally "ox-head") but instead, it seems, as a configuration that is "held by length and length". The sides are 17 (upper width, w_u), 7 (lower width, w_l), 5`10 (first length, $l_1$) and 4`50 (second length, l_2); the area is stated to be 2 bùr = 1``, which follows from the sides if we apply the surveyors' formula, the average length being l = 5` and the average width being w = 12; but since the sum of the widths is not much superior to the difference between the lengths, any attempt to draw the figure in true proportions will show that this formula is highly unrealistic – it actually gives a value that is c. 1.8 times the largest area the sides can contain.

The problem is to find the "middle crossbeam" or transversal (presupposed to divide the lengths in the same proportion) that bisects the area, together with the segments into which the lengths are divided. The model that is used is

270 I use the occasion to make a personal note: as always when I work on this material, I remain deeply impressed by what was done in the thirties by Neugebauer, Thureau-Dangin, Gandz, Vogel, and von Soden – but not least by the profundity of Neugebauer's insights on points where he followed Newton's maxim and "made no hypotheses" that were not needed for his actual project.

271 In principle, even a product may of course do, if only it is not conflated with its resulting number but its factors are kept separate; but this requires that the product have a representation that allows us to see it thus – that is, either as a rectangle or some other concrete representation, or by means of symbolic algebra. But symbolic algebra was not at hand, and the only concrete representation of a product in the present context is exactly as a rectangle (e.g., commercial rates and amounts of silver as in TMS XIII would be meaningless here); therefore the surface with scalings appears to be the only possibility.

272 Based on the transliteration in [MCT, 44*f*]. [MCT, 45] also gives the text of the tablet YBC 9852, in which the lines rev. 7–16 of the present text are copied as a writing exercise.

that of the trapezium, whose bisection we have already seen in VAT 8512 (p. 234); in this case the bisecting parallel transversal does indeed divide the sides proportionally.[273]

Obv.

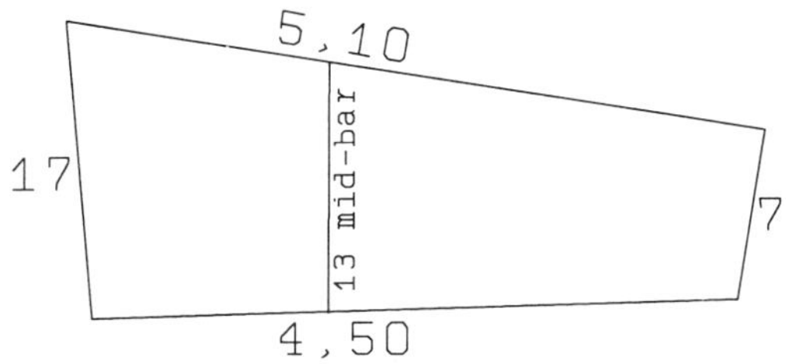

1. If a surface, length length holds, the 1st length 5`10, the 2nd length 4`50,
   šum-ma a.šà uš uš i.gu$_7$[274] uš.1.e 5,10 uš.2.e 4,50
2. the upper width 17, the lower width 7, its surface 2 bùr.
   sağ an.ta 17 sağ ki.ta 7 a.šà.bi 2(bùr)^{iku}
3. 1 bùr each, the surface I have divided into two, the middle crossbeam corresponding to what?
   1(bùr)^{iku}.ta.àm a.šà^{lam} *a-na ši-na* ⟨*a-*⟩*zu-ú-uz ta-al-li qá-ab-lu-ú ki ma-ṣi*
4. The longer length and the shorter length corresponding to what may I posit
   uš gíd.da *ù* uš lugúd.da *ki ma-ṣi lu-uš-ku-un-ma*
5. in order that 1 bùr be bordering? And to the second bùr
   1(bùr)^{iku} *lu-ú sà-ni-iq ù a-na* 1(bùr)^{iku} *ša-ni-im*
6. corresponding to what the longer length and the shorter length may I posit
   ki ma-ṣi uš gíd.da *ù ki ma-ṣi* uš lugúd.da *lu-uš-ku-un-ma*
7. in order that 1 bùr be bordering? The complete lengths,
   1(bùr)^{iku} *lu-ú sà-ni-iq* uš.ḫa *ga-me-ru-ú-tim*
8. both, you accumulate, [the]ir moiety you break:
   ki-la-a-al-le-e-en ta-ka-mar-ma ba-a-[*ši-n*]*a te-ḫe-pe-e-ma*
9. 5` comes up for you. Igi 5` which came up for you you detach:
   5 *i-il-li-a-kum* igi 5 *ša i-li-a-ku*[*m t*]*a-pa-ṭa-ar-ma*
10. (as) to the upper width which 10 over the lower width goes beyond,
    a-na sağ an.na *ša* 10 *e-li* sağ ki.ta *i-te-ru*
11. to 10 the overgoing you raise: 2′ it gives you. ⟨¿Raise to 1``, the complete surface, 2` it gives you?⟩,
    a-na 10 *wa-at-ri-im ta-na-aš-ši-ma* 2 *i-na-an-di-kum* ⟨...⟩

---

[273] Neugebauer and Sachs show in [MCT, 46*f*] that the reasoning of the text cannot be correct with the actual sides; there is no need to repeat their arguments.

[274] I read I.gu$_7$ as a homophonic variant of the usual ì.gu$_7$.

12. You turn around. 17, the upper width, you make hold:
ta-as-sà-ḫa-ar 17 sağ an.na *tu-uš-ta-ak-ka-al-ma*

13. 4`49 comes up for you. From inside 4`49,
4,49 *i-il-li-a-kum i-na li-ib-bi* 4,49

14. 2` tear out: 2`49 the re[s]t.
2 *ta-ḫa-ar-ra-aṣ-ma* 2,49 *a-ḫe-er-*[*t*]*um*

15. Its equalside you take:
íb.si$_8$-*šu te-le-qé-e-ma*

16. 13, the middle crossbeam, comes up for you.
13 *ta-al-lum qá-ab-lu-ú-um i-il-li-a-kum*

17. 13, the middle crossbeam that came up for [you]
13 *ta-al-lam qá-ab-li-a-am ša i-li-a-*[*kum*]

18. and 17, the upper width, you accumulate: [their] moiety [you brea]k:
ù 17 sağ an.na *ta-ka-mar-ma ba-a-*[*ši-na te-ḫe-pe*]*-e-ma*

19. 15 comes up for you. Igi 15 you de[tach,]
15 *i-il-li-a-ak-kum* igi 15 *ta-p*[*a-ṭa-ar-ma*]

20. to 1 bùr, the surface, you rai[se]:
a-na 1(bùr)^{iku} a.šà^{im} *ta-na-*[*aš-ši-m*]*a*

Rev.

1. 2` it gives you. 2` which came you for y[ou],
2 *i-na-an-di-kum* 2 *ša i-li-a-k*[*um*]

2. to 2´ the *arakarûm* you raise[:]
a-na 2 *a-ra-ka-re-e-em ta-na-aš-ši-m*[*a*]

3. 4 comes up for you. 4 which came up for y[ou]
4 *i-il-li-a-kum* 4 *ša i-li-a-k*[*um*]

4. to 2 UŠ you append: 2`4 the longer length.
a-na 2 UŠ *tu-uṣ-ṣa-ab-ma* 2,4 uš gíd.da

5. from the 2 UŠ n° 2 you cut off:
i-na 2 UŠ ki.2 *ta-ḫa-ar-ra-aṣ-ma*

6. 1`56 the shorter length, you make: 1 bùr is bordering.
1,56 uš lugúd.da *te-ep-pe-eš-ma* 1(bùr)^{iku} *sà-ni-iq*

7. You turn around. 13, the middle crossbeam
ta-as-sa-ḫa-ar 13 *ta-al-lam qá-ab-li-a-am*

8. which came up, and 7, the lower width, ⟨you⟩ accumulate,
ša i-li-a-kum ù 7 sağ ki.ta ⟨*ta-*⟩*ka-mar*

9. their moiety you break: 10 comes up for you,
ba-a-ši-na te-ḫe-pe-e-ma 10 *i-il-li-a-ak-kum*

10. igi 10 you detach, to 1 bùr, the surface, you raise,
igi 10 *ta-pa-ṭa-ar-ma a-na* 1(bùr)^{iku} a.šà^{im} *ta-na-aš-ši-ma*

11. 3 UŠ comes up for you. 3 UŠ which came up for you
3 UŠ *i-il-li-a-kum* 3 UŠ *ša i-il-li-a-ak-kum*

12. to 2′ the *arakarûm* you raise you raise:
a-na 2 *a-ra-ka-re-e-em ta-na-aš-ši-ma*

13. 6 comes up for you; 6 to 3 UŠ you append:
6 *i-il-li-a-ak-kum* 6 *a-na* 3 UŠ *tu-uṣ-ṣa-ab-ma*

14. 3`6 the longer length; 6 from 3 UŠ you tear out:
3,6 uš gíd.da 6 *i-na* 3`UŠ *ta-ḫa-ar-ra-aṣ-ma*

15. 2`54 the shorter length, you make hold:
2,54 uš lugúd.da *tu-uš-ta-ak-ka-al-ma*

16. 1 bùr is bordering.
1(bùr)[iku] *sà-ni-iq*

The determination of the "crossbeam" is made in obv. 7–16. The computation starts by finding the average length l, namely, as the "moiety" of the accumulated total lengths l_1 and l_2. Then this length is treated as the perpendicular length l of a right trapezium with widths w_u = 17 and w_l = 7 – see Figure 63. Its igi is "raised" to the difference d between the widths, and the resulting 2′ is "raised" to the total area $A = A_1+A_2$ = 1``. The latter step is forgotten in the text, but the difference in obv. 13*f* shows that 2′ has changed into 2`, which has no other possible explanation.

As usual, this multiplication of a surface by a factor of proportionality $\phi = d/l$ means that the surface has been submitted to a scaling in one dimension, namely, with a factor that changes l into $\phi l = d$, as shown in the lower diagram in Figure 63. This corresponds to the scaling of the triangle into a halved square in VAT 8512 – see p. 238 and Figure 61. The partial areas are changed into $\phi A_1 = S_1$ and $\phi A_2 = S_2$ but remain equal. The area itself is changed into $S = S_1+S_2 = \phi A$ = 2′·1`` = 2`.

In the next step, $\square(w_u)$ is constructed and found to be 4`49; from this, S is "torn out", and the rest is supposed to be the square $\square(c)$ on the "crossbeam" (as can be seen from the taking of the "equalside" in obv. 15). Indeed, since S_1 and S_2 are equal, S is also equal to S_1+S_3, and the rest thus to S_2+S_4, which is precisely $\square(c)$.

On p. 238 it was conjectured, with reference to Figure 60, that the formula for the bisection of the trapezium was found from consideration of two

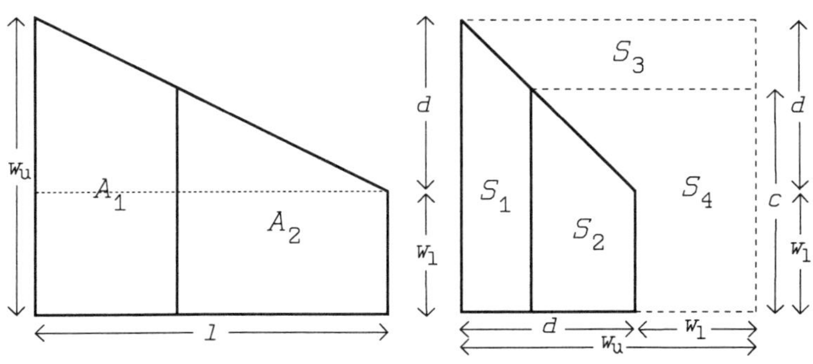

Figure 63. The finding of the crossbeam in YBC 4675.

concentric squares, and then extended by scaling to cases where the angle at the base differs from 45°; the present text confirms this conjecture. We observe that the usual formula $2\square(c) = \square(w_u)+\square(w_l)$ (which we know from VAT 8512) might have been used directly; the recourse to scaling shows both that the text aims at a situation where the correctness of the calculation can be comprehended, and that this situation is the one where the trapezium is cut out from a square.

The rest of the calculation is easy. In obv. 17–19, the average width of the upper area (the "moiety" of the "accumulated" w_u and c) is found to be 15; from an igi-division of its surface (1 bùr) its average length is seen to be 2` (rev. 1). In order to find the real lengths of the upper surface, this is "raised" to a factor 2´ called the *arakarûm*, a loanword derived from Sumerian a.rá.kará. The function in the text is obvious: it is the relative excess of long over average length – a parameter that is common to the upper, the lower, and the total surface because of the implicit condition that the lengths be divided proportionally. The calculation of the ratio is forgotten (maybe once again because of the reappearance of the number 2), but it will have been found as the ratio between (5`10–4`50)/2 = 10 and (5`10+4`50)/2 = 5`. Therefore the excess of the first long length over the first average length is 2´·2` = 4, and the long length itself is 2`4; the short length will be 2`–4 = 1`56. In rev. 7–16 the process is repeated for the second surface.

Some important terminological observations can be made in relation to procedures and concepts. The first concerns the *arakarûm* and its use. We notice that the passages rev. 4–6 and rev. 13–15 are constructed precisely as all other cases where two entities are produced from independently existing average and deviation; the *arakarûm* is thus likely to be *a general term for the ratio between deviation and average* (or, possibly, between "going-beyond" and "accumulation", cf. p. 188).

The etymology agrees acceptably well with this interpretation. a.rá, "step/steps of" we have seen used in several occasions about a factor, for instance, in the composite *aramaniātum* in IM 52301 (see note 242). kará is less clear, but may mean something like "oblique" (the sign is an obliquely oriented GANA, "field", and the sign name indicates this – [SL #105II]). The meaning of *arakarûm* may thus be something like "factor of obliquity".

Another interesting observation can be made by putting rev. 6 and rev. 15 in parallel, remembering also obv. 1. In rev. 15 we see that the longer and the shorter lengths are "made hold", which produces the required surface 1 bùr. Since they are certainly not to be multiplied (the area is found by "raising" their average to the average of 13 and 7), this really means what it says, that the area is contained geometrically between them; thereby the interpretation of the corresponding uš uš i.gu$_7$ in obv. 1 is confirmed.

In rev. 6, what must necessarily be the same process is termed differently, as a "making" (*epēšum*), the term which is translated "proceeding" when referring to the solution of a problem; in general usage, the term may also be used about "building" instead of *banûm* (AHw, 223b); this line is thus a

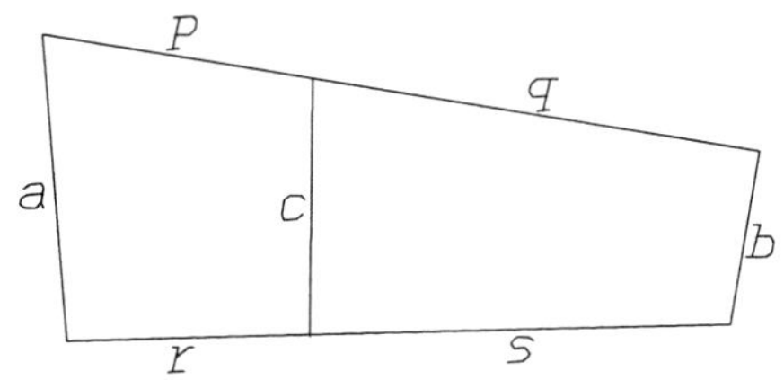

Figure 64. The partition of an irregular quadrangle.

confirmation of the close connection between "making (a length and a width) hold" and "building (a surface)".

Finally, in obv. 8, 18, rev. 9, the "moiety" of an accumulation is spoken of as "their moiety"; as in AO 8862 (above, p. 162), accumulations are thus seen as the aggregates of their constituents, not as the mere resulting numbers.

A brief mathematical note may be added. *Area* is an additive quantity: if an area is divided into two, the total area will be the sum of the component areas. But *computed area* is not necessarily additive: if computations are approximate, only approximate additivity can be expected, and if blatantly wrong formulae are employed, additivity need not result at all. In the present case, however, where the surveyors' formula is quite off the point but nonetheless used for all three areas, additivity holds good. How is that to be explained?

Using the letter symbolism of Figure 64, additivity for the surveyors' formula means (using for simplicity the quadrupled areas) that

$$(a+b)\cdot(p+q+r+s) = (a+c)\cdot(p+r)+(b+c)\cdot(q+s) ,$$

and by rearrangement

$$\frac{q+s}{p+r} = \frac{c-b}{a-c} \quad \text{or, equivalently,} \quad \frac{(q+s)/2}{(p+r)/2} = \frac{c-b}{a-c} ,$$

that is, that the average lengths of the two partial areas are in the same proportion as the consecutive decreases of width. This is evidently the case if we divide a trapezium by a parallel transversal – see, for instance, the left-hand diagram of Figure 63. Since this was the model that was used to find c, additivity followed automatically; but the true distance between the dividing points of the lengths is not likely to be 13 (since the given numbers do not determine the shape, it could happen by accident).

UET V, 864[275]

This text in Sumerian from early Old Babylonian Ur[276] was mentioned in connection with the discussion of BM 13901 #23, the problem about the "four fronts and the area" (see p. 222), as the only other text which provides a square, if not with four syllabic *pāt*, "fronts", then at least with the corresponding Sumerographic sağ 4.bi, "its 4 fronts". The terminology has presented interpreters with some difficulties, in particular because of disagreements about the meaning of the key term *dakāšum*, in part also because of singular use of Sumerian grammatical elements (cf. notes 280 and 281).

Obv.

1. The surface of a confrontation (becoming) a border.[277]
   a.šà LAGAB gú
2. Its surface 4 iku.
   a.šà.bi 4 iku
3. What each have I thrusted forward?
   en.nam.ta *ad-ku-uš*

---

275 Based on the transliteration in [Kilmer 1964], with a correction to line 1 proposed by Kazuo Muroi [1998] (in part a reversal to the first transliteration in [Vajman 1961: 257], to which I had no access when preparing the basic version of the manuscript).

276 Unfortunately, no precise dating can be given because this and other mathematical texts from Ur were seemingly brought to the house from which they were excavated as fill [Friberg 2000: 44*f*]. In the same place a few tablets from Ur III were found, but most datable texts from the location are from the interval 1890 to 1810 BCE.

What can be stated with high certainty from stylistic considerations is that the four mathematical problem texts that were found in the house belong together – cf. p. 352.

277 gú serves as a Sumerogram for *aḫum*, "arm" , "river-bank", "periphery" of a town or a field. "Border" seems an adequate translation.

Kilmer reads line 1 as a.šà es.kár and takes the latter word to be an unorthographic writing of éš.gàr, Sumerogram for *iškarum*, "task", the daily workload of a worker. In the Old Babylonian period, the "task field" may (*inter alia*) be a garden [AHw, 396a]. The sign forms are so close and the variation of each so great in the Old Babylonian period that both readings are possible; in view of the remains of the text, however, Muroi's reading certainly makes better sense. Moreover, the text is not otherwise unorthographic but rather errs by being "hyper-orthographic" – see p. 352.

4. 6`40 the surface, what will be equilateral?
6,40 a.šà en.na ba.si$_8$.e[278]

5. 20 will be equilateral.
20 ba.si$_8$.e

6. . Igi 4, of its 4 fronts,[279]
igi 4 ğál.bi sağ.4.bi

7. detach, 15
ab.te.du$_8$[280] 15′.

8. 15′ steps of 20 you go[281], and then
15 a.rá 20 ù.ub.RÁ

9. 5 you shall see,
5 ì.pàd.dè

10. 5, each I have thrusted forward.
5 ta.àm *ad-ku-uš*

Rev.

11. 20 the length, steps of 5,
20 uš a.rá 5

12. of the thrusting,
di-ki-iš-ti-im

13. you go, and then,
ù.ub.RÁ

---

278 I interpret the final .e as the representative of the suffix /-ed/, whence the translation. Normally it would not be written after a vowel [SLa, §243]; but it does happen, and the repeated uncontracted writing ù.ub instead of the normal u. in the following shows a tendency to make grammatical elements more explicitly visible than normally; cf. note 277.

279 The text has sağ, the Sumerogram which I translate "width" in rectangle problems; I have chosen "front" in the present case because of the parallel with BM 13901 #23 (and also because it may make more intuitive sense in English that a square possesses equal fronts than to find it provided with equal widths).

It should be stressed that the etymological bond between "front" and "confrontation"/"confront itself" is an artefact due to the translation; in Sumerian as well as Akkadian, the terms are wholly unconnected.

280 I suppose .te. is meant as the ablative-instrumental prefix .ta., which agrees would agree well with the semantics of du$_8$/"to detach". ab. will express the impersonal form, cf. [SLa, § 319].

281 ù.ub. is likely to be an unorthodox spelling of the prospective prefix /u/ followed by the pronominal element -.b.-; the usual writing is ub, even though /u/ in isolation is normally ù. The meaning of ù.ub.RÁ is hence "you go, and then", corresponding to the ensuing ì.pàd.dè, "you shall see" [SLa, §253, 255, 411]. ù.ub. (a hyperorthographic ub.) is thus a Sumerian substitute for the Akkadian suffix *-ma* – the use of this Akkadian suffix in texts which otherwise try to write everything in Sumerian seems to demonstrate that the substitution with ub. was a local and recent invention which did not spread.

14. 1`40 you shall see.
1,40 ì.pàd.dè

15. 1`40 steps of 4 you go, and then,
1,40 a.rá 4 ù.ub.RÁ

16. 6`40 the greater surface
6,40 a.šà gu.la

17. you shall see.
ì.pàd.dè

The first interpretation was proposed by Anne Draffkorn Kilmer [1964: 140–142]. She reinterpreted one of the possible uses of *dakāšum*, namely, "to pierce", as "to indent", which she further took to mean "go inwards", namely, by constructing a smaller square concentrically within the given square ("B" in Figure 65). As she points out, this corresponds to the usage in BM 15285, where it can be seen to designate the drawing of a square concentrically within another square. She wondered but found no explanation why the indention could be known to be $\frac{1}{4}$ of the front without being stated explicitly, and appears to have overlooked that the shaded area will not be 6`40 = 400 but only 5` = 300.

In order to overcome this difficulty, and in agreement with what seems to follow from other uses of *dakāšum*, Jöran Friberg [1981a: 58*f*] proposed that the process be one of *outward* expansion, as in "C". The allows him to read gu.la in line 16 as "the great square", in accordance both with the usage in the two-square problems in Str 363 and with lexical lists (Kilmer had been forced to read gu.la as a proper name, never seen in any other mathematical text). Unfortunately, instead of the configuration of concentric squares attested in connection with the term in BM 15285 (a configuration whose special status it has been Friberg's merit to point out) we get something which only with difficulty can be seen as that single "greater" surface of which line 16 speaks (but which is of course a familiar configuration, namely, that of BM 13901 #23; see Figure 57). Friberg is followed by Muroi, who explains [1998: 199] that gú "in our problem [...] probably means four rectangles adjacent to the four sides of a square". None of them comments upon the disagreement with the drawings of BM 15285.

Friberg/Muroi's solution eliminates the assumption that the author of the text committed one of the most elementary blunders ever seen in a Babylonian text. However, if we require that it be meaningful to speak of the resulting configuration as gu.la, "great" or "greater", even though *the area* is not greater, and if we ask for a use of the essential verb which agrees with what BM 15285 clearly presupposes to be a familiar standardized meaning, then we are led to the concentric configuration "D" – which is nothing but the interpretation of BM 13901 #9 as made in Figure 11. The "thrusting" will be the width of the band between the two concentric squares, in agreement with BM 15285.

This last interpretation turns out to make good sense of some details of the text. At first, of course, the very trick consists of cutting the square into four

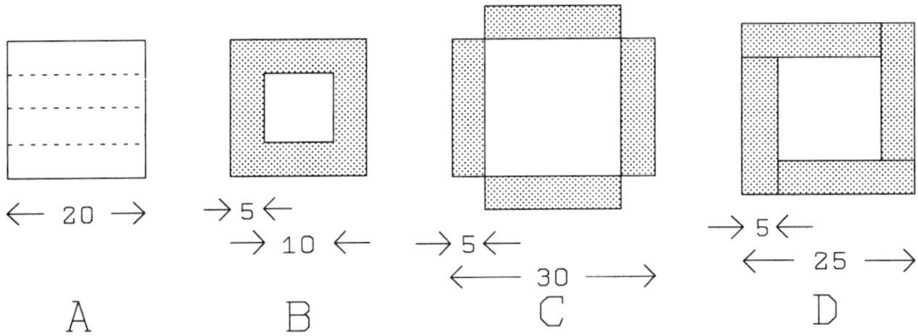

Figure 65. The initial square in UET V, 864, and three interpretations.

equal strips, one for each of *its 4 fronts*, and to thrust *each* of the fronts forwards accordingly. That is why the thrusting has by necessity to be equal to the fourth of the side.[282]

Next, it explains how the *front* can suddenly become a *length* in line 11. Here, indeed, the "front" of the square serves in another function, as the length of a rectangle whose width is the *dikištum*, "thrusting".[283] So it does of course in the Friberg/Muroi interpretatation.

From the terminological point of view, this and another text from Ur (UET V, 859) are remarkable. No other known text from the former Sumerian heartland uses ba.si$_8$ for the plane "equalside" – in fact, the only locality where this usage is found is Eshnunna at the north-eastern periphery (where, however, the term is regularly written in syllabic, "unorthographic" Sumerian). Similarly, the habit of "seeing" results is also foreign to texts from other parts of the former Sumerian core region, as is the reference to "each" side of a square[284] – whether the side be a "front" or a quantity which is identified as "confronting itself". Finally, UET V, 859 contains the phrase ḫé.íb.si$_8$, clearly distinct from the use of ba.si$_8$ for "being equilateral" and an obvious Sumerian calque[285] of Akkadian *limtaḫḫar(ū)*, "let it/them confront itself/each other (as equals)" – in the actual case (treating of the sides of a rectangle) as sides of a square.

282 Muroi introduces a "both" in his translation of line 2 for which there is no correspondent in the text in order to make the text claim the equality of areas.

283 Or, more adequately perhaps, "separation" – namely, the separation of the inner and outer square frame, the width of the border: *dakāšum* may also mean "to split", "to sever", "to separate", which could actually be the basic sense from which the others are derived – that "bulge" on a liver which is referred to as a legitimization of the reading as "outward expansion" may as well have been seen as a splitting of the outer layer from the rest.

284 However, a couple of newly published texts from Nippur (CBS 43, CBS 154–921 [ed. Robson 2000: 39*f*]) reveal the same underlying idea by asking for the side of a square by the phrase *kiyā imtaḫḫar*, referring to the notion of "standing against itself" which we have encountered in BM 13901 no 23 and using the interrogative phrase *kiyā* ("how much each") which always asks for the value of several magnitudes at a time.

285 The customary Sumerian orthography would be ḫé.eb.si$_8$ [SLa, §394].

YBC 8633[286]

The geometrical texts treated in the remains of the present chapter all have a direct or oblique relation to the "Pythagorean rule", as I called it above (p. 197). In the present case the relation turns out to be oblique, perhaps deliberately oblique.

The text is illustrated by a diagram; quite exceptionally, this diagram does not represent the statement alone but the whole procedure. It shows that even this problem deals with a partitioned surface – but here the partitioning belongs with the procedure, not with the statement of the problem. As in real mensuration, the aim is to find a surface by means of subdivision into adequate partial surfaces – but the relation to real field plans with their subdivision into approximate rectangles and approximately right triangles and trapezia is distorted.

Obv.

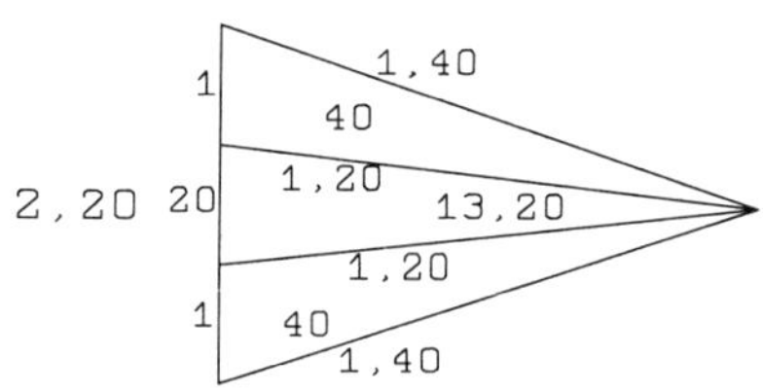

1. A triangle, 1`40 each(?) of both lengths, 2`20 the width. The surface wh[at]?
sağ.dù 1,40 uš.ta⟨¿.àm?⟩ *ki-l[a-l]a-an* 2,20 sağ.ki a.šà *mi-n[u-u]m*

2. You, from 2`20 [of ¿the width?]
at-ta i-na 2,20 *ša* [...]

3. 20 tear out: to the width of the triangle [¿posit? ...]
20 *ú-sú-uḫ-ma a-na* sağ sağ.dù [...]

4. and 2` which you have left, to two break: 1`.
ù 2 *ša te-zi-bu a-na ši-na ḫe-pé-ma* 1

5. 1` the width of the first triangle. 1` the width of the second triangle.
1 sağ.ki sağ.dù *iš-te-en* 1 sağ sağ.dù *ša-ni-im*

6. The second length what?
uš *ša-nu-um mi-nu-um*

7. 20 the bundling to 4 raise, 1`20.
20 *ma-ak-ṣa-ra-am a-na* 4 *i-ši-ma* 1,20

---

286 Based on the transliteration in [MCT, 53].

8. 1`20, the true length, and 1`, the width of the tri[angle]
1,20 uš *ki-nu-um ù* 1 sağ.ki sa[ğ.dù]

9. [t]o two break: 30 to 1`20, the [⸢true⸣? le]nght
[*a-n*]*a ši-na ḫe-pe-ma* 30 *a-na* 1,20 u[š ⸢*ki-nim*⸣?]

10. [ra]ise: 40` the surface of the [first] triangle.
[*i-*]*ši-ma* 40 a.šà sağ[.dù ...]

11. 20 the width of the t[ri]ang[le t]o two [break:]
20 sağ.ki s[ağ].d[ù *a-*]*na ši-na* [*ḫe-pe-ma*]

12. 10 to 1`2[0 the ⸢true⸣? length raise:]
10 *a-na* 1,2[0 uš ⸢*ki-nim*⸣? *i-ši-ma*]

13. 13`20 the surface of the [second] triang[le.]
13,20 a.šà sağ.d[ù ...]

Rev.

1. 1` the width of the triangle to t[wo break:]
1 sağ.ki sağ.dù *a-na š*[*i-na ḫe-pe-ma*]

2. 30 to 1`20, the true length, [raise:]
30 *a-na* 1,20 uš *ki-nim* [*i-ši-ma*]

3. 40` the surface of the thi[rd] triangle.
[4]0 a.šà sağ.dù *ša-a*[*l-ši-im*]

4. 1``33`20 the true surface [...].
1,33,20 a.šà *ki-nu-um* [...]

5. (As to) the bundling of the tra[pezium with diagonal]

ma-ak-ṣa-ru-um ša sa[ğ.ki.gud *ṣi-lip-tim*]

6. 20 to 5 of the diag[o]na[l rai]se[:]
20 *a-na* 5 *ša ṣí-li-i*[*p-*]*ti*[*m i-š*]*i*[*-ma*]

7. 1`40 the dia[go]n[a]l it [gi]ves you:
1,40 *ṣí-l*[*i-i*]*p*[*-ta*]*m i*[*-na-*]*di-kum-ma*

8. 20 to 4, the length, raise: 1`20 the true length..
20 *a-na* 4 uš *i-ši-ma* 1,20 uš *ki-nu-um*

9. 20 to 3 raise: 1 the width of the triangle it gives you.
20 *a-na* 3 *i-ši-ma* 1 sağ sağ.dù *i-na-di-ku*

10. The bundling of a trapezium with diagonal,
ma-ak-ṣa-ru-um ša sağ.ki.gud *ṣí-li-ip-tim*

11. 20 to 5 raise: 1`40 the length.
20 *a-na* 5 *i-ši-ma* 1,40 uš

12. 20 to 4 raise: 1`20 the second length.
20 *a-na* 4 *i-ši-ma* 1,20 uš *ša-nu-um*

13. 20 to 3 raise: 1` the width of the triangle.
20 *a-na* 3 *i-ši-ma* 1 sağ.ki sağ.dù

The surface dealt with in the text is an isosceles triangle, in which the angles at the base are too obviously non-right to allow determination of the surface

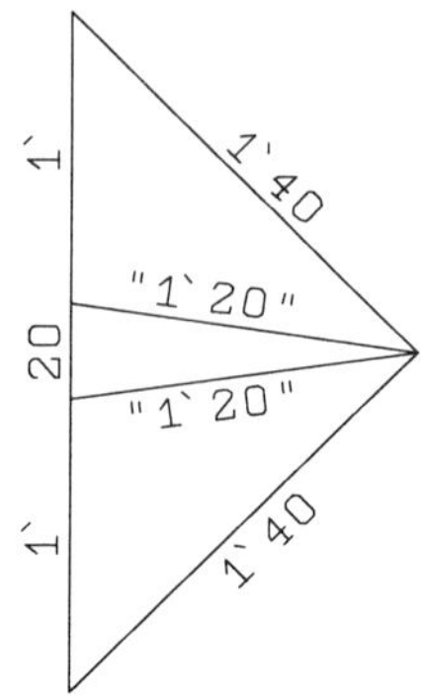

Figure 66. The triangle of YBC 8633 in true proportions

from length and width. Instead, the procedure starts by "tearing out" 20 in the middle of the widths, leaving 1` for each of the outer parts,[287] which thus relate to the lengths as 3 to 5.

This ratio is not enunciated at the present point, but it is the background to obv. 7, where the "bundling 20" is "raised" to 4. What happens is only made explicit in rev. 6–9, and summarized in rev. 10–13.[288] The figure of which a bundle is made is spoken of as a "sag̃.ki.gud with a diagonal", and possesses 3, 4, and 5 as its decisive parameters. Normally the sag̃.ki.gud ("oxhead") is a trapezium, but in the present case it can be seen to be, or to contain as a constituent part, a right triangle with sides 3, 4, and 5, or a rectangle with sides 3 and 4 and appurtenant diagonal 5. Of this model figure a bundle is made, 20 times repeated in each direction, which gives a triangle with length 1`40, "true" length 1`20, and width 1`. This is the triangle which the text supposes to find as "wings" of the original triangle – see Figure 66.

The deviation from the true value of the area (1``23`19) is not exorbitant. The "true" lengths are indeed only c. 12% longer than the common height of the full and the three partial triangles. On the structure diagram it seems even less, but we have no particular reason to believe that the Babylonian teacher was misled by something he would never imagine was made to scale.

Translated into our numerals, the three sides of the triangle are 100–100–140. From this it is immediately evident to us (as also though with less precision from the diagram in Figure 66) that the triangle is very close to being right.The calculator who constructed the problem may have known this too. If he had used the two legs of the triangle as length and width within the

287 Since all numerical information in the problem is homogeneous, the order of magnitude is not determined. I have chosen according to the criterion that a bundle is more easily imagined if it contains an integer number of constituents.

288 The present text is thus a parallel to the didactic Susa texts TMS VII and IX, only with the difference that these give the didactic explanation as introduction, not as an explication given afterwards.

usual procedure, he would have found an area 1``23`20 – 0.02% above the true value instead of 12%.

We cannot be sure that he knew (cf. also p. 387) – all that is needed for the construction of the problem is knowledge of the 3-4-5 configuration, which could be derived from an igi.gub table with proper instructions for use, or could be known as an isolated datum.[289] But at least we may be next to certain that he knew that his supposed "true length" was not quite "true". We must therefore suppose that the problem was constructed with a particular purpose in mind: either to train the *makṣarum* ("bundling") method or to practice the subdivision of inconvenient areas.

A minor terminological observation can be made: IM 55357 (p. 231), when treating of a triangle with the same proportions, spoke of "the length" and "the long length"; in the present case – maybe because "the length" is already occupied, being the "length" of the original triangle and hence the "longer length" of the upper and lower triangles – the same distinction is made by referring to the shorter length as "true" (obv. 8, rev. 8).[290] The two texts come from different localities, but the difference shows at least that none of the texts are expressed in a standard terminology of more than local validity; most likely, the terminology is *ad hoc* and not technical at all.

Db_2-146[291]

Like IM 52301 and IM 55357, this tablet is from Eshnunna; it is roughly contemporary with the former and thus slightly younger than the latter; together with a number of other texts from the same region is provides us with the full question of which the initial accusative "(about a) surface" of BM 13901 #23 is likely to be an ellipsis (cf. note 249).

The question is to find the length (l) and width (w) of a rectangle from the diagonal (d) and the area ⊏⊐(l,w). The solution does not involve the Pythagorean rule, but the dimensions of the rectangle – l = 1, w = 45′, d = 1°15′ – clearly presupposes knowledge at least of the properties of the 3–4–5 configuration; the final proof, moreover, makes full use of the rule.

---

289 It is indeed noteworthy that the text uses these parameters and not 45–1-1,15 (whether to be understood as 45–1`–1`15 or as 45′–1–1°15′), which was the Old Babylonian standard representation of the type in question.

290 Neugebauer and Sachs believe [MCT, 53 n. 150] *kīnum*, "true", to be a writing error for *šanûm*, "second". The repetition of the word makes such a mistake implausible, not least since what is written makes perfect sense.

291 Based on the transliteration in [Baqir 1962: 2].

Obv.

1. If, about a (rectangle with) diagonal, (somebody) asks you
šum-ma ṣí-li-ip-ta-a-am i-ša-lu-ka

2. thus, 1°15 the diagonal, 45´ the surface;
um-ma šu-ú-ma 1,15 *ṣí-li-ip-tum* 45 a.šà

3. length and width corresponding to what? You, by your proceeding,
ši-di ù sağ.ki *ki ma-a-ṣí at-ta i-na e-pé-ši-ka*

4. 1°15´, your diagonal, its counterpart lay down:
1,15 *ṣí-li-ip-ta-ka me-ḫe-er-šu i-di-i-ma*

5. make them hold: 1°33´45″ comes up,
šu-ta-ki-il-šu-nu-ti-i-ma 1,33,45 *i-li*

6. 1°33´45 ¿may your? hand ¿hold?
1,33,45 šu KU.U¿.ZU/BA?

7. 45´ your surface to two bring: 1°30 comes up.
45 a.šà-*ka a-na ši-na e-bi-il-ma* 1,30 *i-li*

8. From 1°33´45″ cut off: {...} 33´45″ the remainder.
i-na 1,33,45 *ḫu-ru-úṣ-ma* {1,}33,45(sic) *ša-pí-il-tum*

9. The equalside of 3´45″ take: 15´ comes up. Its half-part,
ib.sí 3,45 *le-qe-e-ma* 15 *i-li mu-ta-su*

10. 7´30″ comes up, to 7´30″ raise: 56″15‴ comes up
7,30 *i-li a-na* 7,30 *i-ši-i-ma* 56,15 *i-li.*

11. 56″15‴ your hand. 45´ your surface over your hand,
56,15 šu-*ka* 45 a.šà-*ka e-li* šu-*ka*

12. 45´56″15‴ comes up. The equalside of 45´56″15‴ take:
45,56,15 *i-li* ib.si 45,56,15 *le-qe-ma*

13. 52´30″ comes up, 52´30″ its counterpart lay down,
52,30 *i-li* 52,30 *me-ḫe-er-šu i-di-i-ma*

14. 7´30″ which you have made hold to one
7,30 *ša tu-uš-ta-ki-lu a-na iš-te-en*

15. append: from one
ṣí-ib-ma i-na iš-te-en

16. cut off. 1 your length, 45 the width. If 1 the length,
ḫu-ru-úṣ 1 uš-*ka*[292] 45 sağ.ki *šum-ma* 1 *uš*

17. 45 the width, the surface and the diagonal corresponding to what?
45 sağ.ki a.šà *ù ṣí-li-ip-ti ki ma-ṣí*

18. [You, by] your [ma]king, the length make hold:
[*at-ta i-na e-p*]*é-ši-ka ši-da šu-ta-ki-il-ma*

19. [1 comes up ...] may your head hold.
[1 *i-li* ...] *re-eš-ka li-ki-il*

---

292 Thus the hand copy; the ḪA of the transliteration must be a typo.

Rev

20. [...]: 45′, the width, make hold:
[...]-*ma* 45 sağ.ki *šu-ta-ki-il-ma*

21. 33′45″ comes up. To your length append:
33,45 *i-li a-na ši-di-ka ṣí-ib-ma*

22. 1°33′45″ comes up. The equalside of 1°33′45″ ta[ke]:
1,33,45 *i-li* ib.si 1,33,45 *le-[qe]-ma*

23. 1°15′ comes up. 1°15′ your diagon[al]. Your length
1,15 *i-li* 1,15 *ṣí-li-ip-[ta]-ka* uš-*ka*

24. to the width raise, 45′ your surface.
a-na sağ.ki *i-ši* 45 a.šà-*ka*

25. Thus the procedure.
. *ki-a-am ne-pé-šum*

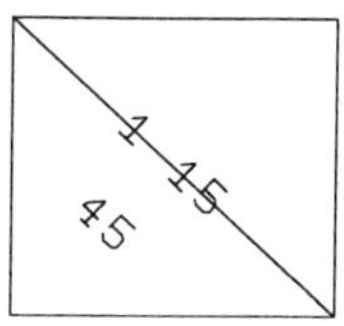

We may follow the procedure in Figure 67. At first the diagonal and its "counterpart" are "laid down" and "made hold". With the resulting 1°33′45″, something is done which involves "the hand" (the last three signs of the line make no certain sense). Possibly, this may just mean that the number should be held at disposition; in view of the formulations of line 11 it seems more plausible, however, that writing on a piece of clay held in the hand is meant (a similar interpretation of the role of the hand in IM 52301 is not excluded, nor, however, positively supported; see p. 213). Next the surface is brought "to two", and then "torn out". This repeats a pattern with which AO 8862 #2 has already presented us: ⊏⊐(l,w) is not in itself part of ⊏⊐(d); but bringing it in

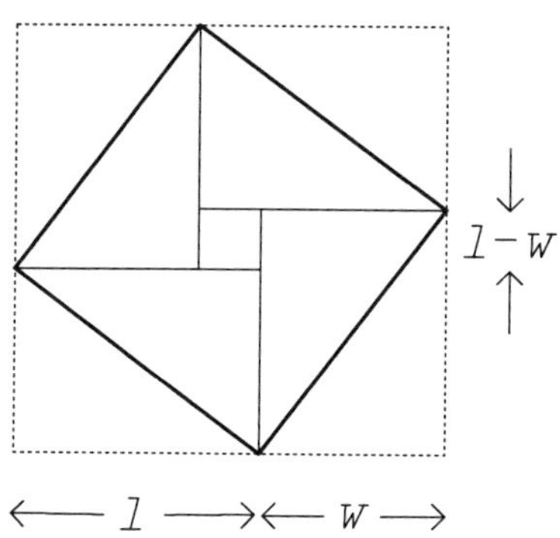

Figure 67. The initial steps of the procedure of Db$_2$-146.

two copies allows a removal; it may be no accident that both texts designate this removal as a "cutting-off".

In itself it does not seem evident that this removal will leave a useful remainder; but the similarity of Figure 67 with a configuration which we have encountered sufficiently often to consider it a familiar standard diagram (in Figures 12, 32 and 35) is helpful. From this diagram it is fairly obvious that the remainder is □(l–w), and its "equalside" thus l–w. Thereby the problem is reduced to finding the sides of a rectangle from the surface and the excess of length over width, and this standard problem indeed provides the pattern for the final part of the solution: the excess is bisected, giving 7′30″ for the "half-part" (*muttatum*, another name for the natural half). The fact that the half-part is "*raised*" to itself suggests that the corner is already in place, and thus that the transformations are thought of as taking place within the same diagram (see Figure 68); but since we are told in line 14 that 7′30″ was "made hold", and in view of the presence of other deviations from customary phrasing and terminology, we should not consider this as certain (but see imminently).

The outcome of the "raising" is characterized as "your hand", and the surface is put above it; this seems to refer to inscription on a pad or device for computation, since no further identification of the addition takes place. The "equalside" of the sum is the usual average side $^{l+w}/_2$; the length is found by "appending" the deviation $^{l-w}/_2$ = 7′30″, the width by "cutting it off" from the "counterpart"; as in IM 52301, "appending" precedes "cutting-off" – and as in BM 85200+VAT 6599 the formulation is elliptic though not as radically.

The calculation is followed by a proof, in which the diagonal is determined by means of the Pythagorean rule, and the surface by "raising" the length to the width. In the determination of d, the squares on l and w are found quite regularly, by making them "hold"; moreover, □(l) is identified in line 21 with the length, which implies that it is really seen as a "confrontation", a square configuration identified with its parametrizing side. It is thus almost certain that "raising" is chosen here for the computation of the surface because the rectangle is already there. This corroborates the analogous

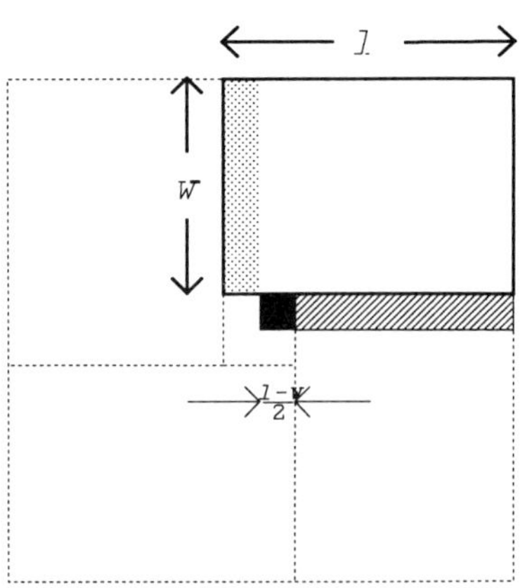

Figure 68. The solution of the rectangle problem in Db_2-146.

interpretation of the use of the operation in line 10. The formulation of the proof will be seen to refer to a formulation of the rule in general terms – "the length make hold", not "1, the length, make hold"

A few supplementary observations connected to Figure 67 will turn out to have implications in later chapters.

Firstly, if the lengths of the four rectangles are prolonged (or, alternatively, if diagonals are drawn in Figure 12), we get the most widespread and probably the simplest intuitive proof for the Pythagorean theorem.

Secondly, if we had "appended" twice the surface to □(d) instead of "cutting it off", □(l+w) would have resulted instead of □(l–w); this would have led us to the other standard rectangle problem, to find the sides from the surface and the sum of the sides. Alternatively, combination of "appending" and "cutting-off" would have given us both l+w and l–w. The former alternative would have been fully in agreement with the habits of Babylonian mathematics, the latter certainly not; both are found in Demotic Egyptian and medieval versions of the problem.

YBC 7289[293]

Obv.

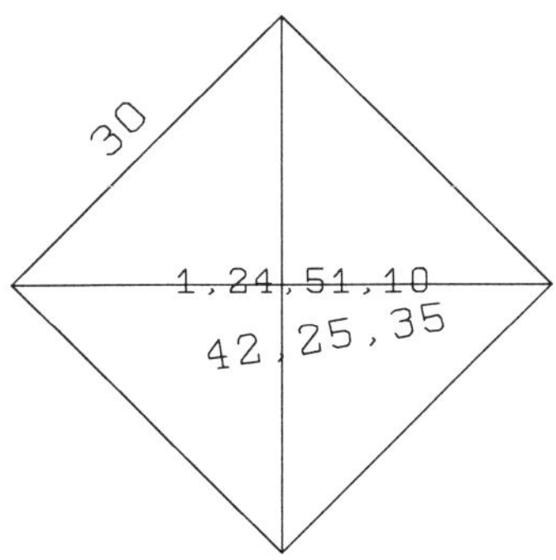

Rev.[294]

The purpose of this diagram with inscribed numbers but no explanatory words is fairly clear. The number 1°24′51″10‴ is a very precise approximation to √2, and identified in the table of technical igi.gub constants YBC 7243, obv. 10 as *ṣi-li-ip-tim* íb.si$_8$, "(the coefficient) of the diagonal of a confrontation[295]"; since it is the ratio between 30 (probably 30′) and

---

293 After [MCT, 42].

294 "a rectangle with an inscribed diagonal, but the numbers are too badly conserved to warrant a restoration of the dimensions" – [MCT, 43 n. 125c].

295 In the context of igi.gub tables, it is virtually certain that both íb.si$_8$ and

42,25,35 (probably 42′ 25″ 35‴), we may conclude that the text finds the diagonal of the standard square □(30′) in agreement with the tabulated coefficient.

Since both diagonals are drawn, we may guess that the diagram is also meant to suggest the naive-geometric argument by which the square on the side can be seen to be twice the square on the bisected diagonal (and, equivalently but not directly to be seen, the square on the diagonal to be twice the square on the side). This is only a special case of the Pythagorean rule, but enough to support the assumption that the rule had been discovered by means of such arguments, for instance in a diagram like that of Figure 67, which is indeed a generalization of a fuller version of the present diagram that is found in BM 15285 #11 – see Figure 69, which confronts this fuller version with the diagram of Figure 67.

A different question is the origin of the coefficient 1°24′51″10‴. In another igi.gub table (TMS III, 31), we find that 1°25′ is the igi.gub *šà* bar.dá šà NIGIN, "the constant factor of the diagonal of the confrontation",[296] which Neugebauer and Sachs only knew from a Seleucid occurrence. For this they gave the explanation that it is found as the mean between the rough approximation 1 ½ and 2÷1 ½. Indeed, if a is an approximation of √2, then 2÷a is another approximation which errs by almost the same amount as a but in the opposite direction. They do not seem to have noticed that the algorithm of "side-and-diagonal numbers" known from classical antiquity[297] ($d_{n+1} = 2s_n+d_n$, $s_{n+1} = s_n+d_n$) also leads quickly to the approximation 1°25′ – starting from 1:1 we continue with 3:2, 7:5, and then come to 17:12 = 1°25.

Neugebauer and Sachs also show that iteration of the procedure they propose (going on from a = 1°25′), results in the approximation 1°24′51″ 10‴... . A problem which they do not discuss is how the division by the irregular number 1°25′ was to be performed; a good guess would be that a convenient regular number close to 1°25′ was chosen. It is not obvious,

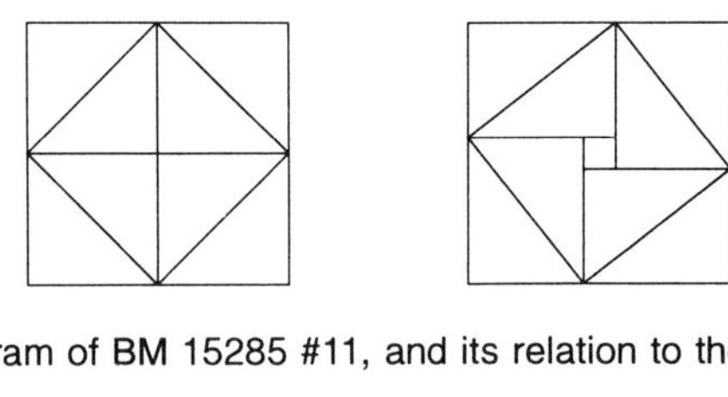

Figure 69. The diagram of BM 15285 #11, and its relation to the diagram of Figure 67.

(presently) NIGIN are to be read as logograms for *mitḫartum* (as íb.si$_8$ in BM 85210).

296 The reading of BAR.TA as bar.dá, literally "cross-bar", was proposed by Muroi [1992: 47].

297 Known from Theon of Smyrna [ed., trans. Dupuis 1892: 70–75], and from Proclos's commentary to Plato's *Republic* [Kroll 1899: II, 393–400]; cf. [Hultsch 1900].

however, how such a number shall be chosen so as to give the right approximation and neither one that is too rough nor one that is too close; this is shown by Fowler and Robson [1998].

It should be mentioned that the algorithm of "side-and-diagonal numbers" ($d_{n+1} = 2s_n+d_n$, $s_{n+1} = s_n+d_n$) also leads to the approximation 1°24′51″10‴. Starting from 1:1 we continued with 3:2, 7:5, and then with 17:12 = 1°25. The next steps are 41:29, 99:70, and 239:169. If the latter ratio is expressed as a sexagesimal number, we get 1°24′51″, and a remainder of $^{21}/_{169}$, which is very close to $^{1}/_{8}$.[298] If we remember that the system was decimal-seximal rather than properly sexagesimal (see note 19), the next place – the tens of the order of thirds – is down by a factor 6. Thus, noticing that $^{1}/_{8}$ is close to $^{1}/_{6}$, we get as the next approximation precisely 1°24′51″10‴, which is indeed the closest approximation at this level.[299]

It may be of interest that Leonardo Fibonacci's *Pratica geometrie* [ed. Boncompagni 1862: 62] finds the side S of a square from the difference between diagonal D and S by means of a diagram which explains the algorithm (the square on D compared with that on S, split as $2s+d$ and $s+d$, respectively, s and d being also side and diagonal of a square). The arguments are wholly in the vein of Babylonian "cut-and-paste geometry with accounting" (see Figure 70): □(S) equals V+VI+VIII+IX. But V+VIII equals IV+VII

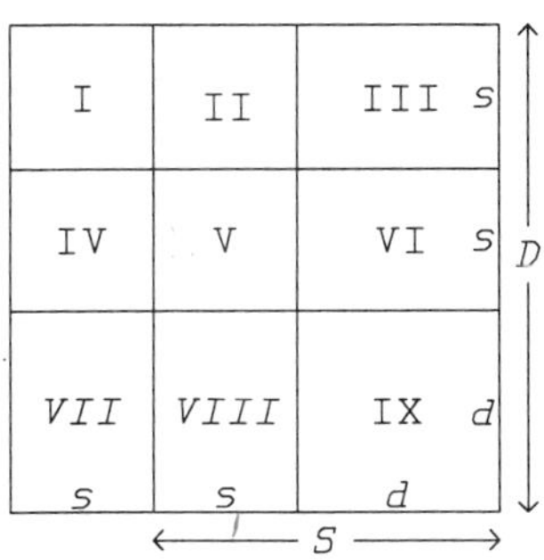

Figure 70. Fibonacci's implicit proof of the side-and-diagonal algorithm.

---

298 Division by an irregular number was certainly not outside the capabilities of the Old Babylonian scribes. Several school exercises from the mid-third millennium train the division of very large numbers by the irregular divisors 7 and 33; the method was stepwise division and conversion of the remainder into smaller units, structurally similar to our long division. See [Høyrup 1982] and [Friberg 1986]. A few tabulations of the reciprocals of irregular numbers are also known (cf. note 50), and even though we do not know about a case where they were used we may be certain that they were computed.

299 Neugebauer and Sachs comment on their conjecture that "the fact that both values for √2 found in our texts are links of the same chain seems to be a rather strong argument in support of our explanation" [MCT, 43]. A strange case, indeed, that the two values belong together in two non-equivalent chains of successive approximations.

and VI equals III; furthermore (since d is the diagonal in $\square(s)$) IX equals twice V and hence I+II. In consequence, $\square(S)$ also equals I+II+III+IV+VII. Therefore $2\square(S) = \square(D)$, whence D must be the diagonal in $\square(S)$.

Another heuristic proof of the rule follows from the construction of a regular octagon by means of two squares as shown in Figure 71. If $\sigma = AC = CD$ and $\delta = CB$, then σ and δ are obviously side and diagonal of the same square; but $AO = DO$, and $DO = AB$ ($DG = BG$, $GJ = AF = AC$, $JO = HF = GC$), whence $AO = OE = \sigma+\delta$, whereas $AE = 2\sigma+\delta$. Thus even $\Sigma = \sigma+\delta$ and $\Delta = 2\sigma+\delta$ are side and diagonal of a square – indeed the inner square of Figure 69, left).

The observation that AB and CE are equal to half of the diagonal gives rise to an easy construction of the regular octagon, in which the oblique square is omitted and the points B and C (and corresponding points on the other sides) are found and connected. This construction is described in a variety of later treatises on practical geometry: the pseudo-Heronian *De mensuris* [ed. Heiberg 1914: 206] – certainly not Hero's work; Abū'l-Wafā''s *Book on What is Necessary from Geometric Construction for the Artisan* [ed., Russian trans. Krasnova 1966: 93]; and Roriczer's fifteenth-century *Geometria deutsch* [ed. Shelby 1977: 119*f*], whose treatment of circles exhibits links reaching back to the Old Babylonian epoch. Roriczer's *Wimpergbüchlein* [ed. Shelby 1977: 108*f*] makes use of the superimposed squares of Figure 71 and shows (though this is not the topic) that Roriczer knew some of their relevant properties. The configuration is also found as an illustration to the determination of the area of the regular octagon in Epaphroditus & Vitruvius Rufus.[300] According to Hermann Kienast (personal communication) it can also be seen to have been

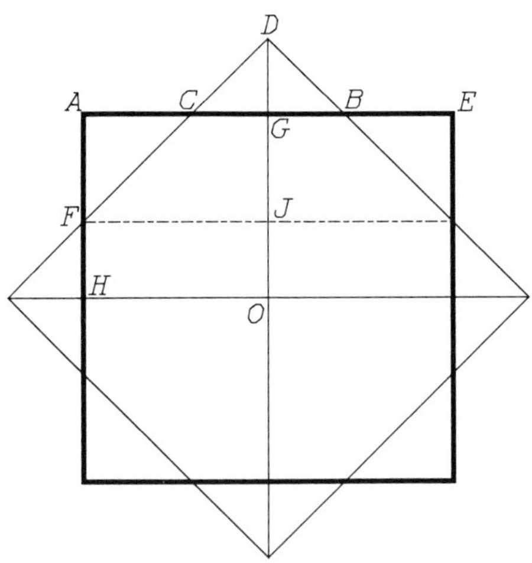

Figure 71. A regular octagon produced by superimposed squares.

300 [Ed. Cantor 1875: 212, Fig. 40]. The text is also in [Bubnov 1899: 539], but the diagram is omitted.

used in the ground plan of the Athenian "Tower of the Winds" from the first century BCE.

All this of course does not prove that the Old Babylonian calculators knew the side-and-diagonal numbers, nor *a fortiori* that they used them in the present case; but is shows that knowledge of the trick could easily follow from the kind of mathematics they *did* engage in.

TMS I[301]

Obv.

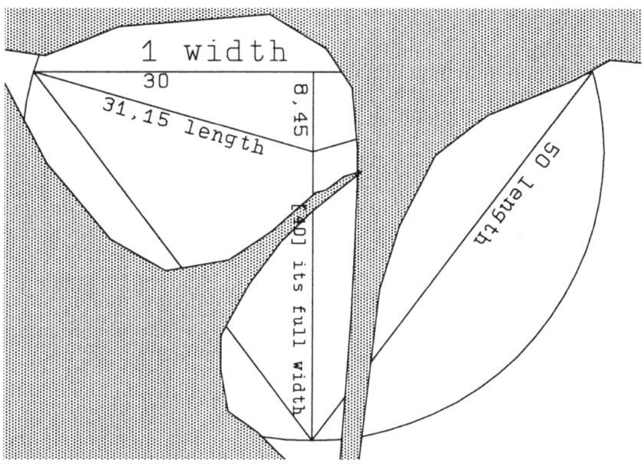

Rev.[302]

The circle on the obverse of the tablet is drawn with a compass, and the sides and height of the inscribed triangle traced after a ruler (and the angle between the height and the base is as right as can be controlled on the photo). What the text does is to find the radius *r* of the circle which is circumscribed about the doubled 30–40–50-triangle – which, as pointed out by Bruins in the commentary, provides us with incontestable evidence that the Susian calculators of the Old Babylonian age were able to conceptualize both the circumscribed circle and the height of an isosceles triangle.

A geometer with Greek training might have found *r* from one of the proportions $2r$:50´::50´:40´ or 40´:30´::30´:($2r$–40´) (see Figure 72), but we possess no evidence that any Old Babylonian calculator would have done a

---

301 Based on the hand copy in [TMS, pl. 1].

302 Numbers, probably intermediate results that are "posited" or "inscribed", mostly heavily damaged and seemingly not connected to the obverse.

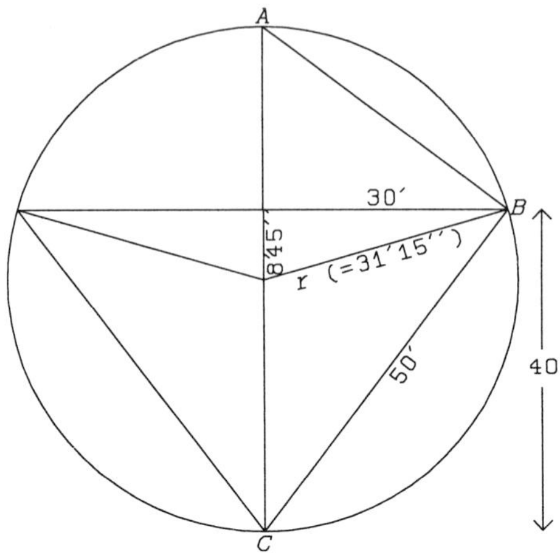

Figure 72. Redrawing of the triangle with circumscribed circle from TMS I

thing like that.[303] The soundest explanation of the procedure – already set forth by Bruins as the only possibility though evidently in numerico-algebraic key – is certainly that the Pythagorean rule was applied. The rule, indeed, tells us that

$$\square(r) = \square(30')+\square(40'-r) ,$$

which can be transformed into

$$\square(r)-\square(40'-r) = \square(30') . \qquad (*)$$

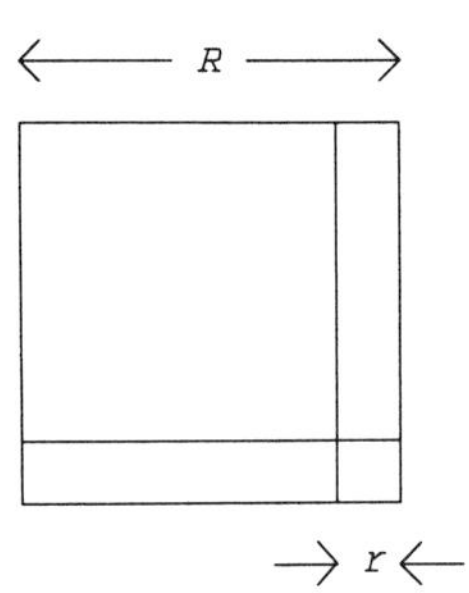

Figure 73. Diagram from which "proto-*Elements* II.7" follows.

303 *If* they realized that $\angle ABC$ is right (which is intuitively obvious for symmetry reasons if only you get the idea to complete the triangle as a rectangle), they *might* have recognized the situation from IM 55357, and thereby have been led to an equivalent of the second proportion (cf. Figure 78 and discussion on p. 274); but in the absence of the dotted lines from Figure 77, it remains highly unlikely that this route was actually pursued.

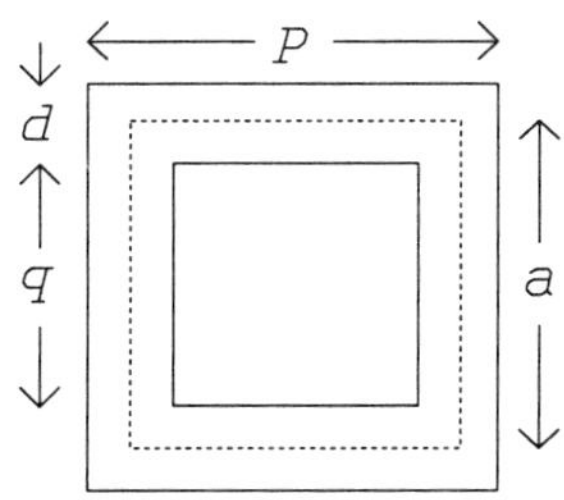

Figure 74. The area of a quadratic border.

As argued (in arithmetical terms) by Robert Whiting [1984: 65*f*], it is virtually certain that it was already known no later than 2200 BCE that

$$\square(R-r)+2\sqsubset\!\sqsupset(R,r) = \square(R)+\square(r) \ ,$$

the same identity as is proved in *Elements* II.7.[304] This rule – which can be argued from Figure 73 by the same kind of arguments as those for the side-and-diagonal algorithm from Figure 70, only simpler – or the parallel rule for $\square(R+r)$ – could be used to resolve $\square(r)-\square(40'-r)$, and we cannot exclude that this was done. Knowing the Babylonian predilection for concentrically situated squares and for average-and-deviation calculations, it seems at least as likely, however, that they would use a rule that can be easily argued from Figure 74:

$$\square(p)-\square(q) = \sqsubset\!\sqsupset(l,d) = \sqsubset\!\sqsupset(4a,d) = \sqsubset\!\sqsupset(2p+2q,d)$$

where l is the mid-length of the border between the squares on p and q, a the average $^{p+q}/_2$ and d the deviation $^{p-q}/_2$; indeed, the latter form of the rule turns up in connections which otherwise are clearly connected if not to the Old Babylonian school tradition then to its closest kin (see below, p. 371).

In any case, the solutions of problems like BM 13901 #9 demonstrate that the Old Babylonian calculators will have had no difficulty in transforming (*) into

$$(2\cdot 40')\times(r-20') = 15'$$

whence it follows that

$$r-20' = 11'15'' \ , \ r = 31'15'' \ ,$$

as indeed stated in the tablet.

On p. 29 it was claimed that "mathematical problems were always constructed backwards from the solution". This text from Susa constitutes one of the extremely few exceptions from that rule[305] – evidently it starts out

---

304 The argument is that a number of school problems about area computation lead to very tedious computations unless this rule is used, in which case they become very simple. It is quite unlikely that the numbers should have been chosen so that this would be the case unless it was done intentionally.

305 Perhaps together with BM 85200+VAT 6599 #15, see above, p. 158 – where the omission of the usual precaution has led to an insoluble problem.

from the 30–40–50 triangle, and then makes use of the familiar algebraic or quasi-algebraic techniques in order to find the radius of the circumscribed circle. The problem turns out to be of the first degree, and therefore it remains true that no single genuine application of the second-degree algebraic techniques was known; no extant text, at least, contains the faintest trace.

VAT 6598 #6–7[306]

Even these two problems presuppose the Pythagorean rule without using it directly; what they do is to show two different approximations to $d = \sqrt{h^2+b^2}$, where h = 40′ nindan is the height, b = 10′ nindan the breadth, and d the diagonal of a door. Since d cannot be found exactly, even the present problems were evidently not constructed backwards from the (non-existent) exact solution.[307]

Rev. I

#6

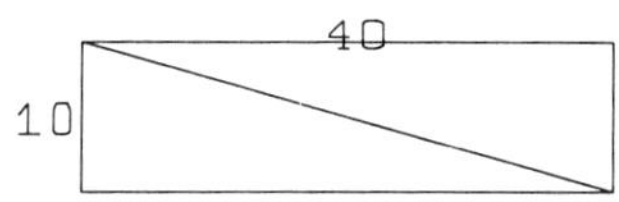

19. A door. ½ (and) 2 kùš the height, 2 kùš the breadth, its diagonal what?
ká ½ 2 kùš sukud 2 kùš dagal *ṣí-li-⟨ip-⟩ta-šu* en.nam

20. You, 10′ the breadth make confront itself, 1′40″ the ground you see.
za.e 10 dagal *šu-tam-ḫir* 1,40 *qà-qá-ra ta-mar*

21. Igi 40′, (of the order of) kùš, the height, detach, to 1′40″ the ground raise,
igi 40 kùš sukud du$_8$.a *a-na* 1,40 *qà-qá-ri i-ši*

---

306 After the transliteration in [MKT I, 279*f*], cf. [TMB, 130]. Christopher Walker has recognized in the fragment BM 96957 a missing part of the tablet, but since both his hand copy and Eleanor Robson's transliteration and analysis of the relevant part were only published in preliminary form in [Robson 1995: 269–280] when the manuscript was prepared, I restricted myself to reproducing the two problems that were indubitably in the public domain at that moment. In Robson's counting, they are #18 and #21.

At the moment when I was making the ultimate corrections the complete tablet *had* been published in [Robson 1999: 231–244]. [Robson 1996] is a proper edition of the brick-problem part of the new fragment, though it is now superseded by the full edition [1999].

307 However, the newly discovered part of the tablet contains problems which make the rather meaningless backward calculation from one side and the approximate value of the diagonal to the known value of the other side! (Only if regarded as control of the numerical computations, the calculations make proper sense.)

22. [2´]30″ you see. ½ of 2´30″ break, 1´15″ you see, 1´15″
[2,]30 *ta-mar* ½ 2,30 *ḫe-pé* 1,15 *ta-mar* 1,15

23. [to 40´, the height, appe]nd, 41´15″ you see, 41´15″,
[*a-na* 40 sukud daḫ].ḫa 41,15 *ta-mar* 41,15

24. [the diagonal. The pr]ocedure.
[*ṣí-li-ip-tum n*]*e-pé-šum*

Rev. II

#7

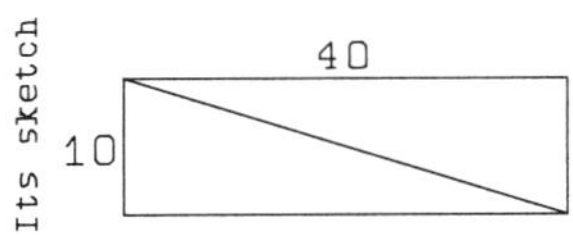

1. 2 kùš the breadth, 40´, (of the order of) kùš the height, its diagonal what? You, 10´, the width,
2 kùš dagal 40 kùš sukud *ṣí-li-ip-ta-šu* en.nam za.e 10 sağ

2. make confront itself, 1´40″ you see, the ground. 1´40″ to 40´, kùš, the height, rais[e:]
šu-tam-ḫir 1,40 *ta-mar qà-qá-rum* 1,40 *a-na* 40 kùš sukud *i-š*[*i-m*]*a*

3. 1´6″40‴ you see, to ⟨2⟩ repeat, 2´13″20‴ you see, to 40´, kùš,
1,6,40 *ta-mar a-na* ⟨2⟩ tab.ba 2,13,20 *ta-mar a-na* 40 kùš sukud

4. append, 42´13″20‴, the diagonal, you see. The procedure.
daḫ.ḫa 42,13,20 *ṣí-li-ip-ta ta-mar ne-pé-šum*

The first problem gives a solution

$$d = h + \frac{1}{2} \cdot \frac{\square(b)}{h} ,$$

where $\square(b)$ is spoken of as *qaqqarum*, "ground", the same term as was used for the base of the "excavation" in BM 85200 + VAT 6599 (see p. 262). No reasons for the formula are given, but there is little doubt that the procedure which we encountered first in BM 13901 #14 was used – see Figure 75: We look for the "equalside" of the surface $\square(h)+\square(b)$, that is, its side when laid out as a square. For this purpose the surface of the smaller square $\square(b)$ is distributed as two "wings" along two sides of the larger square $\square(h)$; at first the area is shaped into a rectangle $\sqsubset\!\sqsupset(h, {}^{\square(b)}/_{h})$, then this rectangle is broken into two rectangles $\sqsubset\!\sqsupset(h, {}^{1}/_{2} \cdot {}^{\square(b)}/_{h})$. If we neglect the absence of the small dotted area, the "equalside" is thus $d = h + {}^{1}/_{2} \cdot {}^{\square(b)}/_{h}$.

Neugebauer gives no explanation of the idea behind this formula in [MKT], but refers to the interpretation in [Neugebauer 1934: 34*ff*]. Here he proposes the same type of iteration for finding $\sqrt{A}$ as for YBC 7289 (see p. 262) – namely, a start from a first approximation $d_0 = h$, from which follows

$$d_1 = \frac{1}{2} \cdot (d_0 + \frac{A}{d_0}) = \frac{1}{2} \cdot (h + \frac{h^2 + b^2}{h}) .$$

The outcome is certainly the same, but in contrast to the geometrical interpre-

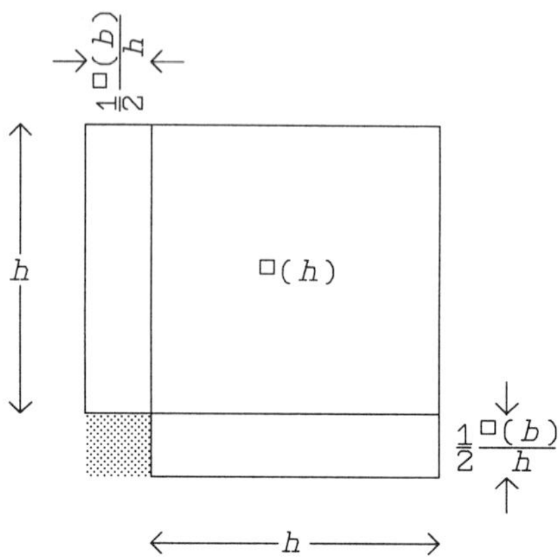

Figure 75. The probable geometrical reasoning behind VAT 6598 #6.

tation it only leads to the actual computational steps of the text after a certain amount of algebraic transformation.

The next problem gives a different approximation, which is obviously nonsensical as it stands, namely, $d = h+2\square(b)\cdot h$. Since this is not a problem but a formula to be used in computation, its addition of a line and a volume is certainly a mistake. It is also much less precise than the formula of #6.[308] In [MKT, 287] Neugebauer reconstructs it as $d = h+\frac{2b^2h}{2h^2+w^2}$, and observes that $2h^2+w^2 = 55' \approx 1$, for which reason the numerical value is not too bad. Again, he refers to [Neugebauer 1934: 36*f*] for a possible explanation. Still with $A = h^2+b^2$ and $d_1 = h+\frac{b^2}{2h}$ this earlier work finds

$$\frac{A}{d_1} = \frac{2h^3+2b^2h}{2h^2+b^2} = \frac{2h^3}{2h^2+b^2} + \frac{2b^2h}{2h^2+b^2}\ .$$

Since $h>b$, $2h^2$ will be much greater than b^2, and $2h^2+b^2$ is therefore close to $2h^2$. Inserting this in the first member (and only there), Neugebauer finds the formula he looks for,

$$d \approx \frac{2h^3}{2h^2}+\frac{2b^2h}{2h^2+b^2} = h+\frac{2b^2h}{2h^2+b^2}\ .$$

He does not comment upon the fact that the error committed when $2h^2+b^2$ is replaced by $2h^2$ is of the same order of magnitude as $\frac{2b^2h}{2h^2+b^2}$ (namely, $-\frac{b^2}{2h}+\frac{b^4}{4h^3}-\ldots$), which is the main reason that the result is so poor.

308 #6 finds 41´15˝, #7 instead 42´13˝20‴. The true value is 41´13˝51‴48⁗ $8^{(5)}$... .

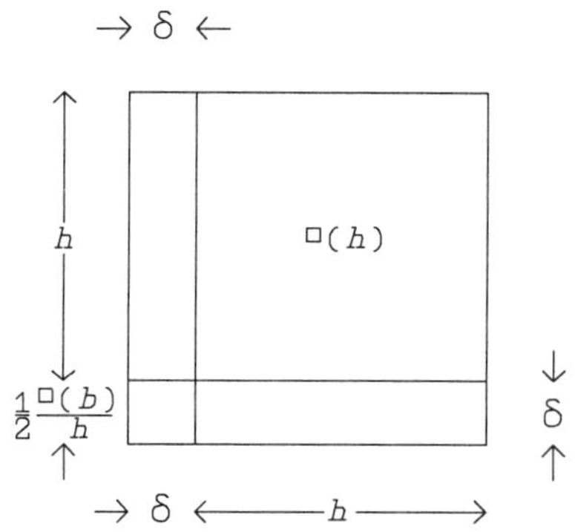

Figure 76. A possible geometrical procedure behind VAT 6598 #7.

It does not invalidate Neugebauer's explanation that it presupposes the author of the text to have committed a blunder (which will certainly not have been elementary before the invention of *analysis infinitorum*); in any case the result as stated *is* so much worse than that of #6 that some serious error must have been committed. But we may still try whether the geometrical interpretation of the method throws new light on it. In #6, the dotted square was simply neglected. Actually, the surface □(*b*) should be distributed not only along the two sides *h* but also in the corner; in total, this surface is therefore equal to a rectangle ⊏⊐(δ,2*h*+δ) – see Figure 76. However, we already know that $\frac{1}{2}\frac{\square(b)}{h}$ is a good approximation to δ, and we may therefore replace the equation

$$⊏⊐(\delta,2h+\delta) = \square(b)$$

by

$$⊏⊐(\delta,2h+\frac{1}{2}\frac{\square(b)}{h}) = \square(b)\ .$$

Since $\frac{1}{2}\frac{\square(b)}{h}$ was found by a "raising" of □(*b*) to igi *h* followed by a "breaking", we may now perform the opposite operations, first "raising" to *h*, which gives the equation

$$⊏⊐(\delta,2\square(h)+\tfrac{1}{2}\square(b)) = \square(b)\cdot h$$

and then "repeating" to 2 (since this is the inverse of "breaking"). Thereby we get

$$⊏⊐(\delta,4\square(h)+\square(b)) = 2\square(b)\cdot h\ . \qquad (*)$$

This would lead to

$$d = h+\delta = h+\frac{2\square(b)\cdot h}{4\square(h)+\square(b)} \quad \text{instead of} \quad d = h+2\square(b)\cdot h\ .$$

Nonetheless, the closeness of the procedure to the familiar techniques of Old

Babylonian mathematics, and in particular the agreement between the operations of #7 and those of #6, makes it plausible that the reconstruction is faithful.

The omission of the denominator 4□(*h*)+□(*b*) cannot be explained by its closeness to 1 – it is 1°48′20, which in this reconstruction explains why the outcome is as far off the mark as it is; it must be due to a genuine mistake – perhaps a mere omission, perhaps a faulty identification of the rectangle in (*) with a broad line ⊏⊐(δ,1).

If the procedure had been brought correctly to an end, and the division by the irregular number performed accurately, the diagonal would have been found to be 41′13″50‴46⁗... . This precision is quite remarkable, though not quite comparable to that of YBC 7289. It should be noted that the procedure is not suited for iteration, since every step is likely to lead to division by an irregular number; in the side-and-diagonal number algorithm (which cannot be generalized straightforwardly to non-square cases), in contrast, everything except the last determination of the ratio takes place in the range of integers.

BM 85194 #20–21[309]

These twin problems deal with a circle, considered as a line which is bent (which explains line 39). The length of the line is 1`, and since the ratio between circular diameter and circumference is supposed to be 20′ (and the inverse ratio between circumference and diameter – our π – thus 3), this leads without any calculation to the assumption that the diameter *D* (the dal, "crossbeam" of line 35 and of line 40, second occurrence) is 20. The coefficient is indeed listed in the tables of igi.gub constants, and is meant there as the value of *D* for circumference *1*, which is precisely the situation here.

In this circle, a chord *c* (even this called dal, "crossbeam"[310]) is drawn with a "descent" *d*. The Pythagorean rule is used to find the relation between *c* and *d*.

309 Based on the transliteration in [MKT I, 148], cf. [TMB, 32].

310 The recurrent identification of the "diameter-crossbeam" as 20 certainly serves to keep the two homonymous entities apart; nonetheless, the use of the same word is a clear trace of teaching in direct confrontation where pointing is possible: "this crossbeam here", "that crossbeam there".

Rev. I

#20

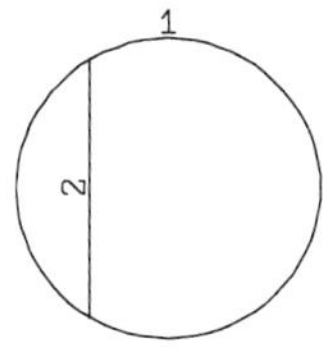

33. 1` the circle. 2 nindan I have descended, the crossbeam what? You,
1 gúr 2 nindan *ur-dam* dal en.nam za.e

34. {...} 2 make hold, 4 you see. 4 from 20,
{za.e} 2 NIGIN 4 *ta-mar* 4 *i-na* 20

35. the crossbeam, tear out, 16 you see. 20, the crossbeam, make hold, , 6`40 you see.
dal ba.zi {*ta-mar*} 16 *ta-mar* 20 dal NIGIN 6,40 *ta*⟨*-mar*⟩

36. 16 make hold, 4`16 you see. 4`16 from 6`40 tear out,
16 NIGIN 4,16 *ta-mar* 4,16 *i-na* 6,40 ba.zi

37. 2`24 you see. (By) 2`24, what is equalside? 12 is equalside,
2,24 *ta-mar* 2,24 en.nam íb.si$_8$ 12 íb.si$_8$

38. the crossbeam. Thus the procedure.
dal *ki-a-am ne-pé-šum*

#21

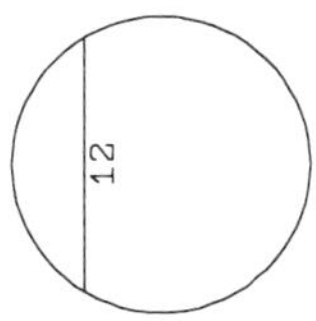

39. If a circle, 1` I have bent,
šum-ma gúr 1 *ak-pu-up*

40. 12 the crossbeam, that which I descended ⟨what⟩? You, 20, the crossbeam, make hold.
12 dal *ša ur-dam* ⟨en.nam⟩ za.e 20 dal NIGIN

41. 6`40 you see. 12 make hold, 2`24 from 6`40 tear out,
6,40 *ta-mar* 12 NIGIN 2,24 *i-na* 6,40 ba.zi

42. 4`16 you see. (By) 16, what is equalside? 4 is equalside. ½ of 4 break,
4,16 *ta-mar* 16 en.nam íb.si$_8$ 4 íb.si$_8$ ½ 4 *ḫe-pé*[311]

43. 2 you see, 2 that which you descended. The procedure.
2 *ta-mar* 2 *ša tu-ur-dam ne-pé-šum*

---

311 Most likely, the original wording will have been
4,16 *ta-mar* 4,16 en.nam íb.si$_8$ 16 íb.si$_8$ 16 *i-na* 20 ba.zi 4 *ta-mar* ½ 4 *ḫe-pé*, which translates
4`16 you see. (By) 4`16, what is equalside? 16 is equalside. 16 from 20 tear out, 4 you see. ½ of 4 break.

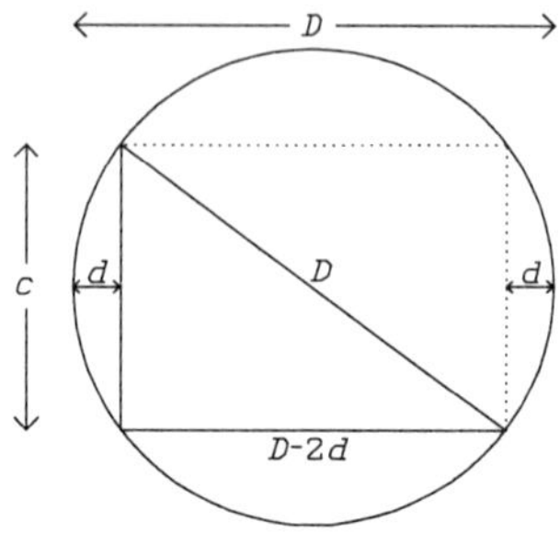

Figure 77. The circle of BM 85194 #20–21, with chord and descent.

In #20 d = 2 nindan is given, and c is asked for. It is used that c and $D-2d$ contain a rectangle with diagonal D, as can be seen in Figure 77 by considerations of symmetry; of course we may also say that c, $D-2d$, and D contain a right triangle, but the Babylonian preference was certainly for the rectangle; moreover, only drawing of the full rectangle makes it intuitively obvious that the relevant angle *is* right. The steps are easily followed: by error, d = 2 is "made hold" and not "repeated".[312] The resulting 4 is "torn out" from D, and c is found as the "equalside" of $\square(D)-\square(D-2d)$.

In #21, the roles of c and d are inverted, both in the statement and the procedure; this time, $2d$ is broken, as it should be – but the preceding passage is confused, seemingly due to overly creative tampering with a better original.

As a mathematical curiosity we may consider Figure 78, in which the diagram of the present problems is superimposed on that of TMS I, which shows how the technique of IM 55357 *could* have been used in TMS I (the shaded triangle is indeed that of IM 55357, with the first subdivision) – but which also makes it clear that this presupposes an indifference to orientations of angles that is not likely to be found where no general angle geometry is at hand and well trained; the quasi-algebraic approach to TMS I remains the only reasonable interpretation. All the diagram tells us

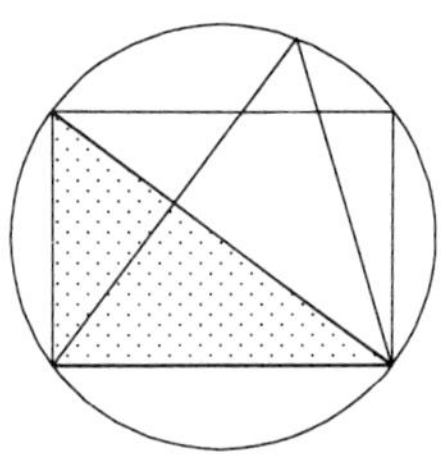

Figure 78. Combination of the diagrams of TMS I and BM 85194 #20–21.

[312] Once more Neugebauer demonstrates by considering this an error that his sensitivity to the details of the texts and the terminology went beyond what he would state explicitly. "Es gehört nicht zu den Aufgaben, die ich mir in dieser Edition gestellt habe, die Konsequenzen zu entwickeln, die sich nun aus diesem Textmaterial ziehen lassen", as he opens the conclusion of [MKT III, 79].

is that Old Babylonian geometry was very fond of the 3–4–5 configuration – so fond that coincidences are bound to occur.

BM 85196 #9[313]

This is by far the earliest occurrence of "the pole against the wall", a problem type that was bond to spread widely during the following three millennia – see [Sesiano 1987]. At times it appears in the present simple version, which ask for nothing but straightforward application of the Pythagorean rule; at times it presents itself in a form that asks for more sophisticated mathematics (we shall encounter this form below, p. 398).

Obv. II

7. A pole, 30′, (that is,) a reed, from [...] ¿its? [...]
[giš]*pa-lu-um* 30 gi *i-na* [...]*-bi-šu* [...]

8. Above, 6′ it has descended, be[l]o[w, what has it moved away?]
e-le-nu 6 *ur-dam i-na ša-a*[*p-la-n*]*u-*[*um* en.nam *is-sé-a-am*]

9. You, 30′ make hold, 15′ you see. 6′ fro[m] 30′ te[ar out, 24′ you see.]
za.e 30 NIGIN 15 *ta-mar* 6 *i-n*[*a*] 30 ba.[zi 24 *ta-mar*]

10. 24′ make hold, 9′36″ you see, 9′⌊36″ from⌋ [15′ tear out],
24 NIGIN 9,36 *ta-mar* 9,⌊36 *i-na*⌋ [15 ba.zi]

11. 5′24″ you see. 5′24″, what ⌈is equalside?⌉ [18′ is equalside, 18′]
5,24 *ta-mar* 5,24 en.nam ⌈íb.si$_8$⌉[18 íb.si$_8$ 18]

12. on the ground it has moved away. If 18′ o[n the g]round,
i-na qá-qá-ri is-sé-a-am šum-ma 18 *i-n*[*a qá-*]*qá-ri-im*

13. above, what did it descend? 18′ make hold, 5′24″ you see,
e-le-nu-um en.nam *ur-dam* 18 NIGIN 5,24 *ta-mar*

14. 5′24″ from 15′ tear out, 9′36″ you see. 9′36″,
5,24 *i-na* 15 ba.zi 9,36 *ta-mar* 9,36

15. what is equalside? 24′ is equalside, 24′ from 30′ tear out,
en.nam íb.si$_8$ 24 íb.si$_8$ 24 *i-na* 30 ba.zi

16. 6′ you see (for what) it has descended. Thus the procedure.
6 *ta-mar ur-dam ki-a-am ne-pé-šum*

In the actual shape, the problem deals with a pole of length l = 30′ nindan (identified as the unit 1 reed) – see Figure 79. At first it stands vertically against a wall; afterwards it is moved to a slanted position, in which its top descends by an amount d and the foot moves a distance s away from the wall.

---

[313] Based on the transliteration in [MKT II, 44], cf. [TMB, 42].

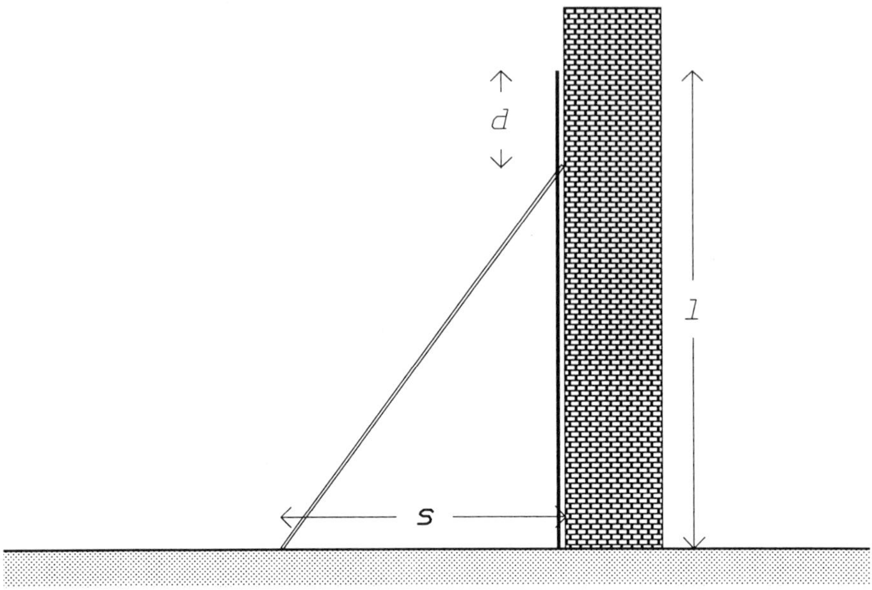

Figure 79. The pole against the wall.

As the previous text, this one presents us with a twin procedure. In lines 8–12, d is regarded as given, and s is found as the "equalside" of $\square(l)$–$\square(l–d)$, in agreement with the Pythagorean rule; in lines 12–16 the reverse calculation is made, $l–d$ is found as the "equalside" of $\square(l)$–$\square(s)$; when this quantity is "torn out" from l, d remains.

Summary Observations

(4.12.96) 1996

Before we leave this chapter on geometry, a few summary observations may be useful.

Let us first look at the techniques that are employed, beyond the basic formulae for area and volume determination. The "algebraic" texts made use of the cut-and-paste technique with completion and of scaling in one and two dimensions; all of these, we have seen, are also used in texts which we will hardly feel tempted to characterize as algebraic but certainly as geometric.

We may try to formulate a notion of "favourite configurations", geometrical configurations that are used to define problems to an extent or in ways that cannot be explained by their practical importance (as can the volume of irrigation canals with sloping sides, etc.). Two categories have turned up in the preceding pages: squares within squares, and "subdivided figures". The former type belongs together with systems of concentric circles within a larger category; the latter can itself be divided into subcategories, particularly important among which are the "striped figures",[314] that is, triangles and trapezia partitioned by parallel transversals, and the special case of the bisected trapezium and its generalization to trapezoids exemplified by YBC 4675.[315]

314 The term was suggested by Jöran Friberg [1990: 558].

315 It will be evident, I hope, that the small selection of problems in this chapter does not, and is not meant to cover, Old Babylonian geometry in any representative way. A large part of the geometric problems we find in the text ask for the determination of areas or volumes, and may in this sense be considered mathematically simple.

However, even if we try to distinguish a particular "sophisticated level", suited (like the algebraic problems) to demonstrate the virtuosity of the reckoner, the selection is far from being representative.

Chapter VII
Old Babylonian "Algebra": a Global Characterization

Algebra?

Neugebauer, Thureau-Dangin, Gandz, and others spoke without doubt about a Babylonian "algebra", and the existence of such a thing was accepted without objections from the 1930s through the late 1960s. It was also accepted that this algebra was numerically based – in [1933] Vogel had proposed a geometric interpretation of AO 8862 #1 (the same as described in Figure 29), but in the end even he accepted the numerical interpretation. In [1936], Neugebauer had set forth the further thesis that the "geometric algebra" of *Elements* II and the "application of an area with deficiency or excess" had been created as a translation into geometry of the findings of Babylonian algebra – a translation that had become necessary if the general validity of the Babylonian results should remain secure after the discovery of incommensurability.[316] Even

---

316 In Neugebauer's formulation:

> "The answer to the question about the historical origin of the fundamental problem of the whole geometric algebra [i.e., the application of an area with deficiency or excess], may now be given without restrictions: on one hand, it arises from the requirement of the Greeks, coming from the development of irrational magnitudes, that the general validity of mathematics should be secured through a shift from the domain of rational numbers to that of general ratios between magnitudes; on the other from the ensuing necessity, *also to translate the results of pre-Greek "algebraic" algebra*.
>
> Once the problem is formulated thus, the rest is fully trivial, and provides

this idea was commonly accepted for decades.[317]

Both the translation thesis and the algebraic reading of the Babylonian texts came under attack around 1970. We shall return below to the translation question and discuss how Neugebauer's thesis must be recast in view of the new evidence; at this point, we shall take up Michael Mahoney's reflections in his essay review of the reprint edition of Neugebauer's *Vorgriechische Mathematik* [1934]. Mahoney, currently absorbed in that algebra which was "a creation of the seventeenth century – AD!" [1971: 375], took advantage of the occasion to ask in which sense "Babylonian algebra" could be taken to be *algebraic*. He proposed a distinction between a mere *algebraic approach* and *algebra as developed from Viète to Descartes*. Algebra proper he took (p. 372) to be characterized by three characteristics: (i) the use of a symbolism which allows us to extract "the *structure* of a problem from its non-essential content", and on which we may operate directly; (ii) the search for "relationships (usually combinatory operations) that characterize or define that structure or link it to other structures"; (iii) abstraction and absence of "any ontological commitments". To this we may add, with Viète, that (iv) algebra, if at all to be characterized as such, should be *analytical*; in fact, without this analytical character, Mahoney's criteria make no sense.

With this delimitation of the concepts, and basing himself on the established arithmetical reading of the texts as represented by the volume under review, he argued that the Babylonian type of mathematics contained only recipes for numerical procedures, and was therefore only a representative of the *algebraic approach*. He made a plea (p. 377)

> to wield Ockham's razor when dealing with Babylonian mathematics and not to assign to the Babylonians any concept, or form of mathematical thought, for which there is no explicit documentation, nor even need.

We might object – Neugebauer had done so in anticipation repeatedly[318] –

---

the *smooth junction of the Babylonian algebra to the Euclidean formulations*" [Neugebauer 1936: 250].

Neugebauer added that this full translation was facilitated by that translation (as he saw things) *of the problems* into length-width geometry which the Babylonians themselves had undertaken.

317 The creation and acceptance (until the onslaught of c. 1970) of the algebraic interpretation and of the idea of a translation from Babylonian into Greek is analyzed in [Høyrup 1996: 7–17].

318 A delightfully aggressive passage is the following comment on a particular sophisticated sequence of fourth-degree problems from a series text [MKT I, 438]:

> Our text material does not reveal how one would confront such problems. However, I must explicitly observe that the cherished expedient, found repeatedly in the literature, namely, to assume that the existence of a common solution $x = 30$, $y = 20$ should indicate that the solution should simply be guessed, is all too cheap. If this were the whole gist of our series texts, then we would have to assume that thousands of problems had been put together for whose solution it had been sufficient, to guess the correct values *a single*

that even when the texts are read as recipes for numerical procedures, these procedures are often so complex that they presuppose coherent mathematical thought; invention by trial and error can be safely disregarded (think of the broken reed in VAT 7532, or the cascade of equations in TMS XIX #2!). It seems more fruitful, however, to explore to which degree and in which sense the new interpretation of the Babylonian supposedly algebraic texts allows us to regard them as constituting *an algebra* according to the four criteria listed. To a large extent, this will be nothing but a recapitulation of scattered observations made on the texts in preceding chapters.

Beginning with (iv), most of the second-degree procedures we have looked at certainly agree with it – we remember Viète's "assumption of what is searched for as if it were given, and then from the consequences of this to arrive at the truly given" (above, p. 33), which is exactly what is done in the cut-and-paste procedures (cf. also the comparison of the naive and the Euclidean approach on p. 98). The many false positions that are used to solve problems or sub-problems of the first degree are also clearly analytical. Even the construction of reference volumes that share some of the required properties of the solution in the treatment of inhomogeneous problems of the third degree is analytical in character, though less directly so than the cut-and-paste procedures.

As to (iii), abstraction and absence of "any ontological commitments", we may notice that the geometry of the standard representation is *functionally abstract*, as observed twice above (pp. 10, 58): a segment may represent a number from the igi-table; it may represent an area as in BM 13901 #10, or a volume as in TMS XIX #2; it is even likely that a school-yard length of 30′ nindan may represent a field length of 30 nindan. It may also represent a commercial rate as in TMS XIII.

As was pointed out, all texts except some of the early ones from Eshnunna and an aberrant passage in one from Nippur use the Sumerograms uš and sag̃ unerringly for the lengths and widths of the standard representation; "real" linear dimensions (the length of a wall, the distance bricks are to be carried, the width of a canal), in contrast, may as well be written in syllabic Akkadian (*šiddum* and *pūtum*) as with the Sumerograms. This suggests strongly that the

> time: an obvious nonsense, which nobody would set forth who had taken the trouble to work through the full material of Babylonian mathematics. Indeed, this not only shows us that the examples of the single tablet from our series texts are often ordered with the most rigorous technical (*sachlich*) order [...], which is only made possible by real insight in the mathematical relationships; we also possess a wealth of other texts, which prove through explicit calculation, that, for instance, the particular biquadratic equations – which are also plentiful in the series texts – were solved absolutely correctly.

It is only to be deplored that such words, deeply buried within one of the most technical sections of an already quite technical edition, were rarely read by those authors of general works on the history of thought, science or mathematics to whom they were addressed.

Old Babylonian authors were explicitly aware of the functionally abstract character of their standard representation.

(i), the use of a symbolism which allows us to extract "the *structure* of a problem from its non-essential content", and on which we may operate directly, is debatable but connected with the question of the standard representation. If we understand "symbolism" as "letter or similar symbolism", we are sure to make the Babylonians fail on this account; however, the gist of the criterion is the *operation on the level of* the symbolism and the distinction between essential structure and non-essential content, and here the very use of the geometric standard representation (whether in drawings to scale, in structure diagrams or in mental geometry) is what allows us to extract from the problem of the broken-reed riddle of VAT 7532, the oil traffic of TMS XIII, and the *igûm-igibûm* problem of YBC 6967 the same essential structure of a rectangle with given area and given excess of length over width, and to extract from the unfinished ramp of BM 85194 #25 the complementary problem with given sum of the sides.

The point where Old Babylonian "algebra" fails indubitably is (ii) the search for "relationships (usually combinatory operations) that characterize or define that structure or link it to other structures". After a few initial attempts to formulate general rules (that these are indeed early we shall see below, p. 383), the Old Babylonian school presented all its insights through paradigmatic examples, from which implicit familiarity with the underlying principles of the problem could be obtained. When it comes to general structures and to the links between structures, no attempt seems ever to have been made to formulate these, nor are traces to be found that they were explicitly searched for.[319]

Old Babylonian "algebra" remained an *art*, not a science, if this is understood as an Aristotelian *epistēme* whose aim is *principles*. On this account, however, any supposed algebra before Viète forsakes, however deep its insights. If we accept to speak of (say) Indian, Islamic, or Latin/Italian medieval "algebra" as *algebra*, then we may safely drop the quotation marks and speak of *Old Babylonian algebra* without reserve.

Nowadays, of course, "algebra" designates a complex, not a single technique or a simple concept; Mahoney's criteria are just as much of an arbitrary choice as the decision to accept that the algebraic art from the Old Babylonian epoch until Bombelli is really an algebra. The technique for solving equations is understood as algebra (the reason that the pre-Viète techniques may also be understood so) – but algebra may also be theory of the solvability of equations (closer to Mahoney's sense), and theory of

---

319 The systematic variation of problems, a distinctive characteristic of the Old Babylonian school, was certainly not meant as a search for underlying structures; to some extent it reflects an interest in probing the tools of the profession, but its main purpose was probably that of training (not least, training the use of the sexagesimal number system). Nonetheless – the surviving texts leave no doubt about this – it led to at least intuitive insight into formal structures and relationships.

groups, etc. *Algebra* is in fact a collective name for a plurality of *algebraic ways of thought*, evidently related either logically or historically with each other, but certainly neither coinciding nor sharply to be distinguished from other ways of thought. The algebraic ways of thought may be said to constitute *today* a kind of Wittgensteinian natural family. *Today* only: if we go back in time, the various components of the family do not belong together (staying with the metaphor, the family was constited through marriage alliances as much as through descent). Thus *Elements* X, which nobody would otherwise identify as a piece of algebra, comes close to modern group theory in its classification of irrational magnitudes and its determination of the relations between the classes.

Equations

Connected to the question of algebra is that of *equations*. Above, this term was used without reserve about many problem statements, in the sense formulated on p. 38: "statements that (the measure of) a more or less complex quantity equals a number", or declarations that "(the measure of) one quantity equals (the measure of) another quantity".

These may not be equations according to the sense a modern theoretical mathematician would prefer to give to that word; the parenthetical references to measures are disturbing.

If so, however, the "equations" of engineers, physicists, and all those who use algebraic procedures to describe real-world phenomena will also be driven into exile: a simple equation like Hooke's law for a spring, $F = -\varkappa d$, expresses that the force F which the spring exerts *if measured in a certain unit* is proportional and opposite in direction to the distance d it has been stretched or compressed from the point of equilibrium, *even this measured in a particular unit*. If any of these units is changed, we get an equation with a different value for the factor $\varkappa$ of proportionality; this can be determined from the original $\varkappa$ and the ratio between the old and the new unit – exactly as the new shape of a Babylonian equation after a change of unit could be calculated by means of a metrological table. In so far, Babylonian "equations" are thus equations in the sense this word is bound to have within any application of algebraic techniques to measurable reality.

As to the equations of the functionally abstract standard representation, the measurement itself has been emptied of reality, and the entities it deals with are functionally identical with their measures; from the functional point of view, these are thus equations even in the mathematician's sense.

Distinctive Characteristics

Deciding that it is more fruitful to count the Old Babylonian technique as a member of the algebraic family than to exclude it does not entail that we should be blind to the differences between Old Babylonian algebra and even its closest analogue in recent times, that is, equation algebra as taught in school. On the contrary, it should be taken as a solicitation to look for *differentia specifica*, the "differences that make the species", beyond those which are too obvious to need further discussion – the absence of algebraic symbolism, the formulation in terms of measurable geometrical entities, etc.

The Given and the Merely Known

The appearance of numbers in certain problem statements that are not used in the solution even though they might facilitate the calculations immensely, but whose inclusion also cause the problem to be overdetermined, and the appearance in procedures of numerical values that are not given in the statement: these peculiarities of the Old Babylonian texts are perhaps those which has caused most bewilderment in modern commentaries. Indeed, when read through the expectations of a modern mathematician they make the Old Babylonian supposed "mathematicians" look as third-rank colleagues or as copyists who did not understand what they were writing. Instead, the ability to use precisely such a set of data that is sufficient to determine the solution and disregard in the procedure those numbers that would make it overdetermined is in fact evidence of mathematical competence on the part of the authors of our texts. It is true that the phenomenon is only directly visible in texts from certain groups; however, its appearance in writing in widely separated groups as well as indirect hints makes it clear that it corresponds to a more widespread characteristic of the oral teaching situation.[320] The sequences of several or numerous problems with the same set of solutions (almost always $l = 30$, $w = 20$ or $l = 30'$, $w = 20'$!) has similar implications.

Noticing the ability of the authors of the texts not to be led astray by the presence of numerical values that are not meant to be used does not explain *why* these values are there. As can be seen in the texts (and as explained in the analysis of these, see pp. 95 and 161), they serve as identifying tags, on a par with those relative clauses which point out that a number n appearing at a certain point is the same as that number n which was "made hold"/ "appended"/"held by the head"/... at an earlier moment.

The "merely known" numbers thus fulfil a function that can also (no doubt more efficiently) be achieved by means of algebraic symbols, or by the Greek letter symbolism in geometrical diagrams. The reason that this has generally been badly understood may be the widespread resolution to read the Babylonian texts as mere recipes in the likeness of FORTRAN programs (without those commentaries which most computer programs allow the programmer to insert in order to remember or make clear to colleagues what a particular step or routine of the program is meant to effectuate).[321] It is

---

320 I have noticed it in the Susa texts; in BM 85200+VAT 6599, belonging to a group presumably to be located in Sippar – see below, p. 332; and in IM 52301, from Eshnunna. With respect to the former Sumerian heartland, Susa can be characterized as south-eastern, Sippar as north-western and Eshnunna as north-eastern periphery. Among texts from the core area, VAT 7521 and Str 367 seemed to reveal a corresponding oral practice.

321 Donald Knuth's reading [1972] of the Old Babylonian procedures as algorithms

definitely easier to recognize which part of *our* mental luggage is absent from the Babylonian mind than to identify ingredients of this foreign thinking that are absent from ours.

"Pedantic Repetitiveness"

Who looks for the first time at VAT 8389 #1 and VAT 8391 #3 (above, pp. 77*ff*) is likely to be struck by the many repetitions, for instance, of the phrase "igi 30´, of the bùr, detach: 2´´" – with the variation "igi 30´, of the second bùr, detach: 2´´", in which the repetition is pointed out quite explicitly to *be* a repetition of exactly the same step (which is hardly a step at all, since the outcome is listed in the igi table).

In Str 367 (p. 239), rev. 2 repeats the prescription "1 and 3 accumulate, 4" which was already in obv. 8. Even here, the same numbers 1 and 3 (namely, the proportionals) are involved in both cases – and once again, the modern reader may get an impression of pedantry and stereotyped duplication.

Actually, the two examples are different in character, and have different explanations. The former indicates, and trains, how to make use of standard tables and scratch pads in order to find the specific rents of the two plots expressed in "standard" instead of "practical" units. The latter example shows that what suggests to us the notion that the Babylonian procedures are nothing but recipes is exactly that they are normally less repetitive than calculations expressed in symbolic algebra.

In order to see that we may compare an algorithmic interpretation of BM 13901 #1 with the corresponding solution of the problem in symbolic equation algebra. Algorithm first:

45´ is given as sum of area and confrontation
1° is posited
broken → 30´
made hold → 15´
appended to 45´ → 1
finding the equalside → 1
tearing out 30´ → 30´, the confrontation

In equation algebra, we get

$$\begin{aligned}
&\square(s) + 1\cdot s = \tfrac{3}{4}\\
\Leftrightarrow\ &\square(s) + 1\cdot s + \square(\tfrac{1}{2}) = \tfrac{3}{4} + (\tfrac{1}{2})^2\\
\Leftrightarrow\ &\square(s) + 1\cdot s + \square(\tfrac{1}{2}) = \tfrac{3}{4} + \tfrac{1}{4} = 1\\
\Leftrightarrow\ &\square(s + \tfrac{1}{2}) = 1
\end{aligned}$$

---

strictly similar to those of computer programming, inspired by extant translations and extant secondary literature, has had much success and contributed to perpetuate the myth that Babylonian mathematics was "empirical" and probably an outcome of trial-and-error experimentation.

$$\Leftrightarrow \; s + \tfrac{1}{2} = \sqrt{1} = 1$$
$$\Leftrightarrow \; s = 1 - \tfrac{1}{2} = \tfrac{1}{2}$$

In the latter description, we see, for instance, the same $\square(s)$ repeated in four consecutive lines, after which the same s occurs twice. This happens for very good reasons: namely, because the symbols are the representatives of the unknown quantities that have to be treated as if they were given (in Viète's explanation of the analytical principle). In a Euclidean proof, in contrast, where the representatives are in a geometrical diagram, the wording of the proof need not be repetitive; nor does a normal Old Babylonian naive-geometrical proof, because the representatives of the unknown quantities are outside the text (whether in an actual or, with the trained calculator, only in an imagined diagram).

The repetition in Str 367, however, is of the same kind as those of our symbolic calculation. Where we would *write* the same binomial $p+q$ (p and q being the proportionals) as components of two different algebraic expressions, the Old Babylonian calculator *describes its calculation* with those given numerical values which serve as their identifying names.

Repetitiveness as such, we see, is no distinctive characteristic of the Old Babylonian texts; what is distinctive (for Old Babylonian as well as modern algebraic texts) is the particular pattern of repetition versus non-repetition – a pattern which in both cases is a consequence of other aspects of the mathematical techniques that were/are used.

Favourite Configurations

On p. 162, the organization of teaching around configurations rather than methods was discussed. Such an organization is not by necessity connected to the predilection for certain configurations – but the algebra texts do show an inclination to formulate problems around a restricted set of favourite configurations, just as the sophisticated geometrical problems bear witness of affection for striped figures and concentric organizations.

To some extent, an apparent foible for a configuration may be an artefact due to the popularity of a particular problem type (cf. imminently). Such may be the case for the trapezoidal field in the broken-reed problems (several variants of the problem from VAT 7532 exist in which the numerical parameters are changed but nothing else).[322] In other cases, a configuration which is somehow characteristic is used as the basis for problems which have to be solved by wholly different methods – for instance, the excavation

322 Even then, however, the configuration might end up being considered interesting on its own account; as we shall see, the bisectable trapezium with widths 7 and 17 serves as foundation for a much simpler problem in UET V, 858, see p. 353.

prolonged downwards or with "ground" and "earth" accumulated, which serves as a pretext for problems of the first, second, and third degree in BM 85200+VAT 6599, but which is also the basis for a sequence of second-degree problems in the catalogue text YBC 4657. Here, we can thus be certain that the configuration in itself counts as interesting.

The simple square and rectangle problems are both so simple and so closely bound up with their function as standard problems within the standard representation that they reveal nothing about the status of the underlying figures; once an algebraic second-technique based on measurable geometry existed, these could hardly fail to appear. Concentric squares could, however; that a whole variety of algebraic as well as non-algebraic problems deals with such systems[323] shows that they enjoyed a particular status.

Favourite Problems

Again, the standard problems to which other more complex problems are reduced are not "favourite problems". Favourite problems and problem types are those which turn up repeatedly in the material but do not "need" to turn up because of the invitations directly offered or constraints produced by the structure and techniques of the prevailing mathematical practice, and which are so peculiar that they can be easily recognized.

The text selections in Chapters III, V, and VI were aimed at exhibiting in breadth the range of Old Babylonian algebra and quasi-algebraic geometry. That some of the problems are indeed favourites was therefore not obvious – the purpose of the chapters would not have been served if (for instance) four broken reed problems with the same structure as VAT 7532 had been included.

But the measurement of the area of a trapezoidal or rectangular area by a measuring reed that undergoes losses in the process *is* a favourite problem – actually two problems, as explained on p. 209. The unfinished ramp seems also to have been a choice problem type – BM 85194 #25–26 are two different versions, and the latter turns up again in BM 85195. Even the oil trade of TMS XIII represents a widespread type: MLC 1842 is a variant which leads to the sum of rectangular sides; the series text YBC 4698 contains a sequence of problems fully analogous to the Susa text [Friberg 1982: 57]; another sequence of similar problem statements are found in the catalogue IM 52685+52304.

The reason for the popularity of these queer problems is probably the discovery that they *could be solved* but only by means of the second-degree technique. The first application of the second-degree technique to a real problem had to wait for more than 2000 years – see p. 206. Finding problems

323 Among the texts that were presented and analyzed above thus many problems from YBC 4714; BM 13901 #8–9 (as elucidated by TMS V, cf. 94); and UET V, 865.

whose appearance would connect them to the professional duty of scribes and which nonetheless asked for use of second-degree algebra will not have been easy; once some clever calculator or teacher had discovered such a problem, there were good reasons to remember it.[324]

The case of problems of the type "accumulation of rectangular length, width, and surface given" (and some other condition) is somewhat different. It is true that the type turns up repeatedly in undressed form; but elsewhere the characteristic structure is found in varying dress: in AO 8862 #7 we are told the sum of men, working days, and bricks (proportional to the product of men and working days) in a brick problem (#4, we remember, is the basic form of the problem, with $l+w$ = ⊏⊐(l,w) as the other condition); in YBC 4668 #A9, we are told the value of $\alpha+\beta+\alpha\beta$ where α and β are the unknown coefficients of a linear problem in l and w; the catalogue text BM 80209 contains several problems about the sum of circular surface S, circumference c, and diameter d.[325] That this asks for application of the second-degree techniques is too obvious to be interesting; in this case it therefore seems to have been the mathematical structure which was intriguing and became a favourite.[326]

"Remarkable Numbers"

Beyond its favourite objects and favourite problems, Old Babylonian algebra was also characterized by its use of a structured set of favourite numbers, numbers which somehow were considered "remarkable"; we may presume that the dedication to this set of numbers would characterize Old Babylonian mathematics more generally, but the systematic organization of algebraic theme texts opens a unique window to this aspect of mathematical thinking.

That a similar category might exist was already suspected by von Soden [1964: 47] – but since one of the deep convictions that a modern mathematical education imparts is that numerical values are immaterial to *the mathematics* of a procedure, the challenge was not taken up.[327]

---

324 This explanation is in need of further elaboration, and should be linked to a discussion of the sociology of practitioners' mathematical knowledge and the role of "supra-utilitarian" knowledge; cf. below, pp. 362*ff*. For the moment, these common-sense considerations will do.

325 It does not seem obvious that this is the same structure – but in the same text group we find a determination of the surface of a semicircle which corresponds to the formula $S = \frac{1}{4}c \cdot d$ for the full circle. If this was used, the structure would be the same apart from a normalization.

326 Originally, as we shall see (pp. 373, 379), the rectangle version of the problem was probably a member of a restricted set of favourite riddles. For the moment, however, we are portraying the algebra of the Old Babylonian school and not its roots.

327 And, as often happens, I only noticed the passage after I had approached the matter for other reasons and written the first draft for [Høyrup 1993a], in which

In order to decide whether a number or set of numbers which recurs often in the texts does so because it enjoys special status one has to rule out rival explanations: Such an explanation could be that it is the easiest choice, given a particular mathematical object – the Pythagorean triples 3–4–5 or 45′–1–1°15′ when we treat of the diagonal of a rectangle, the sides 13–14–15 if want to show how to compute the height of a scalene triangle without ending up in the quagmire of irrationals, etc.; or it may somehow be the outcome of computations and thus not of independent free choice.

Within the confines of Old Babylonian mathematics, such independent and deliberate free choices occur at two levels. Since problems were constructed from known solutions, these had to be chosen – and the choice was free unless is was subjected to mathematical constraints (as the choice of the sides of a rectangle whose diagonal would be used). But the construction of problems would often – not least when whole sequences of problems were produced by systematic variation – involve the choice of coefficients, and even this choice would as a rule be free.[328]

Above, only a small selection of the relevant texts were included, and a complete survey of relevant sources would lead too far astray. The character of the material may be suggested by a complete survey of the problems of BM 13901 (the theme text on squares which we have already exploited extensively).[329] In symbolic translation its problems are (s_i stands for confrontations and Q_i for the corresponding surfaces):

1. $Q+s = 45'$
2. $Q-s = 14`30$
3. $Q-\tfrac{1}{3}Q+\tfrac{1}{3}s = 20'$
4. $Q-\tfrac{1}{3}Q+s = 4`46°40'$
5. $Q+s+\tfrac{1}{3}s = 55'$
6. $Q+\tfrac{2}{3}s = 35'$
7. $11Q+7s = 6°15'$
8. $Q_1+Q_2 = 21'40''$, $s_1+s_2 = 50'$ (reconstructed)
9. $Q_1+Q_2 = 21'40''$, $s_2 = s_1+10'$
10. $Q_1+Q_2 = 21°15'$, $s_2 = s_1-\tfrac{1}{7}s_1$
11. $Q_1+Q_2 = 28°15'$, $s_2 = s_1+\tfrac{1}{7}s_1$
12. $Q_1+Q_2 = 21'40''$, $\sqsubset\!\sqsupset(s_1,s_2) = 10'$

the topic is treated in some depth.

328 Not always: in VAT 8390, the number to which we "repeat" has to be a square. In TMS VII #1, where equality is aimed at, the number of steps has to be a divisor of 50 once l and w are chosen to be 30′ and 20′; in #2, where an excess is the aim, the choice is free.

329 A full transliteration with translation and analysis can now be found in [Høyrup 2001].

13. $Q_1+Q_2 = 28'\,20''$, $s_2 = \frac{1}{4}s_1$
14. $Q_1+Q_2 = 25'\,25''$, $s_2 = \frac{2}{3}s_1+5'$
15. $Q_1+Q_2+Q_3+Q_4 = 27'\,5''$, $(s_2,s_3,s_4) = (\frac{2}{3},\frac{1}{2},\frac{1}{3})s_1$
16. $Q-\frac{1}{3}s = 5'$
17. $Q_1+Q_2+Q_3 = 10\text{`}12°45'$, $s_2 = \frac{1}{7}s_1$, $s_3 = \frac{1}{7}s_2$
18. $Q_1+Q_2+Q_3 = 23'\,20''$, $s_2 = s_1+10'$, $s_3 = s_2+10'$
19. $Q_1+Q_2+\square(s_1-s_2) = 23'\,20''$, $s_1+s_2 = 50'$
20. [missing]
21. [missing]
22. [missing]
23. ${}_4s+Q = 41'\,40''$
24. $Q_1+Q_2+Q_3 = 29'\,10''$, $s_2 = \frac{2}{3}s_1+5'$, $s_3 = \frac{1}{2}s_2+2'\,30''$

In all cases except #8, #12, and #19 where several "confrontations" are involved, their mutual relation is described by their difference. This is mostly done by the expression "*a* over *b*, *d* it went beyond" (translated into symbols as $a = b+d$); in #10, however, we are told that "confrontation to confrontation, the seventh it was smaller" (full text p. 58); in #14 (p. 73), and again in #24, more complex expressions involving one of the "natural fractions" and a "going-beyond" are used.

In some cases, the difference is stated in absolute terms; then it is 10′, 5′, or 2′30″. In others, the relative difference is stated; here, the fractions $\frac{2}{3}$, $\frac{1}{2}$, $\frac{1}{3}$, $\frac{1}{4}$, and $\frac{1}{7}$ are used; when only part of the "confrontation" is subtracted or added, when part of the area is subtracted, and when differences are defined in relative+absolute terms (#3, #4, #5, #6, #14, #24), only $\frac{2}{3}$, $\frac{1}{2}$, and $\frac{1}{3}$ are in use; they thus seem to constitute a particular sub-class, which we may describe as "simple relative differences"; $\frac{1}{4}$ and $\frac{1}{7}$, on their part, belong to a sub-class of "complex relative differences"[330]. Actually, the epithets "simple" and "complex" should not be understood as characterizing the numbers themselves but those variants of problems – "simple" versus "complex" variants – which their use brings about.

Of special significance are #10 and #11. #10 might as well have stated that one "confrontation went $\frac{1}{6}$ beyond" the other, and #11 that "confrontation to confrontation, the eighth it was smaller". They thus demonstrate an *express preference* for the fraction $\frac{1}{7}$, which in #10 forces the author to use a phrase for the subtraction by comparison which he applies nowhere else.

330 The small size of the sample is of course insufficient to make such inferences statistically valid – $\frac{1}{4}$ occurs a single time in the whole tablet! But the conclusions drawn from this paradigmatic example stand firm when the whole corpus is taken into consideration. Among the above texts, YBC 4714, TMS VII, TMS VIII, TMS IX #3, TMS XVI #1, AO 8862 #2 and the sequences from YBC 4668 and YBC 4713 are of particular interest.

In #7, multiples instead of fractions turn up; here, the factors are 7 and 11. This suggest that the denominators of the "complex" category may also occur as factors (which, however, turns out to be true for 7 and 11 only). The complete list of remarkable numbers belonging to the complex partitive-multiplicative domain turns out to consist of 4, 7, 11, 13, 17, and 19.

The "simple" partitive domain turns out already to be fully represented in BM 13901; at most we may say that the "half of the third" of YBC 4714 rev. II 10 represents an extension of the category, and a way to avoid the relative difference $\frac{1}{6}$, which would otherwise be compulsory under the conditions of systematic variation (whereas $\frac{1}{5}$ cannot be dodged and *is* found in rev. I 21).

The preponderance of 10 and its successive halves (in the order of integers or minutes) is probably no expression of an immediate predilection for these *as differences* but rather the result of preferred absolute values of the sides of rectangles and squares – 20 and 30 for the dimensions of rectangles, multiples of 5 if several squares are involved.[331]

We need not ask for the reason that $\frac{2}{3}$, $\frac{1}{2}$, and $\frac{1}{3}$ enjoyed a special status as simple fractions in the school – the fact that special signs were used for them is sufficient to show that this status had its roots far back in time.[332] However, the list of "complex" multiplicative-partitive remarkable numbers may cause perplexity. Most of them are prime numbers – but no hint exists that this was a concept that worried the Babylonians, and the prime number 5 is missing while 4 is an indubitable member of the list. Then all except 4 are irregular – but 14 is also irregular, and no member.

We gain nothing by appealing to number-psychological universals – "the sacred number 7", etc. Even if we believe in their existence (and 7 does play a role in Mesopotamian mythology, but so do several regular numbers as "sacred numbers of the gods"), we would still have to explain why we find 4, 7, 11, etc., only as denominators and factors but never as the sides of rectangles or squares.

Since being prime or irregular is only important if the numbers are used as denominators or divisors, these characteristics are likely to be involved in the explanation. The best guess – and a guess it has to be – is probably that these numbers represent *number in general*. $\frac{1}{4}$ is indeed the first "non-simple" fraction (it is never replaced by $\frac{1}{2}$ of $\frac{1}{2}$), and in this sense 4 is also the first non-small number. As to the others, the condition that a fraction $\frac{1}{n}$ has to be dealt with *as itself* is that it is irregular and prime; $\frac{1}{5}$ could be expressed as 12 (that is, 12′), $\frac{1}{6}$ (as we have seen) as $\frac{1}{2}$ of $\frac{1}{3}$, and $\frac{1}{14}$ as $\frac{1}{2}$ of $\frac{1}{7}$ (as it

331 Even in cases where the side of a square has to be divisible by 7·7 or 7·11 in order to allow a solution (as in BM 13901 #17 and in many problems in the catalogue text TMS V), it is always chosen so as to be *also* divisible by 5 (that is, divisible in its own order of sexagesimal magnitude, which 49 and 1`17 are not).

332 We may of course also agree that these *are* the simplest fractions we can imagine – but the use of successive halves in ancient Egyptian metrology shows that simplicity is not wholly independent of culture and habit. Those who grew up with the metric system may regard $\frac{1}{10}$ as simpler than $\frac{2}{3}$.

would always be in Arabic). That no prime numbers above 20 occur may have to do with linguistics: for ordinal numbers above 20, the corresponding cardinal number seems to have been used [AHw, 70]. This might imply that "the *n*th [part]" for $n>20$ would be indistinguishable from *n*; more important (since a technical terminology can always find subterfuges to avoid such blocks if needed), the linguistic limit will probably have reflected a conceptual boundary, between *concretely imaginable* and *large* numbers which were too abstract to enter conveniently in procedures like those taught in TMS XVI (where $n = 4$) and used in TMS IX #3 ($n = 17$).[333]

"Broad Lines" and "Thick Surfaces"

As we have seen (p. 17), the Old Babylonian volume metrology was based on the notion that surfaces are provided with a virtual height or thickness equal to 1 kùš; this has long been known. What has not been noticed is that this idea shows that the multiplication by "raising", derived originally from volume (and possibly as a special case brick) metrology, can thereby be seen to be an operation of proportionality – "to find the volume on the surface *S* if its real height is *h* instead of *1*." This conforms perfectly with the use of "raising" when technical constants are used: if 3 iku is the surface which a labourer can harrow in *1* day, then what he will harrow if his working time is *d* days will be 3 iku "raised" to *d*. In this form, the igi.gub factor is more easily grasped and explained than as an abstract ratio expressed in the unit $^{\text{iku}}/_{\text{man-day}}$.[334]

As we have also seen, surfaces are determined by "raising" when the computation is not tacitly implied by a rectangle construction. But if the surface of a rectangle ▭(*l*,*w*) is found by "raising" *l* to *w*, and if "raising" is to be understood in the same way, the idea behind the computation will be to find the surface if the width is *w* instead of *1* – that is, the length *l* will have to be a strip or broad line, to be provided with a virtual breadth *1*.

The notion of broad lines is widely diffused in pre-Modern practical geometries,[335] and together with the thick surfaces it explains the Old Babylonian problem statements where sides are "appended" to surfaces and

333 Cf. that the Indo-European languages have developed individual forms for the numbers until 20, which only etymological analysis will resolve into separate components; numbers beyond 20 can be analyzed directly, except in modern Indo-Aryan languages (Gypsy and Sinhalese apart), in which individual forms extend until 100; see [Gvozdanović (ed.) 1992, *passim*], and [Berger 1992: 243–245] for the Indo-Aryan case.

334 When a kid I was indeed taught – and others with me – that density was the weight of one cubic centimetre measured in gram *and therefore dimensionless!* Outrageous from a physicist's point of view, but it could be understood at an age where most had difficulty with considerations of proportionality.

335 Cf. note 76 and [Høyrup 1995a]. We may observe that cloth is still sold according to the system of "broad lines".

surfaces to volumes – those which have made modern workers conclude that Babylonian algebra could not possibly be based on geometry (cf. p. 7 and note 269). However, widespread as it was among practitioners, it was always suspect among theoreticians and in schools, and we should notice the presence of attempts to rationalize it in a number of our texts.

In AO 8862, we remember, sides or fractions of sides were "appended" without worries to the surface, which should only be possible if the sides were already of the same kind as the surface, that is, broad lines – in particular in view of the precision of the text on another ontological or "metacomputational" point: when "cutting off" half of the aggregated sides from a surface of which this entity is not part it takes care to "bring" it first.

In BM 13901 and a number of other texts we find instead that the operation by which sides and areas or surfaces and volumes are added is an accumulation, which allows to combine the *measuring numbers* and not the entities themselves. Moreover, BM 13901 introduces the "projection", precisely that virtual standard breadth with which the broad lines are provided automatically. In TMS IX #1, a different terminology is used – in order to make geometrical sense of the addition of 1 length, the width is provided with a "base"(?) 1 – and #2–3 from the same tablet refer explicitly to the length and width prolonged by 1 as length and width "of the surface 2". In YBC 4714 #30–39, the introduction of an "alternate width" shows us a third way to solve the dilemma.

Even though the broad lines and the thick surfaces are basic, not only for area and volume computation but also for the whole of Old Babylonian second- and third-degree algebra, the status at least of the lines thus turns out to have caused qualms. In some texts (and, we must presume, schools), lines are presumed to *be* broad; in others, they appear to be "lengths without breadth", in agreement with the Euclidean definition (and the modern understanding); they have to be *made* broad by being provided with a "projection" or by some similar device.

Did They "Know" It?

The similarity between Old Babylonian and Modern algebra – even greater in the received interpretation than in the geometric reading of the texts – has led to discussions whether the Babylonians had discovered this or that ingredient of modern algebra or mathematical thought.

As it is formulated, the discussion is often based on a "naive Platonism" which modern metamathematics has left behind since the acceptance of non-Euclidean geometry and the works of (say) Dedekind and Peano. Possibly, God (or Nature, or the constitution of our nervous system, or some interaction between these) gave us the integers, but all the rest is certainly a human construction, as claimed by Kronecker – whether a free construction or one still guided by the conditions of Nature as to what can be adequate still being open to discussion. In any case, no serious philosophy of mathematics will assert today that negative numbers *exist*. Instead it may assert (for instance) that the structure of the domain of positive numbers allows the creation of rules of arithmetic between certain equivalence classes of pairs of positive numbers which makes a subset of these classes isomorphic with the positive numbers – for which reason we may consider the complete set as an extension that beyond this subset comprises a neutral element for addition ("0") and the additive inverse of each positive a ("$-a$").

Asking whether the Babylonians had *discovered*, or had *not discovered*, negative numbers, is thus as meaningful as asking whether they had discovered the Eiffel tower or not; the meaningful questions concerning the Eiffel tower would be whether they made constructions that in one way or the other expressed similar aspirations as those of the illustrious engineer, and whether they had created the techniques of which he made use. Questions whether the Babylonians had discovered this or that mathematical object (etc.) should be reformulated correspondingly – but in such a reformulation they *will be* meaningful, and contribute to the portrait of Old Babylonian algebra.

Zero

Our notion of zero is a compound concept, consisting of two ingredients, one of which is itself composite. One is the idea that the result of a subtraction where nothing is left (2–2, etc.) can be dealt with *as a number* on a par with the outcome of a subtraction where something remains (3–2, etc.); the other is the use of *the cipher* in the place value system, on one hand ("intermediate zero") as a separator which allows us to distinguish 11 from 101, on the other as an indicator of absolute magnitude ("final zero"), which allows us to distinguish 110 from 11. Only the development of a view of a number written in the place value system as $a_n a_{n-1} ... a_2 a_1$ as a sum $\sum_{i=1}^{n} a_i \cdot 10^{i-1}$ (where a_i is one of *the numbers* {0,1,2,...,9}) merges the two or three ideas into a single concept.

The Old Babylonian calculators had no such unified zero. Even their ways to handle an outcome "nothing" shows that they were very far from the idea of treating it as a number. In TMS VII #1, as we have seen (p. 185), "tearing out" 20´ from 20´ left *nothing worth speaking about*, that is, the statement of a remainder was omitted; in another case of "remainder zero" of a subtraction by "tearing-out" (noticed by Kazuo Muroi [1991: 60*f*]), namely, the series text VAT 7537, obv. III 2, the outcome is asserted to be *ma-ti*, "missing" (stative of *maṭûm*) "to be(come) small(er)".

A way to formulate an "outcome zero" of a subtraction by comparison is found in TMS V, rev. II 21: the first entity is said to be *kīma*, "as much as", the other.

It was not to be expected *a priori* that two different concepts of subtraction should lead to a common concept for "remainder nothing" and "difference nothing"; the actual ways to deal with these noughts suggests that the "difference nothing" was at least treated technically, since the formulation coincides with the expression for equality used in the formulations of equations when both sides are concrete entities (see p. 39). The "remainder nothing" seem to be so much out of the way that a shared standard way to speak about it seems not to have existed (but it is not impossible that the creators of the series texts will have encountered it sufficiently often to have developed a terminology like that of VAT 7537, stuffed as these texts are with systematically varied differences).

As to a cipher or digit zero, it was mentioned in note 19 that two Susa texts (TMS XII and XIV) use "intermediate zeroes" to indicate a level of missing tens or missing ones; in the Seleucid epoch, such intermediate zeroes were used more systematically, but now mostly as indications of a missing sexagesimal level. No notation for a "final zero" or "sexagesimal point" that might serve to identify the absolute order of magnitude of a number was ever used. Already for this reason, interpretation of the intermediate zero as a number on a par with other digits is excluded. It was and remained a device to indicate a void place, whether seximal-decimal or sexagesimal.

Negative Numbers

In the mathematical commentaries of [MKT], lines like the following may be found [I, 455]:

$$2(x+y)^2-2,0(x-y) = 1,3,20$$
$$\tfrac{1}{5}(x+y)^2-1,0(x-y) = -1,40$$

These were meant to render the two different ways to formulate the "subtraction by comparison", indicating respectively excess (ugu ... dirig, "over ... goes beyond") and falling short (ba.lal, "it is smaller") as we have also encountered them in BM 13901 #10–11. Neugebauer would speak of the "negative right side" of the lower line, but this word was always aimed at the modern interpretation, not at the Babylonian text. Careful writing, however, is no guarantee of careful reading, and the Babylonian "knowledge of the negative numbers" has become one of the urban legends of the history of mathematics.

Even though the purported evidence is irrelevant, the legend might of course happen to be true. It is therefore of interest to investigate to which extent the Babylonians would make use of techniques or concepts that are similar to those connected to our negative numbers.

A complete survey of the occurrences of comparison expressed by means of *maṭûm* or its logogram lal [336] shows that all belong within one of two structures.

One structure is that of BM 13901 #10–11, of which #10 makes use of *maṭûm* and #11 of *eli ... watārum*. The aim is obvious, namely, to make sure that the difference expressed in relative terms become $^1\!/_7$ in both cases instead of $^1\!/_6$ and $^1\!/_8$, respectively; the choice of one or the other term for comparison is thus determined by the wish to formulate the difference as a remarkable fraction. Due to the existence of two different formulations, two different problems can be formulated around a single fraction.

The other structure is represented by the lines that were quoted in Neugebauer's symbolic translation. In these cases, an external constraint determines the order of the two entities that are to be compared; if the first is larger than the second, it is stated to go beyond it; if it is smaller, the lal terminology is used. Such external constraints are common in the series texts; often, two basic entities A and B (in the example, the square surface on the accumulated sides of a rectangle, and their difference) are multiplied by coefficients α and β, and in a sequence with fixed A and B and changing α and β the result of a comparison of αA and βB is given – in Neugebauer's translation, the difference $\alpha A - \beta B$, may of course be positive or negative, depending on whether αA exceeds or falls short of βB. Because of the compact style of the series texts, where only the variations of α and β are indicated, the order of αA and βB has to be the same in all enunciations.[337]

It should thus be quite clear that the choice of one or the other operation has no more to do with a notion of negative numbers than the distinctions before/after, above/below, larger/smaller, younger/older.

What we do find occasionally in the texts is a marking of certain numbers as *due to be subtracted*. We may compare two parallel phrases from the statement of BM 85200+VAT 6599 #29 and #30:

> The 7th part of that which the length over the width goes beyond, and 2 kùš, that is the depth

and

> The 7th part of that which the length over the width goes beyond, and 1 kùš diminishing (ba.l[al]), that is the depth.

The "and"/*ù* of #29 designate an addition, and the formulation shows that the

---

336 Normally extended grammatically into ba.lal, occasionally replaced by tur in YBC 4714, cf. p. 134.

337 A different type of constrained order is found in YBC 4714 (pp. 111*ff*), where the sides of squares inscribed successively one into the other are listed in all cases in decreasing order. In still other cases, grammatical constraints intervene – if one entity is complex and the other simple, the former has to be described first, as in YBC 4710, rev. II 5–15. See [Høyrup 1993b: 57*f*].

normal role of a number is this construction is supposed to be additive; in #30, the quantity 1 kùš is therefore marked expressly as being subtractive.

Another example is offered by TMS XVI #1 (above, p. 85); in translation, line 1 states

> The 4th of the width, from the length and the width to tear out, 45′.

Lines 3–4 run

> 50′ and 5′, to tear out, posit. 5′ to 4 raise, 1 width. 20′ to 4 raise, 1°20′ you see, 4 widths. 30′ to 4 raise, 2 you see, 4 lengths. 20′, 1 width, to tear out, from 1°20′, 4 widths, tear out, 1 you see [...].

In line 8, finally, we find a reference to the statement

> Since "The 4th of the width, to tear out", it is said to you, from 4, 1 tear out, 3 you see.

In #2 from the same tablet, line 23, one finds

> [...] 45′ you see, as much as (there is of) widths posit, posit to tear out
>
> [...] 45 *ta-⟨mar⟩ ki-ma* sağ ğar ğar zi-*ma*

In #1, the subtractive role of the fourth of the width and its value 5′ is thus pointed out repeatedly; in the formulations of this problem, nothing indicates that this is more than a reference to the role of the quantity/number. In #2, however, the phrase "posit to tear out" is a very strong indication that this role is somehow indicated by the way the quantity is "posited", that is, taken note of materially; *how* this was done remains unexplained, and no other text is of any help – but the place of the number within a calculational scheme is definitely a possibility that is closer at hand than a equivalent of our minus sign.

The category of subtractive numbers may seem to constitute a doubling of the domain of numbers, which is the way negative numbers were conceived for a long time; but this doubling is not really one before the new "numbers" are operated upon as we operate on the old numbers. Not only is any multiplication of a subtractive number by a subtractive number absent from the Babylonian texts, even explicit rules analogous to the identity $a-(b-c) = a-b+c$ are absent.[338] Though the category of subtractive numbers can be formulated within our framework of positive and negative numbers, it is a simpler structure – and no inherent dynamics will "automatically" enforce its extension.

---

338 Old Babylonian accountants will certainly have known (from copiously repeated experience if in no other way) that subtracting c less leaves a remainder that is c greater; however, as long as this is not formulated around the category of subtractive numbers, it is a different matter.

Irrational Numbers?

For a competent calculator it will have been obvious that the calculation of a square root via the procedure shown in Figure 75 will always leave a remainder, and that iteration of the procedure will not bring it to an end. It is less obvious that even combination with compensatory procedures, as shown in Figure 76, though they may improve the precision of the approximation, will never yield the true root. It is not unlikely, however, that calculators who were familiar with the trick of the quadratic completion will have been aware that neither of the two procedures can be definitive.

Nor is it unlikely that those who found the approximations 1°25′ and 1°24′51″10‴ to $\sqrt{2}$ will have known that these *were* approximations – any method which could produce them would have revealed that they were steps in a procedure that might be continued.

Does that mean that the Old Babylonian calculators will have "known" that, for instance, $\sqrt{2}$ is irrational, as sometimes maintained?

Firstly, of course, it may be objected that arguments like these are not pertinent: even if we try to find $\sqrt{9}$ by the methods of Figure 75 and Figure 76 we shall never get to precisely 3 unless this is also our starting point. But this is a logical point that might perhaps have escaped the Babylonians.

Next one has to observe that the Babylonians knew about other arithmetical processes that did not stop – namely, the calculation of the igi of irregular numbers.

There is certainly a difference: the calculation of the igi of an irregular will be cyclical, and the corresponding sexagesimal number periodical; square-roots of non-squares are not. We have no evidence at all, however, that the Old Babylonians were alert to this fact or interested in it;[339] nor is there the least reason that they should have been forced to discover the basis for periodicity (namely, that the number of possible remainders is finite).

We may add that the primary reason that the difference between periodical and non-periodical sexagesimal developments is interesting is that it turns out to be linked to a distinction between rational and irrational numbers which is already established on more obvious grounds. The Greek discovery that certain ratios could be given no (arithmetical) name is thereby dependent on the idea of searching for a common measure for two quantities of the same kind and on the anthyphairesis-algorithm for constructing it. Both were wholly outside

---

339 Among the reciprocals of irregular numbers that are listed in the tablet YBC 10529, only the reciprocal of 1`1 (59′59‴59⁽⁵⁾...) is developed to the point where the periodicity shows up; this is certainly not enough to suggest empirically that they are all periodical. All the reciprocals of irregulars listed in M 10, John F. Lewis Collection, Free Libr. Philadelphia, are interrupted before exhibiting periodicity.

the scope of Old Babylonian mathematical techniques and interests. In the absence of such notions and techniques, a distinction between irrational and other non-finishing numbers (the reciprocals of non-regular numbers) would have been as absurd as a distinction between algebraic and transcendental irrationals within the framework of Greek (or any pre-nineteenth-century) mathematics.

Logograms as "Mathematical Symbols"?

Not rarely, the secondary literature echoes Neugebauer's early suggestion that the logograms of the mathematical texts, in particular those of the series texts, might function "precisely as mathematical symbols" ([1932: 222]; [MKT I, viii]). It may therefore be adequate to repeat and extend the arguments against this view.

In [1842: 302], Nesselmann introduced a distinction between rhetorical, syncopated, and symbolic algebra. Rhetorical algebra, in this classification, is algebra expressed in full words, as the algebra of al-Khwārizmī and its Latin translations; syncopated algebra is written by means of abbreviations, but these can be expanded without difficulty into the full words and sentences of a rhetorical algebra; examples are Diophantos's *Arithmetic*, but also the Italian late medieval and Renaissance algebras with their use of **m** for *meno* and **℞** for *radice*. Symbolic algebra is the kind which is characterized by Mahoney's criterion (i), by the use of a symbolism which allows us to extract "the *structure* of a problem from its non-essential content", and *on which we may operate directly*. Its symbolism may be of the Cartesian kind, but even the medieval Indian use of *schemes* belongs to the category.

As pointed out in the discussion of YBC 4714 (see notes 141 and 152; other texts would allow us to make the same observation), the logograms of the series texts are ambiguous – even more ambiguous than texts written in full syllabic language, although even such texts may use exactly the same phrase in two radically different senses.[340] However, ambiguity which can only be resolved with reference to the text as a whole (often including the numerical parameters) certainly excludes operation on the level of the supposed symbols. We may therefore exclude that the logograms might serve in this function.

340 Thus, in BM 13901, exactly the same sequence of signs appears in obv. II 28 (#12) and rev. I 50 (#19), *mi-it-ḫa-ra-ti-ia uš-ta-ki-il₅-ma*, "My confrontations I have made hold:". In both cases, two "confrontations" s_1 and s_2 are involved – but in the first instance, the construction of ⊏⊐(s_1,s_2) is meant, whereas the second speaks about the separate constructions of $\square(s_1)$ and $\square(s_2)$. Which of the possible interpretations should be chosen in each case is only made clear by what follows: in #12 immediately the resulting 10′, in #19 the phrase a.šà[lam] *a*[*k-mu*]*r*, "T[he] surface I have [accum]ulated".

Even if we read the claim less strictly and conflate the syncopated and symbolic categories (Neugebauer [1932a: 5] probably did not intend it that way, but some of those who echo him might overlook the difference[341]), the ambiguities ruin the claim, since they do not allow translation into full phrases unless the whole context is taken into account. In their writing, the series texts are neither rhetorical nor syncopated, nor symbolic; in a term for which Nesselmann had no use, they are *stenographic* – the written text is as support for memory, and needs to be reconstructed before it can get its full meaning.

As argued on p. 281, the place where Old Babylonian algebra operates directly is on the level of the geometric standard representation and not on the level of reality that is represented: during the operations it can be safely forgotten whether commercial rates, distances and areas measured in terms of broken reeds of unknown length, or *igûm-igibûm* pairs are concerned – just as safely as if the representatives had been letters standing for numbers.

We may add that several representations on which we operate directly cannot coexist, unless they are either alternatives of which only one is actualized, or hierarchically ordered and thus allow further reduction of one representation to the other. The existence of the geometrical standard representation makes a symbolic function for the logograms superfluous.

Overall Organization

Portraits consist of details – but details without the adequate order would be like John Donne's anagram of beauty who "hath all things whereby others beauteous be; for though her eyes be small, her mouth is great; though they be ivory, yet her teeth are jet". We shall therefore end the characterization of Old Babylonian algebra by approaching explicitly some features of its over-all organization of algebraic thought and practice which we have so far left fully or partially aside.

Technical Terminology

At various points (thus pp. 5, 291), aspects of the Old Babylonian mathematical terminology were discussed with reference to the characteristics of technical terminologies; at others it was concluded that particular ways of speaking were probably not technical at all (e.g., pp. 163, 257). This calls for a

341 André Weil, whose competence as a mathematician nobody will doubt, takes care in his defence of the algebraic interpretation of Greek geometry [1978] not to distinguish the two types.

closer investigation of the character of the terminology: *is* it justly characterized as technical, and if not, how should we describe it?

Most likely, technical terminologies in an absolute sense of a pure formal language do not exist; a formal language, if it is to serve as terminology for something outside itself, will be linked to this external field by rules which cannot be formal nor fully fixed. We may say, however, that a terminology becomes technical to the extent that its terms are used without regard for connotations. If we speak about *AD* as the perpendicular on *BC* in both sides of Figure 80, then the connotation of the pending plumb line seems to have been forgotten, and a technicalization has started. If we speak of "dropping" the perpendicular from *A* on *BC* in both cases, technicalization has gone even further. If, on the other hand, it depends on context whether ΔABC is "equal to" ⊏⊐*DE* in Figure 81 (namely, on whether the topic is areas or congruence), then "equality" is only a substitute for an open-ended idea of "sameness", and the technicality intimated by the use of a term with a Latin root is counterfeit. In this case we may speak about a standardized use of everyday language which is less technical than the practice it is used to describe – the practice of Euclidean proofs is fully able to keep the two concepts apart which language conflates.

The terminology of the Old Babylonian mathematical texts is certainly standardized, very standardized indeed; even for a long and fairly varied text like BM 85200+VAT 6599, the total vocabulary if every grammatical form is counted separately consists of 50 words (if only roots are counted, it dwindles to 44). But close scrutiny reveals that technicalization has not gone very far. In order to see that we may look at a few examples.

The first is the use of the term *qaqqarum*, "ground", in the texts BM 85200+VAT 6599, VAT 6598 #6–7, and BM 85196 #9 – all three quoted above. The three texts belong to the same group and are so close to each other that they can safely be taken to represent the same school or professional group. In BM 85200+VAT 6599, the "ground" is used about the base of the excavation (which *is* indeed a "ground" or floor). In VAT 6598, it is used about the square on the base of the door, which has no real existence as a ground; whether the scribe would also speak of the square on the height as a ground is not clear from the conserved part of the text – the word *may* be chosen because the width of the door is measured along the ground, but also in order to avoid the usual term a.šà for a surface because its basic meaning "field" seemed inadequate; in both cases, everyday connotations either of the word itself or of the possible alternative have interfered with the terminologi-

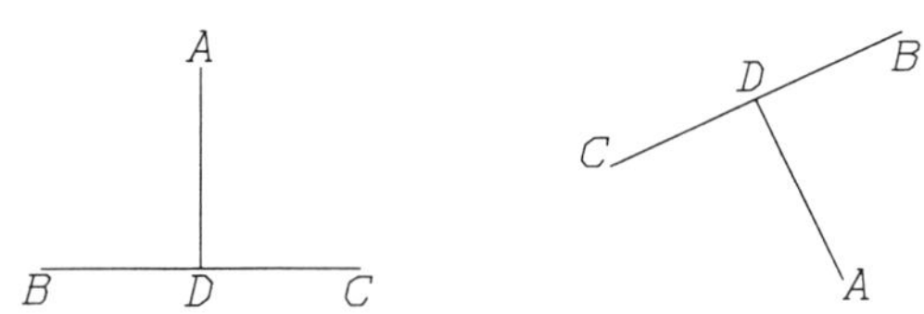

Figure 80. Perpendiculars.

cal choice. In BM 85196 it is used to distinguish that end of the pole that touches the ground. Obviously, the term moves within the whole range of everyday meanings, and only the context constituted by the description as a whole clarifies what is meant technically.

Another striking use of one of the habitual core terms in BM 852300+VAT 6599 was pointed out in note 159, namely, the use of ba.zi, "to tear out", when earth is dug out. As a matter of fact, the spectrum of terms used about the identity-conserving subtraction is generally illuminative. The standard term is certainly *nasāḫum*/zi, "to tear out". We have seen that certain texts replace it by *ḫarāṣum*, "to cut off", which might of course be (and is likely to be) a local variation; more striking is that some texts use *ḫarāṣum* together with other terms – AO 8862 and YBC 4663 together with *nasāḫum*, YBC 4662 with *tabālum*, "to take away" (more specifically, to seize or reclaim, for instance, by legal action). In AO 8862, as we have seen, there is a preference for "cutting-off" when linear entities are dealt with and for "tearing-out" from surfaces; in the other two texts (which belong to a different text group), this tendency is the rule.

The case of *tabālum* illustrates another phenomenon. Toward the end of the catalogue text TMS V, in rev. II 23*ff*, long after the other homogeneous equations (where identity-conserving subtraction is invariably zi), just before systems of two and three concentric squares, comes a sequence of problems which – given the location in the tablet and the uncommon choice of grammatical person – can hardly mean anything but "from a quadratic field, somebody has reclaimed $\frac{1}{3}$ of the area, and what is left amounts to 10` sar". Here, *tabālum* is used as an extra-mathematical term, in order to provide the problem with an apparent real-life dress; but since it does designate an identity-conserving subtractive process albeit a particular one, in other texts it moves into the prescription, where it serves as a purely mathematical term. Other terms manifest comparable shifts between extra- and intramathematical uses, called forth by shared connotations – in BM 85196, rev. II 19, 23*f*, *ḫarāṣum* is thus used about the fully concrete and extra-mathematical "cutting off" a piece from a silver coil with monetary function (the invariable choice of the tablet for the mathematical operation is ba.zi); in VAT 6546, which deals with "cutting off" something from a profit, the similar context suggest the use of the same term – but this happens inside the description of the procedure, where it describes the arithmetical operation.

Analogous observations could be made on other classes of terms – in BM 85200+VAT 6599 #16 and #18 we noticed that what would elsewhere be

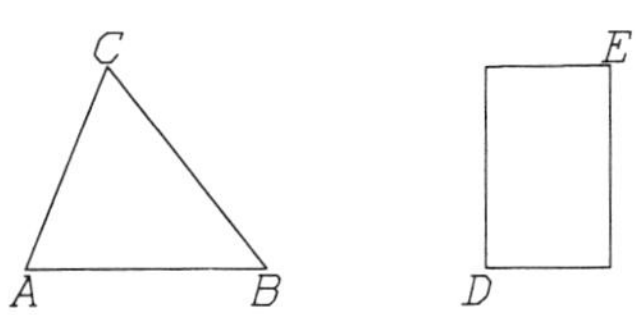

Figure 81. Equal/unequal figures.

an accumulation was termed as an accounting total, probably because of the presence of the *igûm-igibûm*-pair (cf. also p. 20).

This colouring of the vocabulary by concrete connotations never affects the operations which are performed. The standardization of procedures, that is, of mathematical/computational practice, is certainly more outspoken than the standardization of the terminology. In conclusion we may therefore say that the terminology is not truly technical, not even in the relative sense to which this notion has to be reduced according to the above reflections; it is rather a very standardized use of everyday language to describe an extra-linguistic – computational and naive-geometrical – practice which was always *more* standardized than the linguistic description. The linguistic description was thereby analogous to our heuristic explanations in standardized ordinary language of what goes on in those symbolic formulae which with us constitute the level of real technical operation.

Mathematics?

At an early point (p. 8) it was stated that the authors of the mathematical texts were

> teachers of computation, at times teachers of pure, unapplicable computation, and plausibly specialists in this branch of scribal education; but they remained teachers, teachers of scribe school students who were later to end up applying mathematics to engineering, managerial, accounting, or notarial tasks

and that it might be misleading to speak of them as "mathematicians". One may legitimately ask whether it will not by necessity be just as misleading to speak of "mathematics" as of "mathematicians" – and, once this dismal question is asked, to continue and ask whether the very theme of this book – Old Babylonian "algebra" – is not another instance of the proverbial red herring which is drawn across the track of the fox in order to mislead the hounds and the hunters.

It will be convenient to start with the latter question in generalized form: Whether a structuring of the field of knowledge which we speak about as Old Babylonian mathematics in terms of "algebra", "geometry" – and perhaps "theory of numbers", etc. – corresponds to the way the field was ordered by its original workers?

The best evidence we have for their view in this respect is offered by the theme texts. Three of these were discussed above: the "square text" BM 13901, the "excavation text" BM 85200+VAT 6599, and the series text YBC 4714; from two others (the series texts YBC 4668 and YBC 4713) shorter sequences were presented.

BM 13901 is characterized as a "square text" because all its problems deal with one or more squares. Several different methods are used – in particular we notice that #12 reduces a two-square problem to a rectangle problem. But it may be significant that all problems are of the second degree (#12 being a biquadratic); the obvious counterparts of #8–9 in which $\square(s_1)-\square(s_2)$ and $s_1 \pm s_2$ (which appear elsewhere in the corpus, but which reduce to the first degree, cf. p. 267) are omitted, as are simple homogeneous problems (complex homogeneous problems are present). Combining the two observations, we may determine the theme as *non-trivial second-degree algebraic problems about squares with scaling and accounting* (our words, to be sure).

In the excavation text BM 85200+VAT 6599, the excavation configuration provides the manifest definition of the theme; as we saw, problems of the first, second, and third degree occurred together in an order which was determined primarily by the geometric configuration; but we may notice that all solutions fall within a domain that we would call algebraic, even though the solution of the third-degree problems by means of factorization is radically different from the way the first- and second-degree problems are approached. The full range of the theme thus seems to be *algebraic excavation problems* (still our words). The direct computation of volume or diagonal from given dimensions (which would belong in the context if the definition of the theme was provided by the configuration *alone*) is absent.

The series texts are so long that it seems adequate to regard them as built up around an ordered sequence of themes – in particular if we remember that the individual tablet which we happen to possess is only a single member of a complete series. The theme of the first sequence of YBC 4714 (#1–29) seems to have been defined as that of BM 13901; the sequence #30–39, still dealing with two squares, may be a variation, or the beginning of a further extension of the scope of square problems – but still within the range of second-degree algebra with accounting.

The selections from the tablets YBC 4668 and YBC 4713 (the latter of which, as we remember, is virtually identical with the beginning of the former) reflect the contents of these texts as a whole: complex algebraic problems, often derived algebraic problems, about the rectangular configuration ⊏⊐(30,20).

The catalogue text TMS V was referred to repeatedly above, but only for specific points. The text as a whole deals with squares – but it begins with problems where this is only true in the formal sense, since the first sequence has the format "α the confrontation, β of my lengths what", where the variations of β is the pretext for the experiments with composite fractions that were referred to in note 20. Numerically, the "confrontation" (written with the sign LAGAB, which *is* a square frame) and the "length" (uš) are of course identical. Next follow (after a lost section that cannot be reconstructed) problems of the following types (c being the "confrontation", l the length and S the surface; as usually, $p = q+\delta$ corresponds to "p over q, δ goes beyond", whereas $p-q$ represents a "tearing-out"):

$$c+\beta l = \gamma$$
$$c = \beta l+\gamma$$
$$c = \alpha,\ \beta S = ?$$
$$\beta S = \gamma,\ c = ?$$

$$\square(\alpha l)+S = \beta$$
$$S = \square(\alpha l)+\beta$$
$$S+\alpha l = \beta \text{ ("appending")}$$
$$S-\alpha l = \beta$$
$$\alpha c = S+\beta \text{ (ending with } \tfrac{1}{2}c = S)$$

Then come problems dealing with the reclamation of part of a square field (see p. 301), and finally problems about two and three squares which are indicated explicitly to form a concentric system; among these we find both the questions of BM 13901 #8–9 $\square(c_1)+\square(c_2) = \alpha$, $l_1 \pm l_2 = \beta$ and the counterparts $\square(c_1)-\square(c_2) = \alpha$, $l_1 \pm l_2 = \beta$ – not treated there, as we remember.

The theme "square" is thus dealt with in greater breadth than in BM 13901; but the opening is only toward simpler first-degree problems, toward the type $\alpha c = S+\beta$[342], and toward the simpler types of two- and three-square problems; in contrast, problems about geometrical subdivisions of the square as found in BM 15285 (of which #11 is shown in Figure 69, and #24 in Figure 7) are absent.

In BM 15285 (also a catalogue text dealing with squares), on the other hand, all the types of BM 13901, YBC 4714, and TMS V are absent; but many of the subdivisions are more intricate than those shown in Figure 69 and Figure 7, involving, for instance, areas with one or more curved edges. Even though we do not known how the authors would have formulated it, we are forced to recognize that they distinguished a theme "algebraic and quasi-algebraic problems about squares" from the theme "geometric problems about squares".

We may also contrast TMS V to the first part of BM 80209, another catalogue text that deals with squares [Friberg 1981a]. The first two questions run "α, each, stands against itself, the surface what?". Then the diagonal (termed dal, "crossbeam") is asked for; a last question, similar to UET V, 864) concerns the "thrusting" of the square by which two concentric squares are produced (cf. also note 250).

The second part of the same catalogue deals with circles (cf. note 325 and preceding text), with surface S, circumference c, and diameter d. The questions are

$$S = \alpha,\ c = ?$$
$$S+\alpha c = \beta$$
$$S-\alpha c = \beta$$

342 The analogue of our $\alpha x = x^2+\beta$ (α and β positive), the case with a double positive solution. This is the only place in the record where this type appears explicitly as a problem on square side and area; the solution will evidently have followed the pattern of the rectangle problem $l+w = \alpha$, $\sqsubset\!\sqsupset(l,w) = \beta$.

$$S_1+S_2 = \alpha\ ,\ \ c_1 = c_2+10$$
$$S+d+c = \alpha$$

Both sections thus start with the basic problems of and formulae used by surveyors; how exactly the problem of concentric squares should be classified is not clear, but the circle sequence then goes on with algebraic problems (the appearance of the concentric squares in TMS V suggest that the thrusting-problem was counted similarly).

As it will be argued below (p. 371), this text is one of the witnesses of the soil from which the algebra of the Old Babylonian school grew.[343] With this in mind we may characterize BM 13901, YBC 4714, and TMS V as expressions of the systematic elaboration of the short section of "supra-utilitarian" riddles that follow after the useful formulae in BM 80209 (we shall return to this concept below, p. 366; briefly, it refers to mathematical problems and activities that continue utilitarian practice and refer to it but are in fact "pure" – "recreational" problems exemplify the category well). TMS V shows that this elaboration implied the inclusion of first-degree techniques and questions that had not belonged within the original context, and that the resulting combination was still seen as a unit, distinct from the set of questions with which BM 15285 is concerned.

A group of catalogue texts on "excavations" published in [MCT] (YBC 5037, YBC 4657, YBC 4663, YBC 4662) presents us with the same utilitarian beginning – determination of the wages to be paid for the work from the given dimensions of the excavation – and goes on with supra-utilitarian algebraic problems of the same kind as those contained in BM 85200+VAT 6599 (but only of the firsts and second degree, and without *igûm-igibûm*-problems). Another group deals with the excavation or reparation of "small canals" (YBC 4666, YBC 7164); in these, everything is non-algebraic and everything apparently utilitarian; these characteristics are shared by YBC 4607, a catalogue of brick problems.

A final illustrative catalogue text is YBC 4652, containing problems of the following kind:[344]

> I found a stone, (but) did not weigh it; (after) I weighed (out) 8 times (its weight), I appended 3 shekel, one-third of one-thirteenth in 21 steps I repeated, appended: I weighed (it): 1 mina. The head (initial weight) of the stone what? The head of the stone $4\frac{1}{2}$ shekel.

---

343 A similar catalogue dealing with rectangles is YBC 4612. At first come two sequences of five problems, each about a particular rectangle. Each of these sequences starts by the useful computation of the surface from the length and the width, the rest are simple supra-utilitarian problems in which the surface is given together with l, w, $l+w$, or $l-w$. In the end comes further training of the determination of the surface from the length and the width.

344 I quote the well-conserved #20, using the translation given in [MCT, 101], but inserting my habitual standard translations. 1 ma.na/mina = 1` gín/shekel ≈ 500 g.

These problems are clearly supra-utilitarian riddles, formed around the simpler (practical) problem of finding an original value when the result of a given relative or absolute augmentation or diminution is also given. They cannot be solved by means of a single false position, and since the double false position seems not to have been used, they were presumably to be found by means of backward calculation step by step. Here, the cancellation of the step "one-third of one-thirteenth in 21 steps I repeated, appended" may have been meant to be made by a single false position; globally, however, the solutions will not have been algebraic in character.

All in all we notice that the themes of theme texts are always defined *with regard to* a particular mathematical object, configuration, or situation. But they are not defined *by* this object alone. Often, the distinction between the utilitarian and the supra-utilitarian levels is another aspect of the definition. In others, an initial group of utilitarian problems is followed by another group whose members are supra-utilitarian, indicating that the relation between the two was seen precisely as that between a utilitarian basis and a – particular – supra-utilitarian discipline erected on this basis. The utilitarian problems are indeed not followed by a concoction of all kinds of supra-utilitarian problems, as we have seen, but by problems of a particular kind.

This kind may be delimited more or less narrowly – the contrast between BM 13901 and TMS V illustrates this well. But it remains striking that none of the texts combine problems of the kind that were treated in Chapter VI with the algebraic problems of chapters III and V; in spite of the differences between the methods used to treat the first, the second, and the third degree, these kinds of "algebra" appear to have belonged together even in the view of the Old Babylonian teachers. *The notion of an "Old Babylonian algebra", though our construction, is true to Old Babylonian thought.*

Since this is not our topic, we shall not pursue the question whether our identification of a category of "geometrical" problems also corresponds to an Old Babylonian categorization; in the absence of adequate theme texts (of which we only have BM 15285), the answer to the question may be beyond our reach.

There remains the question whether the category of "Old Babylonian *mathematics*" is true to the thinking of its practitioners, or instead *our* imposition of an artificial division on a scribal practice which functioned without such divisions or which was compartmentalized in other ways.

Elsewhere in the historical record, sources used by historians of mathematics provide us with evidence that the category of mathematics was not invariably seen as a necessary category, even if we only look at literate culture and leave aside the ethnomathematical study of non-literate cultures. One instance is Book XIV of the *Greek Anthology* [ed., trans. Paton 1979], which blends arithmetical epigrams, riddles, and oracle prophecies; obviously, the defining category is the *riddle*, not mathematics. Another is the Middle Kingdom Egyptian Papyrus Anastasi I [ed., trans. Gardiner 1911], a "satirical letter" supposedly written by one scribe in order to expose the ignorance of another; and widely used in scribal education. Here, the poor colleague is claimed to be equally ignorant of Palestinian geography, of the calculation of rations, of the determination of supplies for a military expedition, and of the manpower (etc.) that is needed for the construction of a siege ramp. In the implanting of scribal self-awareness, no specific status was given to the category of mathematics.

The Greek mathematicians did possess a concept of mathematics, notwithstanding the views of the compiler of book XIV of the *Greek Anthology*; similarly, the Rhind Mathematical Papyrus and the Moscow Mathematical Papyrus are justly qualified as "mathematical" – none of them mix up the names of Palestinian cities (or spells, medical recipes, or whatever else one might expect from a text like the Rhind Papyrus whose colophon tells the reader that it deals with "all secrets") with their mathematics.

The theme texts certainly do not transgress the boundary between mathematics and non-mathematics; but more decisive evidence that the category of mathematics as a whole is also true to the thinking of the Old Babylonian school teachers is supplied by the anthology texts, which transgress the borders between single mathematical "disciplines" as defined by the theme texts. Texts like BM 85194 and BM 85196 that combine problems on siege ramps with the determination of circular chords and of the distance the foot of a pole moves outwards when its top is lowered by a given amount, or AO 6770 which combines the rectangle problem dealt with above (p. 179) with composite interest, with an arithmetical stone riddle, and with a "broken reed" problem – such texts might certainly be expected to trespass into non-mathematical domains unless this *was* seen as some kind of trespassing. They never do;[345] the absolute maximum of transgression is the rare colophon of AO 8862 invoking the scribal goddess Nisaba (see note 194).

We have to conclude that it *is* meaningful to delimit a particular Old Babylonian domain of knowledge which we may conveniently characterize as "mathematics", and which encompasses both a level of quite utilitarian calculation with appurtenant metrological and igi.gub tables (excerpts of which are sometimes found in texts which otherwise contain problems), and a supra-utilitarian level at which the algebra – but not the algebra alone – is to be found.

We should only remember that this mathematics contains no theorems; what we find written down are tables, didactic explanations and, first of all, problems and procedures for solving them; the "theoretical" insights on which it builds remain implicit and probably often intuitive (whence, strictly speaking, *pre*-theoretical). Babylonian mathematics was always concerned with computation, with *finding the number;* even in texts like TMS I or BM 15285, this remains the central concern. For that very reason, its geometry had to be a

---

345 School tablets where the student has used one side to copy a proverb and the other (during the next lesson when the clay was still humid) as a scratch pad for numerical computations do not count.

geometry of measurable quantities; and probably for that reason, the immense majority of supra-utilitarian problems (and the majority of problems in general, since the supra-utilitarian level was dominant), were algebraic in character, equation algebra being (even when conceptualized via measurable segments and areas) *par excellence* the technique of *finding the number*.

Chapter VIII
The Historical Framework

So far, we have portrayed the mathematical practice of the Old Babylonian scribes and their teachers; inasmuch as this practice is located differently from ours in historical space, the investigation is certainly a contribution to the *history of mathematics* – but since it has not approached the development of the practice in question (not to speak of the motive forces of this development) it has not yet approached the *history of Babylonian mathematics*. The model was that of structural-functional anthropology rather than history.

This history of Babylonian mathematics will be the topic of subsequent chapters; first, however, a brief presentation of the general historical framework may be appropriate. Since the development and orientation of Mesopotamian mathematics was intimately bound up with written administration and the scribal craft, this framework has to include both the development of state and administration and of the scribal profession.[346]

Landscape and Periodization

Mesopotamia – "the land between the rivers" Euphrates and Tigris – can roughly be divided into a northern region around Assur and Ninive; a central region, from Eshnunna and Sippar toward the north to Kish toward the south, the centre of which from the second millennium onward was Babylon (somewhat south of present-day Baghdad); and an extreme South, with the

346 [Høyrup 1994: 45–87] is a more detailed analysis of the interplay between statal bureaucracy, scribal craft and culture, and the transformations of mathematics from the beginnings through the mid-second millennium, with extensive bibliography.

cities Ur, Uruk, Shuruppak, Larsa, and others, characterized by irrigation agriculture. Even in the centre, irrigation was practised from an early date, whereas the North depended on rainfall. The political division between the North – *Assyria* – and the Centre-to-South – *Babylonia* – thus coincides with an ecological split. To the east of Southern Mesopotamia we find the city Susa as the centre of Elam, a region of river valleys between the Zagros mountains. Since the mid-fourth millennium, strong interactions between Elam and Babylonia were the rule. See the map in Figure 82.

Temporally, one may distinguish the *protoliterate* period (c. 3400 BCE to 3000 BCE according to calibrated radiocarbon datings, subdivided into the early *Uruk IV* and the late *Uruk III* phase), in which writing was created in the South, probably in the city Uruk; the *Early Dynastic* period (c. 3000 BCE to c. 2350 BCE), in which the South was divided between competing city-states; and the *Old Akkadian* or *Sargonic* period (c. 2350 BCE to c. 2200 BCE), in which the central and southern region were united in a regional state ruled by an Akkadian dynasty founded by Sargon of Akkad. After an interlude follows the *neo-Sumerian* or *Ur III* period (roughly coinciding with the 21st c. BCE), where the South was the core of an extremely bureaucratic state encompassing also the Centre.

After the collapse of the Ur III state follows the *Old Babylonian* period (c. 2000 BCE to c. 1600 BCE), whose most famous figure is doubtlessly king Hammurapi (1792–1750); in this phase, the Sumerian language disappears as a spoken language, and Akkadian is split into the clearly distinct northern Assyrian and southern Babylonian dialects.

A new societal breakdown followed. At first the masters of the region were the Kassites, a group of warrior tribes. Toward the end of the second millennium BCE, the Assyrian city state expanded and conquered first the

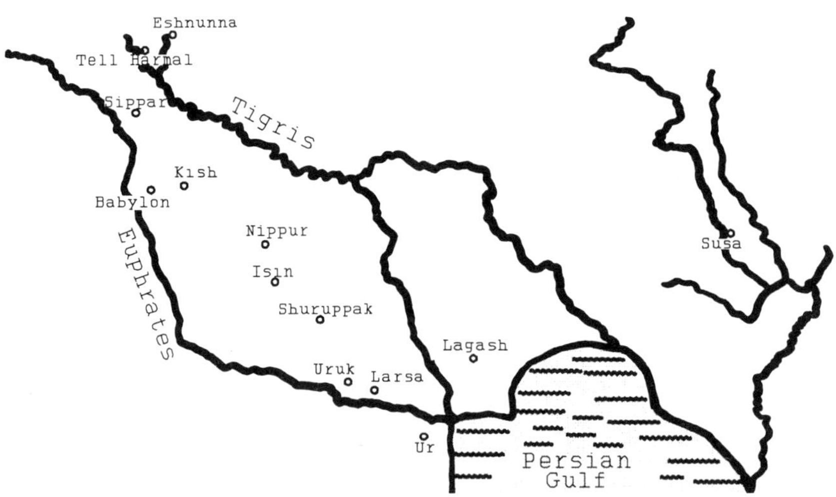

Figure 82. Map of central and southern Mesopotamia, with ancient coast line and main rivers.

whole of Mesopotamia and next the whole Near East. In the final centuries of the Assyrian empire (8th to 7th century BCE), mathematical astronomy may have arisen, but it is only documented directly during the epoch of Persian rule (539 BCE to 331 BCE), which was brought to an end by Alexander's conquest.

After Alexander's death and a brief period of fighting, Mesopotamia fell to one of his generals, Seleucos. The last glow of the ancient Mesopotamian cultural tradition (and its mathematics) falls during the *Seleucid* period and the early part of the Parthian epoch (312 BCE to first or second c. CE).

Scribes, Administration – and Mathematics

The protoliterate writing system – still only partially deciphered – was created during the Uruk IV phase as a tool for the earliest formation of a bureaucratic state[347] headed by a temple institution. The basis for the invention was an accounting system based on small tokens of burnt clay (cones, spheres, disks, pellets, etc.) which will have stood for various measures of grain, for livestock, etc. The system had been used in the Near East since the eighth millennium. Various transformations and extensions of the system had been introduced in mid-fourth millennium Susa in response to needs created by increasing social complexity. These improvements and extensions were taken over by the even more complex Uruk administration, which soon recast the whole complex into a writing system; apparently this was done in one jump, as the result of genuine invention – no traces of development or gradual extension are known.[348]

The best evidence for the link between the token system and writing is constituted by the protoliterate metrological notations, which render the traditional tokens – a sphere as an impressed circle, the cone by an oblique

347 The "state" is understood here as a social system characterized by an at least three-tiered system of control and by extensive specialization of social roles – cf. [Wright and Johnson 1975] and [Høyrup 1994: 48–51].

348 On the token system and its development, see [Schmandt-Besserat 1992]; the book may perhaps not be precise in all details – no pioneering work of this scope ever is; but Schmandt-Besserat has the indubitable merit to have discovered the long-term continuity of this system and to have investigated it in depth and breadth. Cf. the essay reviews by Peter Damerow [1993] and Jöran Friberg [1994] (to the critical points of the latter Schmandt-Besserat has some responses – personal communication).

Obviously, the use of signs for words or concepts and not only for measured quantities constitutes a fundamental change; whether we say that writing was created though a recasting of token accounting system or that this system was part of the inspiration behind the invention is a rather futile discussion if only we remember that the change was sudden and revolutionary; but futile or not is is certainly heated – see, e.g., [Petersson 1991] and [Glassner 2000].

impression of the stylus, the disk as a drawn circle.[349] The integral part of the capacity system for grain is likely to continue the pre-literate system (grain accounting will have been old); a system of sub-units may have been a fresh development[350]. The integral units, with relative values, are as follows:

A small cone: 1
A small sphere: 6
A large sphere: 60
A large cone: 180
A large cone, with impressed small circle: Value?

This system fuses quantity and quality; the picture of the small cone, whose original meaning may perhaps have been a standard basket of grain, is still a standard volume *of grain*; if the grain is malted, the sign itself is modified by addition of small strokes. In the pre-literate period, all metrological symbols are likely to have integrated quantity and quality in this way – a disk marked by a cross may have signified a sheep, etc. With the advent of writing, a separation of quantity and quality became possible, and "2 sheep" would be written as a circle with a cross (a picture of the sheep-token, quality alone) and a numeral meaning 2 (quantity alone). This notation for "almost abstract" number is likely to be a new creation of the protoliterate period, made by adaptation of the grain system to existing spoken numbers (and perhaps extended upwards beyond the range of existing spoken numerals). The signs are indeed the same, but their mutual order and relative values (which in this case are also the absolute values) are different:

A small cone: 1
A small sphere: 10
A large cone: 60
A large cone with impressed circle: 600
A large sphere: 3600
A large sphere with impressed circle: 36,000

Systematic mathematical thinking is visible in the construction of the system: 60 is an enhanced 1; 600 is created by impressing the sign for 10 on the sign for 60; 36,000 is composed in analogous way from 3600 and 10.

---

349 Immediately preceding the invention of writing there is a short phase where tablets were produced on which only metrological notations but no word signs are found. Since the tablets are rarely found in original position, the chronological precedence of these "numerical tablets" is inferred, not independently established; but since the notations are clearly different from those of the protoliterate phase (and reflect experimentation and lack of established conventions rather than different conventions), the inference is convincing; see [Nissen, Damerow, and Englund 1993: 125–130].

350 Evidently, it makes no absolute sense to speak of integral units and subunits in a metrological system; the relative sense is constituted by the way the writing system deals with them.

The characterization of the system as only "almost abstract" refers to its use in the length system: the later usage according to which a number alone would imply the standard unit (thus nindan in length metrology) goes back to this phase. We may therefore presume that even the length system is a fresh invention; almost certainly, the system of area units based on the length system (but with names that point back to pre-existing "natural" irrigation, ploughing, and seeding measures) is an innovation.

Almost certainly innovative are sub-unit extensions of all metrologies and an administrative calendar decoupled both from the natural month and the liturgical calendar.[351]

In this phase, the stratum of ruling managers of the bureaucratic system seems to have been responsible itself for accounting and written administration (and teaching); no distinct group of scribes can be traced. This identity of the stratum of managers with the numerate and literate class is reflected in a complete integration of mathematics with its bureaucratic applications – school texts are "model documents", distinguishable from real administrative documents only by lacking the name of a responsible official and by the prominence of nice numbers. But the integration was mutual: bureaucratic procedures, centred on accounting, were mathematically planned, for instance, around the new area metrology and the calendar.

The third millennium continues the mutual fecundation of administrative procedures and the development of mathematics (in a process whose details we are unable to follow). The scope of accounting systems was gradually expanded, and metrologies were modified intentionally so as to facilitate managerial planning and accounting. At the same time, there was a trend toward "sexagesimalization", increasing use of the factor 60.

Around 2600, however, when a distinct scribal profession emerged, numeracy and literacy were emancipated from the full cognitive subservience to accounting and management. For the first time, writing served to record literary texts (proverbs, hymns, and epics); and we find the first instances of supra-utilitarian mathematics – mathematics starting from applicable mathematics but going beyond its usual limits. It seems as if the new class of professional intellectuals had set out to test the potentialities of the professional tools – the favourite problem was the division of very large round numbers by divisors that were more difficult than those handled in normal practice (see above, note 298).

The language of the protoliterate texts is unidentified, whereas the language of the third-millennium southern city states was Sumerian.[352]

---

351 A broad summary of fourth and third millennium mathematical techniques (including a description of the metrologies) can be found in [Nissen, Damerow, and Englund 1993] – the detailed information about the proto-literate phase can be found in [Damerow and Englund 1987]; for the administrative calendar, see [Englund 1988]. The conjectures about the development of the numerals are mine.

352 The protoliterate script had been logographic, deprived of phonetic and grammatical elements; documents were organized as schemes that corresponded to

Toward 2300, however, the Akkadian-speaking Sargon dynasty conquered the whole Sumerian region, and soon the entire Syro-Iraqian area. Sumerian remained the administrative language (and hence the language cf scribal education) during this "Sargonic" or "Old Akkadian" phase, but new problem types suggest inspiration from a "lay" (that is, non-school), possibly non-Sumerian surveyors' tradition – area computations that are very tedious unless one knows that $\square(R-r) = \square(R)+\square(r)-2\sqsubset\!\sqsupset(R,r)$, and the bisection of a trapezium by means of a parallel transversal (cf. above, pp. 237 and 267).

The 21st century is of particular importance. After the breakdown of the Old Akkadian empire, the new "neo-Sumerian" territorial state ruled by the "Third Dynasty of Ur" (whence the other name, Ur III) established itself in 2112 ("middle chronology"). A military reform under king Šulgi in 2074 was followed immediately by an administrative reform, in which scribal overseers were made accountable for the outcome of every $^{1}/_{60}$ of a working day of the labour force allotted to them according to fixed norms; at least in the South, the majority of the working population was subjected to this regime, probably the most meticulous large-scale bureaucracy that ever existed.

Several mathematical tools seem to have been developed in connection with the implementation of the reform (all evidence is indirect): A new book-keeping system – not double-entry book-keeping, but provided with analogous built-in controls; the sexagesimal place-value system used in intermediate calculations; and the various mathematical and technical tables needed in order to make the place-value system useful (see the description of their use on p. 17).

No space seems to have been left to autonomous interest in mathematics; once again, the only mathematical school texts we know are "model documents". As we shall see (p. 377), this absence of problem texts will not be due to the (bad) luck of excavations.

For several reasons (among which probably the exorbitant costs of the administration) even the Ur-III state collapsed around 2000. A number of smaller states arose in the beginning of the succeeding "Old Babylonian" period (2000 to 1600), all to be conquered by Hammurapi around 1760. Without being a genuine market economy, the new social system left much space to individualism, both on the socio-economic and the ideological level.

---

bureaucratic routines, and did not attempt to render sentences; the grammatical elements and incipient use of phonetic principles that begin to turn up around 2700 are indubitably Sumerian. If we dismiss fanciful ideas about foreign conquerors from Caucasus, Tibet, Thailand, etc. (of which there have been many, but never supported by the least evidence), it seems a natural assumption that the language spoken in the southern region remained unchanged, and that the language of protoliterate Uruk was thus an early Sumerian. Much in the structure of Sumerian suggests, however, that it may have developed locally from a creole spoken by enslaved populations in the late fourth millennium and then taken over by the minority of masters (as often happens in such situations when no metropolis can protect the original language of these). See [Høyrup 1992a].

In the domain of scribal culture, this individualism expressed itself in the ideal of "humanism" (*sic* – nam.lú.ulù, Sumerian for "being human"): scribal *virtuosity* beyond what was needed in practice. This involved the ability to read and speak Sumerian, now a dead language known only by scribes, as well as supra-utilitarian mathematical competence.

The vast majority of Mesopotamian mathematical texts come from the Old Babylonian school (almost exclusively teacher's texts or copies from these, not student exercises as the third millennium specimens). They are invariably in Akkadian, another indication that the whole genre of "humanist" mathematics had no Ur-III antecedents.[353] As we shall see, the adoption of material from non-school environments into the new Akkadian curriculum will have taken place simultaneously in Eshnunna and in the former Sumerian South (perhaps Larsa); a first attempt (with scant repercussions) seems to have been made in Ur.

Inner weakening (which had already led to the loss of the Southern provinces) followed by a Hittite raid put an end to the Old Babylonian state in 1600. Then Kassite warriors subdued the Babylonian area, for the first time rejecting that managerial-functional legitimization of the state which, irrespective of suppressive realities, had survived since the protoliterate phase and made mathematical-administrative activity an important ingredient of scribal professional pride. The school institution disappeared, and scribes were trained henceforth as apprentices within scribal "families". Together, these events had the effect that mathematics vanishes almost completely from the archaeological horizon for a millennium or more (one Kassite problem text and one table text have been found; the problem text seems to derive from the style of the Old Babylonian periphery); metrologies were modified in a way that would fit practical computation in a mathematically less sophisticated environment (making use of normalized seed measures or of systems based on broad lines in area mensuration [Powell 1990: II A–B]; though no longer an object of pride, mathematical administration did not disappear).

Around the "Neo-Babylonian" mid-first millennium, mathematical texts turn up again, for instance, concerned with area mensuration, the conversion

---

353 The sometimes heavy use of Sumerian word signs may mislead on this account, and have misled many workers into the belief that the whole complex of Old Babylonian mathematics was of Sumerian origin. With utterly few exceptions, however, the underlying grammatical structure is Akkadian; in the actual exceptions (see, for instance, note 281), the Sumerian grammar is either wrong and somehow home-made or suspiciously dependent on the grammar book (more so than in the Sumerian literary compositions and grammatical texts of the Old Babylonian age). The use of Sumerograms is no more proof of a Sumerian origin than the Vatican Latin dictionary is proof that the ancient Romans knew about railways, trade unions, and nuclear fission.

In some but not all cases, the denial of Akkadian mathematical innovations has certainly been an expression of anti-Semitism (anti-Arabic and anti-Jewish alike), and the argument of Sumerograms only a welcome pretext.

between various seed measures, and some supra-utilitarian problems of the kind that had once inspired Old Babylonian algebra. This and other features may reflect renewed interactions between the scribal and the lay traditions, which so far cannot be traced more precisely.

One of the Neo-Babylonian texts starts by listing the sacred numbers of the gods before going on with genuinely mathematical topics[354]. This break-down of cognitive autonomy corresponds to what the texts tell us about their owners and producers (such information is absent from the Old Babylonian tablets); they identify themselves as "exorcists" or "omen priests" (another reason to believe that their practical geometry was borrowed from lay surveyors).

A final development took place in the Seleucid era (311 onwards). Even this phase is only documented by utterly few texts: some multi-place tables of reciprocals probably connected to astronomical computation; one theme text; an anthology text focusing on practical geometry; and an unfocused anthology text. The unfocused anthology text shows some continuity with the Old Babylonian tradition (including its second-degree algebra) but also fresh developments (e.g., formulae for $\Sigma 2^n$ and Σn^2) – cf. pp. 389*ff*. The theme text contains "algebraic" problems about rectangles and their diagonals of which only two types (A and $l \pm w$ given) are known from the earlier record, but where even these are solved in a different way. It seems to be a list of *new* problem types, either borrowed from elsewhere or fresh inventions in the area. The Seleucid texts make heavy use of Sumerian word signs, but in a way that sometimes directly contradicts earlier uses. At least to some extent they represent a new translation into Sumerian of a tradition that must have been transmitted outside an erudite scribal environment.

---

354 W 23273, cf. [Friberg 1993: 400].

Chapter IX
The "Finer Structure" of the Old Babylonian Corpus

In the preceding chapters, occasional references were made to specific text groups, but apart from this the Old Babylonian mathematical corpus was treated as an undifferentiated whole.

In many respects this is justified, and not only by our lack of adequate knowledge – but it *is* difficult to group a material if all that is known about the provenance of most tablets is that they were bought at the antiquities market and that the writing seems to be Old Babylonian. Certain differences of terminology (for instance, that some texts make us "see" a result, some make it "come up", and some say that the computation "gives" it) do suggest regional variation, but no evident link between mathematical substance (problem types and methods) and such differences of "dialect" presents itself.

A first attempt to group the texts known by then was made in 1945 by Goetze and published in [MCT, 146–151].[355] He used variations in dialect and orthography (and, to a slight extent, vocabulary[356]) to distinguish six different text groups (texts with too much logographic writing being either left out or ascribed to one of the groups already established if "connected by external appearance and content with other tablets of the group" – [MCT, 149 n.356]). Quotations will illustrate that the argument builds on philological details:[357]

---

355 Strictly speaking, already Neugebauer had tried to define two groups of texts that he felt belonged together – [MKT II, 50], cf. [Høyrup 2000a: 119].

356 Vocabulary, not terminology: he did not distinguish cases where the verb *nadānum*, "to give", refers to the giving of a result from others where it refers to the payment of interest on a loan.

357 The main division is into a "northern" and a "southern" dialect, the former being that of the Hammurapi code and of texts from Dilbat and Sippar, the latter being

1. "This group is certainly to be localized in the South, in all probability Larsa. It employs PI for both *pi* and *pe*, and shows numerous repeated vowels."
2. "This is likewise a southern group. It employs PI for both *pi* and *pe*, but exhibits repeated vowels only sparingly."
3. "This group, likewise southern, is localized in Uruk. It employs BI for *pé*." It includes the Strasbourg texts, VAT 7532, and VAT 7535, already clustered by Neugebauer in [MKT II, 50].
4. "As far as linguistics is concerned, this 4th group cannot be distinguished from the 3rd. [It is] quite clear, however, that here PI is *pi* and BI *pé*. The provenience may likewise be Uruk."
5. "The employment of BI for *pi* [*sic*, misprint for *pí*] and the occurrence of SU make this a northern group".
6. "This group combines northern and southern characteristics. It is slightly younger in date than the other groups. Since by now it seems clear that Akkadian mathematics (as the other varieties of Akkadian writings) originated in the South,[358] the situation is satisfactorily explained when it is assumed that the 5th [*sic*; error for 6th] group comprises tablets based on southern originals, but written and modernized in the North. The southern originals were close to the 1st and 2nd group". The group in question comprises BM 85194, BM 85196, BM 85210, BM 85200+ VAT 6599 and VAT 6598, clustered by Neugebauer, but not the almost exclusively logographic series texts which Neugebauer had believed close to them. The series texts, indeed, are not explicitly included in any of Goetze's groups but mentioned in connection with his group 2.

Since 1945, a number of texts with well-defined provenance have been published. This allows the addition of two new groups,

7. The texts from Eshnunna, dated between c. 1800 and c. 1775.
8. The Susa texts, which according to palaeographic criteria should be dated in the late Old Babylonian period.[359]

---

described mainly on the basis of Larsa texts. Characteristic northern spellings are the following:

ṭá, ṭe₄, ṭì, ṭú, sa, si, su, ás, is, ús, ba, bi, bu, pa, pí, pu.

The corresponding southern sequence is

ṭa, –, ṭi, ṭù, sà, sí, sú, as, is, us, ba, bi, bu, pa, pi, pu.

It is mentioned as self-evident (p. 146) that "texts from other places will probably necessitate the positing of additional 'dialects'", and pointed out (p. 147) that one group of mathematical tablets "probably at home in Uruk" also uses BI as *pé*. To this comes a number of other characteristics, such as a southern preference for phonetic complements of the consonant-vowel-consonant type and a northern preference for the type vowel-consonant.

358 After the discovery of the Eshnunna texts (cf. presently), this belief in the priority of the South has to be given up. It should also be observed that the first law-code in Akkadian was produced in Eshnunna as early as the 19th century.

359 The excavator was too sloppy to care for stratigraphy of tablets and mud brick

Description of the Groups

With this new material at hand, it is possible to approach the problem of text groups anew, and to combine Goetze's orthographic criteria with observations of terminology in the widest sense (vocabulary in the context of function).[360] By necessity, the discussion is very technical; the reader who is not very keen on philological details may skip it and jump directly to "The Outcome", p. 358. If the picture delineated there seems to fanciful, it will always be possible to return to the maze of details that provides the arguments.

Group 7: The Eshnunna Texts

The published Eshnunna texts fall in two subgroups, one of which ("7A") is stylistically very coherent, the other ("7B") certainly not.

Group 7A has good reasons to be coherent – all its texts are from c. 1785/84, all but one were found in the same room, and the last in the immediate vicinity.[361] Among the characteristics, these may be listed:

- All are in "riddle format", starting by the phrase *šum-ma ki-a-am i-ša-al(-ka) um-ma šu-ú-ma*, "If [somebody] asks (you) thus:".
- All introduce the prescription with the formula *at-ta i-na e-pé-ši-ka*[362], "you, by your proceeding". No closing formula is present.
- None of the "logical operators" *aššum* ("since"), *inūma* ("as"), and *šumma* ("if") are used.
- In the many cases where the transition to a new section of the prescription is marked, the phrase used is *na-ás-ḫi-ir*, "turn yourself around".
- The results of calculations are marked by one of the phrases *ta-mar*, "you see", or *i-li-a-ku-um*, "comes up for you", in both cases often combined with an enclitic *-ma* on the verb for the operation[363].

---

structures! Cf. [Robson 1999: 19] and [MCT, 6 n. 28].

360 [Høyrup 2000a] is a detailed investigation of the issue; the remains of the present chapter summarizes the results and draws some further conclusions.

361 IM 53953, IM 53957, IM 53961, IM 53965, IM 54010, IM 54011, IM 54464, IM 54478, and IM 54538 were found together. The last tablet from the group is IM 54559.

In [Høyrup 2000a], I counted Db_2-146 as a peripheral member of 7A because of the riddle format; second thoughts have made me ascribe it to the diffuse rest group 7B.

362 We may observe the use of BI for *pé*, as in Goetze's groups 3-6. The use of BI for *pí* (*pí-ti-iq-tum* and *e-pí-ri-ka*, IM 54011 obv. 2, rev. 2; *ša-pí-il-tum*, IM 54464 rev. 1), however, is only shared with groups 5 and 6.

363 Only IM 54559 (the text not found in the same room as the rest) uses both *ta-mar*

- When results "come up", the interrogative phrase of the question is *mīnûm*, "what"; when they are "seen", we always find *kī maṣi*, "corresponding to what"; only IM 54011 also has *mīnûm* (it contains two problems). The "mixed tablet" IM 54559 has *mīnûm*.
- There is a clear preference for *ḫarāṣum*, "to cut off", over *nasāḫum*, "to tear out"; *nasāḫum*, indeed, only occurs in IM 54464 (rev. 9) in the relative clause *ša ... ta-su-ḫu*, "which you have torn out", the (damaged) reference of which (obv. 10) seems not to contain this verb at all but only an "excess" (*watrum*).[364]
- The quadratic "equalside" is written unorthographically, as íb.si – or indeed íb.se.e (SI=SE), even more indubitably a syllabic writing of the Sumerian pronunciation; in IM 54478, a cubic "equalside" is designated with the orthographic íb.si$_8$. Both are treated as verbs.
- In processes of "breaking" (*ḫepûm*), only the entity to be broken is mentioned explicitly, not the resulting "natural half" (neither as *bāmtum*/ "moiety", nor as šu.ri.a, "the half", nor with the sign $\frac{1}{2}$).
- In backward references to rectangularizations, the relative clause *ša tuštakīlu*, "which you have made hold" is used instead of the noun *takīltum*, the "made-hold".
- In agreement with what is the invariable pattern outside Eshnunna, the lengths and widths of fields are mostly written logographically, as uš and sağ.ki. However, IM 53965 (a "broken reed" problem) has syllabic writings of *šiddum* and *pūtum* throughout.[365]
- In cases whose underlying structure is a rectangle with given area and given difference between the sides, and where the "norm of concreteness" (pp. 58, 174) would make the removal precede the corresponding "appending" (IM 53965, IM 54559), the abbreviated formula from BM 85200+VAT 6599 #24 and #26 is used; "to 1 append, from 1 cut off" (see p. 155). The "norm of concreteness" thus is not observed.

---

and *i-li-a-ku-um*; IM 54464 makes a "raising"-multiplication (*našûm*) "give you" (*i-na-di-na-ku-um*) the result), but then repeats the calculation making the result "come up" (*i-li*), as if a slip had occurred when the scribe submitted an original to stylistic normalization (or tried to conform to a style which was not fully his, if his composition is original); as we have seen, constructions with *nadānum*/sum, "to give", are used regularly in questions about division by irregular numbers, and we may suppose that they had a traditional connection with sexagesimal multiplication.

364 Even this is thus probably the result of a slip, a deviation from a style deliberately striven for, cf. note 363.

365 On the other hand, the *ši-du-um* of IM 54538 (obv. 2) and IM 54011 (obv. 2), respectively a carrying distance and the length of a wall to be built, are not exceptional but correspond to a pattern found elsewhere. Even the dimensions of a field measured by a breaking reed may of course have been seen as "real" and not belonging to the standard representation.

Given the early date of the group it is remarkable that its problems cover the whole range of Old Babylonian mathematical themes: manpower for the carrying of bricks and for the building of an earth wall; a broken reed problem leading to a mixed second-degree equation; combined commercial rates; a complex problem dealing with a rectangle which is reduced to a standard problem about a different rectangle (*viz.*, given area and excess of length over width); and a cubic excavation.

Quite intriguing is IM 53957 ([Baqir 1951: 37], corrections and interpretation [von Soden 1952: 52]):

> If [somebody] asks (you) thus: To $\frac{2}{3}$ of my $\frac{2}{3}$ I have appended 100 silà and my $\frac{2}{3}$, 1 gur was completed. The *tallum*-vessel of my grain corresponding to what?

This may be compared to problem 37 of the Rhind Mathematical Papyrus [trans. Chace et al. 1929: Plate 59]:

> Go down I [a jug of unknown capacity – JH] times 3 into the *ḥeḳat*-measure, $\frac{1}{3}$ of me is added to me, $\frac{1}{3}$ of $\frac{1}{3}$ of me is added to me, $\frac{1}{9}$ of me is added to me; return I, filled am I [actually the *ḥeḳat*-measure, not the jug – JH]. Then what says it?

The coincidences are too numerous to be accidental: firstly, there is the shared use of an ascending continued fraction; these are not only extremely rare in the rich Egyptian record, the RMP example appears to be the only ascending continued fraction occurring at all. To this comes the details of the dress: an unknown measure which is to be found from the process, the reference to a standard unit of capacity, and the notion of filling.

The Egyptian problem is solved in agreement with the normal procedures of Egyptian arithmetic, in a way that depends critically on the fine points of the system of aliquot parts. The Eshnunna solution, on the other hand, is no solution at all but a sequence of operations which only yields the correct result because the solution has been presupposed – what sixteenth-century cossists would call *Schimpfrechnung*, "mock reckoning", a challenge meant to impress and make fools of the non-initiate. Problems of this type turn up regularly in medieval and Renaissance treatises on applied mathematics that draw directly on oral or semi-oral practitioners' traditions[366] – exactly the traditions where rules for practical computation go together with mathematical riddles that seem to refer to practice but rarely have any practical application. Without pursuing the argument for the moment we may infer that the problem has its origin in a similar practitioners' environment (merchants?) in touch with both Egypt and Mesopotamia in the early second millennium, and that it was adopted by both Egyptian and Eshnunna scribes – in Eshnunna preserving the eristic form and purpose, in Egypt transformed into "good mathematics".[367]

---

366 We shall return to the characteristics of oral practitioners' environment below, pp. 362*ff*.

367 Eventually, the transformation into "good mathematics" also took place in Babylonia, where the problem turns up in somewhat altered shape as YBC 4669

This inference fits the common riddle format of the 7A texts. The traces of deliberate stylistic normalization in IM 54464 (see notes 363 and 364, and presently) should warn us, however, against perceiving the texts of the group as nothing but written versions of traditional material. The evidence is delicate, but the scribes (or the scribe) responsible for the production of the ten tablets belonging to the group seem to have aimed deliberately at stylistic demarcation, imitating an archaic (probably Akkadian) model, borrowing part of its material from lay practitioners (at least the quasi-algebraic problems on rectangular and trapezoidal "fields" and the problem on the *tallum* vessel), and eliminating perhaps some of the stylistic features of Ur III computational practice.

The evidence offered by the uneven group 7B corroborates these conjectures. It comprises the texts IM 55357, Db_2-146, and IM 52301 (above, pp. 231, 257, and 213, respectively); furthermore Haddad 104, which contains 10 non-algebraic, practical problems on volumes, work norms, etc.; the catalogue texts IM 52916 and IM 52685+52304 (the "Tell Ḫarmal Compendium"); and the two unpublished tablets IM 43993 and IM 121613, which I know about thanks to the courtesy of Jöran Friberg and Farouk al-Rawi [1994a].

Of these, IM 55357 is at least as old as the texts from 7A, maybe slightly older; the Compendium and the unpublished texts are undated, and the rest some 8–10 years younger than group 7A. The whole group 7 thus comes from a small area and a brief period; the homogeneity and variation with which it presents us illustrates the limits within which a local mathematical culture would move.

In relation to the characteristics of group 7A that were listed, the following observations can be made:

- Db_2-146 and IM 43993 open as riddles; in Haddad 104, problem statements begin by notifying to be the *nēpešum*, "procedure" for a particular object – or, if variants of a preceding type, by *šumma*, "if (however)". In IM 52301, the two problems and the rule in #4 start *šumma*, "if"; IM 55357 begins by stating the object (sag̃.dù, "A triangle"), and IM 121613 starts directly by the data. This exhausts the whole range of variants attested in the complete Old Babylonian material, and adds a possibility that is not attested elsewhere (the initial *nēpešum*).
- All but IM 121613 (which has only "You") start the prescription by the formula "You, by your proceeding" or some variant (as group 7A),[368] and all but IM 121613 (and IM 55357 which is not completed) close by

---

#B4–5 – so transformed indeed that the family likeness with the Egyptian problem is no longer obvious.

368 Strictly speaking, this formula comes before the didactic explanation of the situation in IM 43993 (see note 111 and appurtenant text); the prescription proper opens by *šumma warkatam taparras*, "if you distinguish what is below", a standard phrase for investigating closely [AHw, 831].

saying that this was the procedure (in contrast to group 7A).

- Beyond the uses described above, *šumma* serves to introduce the proof in Db_2-146 and to introduce subsections of the prescription in Haddad 104. Neither *aššum*, "since", nor *inūma*, "as", are to be found.
- Transitions to new sections may be marked by *ta-ás-sa-ḫa-ar*, "you turn around" (IM 43993, after the explanation of the situation); *tu-ur* or *tu-úr*, "turn back" (IM 52301 and Haddad 104); or *na-ás-ḫi-ir*, "turn yourself around" (IM 55357, as in group 7A).
- Results are "seen" in IM 52301, IM 121613, and IM 55357 ("equalsides" are "made come up" in the former two but still "seen" afterwards; in the latter it is a verb); they "come up" in Db_2-146, Haddad 104, and IM 43993 (in Db_2-146, the "equalside" is to be "taken", but it still "comes up").
- In Db_2-146, the interrogative phrase is *kī maṣi*, "corresponding to what" – coupled to results "coming up", against the system of group 7A. The broadly discursive statements of Haddad 104, in which results "come up", mostly use *mīnûm*, "what"; but one asks *kī maṣi*, "corresponding to what"; one has the rare *kiyā*, "how much each";[369] and one contains the exhortation to "give me" the number, not known from elsewhere. In all the other texts, the question is *mīnûm*, irrespective of how results are announced.
- Subtraction by removal is termed *nasāḫum*/zi, "to tear out", in IM 55357, IM 43993, IM 121613, and in the Compendium; in IM 121613 it is actually a reduplicated zi.zi, not found in any published Old Babylonian mathematical text but used in "Šulgi-Hymn B", l.17, in a description of the school curriculum, and thus one of the few terms of whose properly Sumerian origin we may be certain.[370] The other texts (Db_2-146, IM 52301, and Haddad 104) employ *ḫarāṣum*, "to cut off", as does group 7A.
- Except in IM 55357, the "equalside"/"equilateral" is a noun which is "taken" (Db_2-146) or "made come up" (Haddad 104, IM 52301, IM 43993, IM 121613). Except in IM 55357, the writing is always unorthographic/phonetic – sometimes with Akkadian declination; in Haddad 104, IM 52301, and IM 121613 it corresponds to the orthographic counterpart ba.si$_8$, in IM 55357, Db_2-146, and IM 43993 to íb.si$_8$. All occurrences refer to the quadratic, none to the cubic configuration.
- In Haddad 104, IM 52301, and IM 121613, "breaking" does not identify the resulting "moiety"; in Db_2-146 it carries the non-standardized name

369 Elsewhere, *kiyā* mostly serves in connection with sharing between šeš, "brothers" (VAT 8522#2, VAT 6597, YBC 4608; in MLC 1842 it is connected to a problem of combined commercial rates, similar to TMS XIII (see p. 286); in IM 54559 it asks for the sides of the square basis for an excavation, whereas the depth is asked for with *kī maṣi*. In all cases it thus asks for several values at a time (whence my translation).

370 [Ed. Castellino 1972: 32]. zi.zi occurs in the hymn together with gá.gá, which in Old Babylonian mathematics was written g̃ar.g̃ar~*kamārum*, "to accumulate"), and with šid and níg.šid, "counting" and "accounting".

muttatum (see p. 260); in IM 43993, the very operation is designated by the non-standard *letûm*, "to split"; only IM 55357 has the normal *bāmtum*, written BA. IM 52301 "lays down" the "counterpart" of the result, whereas IM 43993 "posits" it. Everything suggests an originally loose usage, which in later texts was straightened, so as to eliminate the unorthodox *muttatum* and *letûm* and use *meḫrum* only about the "counterpart" of an equalside found after quadratic completion

- IM 52301 and IM 121613 employ the noun *takīltum*, the "made-hold"; Db_2-146 and IM 43993, like group 7A, have the relative clause *ša tuštakilu*, "which you have made hold".
- In IM 43993, both the length and the width of the rectangle are written in syllabic Akkadian (*šiddum* and *pūtum*, respectively); in Db_2-146, the length but not the width is in Akkadian; in the Compendium, *šiddum* and uš alternate when referring to the side of a square. Elsewhere, the length is always uš and the width always sag̃.ki.
- In Db_2-146 and IM 52301 (the only texts where the question presents itself), the "norm of concreteness" is not observed. Db_2-146 uses the same abbreviated formula as BM 85200+VAT 6599 (see p. 155), but IM 52301 does not.

According to many criteria, as we see, group 7B can divided further into two; but we also see that division according to different criteria never produce the same clusters. Even the coinciding divisions of group 7A according to the way results are announced and the way values are asked for turns out to have no validity outside this group. This may be taken as an indication that the contrast between the uniformity of group 7A and the heterogeneity of group 7B reflects a deliberate attempt on the part of the creators of group 7A to establish a canonical idiom – which agrees well with the lapses that were pointed out in notes 363 and 364 (even the mixture of question types in IM 54011 is likely to be a *lapsus calami*).

By necessity, the creation of a canon involves choices and exclusion of certain ways as non-canonical. In the case 7A, the most conspicuous choices are: the general use of the riddle format; the exclusion of zi/*nasāḫum*, "to tear out"; the verbal character of the "equalside"; and the coupling between the way questions are asked and that in which result are announced (and the fact that the canon allows two ways for both – the systematic coupling rules out that this can be a case of mere oversight or failing will to define a canon for these aspects of the terminology).

The riddle format points to the lay practitioners' environment and away from the Ur III tradition[371]; the elimination of zi/*nasāḫum* is also a refusal

371 Firstly, this conclusion conforms with from the generally eristic character of oral culture; but it is confirmed by the absence of the format from the rest of the Old Babylonian record apart from the ellipsis in BM 13901 #23 (the "four fronts and the surface", see p. 222), and by its presence in certain medieval sources (see below, p. 370).

of an Ur III terminology, and the adoption of *ḫarāṣum* at least a clear reference to a non-Sumerian background[372].

The choice of a verbal and not a nominal use of íb.si$_8$ is less easily interpreted. In either form, the word is obviously of Sumerian origin, and even (as can be seen from the phonetic spellings) pronounced as a Sumerian word. The use of the verb si$_8$ to express that a line is the side of a square goes back at least to c. 2600 BCE.[373] The occasion on which the word came to designate a thing or a number is likely to have been the age when reckoners would meet it routinely as a value entered in a table of square roots from which it could be "taken" – that is, in the wake of the Šulgi administrative reform.

The first to propose that we might distinguish a stratum of pre-Šulgi mathematics from one of Šulgi mathematics was Eleanor Robson [1995: 204–209], in connection with her work on the technical coefficients. Knowing that the use of the verb si$_8$ as a designation of the square side is old, it is a reasonable assumption that the verb provided with grammatical complement (íb.si$_8$) and accompanied by a case suffix .e, "alongside", predates Ur III; that the nominal use which was discarded by the authors of group 7A represents the Šulgi or post-Šulgi stage is almost certain.

The unorthographic ib.si, ib.sí, íb.se.e, etc., are certainly instances of "colonial spelling", on a par with contemporary unorthographic phenomena like "skeptical" and "color": Eshnunna, though integrated in the system of Sumerian city states since the mid-third millennium, had never been part of the Sumerophone area; it emancipated itself from Ur III already in 2026 or 2025. Outside Eshnunna and Susa, unorthographic spellings of the terms are very rare.

Even the backward references of "seeing" versus "coming up" and of the ways of asking are not evident. But we may start by noticing that nominal "equalsides" may be required to "come up" even in a context where results are "seen" (IM 52301), and that problems whose type points towards Ur III practice (not least the problems of Haddad 104) are predominantly of the "coming-up" type; the obvious riddles (the filling of the *tallum* vessel, two broken-reed problems, most of the second-degree problems on rectangles), instead, "see" the results.

An older group of texts exists where results are also "seen": a cluster of Sargonic school texts about rectangles and squares (mentioned above, note 304 and preceding text – see [Whiting 1984: 59 n.2]). In these, the word is pàd, which we also encounter in UET V, 864 and in a few other texts from early Old Babylonian Ur; IM 55357, however, the only one of the Eshnunna texts to

---

372 In contrast to the frequent use of zi even in predominantly syllabic texts, the Sumerographic writing kud for *ḫarāṣum* is dubious and at least very rare in mathematical texts – indeed, kud is used more commonly for *nakāsum*, "to cut down" (a reed, etc.) and for *ḫaṣābum*, "to break off".

373 See the table VAT 12593 of sides and appurtenant square surfaces in [MKT I, 91].

use a Sumerogram in this function, has igi.dù (unorthographic for igi.du_8). A transmission of the term through a Sumerian tradition is thus quite unlikely; we seem to be confronted with two independent translations into Sumerian of the same idea borrowed from a non-Sumerian (that is, given the historical context, an Akkadian) environment – or rather three, one of which was a dead-end, cf. p. 357.

All in all, results that "come up", if not indications of a real link to the Ur III scribal tradition, are thus likely to have been intended (at least by the stylistically conscious originators of group 7A) to suggest a link to traditional scribal computation; results that are "seen" point toward non-scribal traditions, or will have been meant to do so.

If group 7A had been isolated, a straightforward corollary would now have been that *mīnûm* points back to the scribal past, and *kī maṣi* to lay milieus. Group 7B shows that this, even if it may have corresponded to the ideas of the creator of the nine 7A tablets found together, will have little to do with real history. We may add that en.nam, the Sumerogram used currently for *mīnûm* in most text groups, is wholly absent from the Eshnunna texts.[374]

Group 8: The Susa Texts

In total, the Susa corpus consists of 26 texts, of which TMS VII–XXVI are procedure texts. Since TMS XXVI differs from the others on many accounts, we shall look first at the "typical" group, TMS VII–XXV, which constitute a fairly homogeneous group, with the following characteristics:

- Most texts open by stating the parameters, and only indicate in this indirect way which kind of object is dealt with – thus, if only a "length" and a "width" occur and perhaps an area, the object must be a rectangle, the simplest figure fully determined by its length and its width. If this is insufficient, the object is presented explicitly or (XVIII) by means of a drawing. No *šumma* or other formula occurs.
- The prescription is provided with an opening formula – a terse za.e, "you".

---

374 The text IM 55357 does have a Sumerographic writing of the accusative *mīnâm*, namely, a.na.àm, for which later lexical lists give the equivalence a.na.àm~*mīnûm* [AHw, 655b], and which is also used together with en.nam as the corresponding accusative in the text UET V, 859 (cf. above, note 42).

Analysis of the word (a.na~*mala*, .àm enclitic copula) suggests that the original correspondence is rather with the alternative interrogative phrases *mala maṣi* [AHw, 621b] and *kī maṣi* (see also [SLa, §120]). The coupling to *mīnum* thus seems to be an Old Babylonian construction, brought about by the need to find an equivalent. Even en.nam, rarely used outside the mathematical texts[AHw, 656a], appears to be a new construction.

- Closing formulae only occur in IX #1 (*ki-a-am ne-pé-šum*, "thus the procedure") and #2 (*ki-a-am ak-ka-du-ú*, "thus the Akkadian (method)") and XVII (*ki-a-am ma-ak-ṣa-ar-šu*, "thus its bundling"). Apart from IX #1, closing formulae are thus used when a method with a particular name has been taught.
- Explicit questions are mostly absent from the statement. XII, XVII, and XIX, however, use *mīnûm*, XIII has *kī maṣi* and XIV *mīna* g̃ar, "what may I posit?"
- Results are regularly followed by *tammar*, "you see".
- Shift of section in the prescription is marked with high frequency, and invariably *tu-úr*, "turn back". XII marks a shift within the statement with the corresponding first person singular *a-tu-úr*, "I turn back".
- Some characteristic regularities in the use of logograms can be noted. *kamārum* ("accumulation") invariably occurs as UL.GAR, and *nasāḫum* as a naked zi (not ba.zi), except in XVI #1, where a zi from the statement is quoted as a syllabic infinitive *na-sà-ḫu*. Almost all texts use NIGIN where syllabic texts would have *šutakūlum* or *šutamḫurum*. IX, when quoting a NIGIN from the statement, does so with a syllabic infinitive *šu-ta-ku-lu*. XVII uses syllabic writings of *šutakūlum* throughout for squaring, and XX and XXIII employ *šutamḫurum* systematically in the same function. The "natural half" is written logographically as $\frac{1}{2}$, and *meḫrum* as gaba. The "equalside" appears as an unorthographic íb.si, never as íb.si$_8$; it functions as a verb. uš and sag̃ appear as such, never syllabically nor as sag̃.ki.
- The term *takīltum* is used in four texts (XII, XIX, XXI, XXIV), while the alternative relative clause *ša tuštakīlu* is absent.
- "Intermediate zeroes", we remember, are experimented with in XII and XIV, cf. note 19.
- *alākum*, "to go", is used in many of the texts in a general sense that may be interpreted "do in repeated steps". Often the use is multiplicative (in agreement with the etymology of the term a.rá, "steps of", and the phrase a.rá *N* ^{e}tab, "in steps *N* repeat" used in many of the series texts); in VIII #2 it is used in analogy with the "repeat-append" pattern of the stone riddle YBC 4652 #20 (see pp. 188 and 344).
- The division of a number *A* by an irregular number *B* follows the invariably pattern *mi-na a-na B* g̃ar *šà A i-na-di-na Q* g̃ar, "What to *B* shall I posit that gives me *A*? Posit *Q*".[375]
- Among the "logical operators", *aššum* turns up frequently. Its function is to argue for the calculations to come, either with reference to the statement (quoting a complete phrase or a single word), to general knowledge of the characteristics of the object dealt with (XIV, obv. 9), or to the situation

375 XXIV, 24f; the corresponding but more or less damaged passages in other texts all agree (VIII, 8; IX, 48; X, 20-22; XIII, 13*f*).

that has been established so far. *inūma* is found in IX #2, in complete parallel to the use of *aššum* in #1. *šumma* is absent.

TMS XXVI, the last of the procedure texts, deviates from the shared pattern of the other Susa texts on several accounts. While its problems start as those of the other texts, either by explicit introduction of the object or implicit presentation through specification of the parameters, the prescriptions are not introduced at all. Nor do results have any formal marker, not even a preceding *-ma*. Shifts of section are indicated in the second person present tense, as *ta-ta-a-ar*, "you turn back". Removal is kud, perhaps meant as a logogram for *ḫarāṣum* (but possibly for *nakāsum* or *ḫaṣābum*), not zi/*nasāḫum*. The (plane) "equalside" is a noun and written íb.si$_8$, not íb.si; squaring is gu$_7$.gu$_7$ [Muroi 2001: 229*f*].. a.rá, the term for purely numerical multiplication, is found in several places where other texts have *šutakūlum*, *šutamḫurum* or one of the corresponding logograms (which reminds of AO 8862 – see above, p. 162). The tablet is clearly an outsider in the corpus.

Even though most of the parameters used to characterize the procedure texts belong with the prescription, the two closely related catalogue texts V and VI can be seen also to fall outside the group constituted by VII–XXV. In the present context it is noteworthy that the area of a square is designated a.šà LAGAB (in V) or a.šà NIGIN (in VI); V uses NIGIN as the plural of LAGAB, in a way that is reminiscent of the distinction between a singular LAGAB and a plural *mitḫarātum* in the Tell Ḥarmal Compendium; both speak of the side as uš, as done in the Compendium (and perhaps in VIII, cf. note 212). While the procedure texts use UL.GAR for "accumulation" (*kamārum*), V uses g̃ar.g̃ar; VI, when adding square sides to the area, "appends" them (daḫ), thus perceiving the side as a "broad line", in the likeness of the Tell Ḥarmal Compendium. Both use zi for removal.

Even this group thus consists of a core ("8A") produced according to a fairly strict and deliberately chosen canon, and an assortment of texts that illustrate which other choices had been possible within the local orbit.

In terms of Goetze's linguistic distinctions, the Susa corpus mixes "northern" and "southern" orthography, with "southern" preponderance.[376] However, since Akkadian was anyhow a scribal language in Susa and not the

---

376 BI is used for *pé* (*ne-pé-šu*, IX, 9; *ḫe-pé* in IX, 39 and XV, 2, however, are editorial errors for *ḫe-pe*), and for *bi* (*wa-ṣú-bi*, IX, 32; *qa-bi-a-ku-um*, XVII, 5); the value of PI in *ḫe-pe*/*pi* (*passim*) can therefore be assumed to be *pi*, /pe/ being occupied (cf. also the possible *a-ta-ap-pi-šu*, XXIV, 34). These together agree with Goetze's "Uruk" group. Also "southern" are *sà* (*na-sà-ḫu*, XVI, obv, 8; *ta-na-as-sà*†*-ḫu*, XI, 5; and *sí* (*na-sí-iḫ*, VII, 35, 40, VIII, 9).

"Northern" are *ṭú* (*pu-ṭú*⟨*-ur*⟩, VII, 10 and *passim*; *ta-aq-ši-ṭú*, XIII, 10); *Ca-ya* (*ka-ya-ma-ni*, XI, 19, XIV, 8, cf. XII, 8, 15); and probably *ṭúr* (*pu-ṭúr*, IX, 45, X, 23, XII, 6).

local tongue [Amiet 1979: 202], it is not immediately clear which conclusions should be drawn from this observation.

Groups 6 and 5: Goetze's "Northern" Groups

Neither *tammar*, "you see", nor UL.GAR appear in texts from Goetze's "southern" groups apart from marginal use of the former term in YBC 4662 (to which we shall return)[377] but regularly in his group VI. If we take the Eshnunna texts and group 6 to represent a "northern tradition",[378] the Susa texts clearly belong in the same context; instead of "northern", the use of *tammar* is thus a characteristic of the "periphery" (with respect to the Ur III core area).

Goetze counts the following texts to group 6: BM 85194 (excerpts above, pp. 175 and 272); BM 85196 (excerpt above, p. 275); BM 85200+VAT 6599 (above, p. 137); BM 85210; VAT 6597; VAT 6598 (excerpt above, p. 268); and MLC 1354. Of these, the first four (each of which contains many problems) were already grouped by Neugebauer, who also thought them close to the series texts. They are indeed very close to each other on a number of points which Neugebauer did not mention explicitly but probably used as the basis for his general judgment:

Statements mostly start by identifying the object; BM 85194 and BM 85196 sometimes also give a diagram, and BM 85194 and BM 85210 sometimes (thrice each) start *šumma*. Questions are made explicit, as a rule by means of the interrogative phrase en.nam (logogram for *mīnûm*, "what"), in one problem in BM 85194 and one in BM 85196 by *kī maṣi*. Prescriptions open with the formula za.e, "you", and close *nēpešum*, "the procedure" –

377 Both also occur in MLC 1950, which Goetze connects to group 3 for unstated non-orthographic reasons. Probably he has been inspired by the presence of the introductory formula of the prescription, za.e kìd.ta[.zu.dè], "you, by your proceeding"; however, all "southern" occurrences of this formula (including all other group-3 occurrences) have da instead of ta, which seems to be a characteristic periphery spelling – an instrumental suffix translating *ina* from the corresponding Akkadian phrase; the spelling of the core area, on the other hand, could be a phonetic variation, but might also represent a confusion of cases ("with" instead of "by"); cf. [SLa, § 204].

378 Since Akkadian mathematics can no longer be taken to have arisen in the South (cf. note 358), Goetze's characterization of this group as "based on southern originals, but written and modernized in the North" is no longer credible. The simultaneous use, e.g., of the "northern" *sa* and the "southern" *sà* will therefore have to be explained in a different way. Since supposedly "northern" and "southern" features may coexist within the same line (e.g., BM 85194, obv. II 39, 43), a mixed origin of the material of these texts (many of which *are* very mixed anthology texts) is probably no adequate explanation. As we shall see (p. 332), the texts appear to come from Sippar.

occasionally in BM 85194 and BM 85196, once in BM 85200+VAT 6599 *kīam nēpešum*, "thus the procedure". Shift of section are indicated nig̃ín.na (probably meant as *târum*, "to turn back"), and results are marked *tammar*. All four texts end by indicating the number of sections (*kibsum*, "steps" or "computations") contained, BM 85194 and BM 85196 referring to this number as "sum-total" (šu.nig̃ín).

All four texts use UL.GAR for "accumulation", and ba.zi for removal (BM 85196 has a couple of syllabic forms of *nasāḫum*). *ḫarāṣum*, "to cut off", turns up in BM 85196 #18, but as a physical, not a mathematical operation ("cutting off" a bit of a silver coil). The "natural half" is written ½. íb.si$_8$ is a verb, "to be equalside", in two as well as three dimensions.[379] When divisions by irregular numbers occur (BM 85200+VAT 6599; BM 85210), the format of the question coincides with that of the Susa texts (group 8A), apart from orthography and the use of the logogram sum, "to give", in the former text.

The same features recur in VAT 6597 and VAT 6598.[380] confirming that these two tablets belong together with the four tablets grouped by Neugebauer in what we might call group 6A.[381] MLC 1354, on the other

---

379 Three dimensions only occur in BM 85200+VAT 6599, where íb.si$_8$ produces the side of a cube; the side n of a rectangular prism $n \cdot n \cdot (n+1)$; and the three sides of a general rectangular prism.

An apparent exception to the verbal character of íb.si$_8$ is found in BM 85210, rev. II 15, 40 *ta-mar* 40 íb.si$_8$ *im-ta-ḫar i-na* 40 íb.si$_8$ 20 *bi-ri-⟨tim⟩* ba.zi. Here, however, the term has nothing to do with (the extraction of) a "geometric square root"; instead it designates the side of a square to be constructed, serving as a logogram for the "confrontation"; the passage should thus be translated "40 you see, 40 (as) confrontation confronts itself. From 40, the confrontation, 20, the dividing line(?) tear out". Similarly in rev. I 6*f*, 1 bal íb.si$_8$ *šu-tam-ḫir*, "1 the conversion (as) confrontation make confront itself". We have encountered the same usage in the geometrical catalogue BM 15285 (p. 60), the series texts YBC 4714 (p. 111), and in the procedure text YBC 6504 (p. 174); it is also found in the procedure text Str 363 (two-square problems similar to BM 13901 #14). In both of these procedure texts as well as in the very same problem of BM 85210, íb.si$_8$ in square-root function is also present as a verb.

380 With the variation that VAT 6597 #2 (a division between "brothers") uses the interrogative phrase *kiyā*, "how much each" (cf. note 369).

In VAT 6597, the final segment is destroyed; whether it included a counting of problems is thus unclear. The fragment which Christopher Walker has added to VAT 6598 shows that its colophon is different and includes a name which is likely to be from Sippar [Robson 1999: 240 n. 26].

381 BM 85194, BM 85196, BM 85200+VAT 6599 and BM 85210 (for short: the BM-texts) still constitute one subgroup and VAT 6597 and VAT 6598 ("the Berlin texts") another. Both Berlin texts carry the phrase *rēška likīl*, "may your head hold", (initial magnitude) while it is found in none of the BM texts (it also occurs in groups 7A, 7B, and 8A, and in TMS XXVI). Apart from the occasional use of *šumma*, "if" as an opening phrase (including the opening of a proof in BM 85196, obv. II 12), "logical operators" are rare in the BM texts – *inūma*, "as", occurs

hand, differs on most of them. Its statement tells the object, but ends with the question *mi-nu*, not en.nam. The prescription starts with the full phrase za.e kìd.ta.zu.dè; since the final part of the text is destroyed we do not know whether there was a closing formula; no sections are marked, and results are mostly stated within the structure *-ma ... tammar*. The "natural half" is *bāmtum*, not $\frac{1}{2}$. Neugebauer and Sachs reconstruct a lacuna as g̃ar.g̃ar-*ma*, but UL.GAR-*ma* seems equally possible. Squaring is *šutamḫurum*, used in only half of the 6A texts.[382]

On these accounts (and in orthography, as far as the brevity of the texts and the use of logograms permit analysis), MLC 1354 is closer to MLC 1950 (cf. note 377) than to group 6A. With some caution we may put these two together as group 6B.[383]

For use in the following we may take note of a few further terminological peculiarities of group 6A.

- tab is used regularly as a logogram for *eṣēpum*, "to repeat", in phrases of the type *X ana* 2 tab.ba, "*X* to 2 repeat". But a couple of times BM 85194 combines it with a.rá, *X* a.rá 3 tab.ba, "*X* in 3 steps repeat" (obv. II 44, 50).[384]
- When speaking about squares, BM 85194, BM 85196, BM 85200+ VAT 6599 and BM 85210 all employ the Gt-form *imtaḫḫar*, as *X imtaḫḫar*, "*X* stands against itself", *X* ta.àm *imtaḫḫar*, "*X*, each, stands against itself", or *X* íb.si$_8$ *imtaḫḫar*, "*X* (as) confrontation stands against itself" (cf. note 379).

---

twice in BM 85194. On their part, both Berlin texts, much shorter though they are, use *aššum*, "since".

Many features of the terminology and phrasing shows the catalogue text BM 80209 (above, p. 304) to be a member of group 6A. Its use of *šumma* suggests that it belongs together with the BM-texts.

382 BM 85194 and BM 85196 use NIGIN for squaring; BM 85200+VAT 6599, BM 85210, and VAT 6598 use *šutamḫurum* for squaring but ì.gu$_7$(.gu$_7$) when referring to it in relative clauses. Outside group 6A *šutamḫurum* occurs in one text from group 7A (IM 54478) and in several from group 8A. All these are from the periphery; the only occurrence in a text from the "core" is YBC 6504 (group 1). In general, *šutamḫurum* is thus a periphery but not specifically a group 6 term.

383 A small cluster of four text (VAT 672, VAT 6505, and the palaeographically similar fragments VAT 6469 and VAT 6546) belong together as a group. They have much in common with group 6A, but they also differ from 6A as well as 6B, for which reason it may be adequate to classify them as 6C. In the present context there is no need to analyze them further, nor to argue that they actually form a group.

384 In both cases *X* is a circle diameter, and the result thus the corresponding perimeter. The meaning is therefore likely to be the usual concrete repetition; that it is "gone" in three steps (a.rá) points to the use of *alākum* in the Susa corpus (also reflected in the early Old Babylonian texts from Ur and thus no Susa specialty, see below, p. 342). The phrase itself points to the series texts.

– As in groups 3 and 4, g̃ar.ra is used as an imperative ("posit!"); in the present group it also serves as a precative, "may you posit".

As with the groups 7A and 8A, the uniformity of group 6A is evidence of an effort to create and observe a particular canon. There is little doubt that all 6A-texts come from the same school – nor that this school was located in Sippar, since unpublished tablets excavated on this site share the characteristic vocabulary of the group (Friberg, personal communication) (to which comes the colophon of VAT 6598). MLC 1354 and MLC 1950 are clearly *not* from this school, but sufficiently close to the 6A-texts to be a likely representative of a local style – which agrees well with the fact that MLC 1354 was bought in Sippar [MCT, 150 n. 361].

Goetze's group 5 consists of only 3 texts: YBC 6967 (above, p. 55); MLC 1842 (heavily damaged); and YBC 10522 (a fragment excerpted from the middle of the prescription of a longer procedure text). Their common origin is confirmed by terminological parameters: all use *elûm* for results (mostly in the form *illiakkum*, "comes up for you"); all use *šutākūlum* (YBC 6967 for rectangularization, YBC 10522 for squaring, MLC 1842 undecidable). Both YBC 6967 and MLC 1842 introduce the statement implicitly, by giving the parameters; YBC 6967 opens the prescription with *atta*, while MLC 1842 has *at-ta i-na e-pé-š[i-ka]*. YBC 10522 marks shift of section *tu-úr-ma*, as does probably MLC 1842.

The group is thus clearly related in style to Haddad 104, and in general to that part of the Eshnunna corpus which points to the Ur III tradition as it had been received and transformed in the periphery. A further point of contact with Haddad 104 is the explicit statement in MLC 1842 that a number is to be "inscribed" (*lapātum*) on a calculation tablet.[385] YBC 6967 demonstrates affinity with Db_2-146 by "taking" (*leqûm*) the "equalside" (a lacuna prevents us from knowing whether ba.si_8 or íb.si_8, but in any case a noun) and by "laying down" (*nadûm*) the "counterpart" (*meḫrum*).

But there are differences too, at least in YBC 6967: *ḫarāṣum* is replaced by *nasāḫum*, *takīltum* replaces the relative clause *ša tuštakīlu*. The "natural half" is *bāmtum*, found in only one Eshnunna text (as ba, in IM 55357; Db_2-146 uses the unique *muttatum*). The former two features correspond to the Susa group 8A, the reference to the *bāmtum* does not.

Whether the three texts come from the same school cannot be decided, for this purpose the material is too meagre; but they clearly represent another style than group 6.

385 See [Robson 1999: 30 n.27], correcting [Høyrup 1990: 58 n.83].

Groups 4 and 3: Goetze's "Uruk Groups"

Goetze says explicitly that his groups 3 and 4 cannot be distinguished as far as linguistics is concerned; instead, his separation criterion is terminological (as it turns out, vocabulary rather than terminology, see note 356). The two groups do indeed differ strongly on this account.

Group 4 as constructed by Goetze contains the following tablets: VAT 8389 (excerpt p. 77), VAT 8390 (excerpt p. 61), VAT 8391 (excerpt p. 82), VAT 8512 (p. 234) , VAT 8520, VAT 8522, VAT 8523, VAT 8528, YBC 4186, YBC 6295 (p. 65), YBC 8588, YBC 8600, YBC 8633 (p. 254); VAT 8521 was affiliated to the group for non-linguistic reasons.

As Goetze points out, the phrases *inaddinam*, "it gives me", and *ittaddikkum*, "it gives to you", are used in many of these texts. He does not discuss the contexts in which the terms occur, which might have made him discover that the use in VAT 8521 is extra-mathematical and thus no strong argument for the suggested affiliation.[386]

Even in YBC 6295, the use of *nadānum*, "to give", is not as in the others (and the grammatical forms are not the same). "Since (for) 3°22′30″ the equilateral (the table) has not given to you" (cf. note 91), the "equilateral" is found by the *makṣarum* ("bundling") method from "7′30″ (for) which the equilateral it gives you".

As in groups 6 and 8, the main use of *nadānum* in group 4 is in connection with divisions by an irregular divisor. In YBC 8588 and YBC 8633, however, we find a different use of the term (*it-ta-di-kum* and *i-na-di-ku*, "it gives to you" and "it gives you", respectively, the latter twice), namely, for the outcome of a calculation. In both cases, as already in IM 54464 (group 7A, see note 363), the calculation in question is a "raising" (íl, *našûm*).

Except in these three passages, results are only (if at all) marked as such by a preceding enclitic *-ma* (which of course also serves in the general meaning "and then") – in strong contrast to all the groups so far examined.

As in group 6A, g̃ar.ra is used logographically for the imperative of *šakānum*, "to posit" (*šukun*). The precative *luškun*, on the other hand, is

386 In VAT 8521, indeed, *nadānum* is not used as a mathematical term but about the interest "given" by a capital; the grammatical forms that occur, moreover, all happen to be different from those that turn up in texts from the group when the verb functions as a mathematical term. Since *nadānum* is the standard term in connection with interest, the connection is as tenuous as could be. To make things even worse, the only text certainly belonging to the group and dealing with interest problems – *viz.*, VAT 8528 – does so from the perspective of the creditor and uses *nadānum* and the logogram sum about the *borrowing* of the capital, while interest is "taken" (*leqûm*).

written syllabically, and so is the subjunctive *taškunu*. This corresponds to a general tendency of this group to render "standard forms" logographically but oblique forms in phonetic writing: the imperative of *kamārum*, "to accumulate", is g̃ar.g̃ar, while the corresponding sum is a syllabic *kumurrûm*; most texts also write the nominative *mīnûm*, "what", logographically as en.nam, while giving the accusative *mīnâm* syllabically. The exceptions to the latter rule are VAT 8512, YBC 8600, and YBC 8633, which write *mīnûm* syllabically and which are also the only ones to start the prescription with an introductory formula (*at-ta*). VAT 8521, on the other hand, agrees to the full; we may conclude that Goetze was probably right in including it although his only explicit argument for doing so was vulnerable.

Statements mostly introduce the object implicitly, through specification of the parameters; in some cases, the object is stated explicitly (in VAT 8523 in the first problem, after which the others open *šumma*, "if (however)", as in Haddad 104 and elsewhere; even VAT 8391 uses *šumma* in this way. The interrogative phrase is mostly en.nam/*mīnûm*, but a few cases of *kī maṣi*, "corresponding to what", and one of *kiyā*, "how much each" (VAT 8522 #2, another "brother" problem) occur. Except in the three cases of *at-ta* that were just mentioned, no formula introduces the prescription; closing formulae are totally absent. Shifts of section are not marked. The phrase *rēška likīl* occurs in exactly half of the 14 tablets.

The most common logical operator is *šumma*; apart from the use in the beginning of statements considered as variants, its main function (VAT 8389, VAT 8390, VAT 8391, VAT 8520, VAT 8521) is to open the proof; in VAT 8512, VAT 8522, and YBC 8588 it serves within the prescription to open a new line of reasoning after the establishment of a preliminary result. *aššum* serves as a reference to the statement (VAT 8390), or to the given situation (VAT 8523, YBC 6295).

Removal is *nasāḫum*;[387] in VAT 8389, VAT 8390, VAT 8391, VAT 8512, VAT 8520, VAT 8523, and YBC 8633, the process is said to "leave" (*ezēbum*) the remainder – a word which is used nowhere else in this function.[388] *ḫepûm*, "breaking", is always "into two", and the "natural half" is never mentioned. *šutakūlum* is used for rectangularization, and no other term for squaring or rectangularization occurs; YBC 4186, however, uses the phrase 10 nindan *imtaḫḫar* to state that the base of a prismatic cistern is a square with side 10 nindan, and must have contained a term for squaring in a lacuna in line 10. In the only places where the procedure allows its occurrence (VAT 8512, VAT 8520), *takīltum*, "the made hold", is used instead of the relative clause *ša tuštakīlu*, "which you have made hold".[389]

---

387 A possible occurrence of ba.zi in VAT 8522, obv. II 1c is certainly no subtraction.

388 BM 85196 and VAT 6598, both from group 6, use the Sumerogram íb.tag$_4$. So does the Eshnunna text IM 55357 (group 7B) and a number of series texts (the latter often as a noun, designating the remainder).

389 Neugebauer's restoration of two damaged passages in VAT 8389, obv. II 4 as

íb.si$_8$ is a noun, "the equalside". So is ba.si, "the equilateral" when it is used;[390] the unorthographic writing should be noticed.

Summing up we may conclude that the group is coherent, and that only the agreement between the presence of *atta* and the syllabic writing of *mīnûm* allows us to single out a possible subgroup 4B (VAT 8512, YBC 8600, and YBC 8633) from the main group 4A, consisting of VAT 8389, VAT 8390, VAT 8391, VAT 8520, VAT 8521, VAT 8522, VAT 8523, VAT 8528, YBC 4186, YBC 6295, and YBC 8588. In any case, a common strict canon is observed.

The following texts are included by Goetze in his group 3: Str 362, Str 366, Str 368, VAT 7530, VAT 7531, VAT 7535, VAT 7620, YBC 4608; to these he adds for non-linguistic reasons: Str 363, Str 364, Str 367 (above, p. 239), VAT 7532 (p. 209), VAT 7621 – and finally MLC 1950, which, however, differs from the shared norms of the group on all significant accounts, and which I have therefore moved to group 6B (see note 377 and p. 331). The Strasbourg texts, VAT 7532, and VAT 7535 were already grouped by Neugebauer.

Some of the texts contain only statements (VAT 7530, VAT 7531, VAT 7621) or only statements and answers (Str 364). Leaving these aside for a moment, all procedure texts belonging to the group turn out to share a number of characteristic features.

Firstly, the prescription always opens za.e kìd.da.zu.dè; no shifts of section are marked within the prescriptions,[391] and no closing formulae are present.

Secondly, results are followed in all texts by the phrase ^{in}sum, "it gives"; twice in Str 363 and twice in YBC 4608 we find instead the syllabic *i-na-di-nam*; three of the four syllabic cases concern divisions by irregular numbers.

Questions are always made explicit, mostly with the phrase en.nam, "what", sometimes in Str 362 with *kī masi*, once (in a "brother" problem) in YBC 4608 *kiyā*. In Str 368 and VAT 7535, it is also stated that nu.zu, "I do not know" the entity asked for, while YBC 4608 has the corresponding syllabic *ú-ul i-de*.

---

ta-ki-il-tam is better disregarded, cf. note 374.

390 Namely, about the cubic "equilateral" in VAT 8521 and YBC 6295; and in the expression ba.si 1-lal, "equilateral, 1 diminished", about the side n of a prism $n{\times}n{\times}(n{-}1)$ in VAT 8521. VAT 8528 uses íb.si$_8$ alternating with ba.si in an even more generalized sense (the number of periods in a sequence of successive doublings, mathematically speaking the "dyadic logarithm").

391 VAT 7532 and VAT 7535 ("broken-reed" problems) seem to indicate shifts of section in the statement by the word *a-tu-úr*, as done in TMS XII (see p. 327). But since they separate the measurement of the length and of the width they may also be interpreted concretely, as "I turned [around the corner]".

Other features are shared by all texts whose themes allow them to turn up:

- "Breaking" is never "into two" but a process which brings forth a single natural half – mostly written $\frac{1}{2}$ but *bāmtum* in YBC 4608.
- Rectangularization is absent, and squaring is always $\text{du}_7.\text{du}_7$ (read ZUR.ZUR in [MKT]), referred to in Str 363 with a subjunctive *ša tuštakīlu* and in Str 368 as *ša* $\text{du}_7.\text{du}_7$. The *takīltum*, the possible alternative to these relative clauses, does not occur; but VAT 7532 and VAT 7535 use a different relative clause, *ša tēzib*, "which you have left", referring thus not to the half of the broken entity which was moved around in order to form the other side of a square but to the one that was left in place when the other half was broken off.[392]
- íb.si_8 is a verb: in Str 363, Str 366, VAT 7532, and VAT 7535 this is revealed by the full clause *A*.e *B* íb.si_8 (in VAT 7535, *B* is en.nam), in Str 368 by word order.

Division by irregular numbers is treated with some variation around the pattern *mi-nam a-na A* ḫe.ĝar *ša B* ^{in}sum *Q* ḫe.ĝar, "What may I posit to *A* which gives me *B*? You may posit *Q*" (VAT 7532, rev. 4–5). Logographic equivalences apart, the formula coincides with the first part of the formula from group 4, while the second part of the latter – *Q a-na B* íl *A it-ta-di-kum*, "*Q* to *B* raise, *A* it gives to you" – is always omitted in group 3. Two texts (VAT 8390, YBC 4608) employ the rare locution a.šà *abni*, "a [rectangular] surface I have built".[393]

Most texts from the group make heavy or very heavy use of logographic writing, and often provide the verbal roots with a Sumerian grammatical affix; prepositions, however, always remain syllabic and are never replaced by the Sumerian case suffixes, which allows us to conclude that the language is indubitably Akkadian and neither Sumerian nor some attempted ideographic supralinguistic symbolism; this also follows from the occasional syllabic writing of oblique forms of terms which are elsewhere represented by logograms, and by the phonetic complements on ^{in}sum (*inaddinam*) and ^{e}tab (*ēṣip*, *eṣip* – the latter form is always represented by tab.ba in group 4).

The only logical operator to be found is *aššum*, which introduces an argument by "single false position" in VAT 7532 and VAT 7535 (which are also the only texts that possibly mark a shift of section, *viz.*, in the statements), "since $\frac{1}{6}$ of the original reed was broken off, inscribe 6, let 1 go away, ...".[394] The phrase *rēška likīl* is found only twice, in Str 362 and

392 The same idea is found in the pseudo-Heronian *Geometrica*, ms S, 24.3 (square area plus perimeter equal to 896 [ed. Heiberg 1912: 418]), a problem with evident roots in the surveying tradition to which we shall return (p. 368).

393 It may be significant that the "equalside", when present, is also a verb in the other procedure texts where surfaces are "built", namely, TMS XVII and AO 8862. The last occurrences of surface-building are in the Tell Ḥarmal Compendium and in the catalogue text TMS V, from which the "equalside" is absent.

In IM 52301, where a linear extension is "built", *basûm* is a noun, cf. note 243.

394 *aššum* also occurs twice in YBC 4608, but in the construction *aš-šu X*

YBC 4608.

In spite of certain differences between these procedure texts there is no doubt that they belong together not only for linguistic reasons but also because of their mathematical style, nor that the texts which Goetze included for non-linguistic reasons should be included in the group (apart from MLC 1950). Since Goetze's linguistic criteria turn out to be so certain when membership of the group can be cross-checked with the terminology, we may safely accept even VAT 7530 and VAT 7531. Str 364 and VAT 7621, admitted by Goetze for non-linguistic reasons, are so close in terminology (including the use of Sumerograms which Goetze did not consider) to Str 367 and VAT 7531, respectively, that there is no reason to doubt their legitimate affiliation; Str 364 also shares nu.zu, "I do not know" with Str 368 and VAT 7535 (Str 362 has a corresponding syllabic *ú-ul i-de*), while its use of *ezēbum*, "to leave" (rev. 9) is reminiscent of the function of this term in VAT 7532 and VAT 7535.

The indubitable discrepancies between the texts do not allow us to isolate one or more subgroups; texts which should belong together according to one criterion are always separated by others. The global pattern being much more coherent than, for instance, that of group 7B, the whole group will indubitably represent another search for a canon – in certain respects perhaps in deliberate contrast to the canon of group 4, if both groups are really located in Uruk (but Uruk was so large a city that even outspoken contrasts may have resulted from the internal traditions of schools that coexisted without having any significant communication).

Group 1: The "Larsa" Group

Goetze's group 1, "certainly to be localized in the South, in all probability Larsa", comprises the tablets AO 6770, AO 8862, YBC 4675, YBC 5022, YBC 6504, YBC 7243, YBC 7997, YBC 9852, YBC 9874, and, conjecturally, YBC 9856 and Plimpton 322. Of these, YBC 5022 and YBC 7243 are tables of technical igi.gub constants, and even Plimpton 322 is a kind of table text. In the present connection the only observations to be made on them is that íb.si$_8$ occurs in the latter two in orthodox spelling; that *nasāḫum* turns up in one of the headings of Plimpton 322; but that the orthography of Plimptom 322 points to group 6 with at least as high plausibility as group 1, and that its mathematical substance points to the periphery rather than toward the core (see p. 386). YBC 9852 is a copy of some lines from YBC 4675, deviating from this model only in a few spellings and by two omissions. In the following, we shall therefore concentrate on the texts AO 6770 (excerpt above, p. 179), AO 8862 (#1–4 p. 162), YBC 4675 (p. 244), YBC 6504 (p. 174), YBC 7997, YBC 9856, and YBC 9874.

---

a-ma-ri-i-ka, "in order that you see [i.e., find] *X*", and not as a logical operator.

Beyond the shared orthographic characteristics noticed by Goetze, there is little that keeps the texts in question together as a coherent group. We may notice the absence of logical operators within the prescriptions, and that these never close with a formula; *tammar* is never used with results. Beyond this, the texts have to be discussed one for one.

AO 8862 has always been regarded as one of the earliest Old Babylonian mathematical texts. First come the four "algebraic" problems on rectangular fields. Three problems on brick carrying follow, of which #5–6 deal with proportional sharing of wages, whereas #7 is of the second degree and structurally identical with a rectangle problem that is listed in the Tell Ḥarmal Compendium and solved in the unpublished Eshnunna texts IM 43993 and IM 121613 (group 7B): given $A+l+w$ and $w:l$. Only #1–3 include prescriptions.

#1–4 start by stating the object. The statements are subdivided into sections by means of the words *a-sà-ḫ̮i-ir*, "I turned around", and *a-tu-úr*, "I turned back" (in #1–3 both); the context (and the contrast with the subdivision of the procedures by the unique *lā watar*,"do not go beyond") suggests that they are meant concretely, as "walking around" and "returning to the starting point". The question is *mīnûm*.

The brick-carrying problems (whose topic is familiar from Ur III accounting) follow a different format. They start by stating the work norm and corresponding wage for one man; then follows the word *inanna*, "now", introducing the description of the actual organization of the work;[395] the interrogative phrase is *kī maṣi*.

The prescriptions of #1–3 open with the phrase *at-ta i-na e-pe-ši-i-ka*. As regards the use of tense and person, #1 and #3 follow the "standard format" (see p. 32). Interesting aberrations are found in #2; $A+\frac{1}{2}l+\frac{1}{3}w$ is referred to (II 10) as *ki-im-ra-ti-i-a*, "my things accumulated" (actually a plural, as we remember), which might of course mean that the person who instructs is thought of as identical with the one who stated the problem – if only $l+w$, equally referring to the statement, had not been referred to as "your things accumulated" (II 16). Similarly, "you append" 25′ to 3°25′ (II 27) in order to find the length of a modified rectangle, while "I tear out" (II 32) in order to find its width. No system can be found in the anomalies; they appear to have resulted from the author's failure (or failing intention) to adapt completely to a standard format, which makes the style of actual oral instruction surface.

---

395 Separation of standard information from description of the actual situation is the general function of the term. This standard information may be an igi.gub factor (AO 8862 #5–7, YBC 4673 #2–3, YBC 10772 – all dealing with brick carrying and having the same factor); it may be the rent to be paid per bùr for different fields (VAT 8389 and VAT 8391, *passim*); or the dimensions of an old dike, where the specific information concerns the reparation to be performed (YBC 4673 #14–15).

The only apparent exception is YBC 4669 #B6, where the word has been moved so as to be coupled with the question.

The way rectangularizations are spoken of is also instructive (see p. 163). It may be done by the habitual *šutakūlum*, "to make hold", taken to imply tacitly the numerical computation of the resulting area; but this computation may also be reported separately (by a.rá, "steps of"); and the construction process itself may be referred to instead as "inscription twice" or "breaking". This multiple variation does not fit the editing of an existing written text; it suggests, instead, a scribe who brings into writing the methods of non-scholastic practitioners, in part using their terminology (*šutakūlum*, we may suppose) but then combining this with the terminology for sexagesimal computation, in part describing their procedures in his own words.

This scenario is corroborated by other peculiarities of the text. Firstly, there is the very structure of the basic #1 – in symbolic translation, $A+(l-w) = 3`3$, $l+w = 27$. Addition transforms this into a problem $R = 3`3+27 = 3`30$, $l+W = 27+2 = 29$, where R is the area of a rectangle with sides l and $W = w+2$. No other Babylonian text contains a problem of this structure, nor *a fortiori* the characteristic way to solve it; both the problem type (actually the complementary type $A+(l+w) = \alpha$, $l-w = \beta$) and the trick, on the contrary, turn up again in medieval Arabic sources (for whose pertinence we shall argue below) together with the problem of "the four sides and the square area", whence it was adopted by late medieval Italian and Latin writers.[396] The problem, we may conclude, was certainly present in the lay source tradition, but appears not to have been adopted with much success into the scribe school – probably because the trick, elegant though it be, did not lend itself directly to generalization (exactly as the trick of BM 13901 #23, "the four fronts and the surface"). In a different formulation: the generalized form of the problem $A+\alpha l+\beta w = P$, $\gamma l+\delta w = Q$ asked for different methods – methods demonstrated, for instance, in TMS IX and consisting in the transformation into a rectangular problem in $\Lambda = \gamma\cdot(l+\beta)$ and $\Omega = \delta\cdot(w+\alpha)$. Elegant tricks fit a riddle, but easily turn out to be dead ends in a systematic enquiry.

We may further observe that lines are "appended" to surfaces – "accumulation" is reserved for other purposes. The lines of the geometrical part are thus broad lines, as in the Tell Harmal compendium and TMS VI. The author of the text has felt no need for "projections", "bases", or similar "critical" devices.

Removal is represented by both *nasāḫum*, "to tear out", and *ḫarāṣum*, "to cut off", but with a tendency to distinguish according to everyday connotations. As we have seen, *ḫarāṣum* was important in the Eshnunna corpus, and

---

396 Jacopo da Firenze's *Tractatus algorismi* from 1307, ms. Vat. Lat. 4826, fol 40v–41r, has the original version without swapping of "+" and "–", formulated as a pure-number problem. Jean de Murs, in prop. V.2.26 of his mid-fourteenth–century *De arte mensurandi* [ed. Busard 1998: 187*f*] reduces the rectangle problem $A+(l+w) = 62$, $l-w = 2$ to the problem of four square sides and area, explaining that the sides are broad lines and making a perfect drawing of the situation. Since he does not treat the square problem in question, the whole piece must be borrowed.

nasāḫum apparently eliminated intentionally from group 7A. This purism may perhaps be a school phenomenon rather than a characteristic of the lay practitioners' parlance – an oral culture is more likely to possess a picturesque language rich in concrete connotations than to be purist.[397] In any case, *ḫarāṣum* will surely have been part of the lay parlance, whereas it was soon eliminated from the school tradition (even from the texts from the periphery which on other accounts are clearly related to the early Eshnunna group); on this account, too, the geometrical part of AO 8862 is thus close to the original inspiration – probably much closer than the texts from group 7A.

Two other singularities suggest that the author of the text was not only working at the intersection between a lay tradition to be exploited and a school tradition *in statu nascendi* – that "interface between the oral and the written" which Jack Goody [1987] took as a book title – but also trying to explore the possibilities of this open situation: firstly, sums by accumulation appear as a plural *kimrātum*, "the things accumulated", used explicitly in no other mathematical text[398]. Moreover, the numerical values – which in the perspective of modern mathematics and mathematics teaching are arbitrary and not worth caring about, but which both in the lay tradition (according to whatever testimonies we possess about it) and in the Old Babylonian school were highly standardized – are unusual and even unstable: in #1, the sides are 15 and 12, in #2 they are 4 and 3, in #3 they are 1` and 40.

The way results are indicated has turned out to be an important parameter in the preceding pages. In the present text, as in group 4, a preceding enclitic *-ma* is normally all there is. Twice, however, a result is followed by *inaddikkum*, "it gives you": In II 15, igi 6 g̃ál "gives you" 10′; and in II 20, "breaking" the *bāmtum* from 6°50′ "gives you" 3°25′.

Normally, it will be remembered, *nadānum* when found in texts that for the rest have a different or no marking of results, is connected with the "raising" multiplication. To find the igi may conceivably be seen as belonging to the same domain as "raising", namely, sexagesimal calculation – though with the difference that the multiplications make use of tables but rarely find the result stated directly, while igi 6 is certainly in the table. But "breaking", as distinct from multiplication with 30′, certainly does not belong to the field. Even on this account the present text thus follows a pattern of its own – which would fit a transition between traditions.

In view of the scarce attention which the author has dedicated to terminological uniformity in the geometry part, the striking difference between

397 On solemn occasions, however, the discourse of an oral culture tends to be formulaic – a feature which, when emulated by *literati* (romanticist and others), is easily transformed into purism.

398 The tentative reconstruction of VAT 8512, obv. 6 [MKT I, 341] should indeed be [*ta-w*]*i-ra-tum*, cf. [von Soden 1939: 148]. As we have seen, however, the idea that the accumulation is a plural underlies YBC 4675 (also group 1; see p. 249); it can also be read from the twin text YBC 4662, YBC 4663 – below, p. 347.

the styles of #1–4 and #5–7 is evidence, not only of ultimately derivation from different source tradition but also that these sources were available in still independent form. The distribution of interrogative phrases suggests that the *kī maṣi* would be the standard of the scribal tradition in Larsa (if this is where the text was produced). #7, in which a familiar geometrical riddle problem is translated into a brick-carrying problem, appears to reflect an effort to transfer the techniques of the surveyors' riddles to the domain of traditional scribal computation.

For comparison with other texts from the group, four further observations on the terminology of AO 8862 may be made:

- The remainder after a removal (*ḫarāṣum* as well as *nasāḫum*) is spoken of as *šapiltum*.
- The "natural half" is *bāmtum*.
- *wabālum*, "to bring", is used as in Db_2-146, namely, as a device which makes it possible to "tear out" an entity from another of which it is not part.
- íb.si$_8$ is a verb.

AO 6770 consists of 5 problems of mixed contents; the first, we remember from p. 179, is a rule formulated in general terms and no problem *stricto sensu*; the detailed interpretation of several problems is uncertain, which, however, does not interfere with the present discussion. As the first group of AO 8862, the problems start by presenting the object. #2 and #4 (interest on a loan, and determination of an amount of bitumen by means of an igi.gub coefficient; both traditional scribal computations) ask the question *kī masi*, while #3 and #5 (a riddle on the weight of a stone, almost certainly with a "mock reckoning",[399] and a reed broken in arithmetical series) ask *mīnûm*. #3 and #4 also notify that "I do not know" (respectively *ul īde* and nu.zu) the weight of the stone and the amount of bitumen.[400] In #1 and #2, the prescription opens with the formula *at-ta i-na e-pe-ši-i-ka*. No closing formula is present.

In #1–3, the final result is given "to you" (*i-na-ad-di-ik-kum*, #1) or "to me" (*i-na-ad-di-nam*, #2–3). In #2 (obv. 15), even an intermediate result (after ba.zi) is given "to me"; as in AO 8862 #2, the deviation from the normal "I-you" pattern is noteworthy. Other results are at most marked by a preceding *-ma*.

Removal is represented both by *nasāḫum* (ba.zi in the stone riddle) and by *ḫarāṣum*; the remainder from the latter process is spoken of as *šapiltum*.

YBC 4675, concerned with a (supposedly) bisected trapezoid, was treated on p. 244. The statement opens *šumma* and asks the question *kī maṣi*. The

---

399 As the filling riddle IM 53957, see p. 321. As in #1, though less outspokenly because a particular example is involved, the formulation is in general terms.

400 This usage, we remember, is also found in Str 364 (group 3) and in the broken-reed problems Str 368 and VAT 7535 (group 3) and IM 53965 (group 7A).

prescription carries neither opening nor closing formula. A shift of section is demarcated *ta-as-sà-ḫa-ar*. Results "come up for you" *i-(il-)li-a-(ak-)kum*, except after two "raising" multiplications, where they are "given you", in the nasalized spelling *i-na-an-di-kum*. Removal is *ḫarāṣum*; íb.si$_8$ is a noun and "taken". Thrice, when the *bāmtum* is broken from an accumulation with two components, it is spoken of as *ba-a-ši-na*, "their" natural half – which implies that the accumulation is understood as the plurality of components and not as *one* entity, as also in AO 8862. *šutakūlum* is used (obv. 12) in standard fashion (for squaring), but also in the purely constructive sense of making the opposing sides of a trapezoid "hold each other".

YBC 7997 is a brief text calculating the brick capacity of a cylindrical oven. It starts by stating this object, but has neither explicit question nor formulae of any kind. In several ways it is particularly close to the preceding text: Results "come up" (*i-li-a-am*, "for me", not "you", though other operations are in the conventional second person singular); the final result, coming from a "raising", is "given", without specification of the receiver but with the same nasalized spelling *i-na-an-di-in* as in YBC 4675. But there are also similarities with other texts from the group: As in AO 6770, the prescription attempts to be general and not a mere paradigmatic example (rev. 6, *ma-la i-li-a-am*, "as much as comes up for me" instead of the actual result of the computation).

šutakūlum, "to make hold", appears twice regularly as squaring, after which the resulting areas are "brought" (*wabālum*), in one case to the height, in the other to the factor 5′ (= $^1/_{4\pi}$); both "bringings" thus stand where "raising" would be standard. The resulting two numbers – of which the former is a quasi-volume or quotient volume, *viz.*, a ratio between areas times a height, and the second a circle area – are then taken as the objects of another *šutakūlum*, which is grammatically regular as a rectangularization but is evidently non-standard in its use ("raising" would be the standard choice again, since an operation of proportionality is involved).

YBC 9874 is a short text on the maintenance of a canal. It starts by stating the object, and has neither explicit question nor formulae of any kind. The only observation to be made in the present connection is that the results of the three "raising" operations are "given to you" (*i-na-(ad-)di-ik-ku*). YBC 9856 (included conjecturally by Goetze in the group) contains two problems with answers but no prescriptions, one on work norms and one on proportional sharing (thus traditional scribal computations). Of interest for the present analysis is only that the question is *kī maṣi* (only in #1).

The last text from the group is *YBC 6504* (above, p. 174). While all the others (or at least AO 8862, AO 6770, YBC 4675, and YBC 7997) are mutually connected in one way or the other, this one differs from all of the rest on almost all accounts except Akkadian orthography.

The tablet contains four algebraic problems dealing with the same mutilated rectangle.[401] The object is introduced implicitly, through specification of the parameters; #1–2 leave even the question implicit, whereas #3–4 have a question en.nam. #1–2, on their part, start the prescription with the formula *i-na e-pe-ši-i-ka*, while the prescriptions in #3–4 carry no opening formula. Closing formulae are altogether absent. In contrast to the other tablets from group 1, the present text is predominantly logographic – often in singular ways: results (intermediary as well as final) are followed by the phrase IN.g̃ar, used elsewhere only in the Ur text UET V, 859; "appending" is bi.daḫ (elsewhere occurring only in Str 363, group 3), removal ba.zi (used in groups 4 and 6, in series texts, in IM 55357 (group 7A), and once in AO 6770); for *watārum* serves the truncated logogram SI instead of the normal dirig (=SI.A);[402] the natural half is šu.ri.a, elsewhere the Sumerogram for *mišlum*, the ordinary half; however, *paṭārum*, "to detach", *našûm*, "to raise", and *ḫepûm*, "to break", are syllabic. íb.si$_8$ is a verb when it refers to a (geometric) square root (10.25.e 25 íb.si$_8$). Squarings are *šutamḫurum* in the prescriptions of #1–2 (as in various texts from groups 6A, 7A, and 8A, cf. note 382), while #3–4 have du$_7$.du$_7$ (as most of group 3).

In statements, #1–2 have *X* íb.si$_8$ where #3–4 have *X* du$_7$.du$_7$, both serving as logograms (or at least ideograms) for *mitḫartum* and in the sense of "the square constructed on the side *X*" (as in YBC 4709 and other series texts; cf. also note 379 on a passage in BM 85210). *X* is the expression *mala* uš ugu sag̃ SI, "so much as (that which) the length over the width goes beyond". With the logogram a.na for *mala*, this expression is common in the series texts. Elsewhere this kind of "bracket function" of *mala* is rare; it is found in AO 8862 #1 and #3; in IM 43993 (group 7b, exceptionally with an accumulation and no excess); in VAT 8390 #1 and #2 (group 3); and in BM 13 901 #19 (group 2 according to Goetze, but see below, p. 348). This may not be a mere terminological peculiarity; it rather expresses the fact that problems involving "nested" operations are common in the series texts but rare elsewhere (if not stated sequentially, as in TMS IX # 3 and in TMS XVI). If we add that there are obvious mathematical affinities between YBC 6504 #2, AO 8862 #3, and BM 13901 #19 (all make use of the same standard diagram, the square on the sum of the entities whose difference is spoken of), we may surmise that even YBC 6504 belongs with the rest of group 1 for more than reasons of orthography. If so, however, its position is surely somewhat apart.

---

401 We may notice that the four problems are precisely the basic supra-utilitarian problems on rectangles that are listed twice in YBC 4612 (see note 343) – only with the difference that the area of the rectangle is replaced by that of the mutilated rectangle.

402 Even this is unique, in spite of Bruins's claim [TMS, 54] that zi (=SÍ) in TMS VII, 23, must be a writing error for SI meaning dirig, induced by dictation. zi, "to be torn out", is indeed quite regular, cf. above, p. 181.

Looking back at group 1 as a whole we reach the conclusion that it repeats the characteristics of AO 8862 in larger scale. Few "positive" features connect the texts belonging to the group, but all the more the "negative" characteristic that most deviations from normal usage are concentrated here: that the outcome of an accumulation is regarded repeatedly as a plural; aberrant uses of *šutakūlum*, and conversely the use "breaking" and twofold "inscription" in the normal function of "making hold"; the separate calculation of the surface resulting from a rectangle construction by a.rá; the use of *wabālum*, "to bring", as an all-purpose term; shifts from the second to the first person singular within prescriptions in connection with results "coming up" or being "given". Here we also find attempts to present procedure prescriptions not through paradigmatic examples alone but in general terms – together with IM 52301 #4 (group 7B, see p. 213) the *only* attempts in the whole Old Babylonian record.[403]

Group 1 is not the only place where such deviations from the norm are found. *wabālum* is used in Db_2-146 (group 7a) as in AO 8862, and in the fragment YBC 10522 (group 5) and in CBS 11681 (from Nippur, in mostly syllabic Akkadian) with the function of a "raising"; as we shall see (p. 347), YBC 4662–63 (group 2) handle an accumulation as a composite entity; CBM 12648 (an early Old Babylonian tablet from Nippur in unusually grammatical but seemingly reconstructed Sumerian), uses gu_7[404] not about rectangularization but with triple object, finding a rectangular prismatic volume from length, width, and height. Intrusions of the first person singular in prescriptions are widely spread in connection with division questions but only there.[405]

Precisely this scattered appearance of the peculiar features found densely in group 1 shows that the authors of its texts were not mere bunglers; but they wrote in a situation where it was not clear that the construction of a rectangle should not be spoken of just as "inscription" of its sides, etc. The group as a whole is witness to a period of assimilation of traditions, a phase of creativity which would ultimately give rise to a new tradition governed by rather strict canons – not the same canon everywhere, as illustrated by the difference

---

403 In strong contrast to what the ideals of Greek and present-day mathematics would make us expect, general formulations were thus *not the end point* of the development but *a possibility that was deliberately discarded* in mature Old Babylonian scribe school mathematics, at least from its written expression.

404 As ub.te.gu_7, with a prefix ub.te. which may be composed from /u/ (prospective mark) + /*b*/ (impersonal pronominal element) + /ta/ (ablative prefix) (this analysis is corroborated by UET V, 858, line 7, see note 280). The idea behind the use of the ablative seems to be related to the one which made ba (with the locative element /a/) the prefix for the cubic equilateral, indicating that the process takes place "(out) there" (see note 43): with gu_7 used as a Sumerogram for "making hold", te.gu_7 would mean "make hold, going out (of the plane)".

405 This may, however, have purely grammatical reasons: the precative form ("what may I posit") does not exist in the second but only in the first and the third person singular – see [GAG, §81c].

between the geographically close groups 3 and 4, but still variations on the same pattern and with high local uniformity, as illustrated by the internal homogeneity of each of these groups. Inasfar as one of the traditions to be integrated was that of the lay, semi-oral practitioners, the group as a whole and not only AO 8862 is situated, if not at the "interface between the oral and the written", at least at the interface between semi-oral and literate culture. YBC 6504 may express the attempt to establish new standards – some of them to reappear in group 3, the closest kin (in particular the replacement of the logographic pun ì.gu$_7$.gu$_7$ by the semantically more appropriate du$_7$.du$_7$, if this is the explanation of the choice – cf. p. 23), others unsuccessful and replaced in group 3 by different standards (^{in}sum instead of in.g̃ar, the simple ½ instead of the complex šu.ri.a as logogram for *bāmtum*). As AO 8862 it is engaged in deliberate variation of traditional problem types, but even in this respect (by replacing the simple area by a complex expression) in an idiosyncratic manner.

To some extent, the variations within the group coincide with those of the Eshnunna texts – *nasāḫum*/*ḫarāṣum*, íb.si$_8$ used either as a verb or as a noun designating an entity "coming up" or to be "taken". Even the heterogeneity of the Eshnunna corpus is thus to be explained from an analogous situation, only with the difference that the ten Tell Ḥarmal texts from group 7A express the same attempt to define and stick to a standard as we have found in YBC 6504 and probably in no other group 1 text. The conspicuous differences – not least the total absence of *tammar* from group 1 (in all other respects so eclectic), the different ways to deal with the outcome of a "breaking", and the unmistakeable link between *kī maṣi* and computation in the scribal tradition in group I – exclude that one of the two text groups could be a mere adaptation of texts from the other group (even though mutual inspiration is likely to have influenced the process). Goetze was certainly wrong when maintaining that "Akkadian mathematics [...] originated in the South"; Akkadian school mathematics arose – from synthesis of the adapted Ur III traditions and the lay, probably Akkadophone practitioners' tradition – in parallel processes in the periphery and the core area (and according to all we know about the crystallization of the particular Old Babylonian culture in general, Eshnunna and Larsa are likely foci).

Group 2 – a Non-Group?

As the nucleus of this group, Goetze points to YBC 4662 and YBC 4663, two theme texts on "excavations" (ki.lá) with almost purely logographic statements and fairly syllabic prescriptions, whose "close connection [...] needs hardly any comment" [MCT, 148 n. 354]. To this core he joins YBC 7164, a catalogue of heavily logographic statements and answers on the maintenance of small canals (pa$_5$.sig), omitting – tacitly and for unexplained reasons – not only YBC 4666, of which this latter tablet is a direct continuation, but also

YBC 4657, a similar collection of statements on excavations – actually the statements for YBC 4663, for a missing tablet, and for YBC 4662 (in this order).[406] Finally, since the status of YBC 7164 as a continuation of another tablet reminds him of the series texts, and because the predominantly syllabic procedure text BM 13901 was pointed out by Neugebauer [MKT III, 10] to possess already the systematic order and progression found in the later series texts, even this text is included, though no linguistic clues connect it to the other syllabic texts from the group rather than to groups 3 or 4; as admitted in note 354, "the argument presented may be regarded as circular".

The close connection between YBC 4662, YBC 4663 and, we may add with Neugebauer and Sachs, the statement catalogue YBC 4657, is indeed indubitable. The affiliation of other statement catalogues to the same family – apart from YBC 4666 and YBC 7164 also YBC 4607, YBC 4652, and YBC 5037 – is inherently plausible, not only because of the common style of the statements (an argument which, like coinciding logographic terminology, tends to become circular) but also because of a shared format and similar ductus;[407] most tablets are characterized by an initial group of properly practical problems, after which follow supra-utilitarian algebraic problems (see p. 305). Whether the supposed group is really one therefore hinges on the comparability of BM 13901 with the procedure texts YBC 4662–63. We shall start by describing these latter texts.

The statements start by presenting the object, a ki.lá,[408] and ask the question en.nam, with one appearance of *kī maṣi* in YBC 4662, rev. 2. Prescriptions open with the formula za.e kid$_9$/kìd.da.zu.dè (kid$_9$ in YBC 4663, kìd in YBC 4662); in YBC 4663 they close *ki-a-am né-pe-šu*, in YBC 4662 without any formula. As a rule, results are "given" (*inaddikkum*, varying spellings in both tablets), but in YBC 4662 a solitary *tammar* turns up in rev. 22; in the last problem of the same text (rev. 31–36), *tammar* is used

406 The three procedure tablets correspond to problems #1–8, [#9–18], and #19–28 of YBC 4657. Each sequence constitutes a coherent series of variations on a common basis, while there is a change of theme between #8 and #9 and between #18 and #19; there is also a stylistic break between YBC 4663 and YBC 4662 (see presently); and there is space enough left in the end of YBC 4663 for another problem. All in all it follows that the procedure texts were made first and the statements of the catalogue copied from them.

407 All the tablets in question except YBC 7164 indicate the number of statements contained in larger writing, either at the end or on the edge. YBC 7164, which is also the only member of the group not to insert a line between the single statements, is a companion piece to YBC 4666, and therefore has to go with the rest,

YBC 4612 (referred to in note 343), a catalogue of statements for simple rectangle problems and similar in style, is in a much coarser hand than the others and does not count its problems. Its affiliation with the group cannot be excluded, nor can it however be regarded as reasonably established.

408 At times, this word comes first, at times after the statement of the wage to be paid. But it is always there.

systematically for intermediate results, and *inaddikkum* is reserved for the solution.

As far as mathematical operations are concerned, both tablets employ *ḫarāṣum*, "to cut off", for removal from lines; YBC 4663 employs *nasāḫum*, "to tear out", when areas are involved, while YBC 4662 uses *tabālum*, "to take away". Rectangularization (and a single squaring in YBC 4663, rev. 8) is UR.UR, found nowhere else in the mathematical texts[409], in YBC 4663 alternating with *šutakūlum*. íb.si$_8$ is a noun and "taken", and then when appropriate "inscribed to 2" (*a-na* 2 *lu-pu-ut-ma*). When "detaching" an igi, YBC 4663 alternates between a syllabic *paṭārum* and the logogram du$_8$, while YBC 4662 sticks to du$_8$ alone. None of the tablets obeys the "norm of concreteness", and both use the abbreviated formula "to 1 append, from 1 cut off" which we know from BM 85200+VAT 6599 #24 and a number of Eshnunna texts (see pp. 155, 324, and 346); however, when completing the square, the gnomon area is appended to the complement as in YBC 6967 (group 5), where all relevant texts from groups 6 and 7 append the complement. Both tablets tend to connect *ḫarāṣum* with *ana* instead of the regular *ina* ("cutting off to" instead of "from") and to combine daḫ with a.rá instead of *ana* ("appending step" instead of "to"!). The natural half is written logographically as ½ – in cases where the entity to be broken is an accumulation the whole phrase runs ½ uš *ù* sag̃ *ša* g̃ar.g̃ar^{ru} *ḫepe-ma*, "½ of the length and the width which you have accumulated break" (YBC 4663, rev. 7; YBC 4662 obv. 7). Since other occurrences of *ḫepûm* in the same texts do not always involve a similar description or identification of the entity to be broken (and never any similarly complex description), we are led to a double conclusion: Firstly, the scribes of the two tablets (evidently different scribes, given the systematic divergences on several points) must have worked on the basis of a written source; secondly, this written original must have apprehended the accumulation not as a single entity but somehow as "things accumulated", in the likeness of AO 8862. The rare occurrences of *šutakūlum*, *kī maṣi* and *tammar* suggest that the actual texts result from an attempt to rewrite the contents of the original according to a new standard, in which *šutakūlum* was replaced by UR.UR (perhaps an *ad hoc* construction),[410] *kī maṣi* by en.nam, and *tammar* by *nadānum* in generalized use. The original is also likely to have had the closing formula of YBC 4663, and to have used

---

409 The closest relative is UR.KA, used for squaring in the Kassite text AO 17264. UR itself appears to be semantically close to si$_8$ and *maḫārum* – in the fragment Ist S 428 it is used about the "equalside" of 2 02 02 02 05 05 04, while íb.si$_8$ designates the "equalside" of a factor which is split out, cf. [Friberg 1990: §5.3]. The use for rectangularization, at times with unequal sides, seems somewhat awkward.

410 Such a replacement of the normal verb by an ideogram constructed *ad hoc* and with no precise logographic counterpart might explain that neither the *takīltum* nor the corresponding relative clause nor any other equivalent appears in places where they would be appropriate.

ḫarāṣum – perhaps also for surfaces, since the revised texts diverge on this point. The absence of *tammar* not only from Groups 3–4 but also from group 1 (which otherwise seems to have collected all variant terminologies at hand) shows that the original must have belonged to the periphery type. The non-observance of the "norm of concreteness" points in the same direction.[411]

Even BM 13901 is a theme text, containing "algebraic" problems of the second degree about one or more squares; many of them were presented above. But being a theme text is almost all it has in common with YBC 4662–63. If we leave aside the riddle about "the four fronts and the surface" of #23, statements specify the object only implicitly, and formulate no explicit question. Introductory and closing formulae are equally absent from the prescriptions. Results are at most preceded by an enclitic *-ma* on the preceding operation verb, and as a rule the answer is already involved in the following operation. Statements as well as prescriptions are overwhelmingly syllabic.

Removal is exclusively *nasāḫum*. The outcome of a "breaking" is a syllabic *bāmtum*; when the accumulation of two distinct magnitudes is "broken" (as in #9, above, p. 68), there is no trace in the formulation that its composite nature should be thought of. Rectangularization is *šutakūlum*, and the object of the process is later referred to with the relative clause *ša tuštakīlu*, "which you have made hold". íb.si$_8$ is a verb; when that which "is equalside" is "inscribed", this is done *a-di ši-ni-šu*, "until twice" (a phrase also used in AO 8862), not *a-na* 2, "to 2" as in YBC 4662–63. Unique is the use of a *waṣītum*, "projection", as a rigorous reformulation of the "broad line" (cf. above, p. 292).[412]

The text progresses rather systematically from simple to complex problems, and follows a very homogeneous stylistic canon, quite different however from those pursued in YBC 4662–63 and YBC 6504, and no less from those achieved in groups 3 and 4 and the groups domiciliated in the periphery. It is certainly already fully integrated in a school tradition, well at a distance from that transition between traditions where AO 8862 was located. It is not very likely that the two texts should be equally old (notwithstanding the shared opinion of Neugebauer and Thureau-Dangin). Culturally, at least, BM 13901 is certainly younger, if not with full necessity in terms of chronology.

---

411 Weakly corroborative evidence is provided by the mathematical substance of the texts. The only other text accumulating the base and the volume of a prismatic excavation is BM 85200+VAT 6599 from group 6A.

At variance with what Goetze tries to establish on the basis of unstable spellings, these texts are thus evidence of a diffusion of mathematics from the periphery – probably the north – to the southern core area.

412 The phrase 1 *wa-ṣi-am*, "1 projecting" found in VAT 8391 (see p. 84) and VAT 8528 (obv. 20; both group 4) will probably have the same meaning in a different context.

#23, by being a deliberate archaism (see p. 222*ff*), highlights the distance between the text as a whole and the lay tradition.

Since even spellings do not correspond too well, we may conclude that YBC 4662–63 and BM 13901 do not belong within a common group. The twin text reflects an attempt to organize the material at hand (in the actual case apparently a written text from the periphery) in agreement with a local canon; BM 13901 belongs within an already well developed tradition obeying a wholly different canon – closer to group 4 than to anything else, but still different on many accounts. We may keep the label "group 2" for YBC 4662–63 and associated logographic catalogue texts (certainly YBC 4657, most probably also YBC 4607, YBC 4652, YBC 4666, YBC 5037, and YBC 7164, perhaps YBC 4612). BM 13901 must be considered the only known representative of what might be termed "group ii".

The Series Texts

Neugebauer connected the series texts[413] to the texts of group 6A because of the thematic similarity between the problems of BM 85196 and some of the series texts. The similarity between the excavation texts YBC 4662–63 from group 2 (with the appurtenant catalogue YBC 4657) and the excavation text BM 85200+VAT 6599 from group 6 undermines the argument. As we have just seen, YBC 4662–63 offer evidence that problems did not only travel between schools and regions but were systematically borrowed and adapted to the local canon.

Without including the series texts explicitly in any of his groups, Goetze used them to link BM 13901 with the rest of his group 2, thus presupposing that even they were tied to that group. The argument was risky already in 1945 ("circular", as he admits) and is easily disposed of in the light of texts published since then. It assumes that the systematic order of texts like BM 13901 and YBC 4662–63 and of problem catalogues – with or without a series number – was a local specialty. The existence of the Tell Ḥarmal Compendium and of TMS V–VI (and the group 6 excavation text BM 85200+ VAT 6599, well studied in 1945) shows that this was not the case.

---

413 That is, VAT 7528, VAT 7537, YBC 4668–69, YBC 4673, YBC 4695–98, YBC 4708–15, and A 24194, A 24195.

Friberg [2000: 57] claims it to be an error to count VAT 7528, YBC 4669, YBC 4673, and YBC 4698 among the series texts. It is true that YBC 4669 does not indicate its series number – but this would have been placed in a part of the tablet that has been destroyed (the intact reverse is foreign to the series part); the other three do state it, and YBC 4669 is so close in format to YBC 4673 that they must be presumed to come from the same place, given that the museum numbers indicate that they have been bought at the same moment. Since no arguments are given for the claim, it is not clear on what basis it is made.

In [MKT I, 387*f*], Neugebauer had argued that the series texts would probably come from Kish because of a terminological peculiarity – the supposed use of sig_4 in the sense of "volume" – shared with the catalogue text AO 10822, excavated by Genouillac in 1911 in this place. As pointed out by Neugebauer and Sachs, however, their decipherment of the brick measuring system dissolved the peculiarity into nothing [MCT, 95].

The association with Kish was supported by other arguments – writing and format of the tablets, general terminological agreement of AO 10822 with the series texts VAT 7528, YBC 4669, and YBC 4673, and appearance on the antiques market in the years following upon Genouillac's excavation of Kish. With hindsight, even the format and the terminology turn out to be weak arguments. The lines of the series texts are densely spaced and short, containing some five signs; those of AO 10822 are twice to thrice as long and widely spaced. Terminological similarities do not go beyond what can now be seen to follow automatically from the subject-matter. Taken alone, the acquisition year has little demonstrative force.

Terminology thus seems the only possible cue to the affiliation of the series texts, even though the absence of prescriptions eliminates many of the critical parameters. Happily, the reverse of YBC 4669 contains one brief problem provided with a prescription, in which results are marked igi.du_8, the logogram for *tammar*. The same usage is found in YBC 4673 (rev. III 9, 13, 16, 20). This already points to the periphery, as represented by groups 6, 7, and 8.

A single term in two problems may be suggestive but proves nothing – as we have seen, even the last problem of YBC 4662 uses *tammar*. But other details point in the same direction.

Firstly, there is the use of the term íb.tag_4 for the remainder after a removal in YBC 4668, YBC 4669, YBC 4697, YBC 4710, and YBC 4713. Outside the group of series texts this term is found in BM 85196 and VAT 6598 (both group 6; in the latter, it is a logogram for the verb *ezēbum*, "to leave"), and in IM 55357 (group 7B; verb again).

The phrase *X* a.rá *Y* ^{e}tab is used in VAT 7537, YBC 4668, YBC 4695–97, YBC 4708–4713, YBC 4715, and A 24194–195, and elsewhere only (in the slightly different shape *X* a.rá *Y* tab.ba) in BM 85194, obv. II 44, 50.

YBC 4708 asks repeatedly EN ta.àm íb.si_8, "What, each, is equalside?" Similar references to "each" of the sides of a square are found in BM 85194 and BM 85196 (both group 6); NBC 7934 (not linked to any group); YBC 4607 and YBC 5037 (group 2 catalogues, and thus possibly recasts of "periphery" material in the likeness of the group 2 excavation texts); BM 80209 (group 6, see p. 304); and in UET V, 864, a text from Ur whose terminological resemblance with texts from the periphery were discussed on p. 253. The phraseology seems to belong with the lay surveyors' tradition, which carries it into the Middle Ages.

But the series texts certainly do not belong to any of the periphery groups established so far. They never employ UL.GAR for *kamārum* nor NIGIN for *šutakūlum*/*šutamḫurum*, which sets them well apart from groups 6 and 8A; the sophistication of many of the series texts – regarding mathematical substance as well as the pluridimensional variation of statements – shows them to belong to a more mature phase of scholastization than the Eshnunna corpus; we may also observe the absence of kud (used logographically for *nakāsum*, for *ḫaṣābum*, and perhaps for *ḫarāṣum*) from the series texts.

The distribution of a few other terms and expressions may serve to complete the picture:

ba.zi, the normal writing of the removal in the series texts, is used in group 6A; in IM 55357 from group 7A; and in AO 6770 (only the stone problem) and YBC 6504 from group 1.

nu.zu, "I do not know", is used in connection with the question in YBC 4668, YBC 4673, YBC 4698, YBC 4710, and YBC 4713. Outside the series text group the same phrase or its syllabic equivalent is used in Str 362, Str 364, Str 368, VAT 7535, and YBC 4608 (all from group 3); further in IM 53965 (group 7A), a *tammar*-text dealing with a broken reed and very close to Str 368; and in AO 6770 (group 1), the stone problem and in one other place. In TMS XX, 5, and TMS XXV, 4, the relative clause *šà la ti-du-ú*, "which you do not know", is used as an identifier within the prescription.

The phrase a.na uš ugu sag̃ dirig, "as much as that by which the length exceeds the width", is used in YBC 4668, YBC 4697, YBC 4709, YBC 4711, and YBC 4713. The same phrase in more or less syllabic writing turns up in groups 1 and 3, with a single similar expression in BM 13901 (=group ii). The Susa texts (group 8A) reduce it to dirig.

We may conclude that the series texts are less closely related to group 6A than believed by Neugebauer; that they will have been produced somewhere in the peripheral orbit – that is, outside the ancient Ur III core area. If we look at the problem types where nu.zu and a.na uš ugu sag̃ dirig and their syllabic equivalents turn up in groups 1 and 3 (broken-reed and stone riddles, etc.) we may also infer that the series texts, in spite of their sophistication and highly technical language, were produced in a place where the riddle tradition was closer to the surface than in the school where (e.g.) group 6A was produced and used.[414]

---

414 Close connections, if not specifically to the riddle tradition then at least to the practical background of Old Babylonian mathematics, are also suggested by the fact that even those series texts that contain "algebraic" problems on rectangles indicate the units explicitly, and that the normal dimensions of the rectangles are 30 nindan×20 nindan, not the school-yard dimensions 30´×20´ that are preferred elsewhere (thus in BM 13901, group ii, and in various Susa texts). This is no absolute rule, it is true – YBC 4668 #A35–47, YBC 4695, and A 24195 are of the 30´×20´ type, unless we make the rash assumption that all of these, like the unique sequence #B7–10 of YBC 4668, take advantage of the floating-point character of the numerals and make formal additions without regard for the proper order of magnitude.

Many of the series texts, finally, deal with genuinely practical problems and

Old Babylonian Ur and Nippur

When writing the basic version of the manuscript, I did not have access to [Vajman 1961], in which four mathematical problem texts from early Old Babylonian Ur were published, but only to the publication of UET V, 864 in [Kilmer 1964] and the further commentary to the text in Friberg 1981a. Since then I have not only got hold of the original publication but also received Muroi's analysis [1998] of two of the texts and Friberg's preliminary analysis of all four [2000]. Moreover, whereas I could only refer by then to a single text from Nippur (in seemingly reconstructed Sumerian), Eleanor Robson [2000] has now published a number of mathematical texts from Nippur, including a number of problem texts in predominantly syllabic Akkadian.

Part of the information deriving from these texts has been included above, but a brief summary of the characteristics of each of these groups will be useful.

The whole group from Ur shares several of the characteristics of the text UET V, 864, as discussed above on pp. 250–253. Most striking is the language: apart from the *dakāšum* and *dikištum* of No. 864 and the loanwords *-ma* and *u* they are written in Sumerian and not in Akkadian, and indeed in a Sumerian which is not only much richer in grammatical elements than normal Old Babylonian mathematical texts (even the Strasbourg texts of group 3) but also to agree better with the grammar book. However, they seem exactly to have been constructed "grammar book in hand": the grammatical elements are not contracted as usually in genuine Sumerian, *except in expressions which we know from other sources to have been well established:* indeed, the "remainder" is íb.tag$_4$ and not ì.ib.tag$_4$, even though ì.ib. is the perfect parallel to ù.ub..

All texts except No. 121 also "see" results, which is unique among texts from the Ur III core area (and if anything, Ur is really core!). Among Old Babylonian texts, they are the only ones to use the same term pàd as the Sargonic school texts (see p. 325), but followed by the prospective /ed/ ("you shall see"). The preceding verb may either carry a borrowed Akkadian *-ma* or a Sumerian prefix ù.ub.

Instead of being seen, results may be provided with the enclitic copula .àm, "it is". In UET V, 859 (as in Db$_2$-146), a.na.àm is used for the accusative of "what" (better, since the text is in genuine Sumerian, for the unmarked case).[415] In the nominative (better, the ergative), en.nam is

---

not with artificial "algebraic" matters – perhaps to a larger extent than other texts groups apart from 2, 6, and 7.

415 It is uncertain whether sag murub gi.na a.na gi.na.àm in UET V, 858, line 9, means "the true middle width, how much, true, is it", or it should be read sag murub gi.na a.na{gi.na}.àm, thus using a.na.àm as a nominative. The former

used, also in connection with ba.si$_8$.[416]

The connection in which a.na.àm turns up is a division question, a.na.àm bùr in.gar-*ma* saḫar ì.pad.da 6 in.gar-*ma* saḫar ì.pàd.dè, "what as depth do I posit: the earth shall be found? 6 I posit: the earth shall be found". We notice the form in.gar, found elsewhere only in YBC 6504, with which the use of šu.ri.a for the natural half is also shared. We further observe the absence of "giving", elsewhere compulsory in the division question (apart from group 7, where it seems, however, to have been known but deliberately avoided at least in subgroup A, cf. note 363; in group 7B, no divisions by irregular numbers occur).

No pronominal suffixes are used, but the Akkadian *ad-ku-uš* of No. 864 shows that the habitual pattern of grammatical person was intended; the statements have no opening formula but end with an explicit question. The prescriptions have neither opening nor closing formula. "Logical operators" are absent.

There are some similarities (beyond the "seeing" of results) with group 6 and the series texts. Firstly, the use of íb.tag$_4$ for the remainder (found once properly and once as a verb in group 6; also found in some series texts and, as a verb, in IM 55357 from group 7B); secondly, the expression *X* a.rà *Y* ù.ub.RÁ, "*X* steps of *Y* you go", which is structurally similar to the phrase *X* a.rá *Y* ^{e}tab/tab.ba, "*X* steps of *Y* repeat"; thirdly, the reference to "each" of the sides of a square. The use of íb.si$_8$ for the square configuration (i.e., as a logogram for *mitḫartum*) also recurs in the series texts as well as in YBC 6504. The use of ba.si$_8$ about the plane equilateral is shared only with group 7.

The mathematical substance within the group ranges widely. UET V, 121 #1 treats of a division among "brothers", where the shares are in geometric proportion and the quotient is known; #2 (dressed as a problem about flocks and shepherds) divides 1```1``1`1 by 13`13, by 13, and by 7.[417]

UET V, 858 deals with the bisection of a trapezium with upper width 17 and lower width 7. Even this has roots centuries back (see p. 237). It is interesting, however, that the Ur problem simply takes the ratio in which the length is divided for given (it does not even state it but takes it for silently granted). This enables it to solve everything by means of considerations of proportionality – there is no indication whatsoever in the text that the scribe

---

reading is undoubtedly more strange than the latter.

416 Evidently, the authors consider si$_8$ as an intransitive (or "one-participant") verb. The surface corresponding to the equalside is not mentioned; if it had been, the suffix .e would be understood as a locative-terminative indication, cf. note 42.

417 As observed by Friberg [2000: 36], this has no parallel in later Old Babylonian texts but is related to the divisions by irregular divisors in third-millennium texts from Ebla and Shuruppak. He also points out that the most astonishing implication of this text is that the scribes knew that these funny numbers divided, and explains with a parallel in his analysis of the Ebla division [Friberg 1986] how it could be discovered.

knew the equality between the square on the "true mid-width" (sag murub gi.na) and the average square on the widths.

UET V, 859 #1 deals with a cubic excavation,[418] where the square "surface of the ground" and the volume are known; the challenge (actually not met too well in the calculation) consists in handling the metrological conversions. #2 asks for the dimensions of a cube if the base and the volume are given; the problem is obviously overdetermined, but since no calculation is made we cannot see whether one of the parameters was intended to be "merely known". UET V, 864 was reproduced and discussed above (p. 250).

Absent are "algebraic" problems – that is, problems making use of a quadratic completion or similar methods; the closest we get to cut-and-paste geometry is UET V, 864. In view of the central importance of this problem type in later Old Babylonian mathematics this is striking even for so small a sample.

The newly published Old Babylonian problem texts from Nippur (CBS 43, CBS 154+921, CBS 165, CBS 11681, CBS 19761) are in "normal" syllabic Akkadian with interspersed logograms.[419] All that is known about their provenance is that they are from Nippur and of Old Babylonian date. Most are sadly damaged. In this small lot, "second-degree algebra" is well represented: CBS 43 and CBS 154+921 constitute (or are part of a larger) catalogue similar to TMS V and TMS VI and to the Tell Harmal Compendium IM 52916, IM 52304+52304;[420] CBS 165 has the same character (the former two deal with squares, the latter treats of rectangles); CBS 19761 contains problems about two squares, probably similar to BM 13901 #8–9; only CBS 11681 deals with a simple cubic excavation finding first the volume from the sides (with the familiar dimensions 30´ nindan and 6 kùš), and next goes the other way; in both cases, the factor 12 is applied wrongly, with the consequence that the scribe gets back to the original value. It may be observed that the area is expressed in absolute, not place value number. For the backward calculation, a reference volume is formed: "1, length, 1, width, 12, depth, which you do not know, you inscribe".

Among the characteristics of these texts, the following can be mentioned:

The square configuration is *mitḫartum* – in CBS 19761 once with logogram íb.si$_8$ (cf. note 47), in CBS 43 and CBS 154+921 written lagab but with phonetic complements that allow identification. CBS 19761 refers once in syllabic writing to the "fronts" (*pāt*) of the square, CBS 43 and CBS 154+921

418 This is not told, but the parameters are those with which we are familiar, cf. note 64.

419 As mentioned repeatedly above, CBM 12648 is in (seemingly reconstructed) Sumerian; since nothing suggests that it should form a common group with the syllabic texts, I disregard it in the present connection.

420 These catalogues from Eshnunna, Susa, and Nippur, and no other texts in whole Old Babylonian corpus, speak of the "length" of a square; all of them also treat this length as a "broad line".

refer to the side (singular) as uš, "length", but quite exceptionally provided with a phonetic complement *ia*, "my"; however, when asking for the side of the square the question is *kiyā imtaḫḫar*, "how much, each, confronts itself", again leaving no doubt that the sides are seen as a plurality.

CBS 19761 asks a question *kī maṣi*; in CBS 165, Robson restitutes en.nam everywhere, but as far as I can see on her hand copy without basis in extant parts of the tablets.

In the catalogues CBS 43 and CBS 154+921, lengths are "appended to" and "torn out from" the surface; they are thus treated as broad lines.[421] Strikingly, "tearing-out" is invariably *ina* šà, "from inside"; in contrast, "appending" is always simply *ana*, "to".

The mainly logographic catalogues CBS 43 and CBS 154+921 employ ba.zi; in CBS 19761, however, removal is *ḫarāṣum*, "to cut off", as it often is in groups 1 and 7.

Rectangularization is probably *šutakūlum*; *wabālum*, "to bring", serves instead of "raising" (both in CBS 11681; *wabālum* probably also in CBS 19761). Results "come up"; occasionally they are announced by a preceding enclitic *-ma*.

The reverse computation in CBS 11681 opens *šumma*, "if (instead)"; no other opening formulae are conserved, but this absence is inconclusive because of the damages. Both prescriptions in CBS 19761 open *i-na e-pe-ši-ka*, "by your proceeding". Closing formulae are absent, as are logical operators beyond the use of *šumma* as an opening formula (anyhow, neither formulae nor logical operators have any space in the three catalogue texts).

In view of their Sumerian language, their early date and the city in which they were found, it seems near at hand to suppose that the texts from Ur represent the missing link between the mathematics of late Ur III and that of the later core area; the features which they share with the peripheral groups and which were supposed above to have been come from the lay tradition would then rather have to be understood as reminiscences of the Ur III style.

However, closer analysis of the place of the Ur texts within the total corpus show that they cannot represent a stage in a development that later unfolded in the mathematics of Nippur and groups 1–8. Firstly, the total absence of the suffix .àm as a marker of results from all later texts would be hard to explain – not least if BM 13901 really shares the idea but reinvents the terminology by misapplying the ergative suffix .e, as proposed on p. 52. Secondly, if the southern text groups were inspired by the Ur texts, the conspicuous features which the latter share with the periphery groups ("seeing", the reference to "each" side of a square, multiplication in "going steps", the use of ba.si$_8$ about the plane configuration) would certainly also be

421 The term for appending is daḫ; there is one syllabic imperative *ku-mur*, "accumulate"; but since it goes together with the preposition *ana*, "to", it is likely to be a *lapsus calami* – or perhaps, as we shall see (note 472), a result of metatheoretical bad conscience!

prominent in texts from groups 1-4. Further, if the Ur use of Sumerian pre- and suffixes corresponded to Ur III habits, it would certainly have had some influence in the later texts from the core. All we find, however, in the use of in.g̃ar in YBC 6504, which in itself, as we have noticed, represents a terminological dead end. This text also uses šu.ri.a about the natural half, íb.si$_8$ as a logogram for *mitḫartum* and (shared with many later texts but with no others from group 1) en.nam for *mīnûm*; but the differences are no less conspicuous (ba.zi for "tearing out", íb.si$_8$ also for the equilateral, the actual use of in.g̃ar, etc.), and even YBC 6504 is therefore not likely to represent a direct descendant from the style of the Ur group. The only component of the later terminology that *could be* a legacy from the Ur texts is en.nam, Sumerian or rather pseudo-Sumerian for "what".

Instead we have to see the texts from Ur as an independent expression of the transformation of mathematics that took place in the early Old Babylonian phase; part of the material that was used may well have come from (or through) the Ur III tradition – not least the strange divisions, but also the division among brothers and the excavation.

However, the use of the Akkadian terms *dakāšum* and *dikištum* is already sufficient to show that non-Sumerian surveyors' mathematics was also part of the inspiration; such elements of the approach to surveying mathematics which turn up only marginally or not at all in later texts from the core area (the reference to "each" side of the square, the "standing against each other", and even the notion of "seeing" results) will also have been part of the surveyors' heritage – a part which other schools in the core area choose *not* to take over when adopting the surveyors' riddles (but of whose existence they will have been aware, as demonstrated by BM 13901 #23).

The use of Sumerian and the minimal use of indicators of the structure of problems are likely to be linked with the place in which the Ur texts were produced. Ur, reduced from imperial capital to a provincial town in the Larsa Kingdom, would certainly not be the place where demonstrations of cultural rupture with the Sumerian past were in favour; that supra-utilitarian problems do turn up may tell us that the rupture and the rise of scribal ideological autonomy was first of all results of a spontaneous process, and that the deliberate accentuation in Larsa and Eshnunna was secondary and derived (as one would anyhow have to expect).

The features which the texts from Ur share with group 6 might also be taken as evidence that these latter texts reflect the emigration of scholars after the rebellions and collapses in the South in 1739 BCE; Friberg [2000: 66] argues so, and points out that these political events had been presupposed by Goetze in his description of the group 6 texts. Even on this account, however, closer inspection of the evidence forces us to give up this idea at least as far as mathematical substance is concerned.[422] Even if the phrase *X* a.rà *Y*

422 Whether an influx of scholars from the south influenced orthographic habits in late Old Babylonian Sippar is a different question which there is no reason to approach

ù.ub.RÁ, "*X* steps of *Y* you go" were cleansed for some reason of the verbal prefix and provided instead with an Akkadian phonetic complement or a Sumerian imperative suffix, there would be absolutely no reason to change RÁ into tab, "to repeat". A divergence of this type is much more likely to arise from two independent translations (into Sumerian and Sumerogram, respectively) of the same Akkadian expression (the *alākum* known from the Susa texts, "to go"[423]); this would simply be a parallel to the adoption of the notion of "each" side of the square, the "standing against each other", and of results being "seen". íb.tag$_4$ is too unspecific and too traditional to serve as evidence for a particular link; all that can be concluded from the use of íb.si$_8$ for the square figure in the Ur texts as well as YBC 6504, in various series texts, in BM 15285, in one text from Nippur and in one from group 6 (BM 85210) is that even this may be Sumerian usage. It is also difficult to see why features of early texts from Ur should be borrowed into the Sippar tradition but leave almost no traces in the texts from groups 1–4.

It may still be asked whether the "seeing" of results in the Ur texts is a translation from Akkadian *tammar* or a continuation of the use of pàd in Sargonic problems. Since pàd may simply have been the closest Sumerian equivalent (irrespective of the choice of the composite igi.du$_8$ in IM 55357 and the two series texts YBC 4669 and YBC 4673[424]), coincidence is no proof of continuity; moreover, the Sargonic problems ask their question in a different way ("its" – see below, p. 378), which shows that the general format in which problems were formulated was not transmitted from the Old Akkadian school to the school of early Old Babylonian Ur.

The Nippur texts may give rise to two general observations. Firstly, they demonstrate that the type of catalogue text known until now from the "Tell

---

in the present context. I have my doubts: the equally "mixed" orthography in the Susa texts (cf. note 376) shows that the idea of a mixture may reflect a mistaken notion of what should be the "pure" types.

423 The use of the characteristic "repeat-append" structure in the stone riddles of YBC 4652 (see p. 305), written a.rá *n* e.tab bí.daḫ, shows that tab does indeed fulfill the role of *alākum*.

424 The use of a composite verb, "to open the eye", would have the disadvantage in the context of grammatically correct Sumerian that the entity seen would have to occur in an oblique and not the neutral ("object") case and therefore need a case suffix. In Eshnunna, this will have been no trouble, since the composite verb is written unorthographically and therefore hardly understood in grammatical detail. In the two series texts, the writing is orthographic, but their language makes no pretence to be grammatical.

We may observe that IM 55357 is the only text outside the Ur group which uses a.na.àm as an interrogative phrase. But it uses íb.si$_8$ (orthographic!) and not (as the Ur texts as well as all other Eshnunna texts) some orthographic variant of ba.si$_8$ for the plane equilateral. All in all, IM 55357 thus seems not to be linked directly to the Ur texts, the common formulation of the question notwithstanding (after all, the term is good Sumerian and no specifically mathematical locution [SLa, §120]).

Harmal Compendium" IM 52916, IM 52304+52304, and from the Susa catalogues TMS V and TMS VI was indeed more widespread. Secondly, and of greater importance for our understanding of the general development, they confirm that the notion of "each" side and of the plural "fronts" of a square, though absent from groups 1–4 (except the folkloristic citation in BM 13901 #23) was present in the area (as it had to be the case if the citation in BM 13901 #23 should make sense). It is therefore also likely to have been present within the horizon of the Ur scribes of the nineteenth century, as presupposed in the preceding argument.

Summarizing

On the whole, the inclusion of new parameters has confirmed Goetze's original categorization, moving a few tablets from one group to another or away from the established groups;[425] it has also allowed us to distinguish a couple of subgroups. As demonstrated by the series texts and BM 13901, it does *not* allow us to link all tablets to the existing groups 1–8, Ur, and Nippur. One reason for this is probably that the major groups come from very few very specific locations – tablets that are as alike as the sequence VAT 8389–8391 and which were bought in the same lot (as is evident from the museum numbers) are likely to belong as closely together as the group 7A texts – and comparison of the 7A Tell Ḫarmal texts with the 7B texts from the same city shows that even the same locality might produce texts of widely diverging character during the same decades. Many of the remaining texts may simply have come from other locations. But another reason is that none of the terminological parameters determines anything when taken alone, and that texts containing one or two problems are likely to make use of only a few of the interesting operations.

Discussion of the single isolated texts will contribute nothing significant to the understanding of the historical processes in which Old Babylonian algebra participated and was shaped. We shall therefore go on with a delineation of the landscape which emerges from the haze of details.

The Outcome

The most important aspects of the picture that results from the examination of the characteristics of the texts groups are the following:

---

425 Conversely, the gross agreement with a categorization based on orthographic criteria confirms the correlation between our terminological parameters and the origin of tablets.

- The political division of the Babylonian area into what had once been the Sumerian core area and a periphery which had only been subjected to Ur III rule for a relatively brief period is reflected in a similar division of the Old Babylonian mathematical corpus, where group 5 (not located), group 6 (almost certainly Sippar), group 7 (Eshnunna), and group 8 (Susa) represent the periphery and group 1 (Larsa?), groups 2 and ii (not located), group 3 (Uruk), and group 4 (Uruk?) together with Nippur the core. However, the early texts from Ur, no doubt belonging to the core area, turn out to challenge this simple dichotomy.
- The specific body of Old Babylonian mathematics resulted when the tradition for scribal computation in the wake of the Ur III tradition was fused with one or more non-scribal ("lay") traditions whose most important contribution was a proto-algebraic geometrical area technique, but to which other mathematical riddles will also have to be traced. Since non-school traditions are likely to have been carried by specific professions or crafts, it seems most likely that riddles about the filling of measures or about shrinkage and increase of weights came from another tradition than riddles about area measurement. The fusion was the outcome of deliberate amalgamation, no spontaneous and accidental aggregation of disparate elements.
- The resulting body of mathematics is sufficiently homogeneous to demonstrate the existence of some kind of formal or informal coordination; but also inhomogeneous enough to tell us that the fusion of traditions took place in several centres simultaneously; the process is reflected both in group 1 (Larsa?) and group 7 (Eshnunna). The Sumerian texts from Ur seem to be evidence of a third centre and may precede the creation of the Akkadian mathematical tradition proper (see presently); but they appear to have had slight or no later repercussions. Even though group 1 seems to represent the beginning and group 7 a slightly later phase of the creation of Akkadian mathematics, group 7 represents a local undertaking and no local transposition of the Larsa innovations (nor group 1 a transposition of an earlier phase of the Eshnunna process). Nothing proves that these centres were alone, but what else we know about the role of Larsa and Eshnunna in the formation of the specific Old Babylonian cultural complex suggest that they will have had the leading roles. The general division of the corpus into periphery texts and core area texts also suggest that only two centres were important.[426] In Eshnunna, the process was well

426 With the proviso that the small group 5 might represent a relatively autonomous development, produced in an environment where the role of the traditional scribal component was more important than in Eshnunna.

It might, however, also represent a branching of the core group which happened to be produced in an environment using a different orthography: the texts from Ur seem to show that the "seeing" of results was familiar even in the core area; this is confirmed by the oblique (and seemingly more acceptable) use of the term, "in order that you see [the result] *X*" in the group 3 text YBC 4608 – see

under way around 1800 BCE (which means that the beginning may coincide with the creation of the first law code in Akkadian, precisely in Eshnunna); we may presume the Larsa development to have been at least as early.

- Everywhere, the new scribal schools tried to create a precise canon for the format in which mathematics should be presented (without agreement between schools about what the canon should be). In some cases, the distinct internal uniformity of certain text groups or subgroups reflects the existence of such canons; in others, single texts or clusters of texts offer evidence of the effort to bring forth a canon.[427]
- The existence of a local canon would not prevent the adoption of material borrowed from a school or region obeying a different canon. Single characteristic problem types may also turn up in several groups with a characteristic format that overrules aspects of the local canons, probably because type and particular formal characteristics were borrowed together.
- An easily perceptible characteristic of most periphery groups (all but group 5) is the use of *tammar*, "you see", when results are announced, borrowed from the lay traditions. The term is absent from all core groups (only the texts from Ur have the Sumerian equivalent pàd); some of them tend to use sum/*nadānum*, "to give", which appears to have been originally connected to computation in the sexagesimal place value system and the use of tables. Since it is not used in the Ur texts, it might have developed between the creation of the Ur corpus and the early Larsa texts – which would be an argument in favour of an early dating of the Ur texts, perhaps close to 1900 BCE.[428]
- Also derived from the scribal tradition is the idea that results "come up". In view of the invariably Akkadian writing we can be fairly certain that the latter usage is a post-Ur III Akkadian contribution to this tradition – probably made in the periphery or the northern part of the core (Nippur), since the term is absent from all texts from the core area with the exception of three group-1 texts and the Nippur texts.[429]

---

note 394. But if the usage was indeed well known, then avoiding it looks like a very particular stylistic choice, one that was not likely to be made fully independently in several centres. *If* this argument stands, Ur and Larsa will have been the only centres in the core, and group 5 as well as core groups 1–4 will have been in debt to the Larsa foundations.

427 As we shall see (p. 382), canons were not only concerned with formats but also with "critique".

428 It cannot be wholly excluded, however, that the authors of the Ur texts avoided an existing term with purpose, judging that two different ways to mark results (pàd and .àm) was ample. But even in this case, only a fresh innovation is likely to have been refused.

429 The only Old Babylonian text which has a Sumerographic writing of "coming up" is the Nippur text CBM 12648, which appears to construct Sumerograms for everything, and which has íb.si$_8$ x e$_{11}$.dè, "make the equalside of x come up".

- In Larsa, the interrogative phrase *kī maṣi*, "corresponding to what", was clearly felt to be tied to traditional scribal computation. In Eshnunna, this appears not to have been obvious around 1785 BCE. We thus seem to be confronted with another post-Ur III Akkadian innovation, this time diffused from the core area, however. The absence of en.nam (and en.na) from the Eshnunna texts and its presence in those from Ur suggests that even this way to ask may have been devised in the South but at an earlier moment. a.na.àm, present also in Eshnunna, is traditional Sumerian; ephemeral experiments with its use in Ur and Eshnunna may be independent.
- Though with evident Sumerian roots, the use of íb.si$_8$ and ba.si$_8$ as verbs is not part of the tradition of scribal computation as descending from the Ur III tradition, whose recurrent use of tables had led to a reconceptualization as a noun. The verbal usage is evidence of pre-Ur III interactions between the lay surveying tradition and Sumerian culture.[430]
- The post-Ur III scribal tradition seems to have developed local forms already before the merger with the lay traditions took place: in Eshnunna, variants of ba.si$_8$ are used nominally for the quadratic "equalside"; in Ur, with verb function. In all other texts from both core and periphery, ba.si$_8$ with variants is reserved for the cubic "equalside" and generalized uses.
- Part of the terminology of the surveyors' tradition had no generally established Sumerographic counterpart. In such cases, Sumerographic equivalents might be constructed in more or less fanciful ways (the pun ì.gu$_7$.gu$_7$, "to make eat each other", for *šutakūlum* is an extreme example); closest equivalents might be chosen, as $\frac{1}{2}$ or šu.ri.a, both "normal halves", as Sumerograms for *bāmtum*, the "moiety"; or the Akkadian term itself might tend to be replaced by another one which already possessed an established equivalent (as *ḫarāṣum*, "to cut off", very common in groups 1 and 7, is replaced by zi/*nasāḫum*, "to tear out", in all later groups except group 2).

430 Whether pre-Šulgi scribal culture or non-scribal Sumerian surveying cannot be decided, not least because it is not clear whether the lú.éš.gíd, "rope-spanner" (that is, "surveyor") referred to in Sargonic contracts is a scribe or not (since a "scribe" is also mentioned, he is probably not). See [Krecher 1973: 173–176].

Chapter X
The Origin and Transformations of Old Babylonian Algebra

Practitioners' Knowledge and Specialists' Riddles

Several offhand references were made in the preceding chapters to "riddles", to "practitioners", and to "lay surveyors". In order to make these considerations meaningful and fruitful, a somewhat more systematic discussion may be needed.[431] The following applies to the situation as it looked (with a few exceptions, mostly regarding the Islamic world) until the seventeenth century, that is, to a situation where the knowledge system of a practical profession or craft was much more autonomous than today. As is well known, the relation between theoreticians' and practitioners' knowledge began to change in the late Renaissance, and was wholly transformed by the nineteenth-century advent of the modern engineers' professions; in consequence, today a large part of the practitioner's knowledge (though still far from all of it) is *applied theory*, which complicates the relation between the two types of knowledge without wholly abolishing their distinctive characteristics.

The gist of *practitioners' knowledge* is, by definition, *know-how*, not *know-why* – in Aristotle's terminology "productive knowledge", not "theoretical knowledge".[432] This certainly does not mean that practitioners can have no knowledge "of principles" (Aristotle again) but that the orientation of the practitioners' knowledge system differs from that of the theoretical system. The distinction between these two orientations is of general validity,

[431] With some changes, these reflections draw on ideas first presented coherently in [Høyrup 1990a] (slightly revised reprint in [Høyrup 1994: 23–43]).

[432] *Metaphysica* 982a1 [ed. Tredennick 1933: I,3].

but has particular implications for mathematics; what follows is therefore specifically adjusted not only to the pre-Modern period but also to the domain of mathematics, whose "practitioners" in pre-Modern times were surveyors, master builders, calculators – and, with mathematical practice as only part of the professional duties of the craft, "scribes" and "clerks".

Beyond this distinction between possible orientations of knowledge systems, "productive knowledge" itself may be divided into two types according to the main ways in which it is transmitted. One way is within a master-apprentice network, through on-the-job training; the type of knowledge which results may be labelled "subscientific" (explanation follows). The other is through institutionalized schools, where training is separated from actual practice and taken care of by teachers whose direct connection to the practice for which they prepare is reduced or tenuous; the outcome may be labelled "scholasticized knowledge".

The choice of the term "practitioners' knowledge" instead of "practical knowledge", with reference thus to a social group and not to its practice, is deliberate. Practitioners' knowledge, whether subscientific or scholasticized, is not to be identified with "practical knowledge" alone. The difference has to do precisely with the influence of the *social* systems which carry the knowledge in question.

The larger part of the practitioners' fund of knowledge *is* evidently applicable in practice, at least according to the convictions of the environment within which they function. As far as this part is concerned, *problems – viz.*, the problems which the craft or profession is supposed to deal with – are fundamental, and appropriate techniques have been developed which allow it to handle these problems. But the *training* of future practitioners, whether effected in a school or done on the job, will have to start from simpler tasks than those taken care of by the master – in part from tasks which have been prepared with the special purpose of training the techniques which the apprentice should learn but which have no direct practical relevance; here, *techniques* or *methods* are thus primary, and problems are secondary, derived from the techniques which are to be trained. Anybody familiar with school books on arithmetic will recognize the situation, and scholasticized systems are indeed those where problems constructed for training purposes dominate. Since the genuine practice of the school teacher is *teaching*, the school may end up teaching techniques and methods which have no purpose outside school – and, depending on the general style of the school, perhaps giving reasons for these techniques and methods which come close to investigation of "principles". In any case the school situation will favour systematic drill and – since this is the only way systematic drill can be achieved in mathematics – systematic variation (making *the same* calculation ten times would be meaningless, whereas trying ten times to build a chimney with the same unvaried characteristics is not).

Apprenticeship-based systems, on their part, will tend to train as much as at all possible on "real" albeit simple tasks; since the teacher is first of all a master of the craft who just happens in a given moment to instruct an apprentice, there is no risk that methods will be taught without a purpose. This purpose, however, is still not always practical utility.

Scholasticized systems often make some use of so-called "recreational" problems as a means to create variation. The genuine basis, however, for the invention and spread of these problems – problems that distort everyday settings so as to create a striking or even absurd situation – is the subscientific knowledge system.

This claim – as well as the character and function of the recreational problems – may be illustrated by three examples, picked from an Indian, an Arabic, and a Carolingian source:

(i) Two travellers saw a purse containing money (dropped) on the way. One of them said (to the other), "By securing half of this money (in the purse), I shall become twice as rich (as you)". The other said, "By securing two-thirds I shall, with the money I have on hand, have three times as much as you have on hand". What are the moneys at hand, and what the money in the purse?[433]

(ii) A duck for 5 dirham, 20 sparrows for 1 dirham, a hen for 1 dirham, and so on; you receive 100 dirham, or more or less, and it is said to you, "Buy yourself in total 100 fowls, or more or less, of these sundry kinds".[434]

(iii) A paterfamilias had a distance from one house of his to another of 30 leagues, and a camel which was to carry from one of the houses to the other 90 measures of grain in three turns. For each league, the camel would always eat 1 measure. Tell me, whoever is worth anything, how many measures were left.[435]

Problem (i) is indeterminate, and any set ($11t$, $13t$,$30t$) for the two amounts at hand and the contents of the purse will solve it; Mahāvīra does give a rule for finding one solution (the problem is part of a sequence of systematic variations – the treatise is already well at a distance from the oral origins), but gives no explanation, nor does he waste a word on the existence of other solutions.

Even (ii) is indeterminate, with two free parameters. Abū Kāmil, from whom it is quoted, explains that it is

> a particular type of calculation, circulating among high-ranking and lowly people, among scholars and among the uneducated, at which they rejoice, and which they find new and beautiful; one asks the other, and he is then given an approximate and only assumed answer, they know neither principle nor rule in the matter[436]

– after which Abū Kāmil shows himself to be a mathematician by deriving the complete set of integer solutions, at first to the duck-sparrow-hen problem, then to other variants involving three, four, or five kinds (after having pointed

---

433 Mahāvīra, *Gaṇita-sāra-sangraha* VI.242–243 [trans. Raṅgācārya 1912: 157].

434 From the German translation in [Suter 1910: 101]. The phrases "and so on" and "or more or less" refer to the existence of other variants.

435 *Propositiones ad acuendos iuvenes*, 52 (version ii) [ed. Folkerts 1978: 74].

436 *Book of Rarities in Calculation*; from the German translation in [Suter 1910: 100].

out that if fractional solutions are allowed, there would be no other end to answering than the death of the answerer).

The obvious answer to (iii) is that all the grain will be consumed in transport. Instead the text proposes that the camel make only 20 leagues with the first 30 measures, leave 10 measures and return, carry another load of 30 measures 20 leagues, return and bring the last, take up the deposit after 20 leagues, and thus bring 20 measures to the final destination. The solution is not optimal – a first intermediate station after 10 leagues and a second after another 15 leagues would permit a transfer of 25 measures.

If judged as mathematics, the first and third problem together with their solutions are thus no less to be censured than the common version of the second as judged by Abū Kāmil. But since such problems have survived for centuries or millennia and pop up everywhere between China and Aachen, this cannot be the historically relevant gauge. Knowing *how* to find an answer without knowing much about *why* this would be the answer must have been fully satisfactory. This already shows the affinity of the problems with the practitioners' knowledge system.

The conventional classification of the problems as "recreational" is equally off the point. It is obvious from the formulations that the problems are riddles, not pieces of mathematical research in even the vaguest sense. But it is also obvious ("Tell me, whoever is worth anything ...") that these riddles are not meant for fun but as challenges: these *riddles for specialists* share with other riddles that eristic character which distinguishes oral cultures in general.[437] They are no more "recreational" than the potentially lethal riddle of the Sphinx. They served as a means to *display virtuosity*, and thus, on one hand, to demonstrate the status of the profession as a whole as consisting of expert specialist, and, on the other, to let the single members of the profession stand out, and discover themselves, as accomplished calculators/surveyors/... . At this point it should be noticed that subscientific systems, *qua* their decoupling from school systems, are oral in character even when functioning within societies where a literate class exists; scholasticized systems, in contrast, have always been geared to writing.

The phrase "Tell me, whoever is worth anything ..." was found in a written source; all we can know about oral cultures in former times comes via their reflection in literate culture. But precisely the written sources that so to speak present the "first generation", and not the mature development of a written tradition, are those that often contain phrases of this type, or present a problem or technique as a way to impress the non-specialists who do not understand.[438]

---

437 See, for instance, [Ong 1982: 43*ff* and *passim*].

438 "... to find out, not without exceptional amazement of the ignorant, how many penning, creutzer, or other coin somebody possesses", in Christoph Rudolff's words [1540]. In the pre-Modern world, indeed, subscientific knowledge was neither "folk" nor "popular" knowledge, but a possession of the few to a

The function of the "recreational" problems – to allow the *display of virtuosity* and thus to demonstrate the expert status of the profession as a whole and of its single members – puts certain constraints on the problems that explain their character. They must arouse immediate interest, which explains the "recreational surface": if the camel seems to devour exactly *everything* in the process, then the expert solution allowing a net transfer is impressing; a less striking formulation might provoke the reaction "so what?". The problems, furthermore, must appear to belong to the domain of the profession – mentioning that Friedrich Nietzsche also tried his hand as a composer (and made pretty well) is more likely to be intended to intimate that even in philosophy he will have been, a gifted epigone than to raise his philosophical prestige; similarly, skill in singing does not enhance the *professional* prestige of an accountant, does not demonstrate professional valour. According to their *form*, their dress, the problems have to be practical. But they must also be more difficult than the tasks that any average bungler in the profession performs without difficulty. This, together with the quest for the striking or absurd, is the reason that the problems are *pure in substance*, i.e., separated from real practice – and more truly so than the simplified problems of school teaching which prepare this practice. The two characteristics together are what I have summed up in the term "supra-utilitarian".

Like school problems, however, recreational problems are determined from methods, *viz.*, from the characteristic methods of the profession; often, moreover, and in the likeness of other riddles, from a peculiar trick (like the intermediate stop and return of the camel) that will be known within the subculture of the profession but not outside.

Such tricks will often not be generalizable; often, moreover, as Abū Kāmil and other "scientific" mathematicians from the Islamic Middle Ages assert polemically, the practitioners using them would not know why their tricks worked; at times they do not work at all *as mathematics* but only because of numerical coincidences or because the result has been presupposed, as in the stone riddle of AO 6770 and the filling problem of IM 53957 (above, pp. 341 and 321, respectively). Even more clearly than the indeterminate problems, such examples show that the purpose of posing and solving the problems cannot be to provide insight; the supra-utilitarian level of subscientific knowledge is thus neither a direct nor an indirect underpinning of practice, nor is it a critical reflection on the principles underlying practice; in this respect supra-utilitarian knowledge differs fundamentally from what Aristotle and al-Fārābī would speak of as theoretical knowledge and from what we would call science.[439]

---

significantly higher degree than scientific knowledge today.

439 There is another point in the term "sub-scientific" which I shall not pursue here, namely, that the subscientific knowledge systems have served as inspiration in the development of "scientific" mathematics.

Supra-utilitarian and theoretical knowledge also differ with regard to the primacy of problems versus methods. Greek mathematics may be chosen to represent the defining ideal type of "theoretical knowledge" (which is certainly not respected by everything that parades as theoretical activity, but this is a problem that does not concern the present discussion).[440] Greek mathematics developed around problems. Famous are the three "classical problems", the trisection of the angle, the doubling of the cube, and the quadrature of the circle, and their history is indeed illustrative: when they were formulated as geometrical problems,[441] no theoretically acceptable methods were known that would allow their solution; as long as ancient mathematics remained alive there was a continuous effort to solve them by means of methods more satisfactory than those found by earlier workers.[442] But we may also look at the theory of irrationals. The first discovery of irrationals led to the problems *how to construct* according to a general scheme lines which are not commensurate with a given line (or whose squares are not commensurate with a given square); *how to classify* magnitudes with regard to commensurability; and *which are the relations* between different classes of irrationals? The first problem is the one which, according to Plato's *Theaetetus* 147D, was addressed by Theodoros; further on in the same passage, Theaetetos makes a seemingly first attempt at the second problem; *Elements* X, finally, is a partial answer to all three problems.

In this sense, the *problem* is primary in theoretical science, as in genuine practice, and the method secondary and derived from the requirements of the problem; therefore theoretical science may serve as an underpinning for practice, whether this be its aim or not, whereas the supra-utilitarian level of the practitioners' knowledge system rarely lends itself to this use, even if it may provoke the theoretician's critical scrutiny of *why* and *under which conditions* it works, and thus be *transformed itself into a problem* (as illustrated by Abū Kāmil's mathematical investigation of the problem of the 100 fowls).

---

440 When mathematics *per se* (or any other science) has become a profession, workers in need of a dissertation subject or of another item in the list of publications will easily end up looking out for problems which are likely to be solved with the methods with which they are already familiar. But this does not contradict the general observations: *qua* profession members they are *practitioners* of the craft, and may either intend to demonstrate their membership (the thesis) or feel constrained to do what they are paid for.

441 Thus not as everyday (practical) problems – cf. Aristotle's polemics against various sophists' approaches which simply miss the distinction between the two categories and thus permit trivial solution (*Analytica posteriora* 75^b40-76^a3; *De sophisticis elenchis* 171^b16-22, 172^a3-7; *Metaphysica* 998^a1-4).

442 See [Heath 1921: I, 218-270].

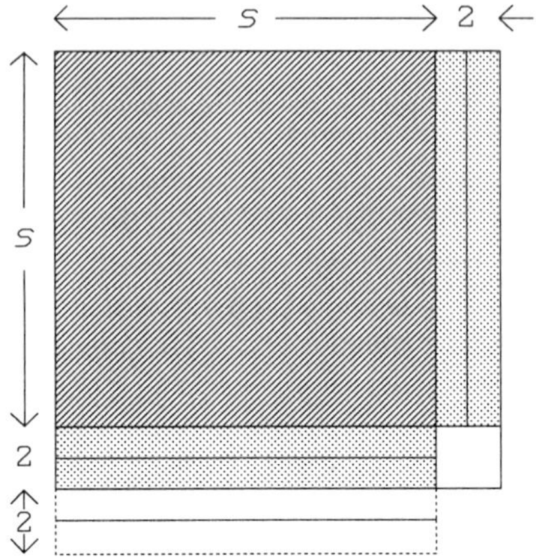

Figure 83. The procedure described in *Geometrica* 24.3.

A Long and Widely Branched Tradition: the Lay Surveyors

The preceding deliberations concerned (mathematical) school knowledge and non-school or "lay" practitioners' knowledge in general. Their relevance for the emergence of the Old Babylonian algebra was already suggested by the scattered characterization of certain problems as "riddles"; but it only becomes obvious when links can be established between the Old Babylonian school discipline and a lay tradition.

We shall start by looking at the further occurrences of the problem dealt with in BM 13901 #23 (above, p. 222). Earliest of these is the appearance as §3 in an anonymous Greek treatise which Heiberg has included as Chapter 24 in the pseudo-Heronian *Geometrica* [ed. Heiberg 1912: 418]:[443]

> A square surface having the area together with the perimeter of 896 feet. To get separated the area and the perimeter. I do like this: In general [καθολικῶς, i.e., independently of the parameter 896 – JH], place outside (εκτίθημι) the 4 units, whose half becomes 2 feet. Putting this on top of itself becomes 4. Putting together just this with the 896 becomes 900, whose squaring side becomes 30 feet. I have taken away underneath (ὑφαιρέω) the half, 2 feet are left. The remainder becomes 28 feet. So the area is 784 feet, and let the perimeter be 112 feet. Putting

[443] The treatise is clearly *not* by Hero, and was never claimed by Heiberg to be so; nor is it connected to the rest of ms S of the *Geometrica* – see below, note 494, and [Høyrup 1997: 77 n.29].

> together just all this becomes 896 feet. Let the area with the perimeter be that much, 896 feet.[444]

The next known occurrence is in an Arabic text which was translated in the twelfth century by Gerard da Cremona as *Liber mensurationum* [ed. Busard 1968] and ascribed to an otherwise unidentified Abū Bakr "called Heus". The Arabic original is likely to reflect the state of the art of c. 800,[445] but the text used by Gerard is obviously a revised version, and we cannot be certain whether Abū Bakr was the original author or responsible for a sometimes unfortunate revision.[446] For convenience I shall use the name henceforth as it if were that of the original author.

Problem 3 of the treatise runs as follows:

> And if he [a "somebody" introduced in #1] has said to you: (Concerning a square,) I have aggregated its four sides and its area, and what resulted was 140, then how much is each side?
>
> The working in this will be that you halve the sides which will be two, thus multiply this by itself and 4 result, which you add to 140 and what results will be 144, whose root you take which is 12, from which you subtract the half of 4, what thus remains is the side which is 10.[447]

---

444 Heiberg does not grasp the geometrical procedure that is described, for which reason his commentaries are misguided, imputing the faulty understanding on the ancient copyist.

As with the Babylonian texts, my translation is meant to be pedantically literal.

445 The reasons for this dating are the following:

- The text often gives a "normal" solution and an alternative by means of *aliabra* (*al-jabr*); this kind of synthesis between approaches is characteristic of Islamic science from the early ninth century onward, but not known from earlier times.
- The *al-jabr* to which it refers is not al-Khwārizmī's treatise; since its use of the key terms *al-jabr* and *al-muqābalah* belongs to an earlier phase of the development of the terminology, it is at least in the pre-al-Khwārizmīan tradition.
- A "square" is spoken of in the Latin text as *quadratum equilaterum et orthogonium*. The Arabic term *murabbaʿ* will therefore have been understood in the original sense of a "quadrilateral", whereas al-Khwārizmīan and post-al-Khwārizmīan algebras use it without further qualification as "square" (with some slips in al-Khwārizmī's own text).

These arguments do not exclude a post-al-Khwārizmīan date for the work – the author may have worked in an environment where the influence of the court scientists from Baghdad had not made itself felt. But even if this should happen to be the case, the treatise is good evidence for what was available before these scientists had set the scene anew.

446 In any case, Abū Bakr was the name of the first Caliph, and therefore very common in every part of Sunni Islam. It is no more useful for identification than a nude John/Johann/Jan/Giovanni/Juan/Jens/Hans in the Christian world.

447 [Ed. Busard 1968: 87]. Since Gerard was a most precise translator, my pedantic English translation can be supposed to be very close to the lost original.

We notice that this text not only has the riddle format, the same distribution of grammatical person and tense as the Old Babylonian texts, "its four sides" and a reference to "each" side, but also the sides before the area and the same solution 10 as the Old Babylonian text (apart from the order of sexagesimal magnitude). We may also notice the distinction between the symmetric process of "aggregation", corresponding to the Babylonian "accumulation", and the asymmetric "addition to", corresponding the Old Babylonian "appending" (the distinction is systematic even in the *Liber mensurationum*). The procedure is evidently the same as that of *Geometrica* 24.3 (Figure 83), and indeed the Babylonian standard procedure for "square area and sides").

In Abraham Bar Hiyya/Savasorda's early twelfth-century *Liber embadorum* we find:

> If, in some square, when its surface is added to its four sides, you find 77, how many cubits are contained in the surface? Taking the half of its sides, which is two, and multiplying it with itself, you find 4. If you add this to the given quantity, you will have 81, whose root, which is 9, you take; and when you subtract from this the half of the addition that was mentioned already, 7 remain. This is the side of the square in question, whose surface contains 49.[448]

The solution is followed by a proof, which uses *Elements* II.6.

Further, in Leonardo Fibonacci's *Pratica geometrie* [ed. Boncompagni 1862: 59], from 1220:

> And if the surface and the four sides [of a square] make 140, and you want to separate the sides from the surface. [...].

Even here, a proof based on *Elements* II.6 follows.

In Piero della Francesca's *Trattato d'abaco* [ed. Arrighi 1970: 122] from c. 1480 we find another version:

> And there is a square whose surface, joined to its four sides, makes 140. I ask what is its side. [...].

Finally, Luca Pacioli's *Summa de arithmetica* from 1494[449] contains the problem:

> And if the 4 sides of a square with the area of the said square are 140. And you want to know how much is the side of the said square. [...].

Elsewhere in the *Pratica*, Fibonacci uses Gerard's translation of Abū Bakr; when doing so, he copies word for word, changing only the grammar at certain points. The wholly different formulation of this problem (not least the idea of

448 II.12 in Plato of Tivoli's Latin translation [ed. Curtze 1902: I, 36]. The free Catalan translation in [Guttmann (ed.) and Millàs y Vallicrosa (trans.) 1931: 37], which goes back to a different recension of the work, here only differs from Plato's text on points where Millàs y Vallicrosa has introduced modern terminology.

449 I translate from the second edition [Pacioli 1523: II, fol. 15r].

separation, shared with ps-Hero, cf. below, p. 407) therefore suggests the use of a different (and independent) source.

Even Piero's version is independent of both Abū Bakr and Fibonacci; this is seen more clearly from his geometrical demonstration, which is rather of the "naive" kind and somewhat clumsy, and which in the actual form may be of his own making (see p. 416).

Pacioli depends on Fibonacci.[450] However, he must have supplementary information, since he knows that the sides "should" come before the area; a passage where he has a better (but still not quite correct) version of a problem than Abū Bakr/Gerard shows that this treatise cannot be his source for the return to the original formulation.[451]

Also interesting is the problem where the sum of circular surface S, circumference c, and diameter d is given. This problem was found in the catalogue text BM 80209 (see p. 287), in the form $S+d+c = \alpha$; with the order $d+c+S = \alpha$ it turns up in two of the treatises which Heiberg aggregated as *Geometrica* (Chapter 24 once more, and mss A+C [ed. Heiberg 1912: 380, 444, 446]), and again in ibn Thabat's *Reckoners' Wealth* – an Arabic handbook for practical reckoners from c. 1200, in which it is also explained that the area between two concentrically positioned squares is the mid-length of the border times the width [ed., trans. Rebstock 1993: 113*f*, 119] (cf. above, p. 267, and Figure 72).

The order of members in the Old Babylonian text is already the normal order of the school, in which areas precedes lines; the Greek and Arabic cases have the linear extensions first, as the riddle of the "four fronts and the area", and among the linear magnitudes the diameter before the circumference. The order $c+d+S = \alpha$ is found in Mahāvīra's *Gaṇita-sāra-sangraha* VII.30 [ed., trans. Raṅgācārya 1912: 192].

The order of members is significant and sociologically informative by pointing to a subtle difference between the riddle and the school problem. A riddle will start by mentioning what is obviously or most actively *there*, and next introduce dependent entities – in the riddle of the three brothers (protectors and potential rapists) and their sisters (virtual victims) the brothers come first; in the case of somebody encountering a group of people these first, next their double, etc.[452] In systematic school teaching, the order will tend to be determined by internal criteria, for instance, derived from the method to be applied. The typical school problems will therefore mention the area before the side – in the solution, the area is drawn first, only afterwards will a rectangle ⊏⊐(1,s) be joined to the area in order to represent the numerical

450 The link is indirect, since Pacioli draws upon a (so far unpublished) fifteenth-century Italian version of Fibonacci's *Pratica* – cf. [Picutti 1989].

451 He may share this source with Jean de Murs, whose solution of the same problem (see note 396) explains Pacioli's illegitimate shortcut.

452 Both examples are picked from the *Propositiones ad acuendos iuvenes* (Nos. 17 and 2, respectively) [ed. Folkerts 1978: 54, 45].

value of the side. The surveyors' riddle, on the other hand, will start with what is immediately given to surveying experience, that is, by the side; the area is found by calculation and hence derivative, and therefore mentioned last.

In Greek and Arabic practical geometry, the fundamental parameter for a circle was the diameter; both the circumference and the area were derived entities. The *Geometrica* as well as the *Reckoners' Wealth* thus follow what would be the typical riddle order of their own epoch. Even for Mahāvīra (VII.19), the basic parameter is the diameter; in Old Babylonian geometry, however, the basic parameter had been the circumference c, d being found as $20'\cdot c$ and S as $5'\cdot\Box(c)$.[453] Though Mahāvīra is coeval with al-Khwārizmī, his problem thus conserves a form that goes back to the early second millennium BCE.

Related to the idea of "the four fronts" of a square is the totality of sides of a rectangle – either the length and the width, or both lengths and both widths. In the Old Babylonian corpus, we have encountered it several times:

- AO 6770 #1 (p. 179) as well as AO 8862 #4 (p. 169) deal with the situation where the accumulation of length and width equals the area.
- The type "accumulation of rectangular length, width, and surface given" turns up so often that it was counted above as a "favourite problem" (p. 287): In TMS IX #2–3 (p. 89), in the unpublished Eshnunna texts IM 43993 and IM 121613, and (in combination with the equality of the accumulation of the sides with the surface) in AO 8862; moreover, in various dresses, in AO 8862 #7, in YBC 4668 #A9, and in BM 80209.

In the classical world, none of the types turn up in properly mathematical texts. But the pseudo-Nichomachean *Theologumena arithmeticae* mentions that the square $\Box(4)$ is the only square that has its area equal to the perimeter (see [Heath 1921: I, 96]), whereas Plutarch (*Isis et Osiris* 42 [ed., trans. Froidefond 1988: 214*f*]) relates that the Pythagoreans knew 16 and 18 to be the only numbers that might be both perimeter and area of a rectangle – namely, $\Box(4)$ and ⊏⊐(3,6), respectively.[454] Mahāvīra's *Gaṇita-sāra-sangraha* deals with the square case in VII.113½ and with the rectangular case in VII.115½.[455].

Other problems with which we are familiar from the Old Babylonian corpus also turn up in the same sources: the rectangle with given area and

453 The underlying idea is that the circle is a "bent line" – cf. p. 272. If the circle is the cross-section of a massive cylinder, the entity which is most easily measured *is* evidently the thread stretched around it.

454 A generalized and arithmeticized version of the problem (to find two numbers whose product has a given ratio to their sum) turns up in Diophantos's *Arithmetic* as I.14, II.3, and lemma to IV.36. The corollary to I.34 refers to the corresponding determinate problem where the ratio between the numbers is also given. If they had been alone, these coincidences might have been accidental; the attested interest in the simple version in Neopythagorean environments strengthens the hypothesis that the Diophantine version *is* a generalization.

455 [Ed., trans. Raṅgācārya 1912: 221]. Elsewhere (p. 224), Mahāvīra treats the case where the rectangular area and perimeter are given separately.

given sum of or difference between the sides, the sum of a square side and the area or the corresponding difference. Regarded as mere mathematical structures these are too simple to prove the existence of a link; but the vicinity of more characteristic problem types and the use of the customary format[456] makes independent reappearance highly unlikely.

Final pieces of the puzzle are furnished by *Elements* II and Diophantos's *Arithmetic*. *Elements* II.9–10 examine the cases $\square(p)+\square(q) = \alpha$, $p\pm q = \beta$, corresponding to BM 13901 #8–9; this is noteworthy because the propositions are never referred to afterwards – cf. below, p. 401. Diophantos's *Arithmetic* I.28–29 treats the two problems $\square(p)\pm\square(q) = \alpha$, $p+q = \beta$. It is striking that these, like the two "rectangle problems" I.27 and I.30, $\sqsubset\!\sqsupset(p,q) = \alpha$, $p\pm q = \beta$ (all four problems translated into arithmetical form), use the familiar method of average and deviation: in I.1–13 (simple first-degree problems), one of the unknown numbers is routinely identified with the *arithmós*; in I.15–25 (undressed "recreational" first-degree problems – "give and take", "purchase of a horse", etc.), a particular choice adapted to the actual case is made.

An inventory of the problem types that are shared between the Old Babylonian algebraic corpus[457] and the various later sources results in the following list:

- On a single square with side s and area $\square(s)$ (${}_4s$ stands for "the four sides"):

$$s+\square(s) = \alpha$$
$${}_4s+\square(s) = \alpha$$
$$\square(s)-s = \alpha$$
$$s-\square(s) = \alpha$$

- On two concentric squares with sides s_1 and s_2:

$$\square(s_1)+\square(s_2) = \alpha\ ,\quad s_1\pm s_2 = \beta$$
$$\square(s_1)-\square(s_2) = \alpha\ ,\quad s_1\pm s_2 = \beta$$

- On a circle with circumference c, diameter d, and area A:

$$c+d+A = \alpha$$

---

456 Beyond the striking features of the format of the *Liber mensurationum* that were already mentioned, two others of may be listed:
   - regularly, it is stated that an intermediate result X is to be "kept in memory" (*X que memorie commenda*).
   - at times, references to the statement are made within the prescription, with the phrase "because his speech was" (*quoniam sermo eius fuit*) followed by a quotation, and similarly.

457 The restriction to the algebraic corpus excludes the simple determination of a diagonal or area from the sides – but such calculations are anyhow too simple to serve as arguments for links.

– On a rectangle with sides l and w and diagonal d:

$$\sqsubset\!\sqsupset(l,w) = \alpha\ ,\ \ l = \beta \text{ or } w = \gamma$$

$$\sqsubset\!\sqsupset(l,w) = \alpha\ ,\ \ l{\pm}w = \beta$$

$$\sqsubset\!\sqsupset(l,w){+}(l{\pm}w) = \alpha\ ,\ \ l{\mp}w = \beta$$

$$\sqsubset\!\sqsupset(l,w) = \alpha\ ,\ \ d = \beta$$

$$\sqsubset\!\sqsupset(l,w) = l{+}w$$

So far, this might look as a strong argument that Indian, Greek, and Islamic mathematics built (*inter alia*) on the legacy of Old Babylonian algebra. But the argument is deceptive. The problems in the list share a common characteristic: all their coefficients are not only integers but *natural*. That is, they correspond to what is *really there* in the terrain: *the* area, not some multiple or fraction; *the* side, or *the* sides; etc. That systematic variation of which a text like BM 13901 bears witness (see the survey on p. 288) is fully absent (not to speak of experiments like those of YBC 6504 or of the series texts). Moreover, the Babylonian occurrences of the most characteristic shared problems are concentrated in groups 7B and 1 – those which seemed to be located at the border between the lay and the school tradition; "the four fronts and the surface" appeared in BM 13901 but as a glaring citation of non-school usage. Even the riddle format of the *Liber mensurationum* belongs in group 7, and (as ellipsis) in the citation of the non-school format in BM 13901 #23

If later traditions had really borrowed the algebra that was developed in the Old Babylonian school, this striking selection would certainly not have occurred.[458] Nor would Abū Bakr and Mahāvīra have had any reason to speak of linear magnitudes before the area, this order being not only extremely rare in the school corpus but also bound to quite particular situations.

It is an elementary rule for the construction of a stemma that similarities between *A* and *B* that are too systematic or too characteristic to be random are due either to descent of *A* from *B*, descent of *B* from *A*, or to descent from a common ancestor. If this rule is applied to the actual case, the later traditions must have borrowed what they have in common with the Old Babylonian school texts from a source that also inspired the Old Babylonian school. As we shall see presently, the Ur III school can be safely excluded. This leaves as the only possible common ancestor a non-school or "lay" tradition – which of course fits both the riddle format and the riddle order of the texts.

---

458 It should be noticed that the circle problem $c{+}d{+}A = \alpha$ is non-normalized, irrespective of whether c or d is taken as the basic parameter; the selection is thus not dictated by the difficulty of treating the non-normalized case (which anyhow would not exclude problems that added a square area and twice the side).

The Sumerian School: the Vocabulary as Evidence

In Chapter IX, in particular in connection with the discussion of groups 7 and 1, various kinds of evidence were mentioned that suggested a non-scribal origin for the algebra. This may be corroborated by an analysis of the terminology.

Our information about the mathematical terminology of the third millennium is scarce. We know that the verb si_8 was used to express that a segment *l* was the side of the corresponding square area at least since c. 2600 BCE (see note 373); uš, "length", sağ, "width", and $ašà_5$ (=GÁN = IKU), "surface", can be followed back to 2400 BCE;[459] the use of the phrase igi *n* ğál for *n* = 3, 4, and 6 is also documented since c. 2400 BCE (cf. p. 28).

The only mathematical documents from the Ur III period that contain terms for mathematical *operations* are the tables of reciprocals and of multiplication, of which the former use igi *n* ğál and the latter a.rá, "steps of" – *if* any of the extant specimens are really of Ur III date, which is difficult to establish with certainty. In any case the stable and invariably Sumerian terminology of the tables allows us to conclude that these terms will have been used already in Ur III; the same conclusion may be drawn about the terminology of the tables of square and cube roots, $íb.si_8$ and $ba.si_8$.[460]

The other mathematical documents from the epoch, accounts and model documents, only give results, and tell neither the details of calculations nor the terminology in which these were spoken about.

As mentioned in note 370, a hymn in the praise of King Šulgi relates that the scribal school is a place where zi.zi.i ğá.ğá are learnt together with šid, "counting", and níg.šid, "accounting". The use of the reduplicated form zi.zi.i may depend on the context (description of a habitual practice and not of the single operation); ğá.ğá and either zi or zi.zi may therefore be presumed to have been the standard terms for addition and subtraction in the Ur III school. ğá.ğá is the *marû* (approximately = imperfective/durative) stem of ğar, "to place" [SLa, 305], later used logographically for *šakānum*, "to posit"; in good agreement with the meaning of *kamārum* for which ğar.ğar is

459 Texts in [Allotte de la Fuÿe 1915: 124–132]. For non-rectangular fields, these surveying texts distinguish uš and uš 2.kam, "2nd length", and sağ an.na and sağ ki.ta, "upper" and "lower width". The equality of (e.g.) lengths is expressed uš si_8. a.šà is used about the area in Sargonic texts [Whiting 1984: 69].

460 The aberrant use of $ba.si_8$ in Eshnunna and Ur, it is true, could suggest that the distinction which all other text group upholds between the two is a secondary development, and perhaps that the form originally connected with the function as a verb was $ba.si_8$. In Eshnunna, it might then have displaced $íb.si_8$ even when used as a noun; in other groups, the term of the tables might have got the upper hand.

used logographically in the Old Babylonian age (namely, "to place in layers, to accumulate"), g̃á.g̃á may thus be understood as "ongoing placing". zi.zi is the *marû* stem of zi (better reading zìg), "to rise, to stand up" [SLa, 322], and may perhaps be understood as "take up from" – not too far removed from *nasāḫum*, "to tear out", for which zi is used logographically in Old Babylonian texts, nor too close, however.

Other terms are not mentioned in the text, not even a term for multiplication, even though multiplication was certainly a cornerstone in the accounting system.

We may approach the question from a different angle and ask whether the use of Sumerograms in the Old Babylonian mathematical terminology informs us about preceding Sumerian usages.

A few terms are written invariably (or almost so) with Sumerograms: uš, sag̃ (including sag̃ an.na and sag̃ ki.ta) and a.šà, when the "lengths", "widths", and "surfaces" of quadrangular and triangular fields are meant.[461] The only exceptions are found in the citation of non-school usage in BM 13901 #23, in one text from Nippur and in a few belonging to the Eshnunna corpus; but even here they are rare. a.šà is regularly provided with an Akkadian phonetic complement (indicating the pronunciation *eqlum*), uš and sag̃ never except for a possessive suffix *-ia*, "my", in the Tell Harmal compendium and its cognates.

A few other terms occur alternatingly as Sumerograms and as Akkadian loanwords, which indicates that they were really spoken with the Sumerian phonetic value: igi/*igûm*[462] (with igi.bi/*igibûm*), and íb.si$_8$/ib.sí/ba.si$_8$/ba.se.ė/*basûm* (with still other unorthographic spellings).

Both of these categories, though used in the algebraic texts, have their roots in a much simpler and much more utilitarian calculational practice. In contrast, the rest of the terms of the algebraic texts are sometimes written logographically, sometimes in syllabic Akkadian – except that a number of essential terms have no Sumerographic equivalent at all, or no suitable equivalent.

This applies first of all to the "logical operators" *šumma*, "if", *aššum*, "since", and *inūma*, "as". The interrogative phrase *mīnûm*, "what", *is* often written with a logogram en.nam, which, however, seems to be an *ad hoc* construction, first seen in the texts from Ur and in YBC 6504 from group 1 (but in no other texts from the early groups 1 and 7); a.na.àm, used for the accusative *mīnâm* in IM 55357 from group 7 and in UET V, 859 from Ur,

---

461 As mentioned on p. 224, however, *šiddum* may replace uš when the length of a wall or a carrying distance is meant; similarly, sag̃ may occur as rēšum, "head", when an initial value is intended or an intermediate result is to be kept in memory.

462 Again, certain Eshnunna texts (thus Haddad 104) are exceptions to the rule and use *pa-ni*, "in front of" – cf. p. 28. But since this is a manifest "scholars' folk etymology", it only confirms the genuine Sumerian origin of the term, which is anyhow well established on direct evidence.

seems to be an experimental borrowing from general (non-mathematical) Sumerian, where it means "what is (it/the reason that)". The other interrogative phrases *kī maṣi*, "corresponding to what", and *kiyā*, "how much each", have no equivalent, even though *kī maṣi* is obviously linked in the core area to computation types that were common in Ur III accounting. The "algebraic bracket" *mala*, "so much as", may be replaced by the interrogative pronoun a.na, "what", but only in the elliptic and late series texts; the same restriction holds true of the use of gin_7.(nam) (the Sumerian equative suffix) as a logogram for the "indication of equality" *kīma*, "as much as".

Among the ways to announce a result, *tammar*, "you see", is only written in exceptional cases with a Sumerogram (pàd in early Ur, igi.dù in IM 55357, igi.du_8 in the series texts YBC 4669 og YBC 4673 – in all cases seemingly a translation from Akkadian); *elûm*, "to come up", has no Sumerographic equivalent at all when used about results, even though it is primarily linked to typical Ur III computations (when used instead of *našûm*, "to raise", it may be written nim); only *nadānum*, "to give", apparently connected originally to sexagesimal multiplication, may be written sum. The enclitic particle *-ma*, often used as a minimal marker of a result, is Akkadian and is often found in texts which otherwise are heavily Sumerographic; the alternative hypergrammatical Sumerian prefix ù.ub. is likely to be an invention of the Ur group since it is never repeated in later texts.[463]

Some terms are only provided with an improper Sumerogram – thus in particular *bāmtum*, the "moiety". When written syllabically, it is kept apart from the normal half, whether written $\frac{1}{2}$ or *mišlum* (thus very clearly in Str 367, see p. 239); when the Sumerograms šu.ri.a or $\frac{1}{2}$ are used, this distinction is lost. Thus also *ḫarāṣum*, "to cut off", which (if it has any mathematical logogram at all), borrows kud from *nakāsum*/*haṣābum*, "to cut down/break off". The same holds good of *šutamḫurum*, "to make confront itself", which (apart from the rare use of íb.si_8) shares its Sumerograms with *šutākulum* – and even the pun ì.gu_7.gu_7 with its totally misleading semantics of "eating" may be characterized as an inadequate Sumerogram for *šutakūlum*, "to make hold".

bāmtum is an essential concept in the second-degree algebra, as are *šutakūlum* and *šutamḫurum*; *ḫarāṣum* is more used in the early groups 1 and 7 than *nasāḫum*, "to tear out". The lack of proper Sumerograms for these is in itself a sufficient argument for the non-Sumerian origin of the second-degree algebra. The absence of adequate equivalents for the whole metalanguage (logical operators, interrogative phrases, announcement of results, algebraic parenthesis and indication of equality) shows that the whole *discourse of problems* was absent from the legacy left by the Ur III school.[464]

463 The closest we ever get (and the only time we get close) is the verb ub.te.gu_7 in the Sumerographic Nippur text CBM 12648 – cf. note 404.

464 An unexpected and, even at the distance of 4000 years, atrocious testimony of one of the most oppressive social systems of history – no modern despotism (except

Such a school, with no space for training problems nor for supra-utilitarian pursuits (not to speak of genuine intellectual interest), was certainly not the cradle of Old Babylonian algebra: no wonder that even the continuation of the neo-Sumerian computational tradition in the beginning of the Old Babylonian period (as reflected, for instance, in the Ur texts, in the brick carrying problems of AO 8862 and in Haddad 104) had to innovate and develop a terminology for problem formulation – the legacy alone left no space for that virtuosity and "humanism" which was the core of the emerging scribal culture.[465]

The "Surveyors' Proto-algebra"

The introduction of problems in the Old Babylonian school was a *re*introduction. Supra-utilitarian problems turn up in the record around 2600 BCE, at the same time as the autonomous scribal profession – cf. p. 313 and note 298; they ask explicitly for a magnitude *X* by the phrase *X*.bi, "its *X*". Of particular interest for our present discussion is a group of problems on squares and rectangles from the Sargonic epoch (c. 2200 BCE or earlier; see [Whiting 1984]). Those on squares seem to use the rule

$$\square(R-r)+2\sqsubset\!\sqsupset(R,r)=\square(R)+\square(r)\ ,$$

(cf. note 304); in those on rectangles, the area and one side is given, and the other is asked for. Results are "seen" (pàd). Questions are still asked by .bi.

The rectangle problems are of the same type as the first supra-utilitarian problems of the Old Babylonian catalogue text YBC 4612 (cf. note 343), given $\sqsubset\!\sqsupset(l,w)$ and either *l* or *w*. It is therefore striking that the other members of

the Red Khmers who abolished schools altogether) ever suppressed independent thinking to the point that school teaching gave up the use of problems. It corresponds all too well to this commentary to the yearly balance of a scribe, who had squeezed the equivalent of $7420\frac{1}{6}$ female labour days less out of his assigned labour crew than he was expected to [Nissen, Damerow, and Englund 1993: 54]:

> From other texts we know what drastic consequences such continuous controls of deficits meant for the foreman and his household. Apparently, the debts had to be settled at all costs. The death of a foreman in debt resulted in the confiscation of his possessions as compensation for the state. One consequence of such a confiscation was that the remaining members of the household could be transferred into the royal labor force and required to perform the work formerly supervised by the deceased foreman.

Whether it was planned or not, Ur III seems to have achieved exactly what Orwell's [1954: 241] Newspeak was meant to effectuate: "to make all [unauthorized] modes of thought impossible" – at least among those overseer scribes who constituted the "Outer Party" of the statal system.

465 Cf. p. 315 on the new scribal culture and p. 360 on innovations.

this fixed set are missing, those in which $\sqsubset\!\sqsupset(l,w)$ and either $l+w$ or $l-w$ are given.

Also Sargonic is a tablet with a bisected trapezium – see [Friberg 1990: 541]. The tablet carries no verbal text, it only shows the trapezium and the sides; from the Old Babylonian texts, however, we know that the procedure would be explained in terms of "making hold", and that the argument would be made by scaling from the simple case where the trapezium is part of the border between two concentric squares (see pp. 238 and 247).

The links between these Sargonic texts and the basis for Old Babylonian are not to be doubted; the use of *tammar* for "seeing" in the Old Babylonian texts shows that the transmission has taken place outside the Sumerophone school environment (on the rare logographic writings, see p. 357).

The absence of problems with given values for $\sqsubset\!\sqsupset(l,w)$ and either $l+w$ or $l-w$ (even though their later perpetual partners are found) is a strong suggestion that the device of the quadratic completion had not yet been invented in the Old Akkadian period; the absence from the Ur corpus (which does reflect the bisection of a trapezium, without demonstrating understanding) could suggest an even later discovery. Since exactly this ruse appears to be spoken of as "the Akkadian (method)" in TMS IX (see p. 94), it is a well-supported assumption that it was discovered in a lay, Akkadian-speaking surveyors' environment, at first as a mere trick comparable to the intermediate stop of the camel (see p. 365). It will soon have been discovered that it could be used for several problem types, and when the new Akkadian scribe school adopted the stock of proto-algebraic riddles and made it the starting point for its creation of a genuine algebraic discipline, this stock will already have encompassed the problems listed on p. 373 – but probably also other problems that turn up in later treatises but were eliminated by the school, and which are therefore not present in the intersection between the Old Babylonian corpus and the later traditions (we should notice that the square problem ${}_4s+\square(s) = \alpha$ has only survived in the school corpus as a citation of non-school material):

On a single square with side s, "all four sides" ${}_4s$ and diagonal d:

$$\square(s)-{}_4s = \alpha$$

$$\square(s) = {}_4s$$

$${}_4s-\square(s) = \alpha$$

$$d-s = 4$$ [466]

466 The problem $d-s = \alpha$ is only found in the *Liber mensurationum*, where α is successively 4, 5, and 4, and with Fibonacci followed by Piero and Pacioli (α = 6). In Abū Bakr's first instance, the exact solution is found ($s = 4+\sqrt{32}$; Fibonacci, etc., also exact, find $6+\sqrt{72}$); Abū Bakr's case $\alpha = 5$ only refers to the preceding method; his third case differs from the first solely by speaking of the excess of the diagonal over "each of the sides" ($d = s+5$) and not of subtraction proper.

Abū Bakr's solution seems to depend on the side-and-diagonal numbers: expressing s and d in terms of the previous stage, $s = \sigma+\delta$, $d = 2\sigma+\delta$ we have $\sigma =$

On rectangles with length l, width w, "all four sides" ${}_4s$ perhaps also these:[467]

$$\sqsubset\!\sqsupset(l,w) = {}_4s$$

$$\sqsubset\!\sqsupset(l,w)+{}_4s = \alpha,\ l = w+\beta$$

and possibly versions of the problems listed on p. 373 that involve "both length" and "both widths".

Yet regardless of these extensions, the stock remains a strictly limited stock of riddles, with no variation of coefficients: ontologically speaking, "all four sides" is not the same thing as "four times the side", and it does not invite to further variation. The riddles, moreover, are really *geometrical*: the entities that are used in the procedure are really those lengths, widths, and surfaces which the problems treat of, there is nothing like a functionally abstract representation. None of the later sources that inform us about the lay tradition contains anything like BM 13901 #12 (p. 71), in which a surface was represented by a line. The method is analytical, and in the context of the Old Babylonian school it became the backbone of a discipline which we may reasonably regard as "algebra". In itself, however, the riddles and the technique by which they were solved constituted neither discipline nor algebra; in view of what they gave rise to we may characterize them as *proto-algebra*.

Scholastization

What we encounter in the Old Babylonian tablets (even in text groups 1 and 7) is already very different. We are therefore not able to follow the details of the process and to ascertain which steps came first and which came later; yet we

---

4, $\delta = \sqrt{(2\sigma^2)} = \sqrt{(2\cdot(d-s)^2)} = \sqrt{(2\cdot 16)} = \sqrt{32}$. Fibonacci's solution (repeated by Pacioli) depends on the diagram shown in Figure 70. Piero uses algebra.

Abū Bakr's repetition of the same problem in versions which differ only in their choice of subtractive operation makes it unlikely that it was invented by the competent mathematician Abū Bakr; he will rather have taken over a familiar problem whose current solution is likely to have been derived from a practical assumption that the diagonal in the 10×10-square be 14; at some moment, a copyist may then have inserted the other version that was around. The assumption that 14 is the diagonal that corresponds to the square side 10 may be as old as the routine reference to the square in question. We have certainly no means to know whether it served as the *pretext for a problem* much before Abū Bakr's times, but in view of the not unlikely use of the side-and-diagonal numbers already in Old Babylonian times (see p. 263) the assumption is not unreasonable.

467 The interest in "all four sides" of a rectangle turns up in certain Greek sources and with Mahāvīra; however, they are not found in Abū Bakr or other Arabic or derived Latin or Italian sources, which might indicate that they represent a late reformulation or trivial expansion of the stock.

may identify a number of changes which taken together created the mature algebraic discipline:

- The two basic tricks (completion and scaling) achieved a new status as widely applicable *methods*. This change will already have been under way in the case of the completion trick, which serves in many of the riddles. Scaling, on the other hand, was only of use in the circle problem, and was thus certainly an isolated trick in the original context; in the texts analyzed above it has become a main pillar.
- Applications of the new technique were found in what seem to have been a process of active experimentation. Some of those that were found were obvious: *igûm–igibûm*-problems as applications of the basic rectangle problems, the sum of men, days, and bricks in AO 8862 #7. Others are truly astonishing: combined rate-problems in the style of TMS XIII (listed already in the "Tell Ḫarmal Compendium" IM 52916, IM 52304+52304); broken-reed problems, both versions of which (see p. 209) are found in the early groups;[468] and the unfinished ramp (only found in the late group 6). An important aspect of this effort is the transfer of the surveyors' technique to domains that belonged within the tradition of scribal computation (the contrast between the surveying problems AO 8862 #1–4 and the brick-carrying problems #5–7 is a beautiful illustration).

  At times the experiments would lead to the formulation of problems that could not be solved by the inherited methods; this is the case of the third-degree excavation problems of BM 85200+VAT 6599, which are obvious calques of the (square or rectangular) area-and-sides problems but turn out not to be solvable by any calque of the completion method.

  Other experiments would lead to reducible higher-degree problems; we may think of BM 13901 #12 and TMS XIX #2. Both instances exemplify the particular inner-geometric variant of the representation principle in which the abstract "space" of line segments is generalized so as to encompass rectangular areas and rectangular prismatic volumes.
- The "natural" pseudo-coefficients "all four" and "both" were expelled, probably as uninteresting, and replaced by a true variation of coefficients (and by other kinds of systematic variation, of which the series texts provide the high point). A particular reason for this may have been the function of the algebraic problems: apart from being supra-utilitarian ways to display virtuosity and scribal "humanism" at a higher level than the traditional riddles, they will have been a useful pretext to train computation in the sexagesimal place value system.

---

468 One might perhaps expect problems of this type to have been devised in the lay surveyors' environment – but since nothing similar is found in *any* of the later sources, this *a priori* expectation seems to be unwarranted. The computations that are involved, we may add, are probably so complicated that a non-literate or semi-literate environment would not find the problems attractive, and therefore not invent them.

Connected to the variation of coefficients was the transformation of linear equations and the "accounting" method for finding coefficients, none of which have any place within the group of traditional riddles (apart, again, from one isolated and simple application in the circle problem).

- Expelled were also non-generalizable tricks like that used in BM 13901 #23,[469] mock solutions as that of IM 53957 (see p. 321), and non-solutions as the postulate that the side of the square is 10 if $d = s+4$ (if this was really part of the heritage, which remains a hypothesis).
- The citation in BM 13901 #23 suggests that the lay favourite value for side of the square was 10. The same can be seen in the *Liber mensurationum* and onwards, but also in a number of other later sources: in al-Khwārizmī's algebra, the archetype of the irrational square root is $\sqrt{200}$, the diagonal of □(10); in Hero's *Metrica* and in *Geometrica*/mss A+C and a number of other works, 10 is the habitual side of regular polygons (but not in other cases where the choice is free, which shows it to be really the preferred value in *this* situation)[470]. In the rest of the Old Babylonian corpus, the preferred square side is 30´ (or 30) – and in TMS II this is also the side of the regular hexagon and heptagon. The adaption of the lay heritage to the school will thus also have entailed a numerical normalization, in which the number 10 – round and striking among speakers of Akkadian with its decadic number words, much less so to users of the sexagesimal system – was replaced by 30´ (when located in the "school yard" order of magnitude). Apart from the simplicity of the value (30´ = $\frac{1}{2}$) it may have played a role that 30´ n i n d a n is identical with the unit 1 g i (or "reed"), a standard unit in third-millennium Sumerian mensuration.[471]

  We do not know how far the later predilection for 6 and 8 as sides of a rectangle (with diagonal 10) goes back in time; so much seems certain, however, that the Old Babylonian predominance of 45´, 1, and 1°15´ (or 45, 1`, and 1`15) is a consequence of the adaptation to sexagesimal reckoning, and thus another instance of numerical normalization.
- The "appending" of lines to surface, that is, the presupposition that lines are virtually broad, is only found in groups 1 and 7 and in those catalogues from Eshnunna, Nippur, and Susa which refer to the side of a square as its "length"; in other texts the "broad line" can still be shown to be present as a conceptual substructure (cf. p. 291), but it never appears explicitly. It follows that the notion of broad lines must be a heritage from

469 Most likely, the original solution to the problem consisted in filling out all four corners, as suggested by the first geometric demonstration in al-Khwārizmī's *Algebra* (see Figure 88). But this makes the trick neither more nor less suited for generalization.

470 See [Høyrup 1997: 90*f*].

471 I owe this observation to Marvin Powell (personal communication). We notice that 30´ is also the initial length of the broken reed in VAT 7532, and the length of the pole in BM 85196 #9.

the lay surveyors, and that the school could not accept it as such but only after a "critical" reinterpretation. One – widespread – articulation of this critique is the replacement of "appending" by "accumulating", that is, by a statement of problems as dealing with the arithmetical sum of measuring numbers and not with a supposedly concrete combination.[472] Another articulation, much less widespread, is the rationalization of the notion by means of a *waṣītum*, "projection", (BM 13901), of a "second" or "alternate" width (YBC 4714), or of a KI.GUB.GUB, "base" (TMS IX).

But the elimination/rationalization of the "broad line" is not the only instance of critique; another one is the introduction in some text groups of the "norm of concreteness" (pp. 58, 174, and elsewhere), according the which an entity has to be at disposition before it can be appended to another entity. Since this norm is not observed in the early groups (nor in group 6), the "concreteness of thought" which it presupposes is no indication of "primitivity" or failing capability to think in abstract terms; it is the consequence of a *metamathematical choice*, and thus comparable to the eviction of fractions from Greek theoretical arithmetic and to Viète's insistence on homogeneity.[473]

- In the early phase (perhaps as experiments, but in view of the simultaneous occurrence of the phenomenon in groups 1 and 7 and the reappearance in Late Babylonian texts characterized by readoption of lay material most likely in imitation of a lay practice) we find a number of attempts to formulate rules in general terms and not through paradigmatic examples. Later such attempts were given up, probably because they are pedagogically inefficient (in view of the difficulties with which the general formulations present modern workers, this choice will have been fully justified – the terminology was too geared to the numerical example to be unambiguous when deprived of this support).[474]
- Finally, as we have seen in Chapter IX and as summed up on p. 360, a number of schools undertook to develop a canonical format in which mathematics should be formulated; even when borrowing material from

472 This is beautifully illustrated by the faulty use of *kamārum* instead of *waṣābum* in CBS 43 (see note 421): it seems to be due to bad conscience, to the scribe's awareness that the addition of sides to areas *ought to* be spoken of in this way.

473 Viète's case is indeed a close parallel to the introduction of the "projection", and a critique of the use of broad lines in the algebra of his epoch. A few decennia before Viète, Pedro Nuñez [1567: fols. 6r, 232r] had explained that the "roots" of algebra were indeed such broad lines, to be understood as rectangles whose width is "la unidad lineal".

474 Reversely, the general rule is an adequate tool for an oral tradition, being more easily remembered mechanically and transmitted faithfully than the full paradigmatic example; explanations and examples could then be improvised once the master knew what was meant by a possibly ambiguous rule (a parallel is offered by the relation between fixed formulae and relatively free use of these by the singer in oral epic pietry, see [Lord 1960: 99–102 and *passim*]).

elsewhere, they would adapt it to this local canon (with just enough lapses to allow us a glimpse of the process – fully successful adaptations may be plentiful, but cannot be identified).

Together, these features show us what happened when the supra-utilitarian knowledge of the lay surveyors was "scholasticized". We find that repetition with variation that is the inherent inclination of any school system – but often with much more system in the variation than in other schools, in agreement with that particular passion for order that had characterized the Mesopotamian school since its fourth-millennium beginnings (cf. p. 203). We finally find a systematic search for *new applications* of the techniques at hand which might turn out to ask for the creation of *new techniques*; this applies at least to the third-degree excavation problems, but probably also to some higher-degree problems in the series texts where we do not know how they were meant to be approached. According to all we know, the purpose remained that of testing the reach of professional tools and to demonstrate professional valour (most of all perhaps the professional valour of the teacher); but the effect sometimes comes close to that of a genuine scientific exploration. These experiments and this interest in the supra-utilitarian level express the particular ideals of the Old Babylonian age and its school, and stand in strong contrast to the suppressive conditions of Ur III scribal life as revealed in the absence of the very possibility to formulate a school culture where the student should think for himself.

Another contrast to Ur III conditions that was made possible by the cultural climate of the Old Babylonian epoch is the appeal to the understanding of the student which the didactic Susa texts (TMS VII, IX, XVI) present us with in writing, and which other texts (for instance, YBC 8633) suggest was made orally elsewhere. At a higher level, this same search for intelligibility made the authors of the texts go beyond the intuitive transparency of the naive proto-algebraic technique and prompted them to rephrase it "critically".

At an early point (p. 8) it was formulated that the authors of the mathematical texts were "teachers of computation, at times teachers of pure, unapplicable computation, and plausibly specialists in this branch of scribal education; but they remained teachers", and that it would be misleading to speak of them as "Babylonian mathematicians" unless we "take care to remain *very* aware of this difference". We may now formulate in which sense it is *not* misleading to speak of them as mathematicians: *Not* because they were concerned with number and measurable quantity; *not* because they performed numerical computations; and *not* because they based these computations on rules which *we* derive by mathematical considerations – the reasons for which the characterization is currently used, and which are indeed the only reasons that could be given as long as the texts were understood as mere prescriptions of numerical procedures. They were mathematicians in the sense that they explored the possibilities of their conceptual and technical tools, because their pedagogical method made them aim at system and at comprehension of principles, and because the aspiration to make their procedures transparent and comprehensible made them strive to develop clear formats and caused them to engage in critique.

An Aside on the Pythagorean Rule

To historians of mathematics, the most striking facet of the new view of Babylonian mathematics that emerged in the 1930s was certainly the algebra. Until then, indeed, algebra had been supposed to be an invention of medieval India and Islam, somehow anticipated in Diophantos's *Arithmetic* and the "geometric algebra" of *Elements* II; algebra, moreover, was known by them to be the key ingredient in the transformation of ancient into Modern mathematics. When Neugebauer's idea was added that the geometry of *Elements* II was a translation of age-old algebraic Babylonian results, the outcome was certainly revolutionary.

To a general public, unburdened by prejudice about the origin of algebra and more impressed by the difficulty of school geometry than by the manipulation of simple equations – not least, thus, to the general public of Assyriologists – it was (and remains) more striking that even the theorem of Pythagoras appeared to have been known in the Old Babylonian period. After all, the theorem was linked to Greek mathematics not only by its name but also by the familiar anecdote according to which Pythagoras had sacrificed one hundred oxen to the gods because they had granted him the discovery.

Since the preceding chapters shed some light on the possible origin of the rule (which could be no "theorem" in the Old Babylonian context, cf. p. 197), it may be worth summing up what can be said.

Disregarding the special and simple case of the diagonal of the square (represented above by YBC 7289, p. 261), the rule is used in a total of nine texts, seven of which were analyzed above:

- Db_2-146, the rectangle problem with given area and given diagonal (p. 257).
- TMS I, the isosceles triangle inscribed in a circle (p. 265).
- TMS XIX, the first- and the eighth-degree problems about a diagonal (p. 194).
- VAT 6598 #6–7 (now VAT 6598+BM 96957 #18 and #21), the diagonal of the door (p. 268).
- BM 85194 #20–21, the determination of a circular chord from its arrow and vice versa (p. 272).
- BM 85196 #9, the pole against the wall (p. 275).
- YBC 8633, the subdivision of an isosceles triangle (p. 254).

The last two texts are

- TMS III, a table of technical igi.gub constants, which gives 1°15′ as the "constant for the diagonal".
- Plimpton 322, a tabulation not exactly of Pythagorean triples $a - b - c$ but of $\dot{?}?? - \bar{c}^2 - b - c$, ??? standing for one or (probably) more lost columns, $\bar{c} = c/a$. All pairs $(\bar{b},\bar{c}) = (\frac{\tau-t}{2},\frac{\tau-t}{2})$ are listed for which $\sqrt{2}-1 < t < \frac{5}{9}$, t being the ratio between two regular numbers no greater than 125, $\tau = 1/t$. The headings of the columns show that the numbers are understood as having to do with the [length,] width, and diagonal of a rectangle. For the rest, the purpose of the table is an enigma, and none of the explanations suggested so far seem plausible.[475]

The distribution of these texts on groups is striking. Db_2-146 belongs to group 7B (Eshnunna) and is the oldest of all. TMS I, III, and XIX are from group 8 (Susa); VAT 6598, BM 85194, and BM 85196 belong to group 6A (Sippar). Goetze ascribes Plimpton 322 very tentatively to group 1, saying that it "may or may not belong here" [MCT, 147 n.353]. His only argument is the spelling *[in-]na-as-sà-ḫu-ú-ma*, which may indeed just as well (if not better) point to group 6. YBC 8633, the only text that misuses the rule, is also the only text that certainly comes from the core area (possibly Uruk). Whereas most other problem types occur both in the periphery and in the core, computations that involve the Pythagorean rule are thus conspicuously linked to the periphery, where the influence of the lay tradition seems to have been continuous.

For chronological reasons, only Db_2-146 can be informative about the origin of the rule, all the other texts being later. What it tells us is, firstly, that the standard representative of the rectangle with proportions 3:4:5 was already 45′:1′:1°15′ in the early eighteenth century BCE; secondly, in the proof, that the rule was known perfectly well (and that it was formulated exactly as a general *rule*, cf. p. 261).

The solution itself does not make use of the rule; however, as pointed out on p. 261, the diagram used in the solution (see Figure 67) is the basis for the most widely diffused cut-and-paste proof of the rule. If the rule was not known already, it will have been very easy to discover it in connection with the solution of this problem – in particular if also the alternative solution which adds $2A$ to $\square(d)$ instead of subtracting was known. That both solutions were in use is indeed quite likely – the solution of the present tablet turn up

---

475 See [Friberg 1981], whence the preceding analysis is borrowed. Friberg's own proposal – that the table was meant to provide parameters from which second-degree equations could be constructed – though not impossible does not fit the Old Babylonian habit of constructing problems from known very simple solutions. A new perspicacious analysis is [Robson 2001].

again in Savasorda's *Liber embadorum*, whereas Abū Bakr (and Fibonacci) have the additive variant.

Indirect and not fully conclusive arguments can be given that knowledge of the rule does not predate the invention of the quadratic completion. One is that no earlier text suggests use or knowledge of the rule; but this may be a pure accident – only a single text proves that the bisection of the trapezium was known in the Sargonic era. The other is the apparent absence of the rule (except the improper use in YBC 8633) from the core area, which suggests that the discovery was made too late for it to be at hand when the Larsa adoption of surveyors' material took place (whereas the diagram of Figure 12 certainly was, being used in the two group-1 texts AO 8862 and YBC 6504). This may be the reason that the Pythagorean rule never acquired any prominent status in the core region, even though other procedures based on Figure 12 are encountered often (thus in AO 8862, in YBC 6504, in BM 13901). This argument has some force, but alternative reasons for exclusion of the rule cannot be ruled out. It remains a substantiated hypothesis that the rule was discovered around the nineteenth century BCE (or reached the Babylonian periphery from outside at that time), and a mere conjecture that the discovery was made in connection with the solution of the problem of Db_2-146.

The Later Phases

After the disappearance of the school institution, everything mathematical vanishes from the archaeological horizon. From the Kassite period, a single table of technical constants[476] and a single procedure text AO 17264 has come down to us. The topic of the latter is not algebraic but the division of a trapezium between "brothers" into six strips of which the first and the second are equal, the third equals the fourth, and the fifth equals the sixth. The tablet was reported by the dealer to be from Uruk, but the terminology belongs to the peripheral family without being derived from any single group of those discussed above – see [Høyrup 2000a: 159*f*]. As Lis Brack-Bernsen and Olaf Schmidt conclude their analysis of the text [1990: 38], the problem

> is beyond the capability of Babylonian mathematicians, and it looks as if they have given up in despair in their attempt at solving this problem and just given some meaningless computations that lead to a correct result.

In other words, what the text offers is a mock solution, the first since those of the stone riddle in AO 6770 and the filling problem of IM 53957 (counting YBC 6504 #4 as a fallacy and not as a deliberate burlesque). Like these, the text is a piece of sham mathematics – all that remains of the stringency and

---

476 Published in [Kilmer 1960]. The date may actually be even later, cf. [Robson 1999: 26 with n.19].

creativity of the Old Babylonian mathematical school is the higher level on which the fraud is perpetrated.

After this, the first mathematical texts we know about are from the Late Babylonian epoch[477] – sixth to fourth century BCE. Of these, W 23291-x was published in [Friberg, Hunger, and al-Rawi 1990a]; W 23291 was analyzed recently in [Friberg 1997]. W 23273, a metrological table that starts by telling the sacred numbers of the gods [Friberg 1993: 400], is only interesting in the present context because of this beginning; it shows indeed that the cognitive autonomy that had once characterized Old Babylonian mathematics (see p. 307) could not be upheld in an environment where the scribes of mathematical texts presented themselves as "exorcists" (thus the colophone of W 23291-x).

W 23291-x and W 23291, instead, are of real interest. Since they are available in adequate conformal translation, I shall only sum up the essential conclusions that may be drawn concerning the continuation of the algebraic tradition.

Both texts use new area metrologies whose units are *not* the squares on the basic length unit. One type is constituted by the "reed metrologies" (one based on the Sumerian and Old Babylonian gi, "reed", of 6 kùš, the other on the Neobabylonian gi which consisted of 7 kùš). In both cases, an area is measured as the length of the corresponding broad line with breadth 1 gi. The other is the seed metrology, where the area is measured by the amount of grain needed to seed it (according to the standard that 100 silà would seed 3·□(100 kùs)).[478] If any doubt should persist whether the Old Babylonian "surfaces" are meant as such or as metaphors for arithmetical products, no such doubt is permitted by the Late Babylonian texts.

W 23291-x contains various area measurements. None of them are algebraic, nor supra-utilitarian in other ways; of some interest, however, is a problem where the area of the band between concentric circles is determined as the product (in this historical phase always termed a.rà, "steps of") of the width of the band and its mid-length.

W 23291, in contrast, contains a number of proto-algebraic problems: To find one side of a rectangle when the area and the other side are given (since the area is given in seed measure, this is more than a simple division problem); to find the sides when the area and the sum of the sides or their difference are given; to find the sides of two concentric squares when the area of the intermediate band and the width of this band (the "thrusting" of UET V, 864) are given. All, as we see, belong to the original stock of surveyors' riddles. Nothing is connected to the innovations that were introduced by the

---

477 In established terminology, the "Late Babylonian" epoch includes the periods of Chaldaean, Persian, Seleucid, and Parthian rule, from c. 625 BCE onward. For convenience I shall use the term in the sense which in full formulation would be "pre-Seleucid Late Babylonian".

478 Cf. discussion in [MCT, 143–145].

Old Babylonian school; what we see is really a *new* adoption of lay material into the scholarly environment on the conditions of current metrology.

This time, everything remains geometrical (to the extent that the limited text material at our disposal allows us to know); no attempt at representation is undertaken. An interesting new format is created. After stating the problem (in a concrete, paradigmatic example) comes the phrase mu nu zu.ù, "since you do not know",[479] after which follows a rule formulated in general terms. Then come two versions of the actual computation, one made under the condition *šumma* 5 *ammatka*, "if your cubit is 5´", that is, if the nindan is taken as the basic unit and the cubit or kùš (*ammatum* in Babylonian) thus 5´, the other under the assumption that "your cubit is 1", that is, with the kùš as the basic unit. Thanks to the combination with concrete examples, the abstract rules are more comprehensible than those which we find in the Old Babylonian groups 1 and 7.

The basic terminology shows some continuity with what we know from the Old Babylonian texts. igi and igi.gub appear as loanwords, *igûm* and *igigubbûm*; for squares, "each" side is referred to. The general rule ends by telling that "you see" the quantity which was asked for. When spoken of in the general rules, multiplications may be "going" (RÁ) or "raising" (*našûm*/íl); in the actual computations it is a.rá, "steps of". "Accumulation" is g̃ar.g̃ar (not UL.GAR as it is in the Kassite problem text), appending is *waṣābum*/daḫ. *nasāḫum*, "to tear out", is still found in syllabic writing, but may also be expressed by the Sumerogram nim, "to lift", which in the Old Babylonian corpus serves as an alternative Sumerogram for *našûm*, "to raise", or for the synonymous *elûm*. Such changing functions of the same Sumerian word show that the Sumerogram had not been conserved in its technical use. When scholar-scribes regained interest in mathematics they therefore reinvented Sumerograms according to (what they understood to be) the general meaning of Sumerian terms. In the actual case they happened to choose a term whose meaning was very close to the sense in which zi, the Old Babylonian logogram for *nasāḫum*, had been used in the Ur III school (see p. 376).

Seleucid Procedure Texts

Three procedure texts of Seleucid date are known: VAT 7848, AO 6484, and BM 34568. Of these, AO 6484 tells in the colophon to be written by the

---

479 Thus *aššum*, "since", which had no Sumerographic equivalent in the Old Babylonian period, is now written mu. It is true that the phrase when taken alone might also mean "the name I/you do not know", which would fit the use of "name" in the Susa texts in the sense of numerical value or identification; but since the phrase follows after an explicit formulation of the question, this reading seems to be excluded.

astrologer-priest Anu-aba-utēr, who lived in the early second century BCE [Hunger 1968: 40 #92 and *passim*]; the colophons of VAT 7848 and BM 34568 have been destroyed, but according to their whole style they will have been produced in the same environment of exorcists and astrologer priests as AO 6484 and W 23291.

VAT 7848 is clearly related to the Late Babylonian tablets dealt with in the preceding section (but contains no supra-utilitarian matters). For areas it uses seed-measure; the conserved problems find a diagonal from the sides 45′ and 1;[480] the area of an isosceles trapezium, where the height is found by means of the Pythagorean rule (W 23291 find the areas of isosceles triangles similarly); the area of the circle as 5′·□(*c*); the volume of a cube ⧈(10 kùš); and another volume.

In part, AO 6484 presents us with the same picture. It finds the area of a square expressed in seed measure, the area of an isosceles triangle (still in seed measure), the side of a square from the diagonal 10 kùš (multiplying by 42′30″[481]); and it determines the volumes ⧈(*a* kùš), *a* = 1, 2, 3, 4, and 5.

However, AO 6484 also contains other, supra-utilitarian matters: the sums

$$\sum_0^9 2^i \quad \text{and} \quad \sum_1^{10} i^2 ,$$

found nowhere else in the Babylonian record; the rectangle problem

$$l+w+d = \alpha \ , \ \sqsubset\!\sqsupset(l,w) = \beta$$

(without procedure; more on this problem type imminently); and a sequence of *igûm-igibûm* problems, one of which runs as follows:[482]

Rev.

15. The *igûm* and the *igibûm*, 2°3′, steps of 30′ go: 1°1′30″ [1°1′30″ steps of 1°1′30″ go: 1°3′2″15‴].
igi *u igi-bu-ú* 2,3 GAM 30 RÁ-*ma* 1,1,30 [1,1,30 GAM 1,1,30 RÁ-ma 1,3,2,15]

16. 1 from inside diminish: remaining 3′2″15‴. What steps of what [may I go so that: 3′2″15‴?]
1 ta *lib-bi* lal-ma *ri-ḫi* 3,2,15 *mi* GAM *mi* [*lu*-RÁ-*ma* 3,2,15]

17. 13′30″ steps of 13′30″ {...} g[o:] 3′2″15‴. 13′30″ [to 1°1′30″ join: 1°15′ the *igûm*].
13,30 GAM 13,30 {GAM} r[á-*m*]*a* 3,2,15. 13,30 [*a-na* 1,1,30 tab-*ma* 1,15 igi[ú]]

---

480 Neugebauer and Sachs restore the lost identification of the object as a "triangle", and hence speak of a "hypotenuse". They give no reason that the text should share the Greek preference for the right triangle and not the normal Babylonian predilection for the rectangle.

481 This is half of the usual coefficient 1°25′ for the diagonal (see p. 262). Though not known from any extant igi.gub table it seems, however, to be considered as a coefficient on its own since no calculation is made.

482 Based on the transliteration in [MKT I, 98].

18. 13′30″ from 1°1′30″ diminish: 48 the *igibûm*.
13,30 ta 1,1,30 lal-*ma* 48 *igi-b*[*u-ú*]

This problem, and three more of the same kind, are important for understanding the historical process. They are, indeed, *the only certain traces in later sources of that algebra which was created by the Old Babylonian school* (which implies that they themselves left no traces in later traditions). Their existence shows that at least this problem type, corresponding to the simple rectangle problem where the area and the sum of the sides is given, survived within an environment where tables of reciprocals were in use and the igi was spoken of with the loanword *igûm*. A Sumerographic innovation shows, however, that this environment will not have been too literate in Sumerian: tab, which in the Old Babylonian texts serves as a logogram for *eṣēpum*, "to double, to repeat (until *n*)", now stands for the identity-conserving addition. Once again we see that the scholar-scribes, when returning to mathematics and expressing it in their usual way, invented new Sumerographic equivalences on the basis of the general semantics of Sumerian words; since tab means "to be/make double", "to clutch/clasp to" [SLa, 318], it may indeed serve for *waṣābum*, "to append" (or *ṭepûm*, which had largely replaced it in the Seleucid epoch) just as well as for *eṣēpum*.[483]

Among other terminological innovations, we may note the use of the sign GAM (two oblique wedges drawn in line) in the function of a.rá, "steps of"; and the way a square root of *a* is asked for, namely, by the purely arithmetical question "what steps of what may I go so that (I get) *a*". In the other major Seleucid procedure text, BM 34568 (see imminently), we observe a continued reference to magnitudes which you "do not know", present in the Old Babylonian corpus in Nippur, in groups 1, 3, 6, 7, 8, and in the series texts, and again in the Late Babylonian texts.

BM 34568

By far the most interesting Seleucid text is the theme text BM 34568, which deserves a treatment of its own; most of what it contains is certainly supra-utilitarian. I transliterate and translate some significant extracts:[484]

---

483 The equivalence is invented *ad hoc* for use in the mathematical texts; outside mathematics, tab appears never to have been used logographically for any of the two additive terms.

484 Based on the transliteration in [MKT III, 14–17], with correction [von Soden 1964: 48a].

Obv. I

#1

1. [4 the length, 3 the wid]th, ⸢what the diagonal?⸣ Si[nc]e you do not know, ½ of your length
[4 uš 3 sa]ğ.ki ⸢en bar.NUN⸣ m[u] nu.zu^{ú} ½ *šá* uš^{ka}

2. [to] your [w]idth you join: That is it. 4, the length, steps of 30′ go: 2.
[*a-na* s]ağ.ki^{ka} tab-*ma šu-ú* 4 uš GAM 30 RÁ-*ma* 2

3. [2 t⸣o 3 you join: 5. 5 the diagonal. The th[i]rd of your width
[2 *a*⸣-*na* 3 tab-*ma* 5 5 bar.NUN *š*[*a*]*l-šú šá* sağ.ki^{ka}

4. [t]o your length you join: that is the diagonal. 3, the width, steps of 20′ you go, 1.
[*a-*]*na* uš^{ka} tab-*ma šu-ú* bar.NUN 3 sağ a.rá 20 RÁ 1

5. [1] to 4 you join: 5. 5 the diagonal.
[1] *a-na* 4 tab-*ma* 5 5 bar.NUN

#2

6. [4 the le]ngth and 5 the diagonal, what the width? Since you do not know, 4 steps of 4,
[4 u]š *ù* 5 bar.NUN en sağ mu nu.zu^{ú} 4 GAM 4

7. 16. 5 steps of 5, 25. 16 from 25 you lift:[485] remaining 9.
[1]6 5 GAM 5 25 16 ta 25 nim-*ma ri-ḫi* 9

8. What steps of what may I go so that 9?[486] 3 steps of 3, 9.
mi-nu-ú GAM *mi-ni-i lu*-RÁ-*ma lu* 9 3 GAM 3 9

8a. 3 the width
3 sağ

---

485 On earlier occasions – [Høyrup 1990: 345] and elsewhere – I have followed Neugebauer's interpretation of *a* ta *b* nim, taking ta seriously as a Sumerian ablative case suffix, which would make the phrase mean "ascend from *a* to *b*", that is, "count the difference from *a* to *b*". Comparison with the subtractions in AO 6484 (*a* ta *lìb-bi b* lal, "*a* from inside *b* diminish") and W 23291 (*a ina b* nim, "*a* from *b* lift" – cf. p. 389) shows that this reading as good Sumerian is mistaken. ta is no suffix but simply serves as a logogram for *ina*; the subtraction has not become arithmeticized but remains one of concrete removal.

Evidently, the ensuing reference to what "remains" should have make the counting-interpretation suspect. I remember to have been worried, but – as it turns out – not worried enough.

By analogy, it seems reasonable to read šè as an independent logogram for *ana* and not as a terminative case suffix. This does not change the interpretation, but makes more intuitive sense. For instance, in obv. II 5, a reading as a suffix implies that the length is found by joining the width to the excess of length over width; the reading *ana* instead makes us join the excess to the width.

486 My choice of the first person singular in these questions emulates the Old Babylonian division question and is unsupported by the text, the writing of the verb being logographic. It is true that the genuine precative forms are only found in the first and the third person, but the present construction might correspond to the Late Babylonian forms described in [GAG, §81e].

#3

9. The diagonal and the length I have accumulated: 9. 3 the width. Wh[at the length] and the diagonal. Since you do not know,
bar.NUN *ù* uš ĝar-*ma* 9 3 saĝ e[n uš] *u* bar.NUN mu nu.zu$^{\acute{u}}$

10. 9 steps of 9, 1`21, and 3 steps of 3, [9. 9] from 1`21 you lift:
9 GAM 9 1,21 *ù* 3 GAM 3 [9 9] ta 1,21 nim-*ma*

11. remaining 1`12. 1`12 steps of 30´ you go[: 36.] 9 steps of what
ri-ḫi 1,12 1,12 GAM 30 RÁ[-*ma* 36] 9 GAM *mi-ni-i*

12. may I go so that 39? 9 steps of 4 you g[o:] 36. 4 the length.
lu-RÁ-*ma lu* 36 9 GAM 4 R[Á-*ma*] 36 4 uš

13. 4 ⟨from⟩ 9 you lift: remaining 5. 5 the diagonal.
4 ⟨ta⟩ 9 nim-*ma ri-ḫi* 5 5 bar.NUN

#5

19. [1` the length, 32 the width, what the diagonal? Sin]ce you do not know,
[1 uš 32 saĝ en bar.NUN m]u nu.zu$^{\acute{u}}$

20. [1` steps of 1`, 1``. 32] steps of 32, 17`4. You accumulate: 1``17`4.
[1 GAM 1 1 32] GAM 32 17,4 ĝar.ĝar-*ma* 1,17,4

21. [What steps of w]hat may I go so that 1``17`4?
[*mi-nu-ú* GAM *m*]*i-ni-i lu*-RÁ-*ma lu* 1,17,4

22. [1`8 steps of 1`8, 1``]17`4. 1`8 the diagonal.
[1,8 GAM 1,8 1,]17,4 1,8 bar.NUN

#6

23. [1` the length and 3]2 my width. What the surface? Since you do not know,
[1 uš *ù* 3]2 saĝ-*a* en a.šà mu nu.zu$^{\acute{u}}$

24. [1` steps of 32 you g]o: 32`. 32` the surface.
[1 GAM 32 R]Á-*ma* 32 32 a.šà

Obv. II

#9

1. The length and the width I have accumulated[:] 14, and 48 the surface.
uš *ù* saĝ ĝar-[*m*]*a* 14 *ù* 48 a.šà

2. Since you do not know, 14 steps of 14, 3`16. 48 steps of 4, 3`12.
mu nu.zu$^{\acute{u}}$ 14 a.rá 14 3,16 48 GAM 4 3,12

3. 3`12 from 3`[1]6 you lift: remaining 4. What steps of what
3,12 ta 3,[1]6 nim-*ma ri-ḫi* 4 *mi-nu-ú* GAM *mi-ni-i*

4. may I go so that 4? 2 steps of 2, 4. 2 from 14 you lift: remaining 12.
lu-RÁ-*ma lu* 4 2 GAM 2 4 2 ta 14 nim-*ma ri-ḫi* 12

5. 12 steps of 30´, 6. 6 the width. 2 to 6 you join: 8. 8 the length.
12 GAM 30 6 6 saĝ 2 šè 6 *te-tip-pi-ma* 8 8 uš

#10

6. The length and the width I have accumulated: 23, and 17 the diagonal. Since you do not know,
uš *ù* saĝ ĝar-*ma* 23 *ù* 17 bar.NUN mu nu.zu$^{\acute{u}}$

7. 23 steps of 23, 8`49. 17 steps of 17, 4`49. 4`49 from
23 GAM 23 8,49 17 GAM 17 4,49 4,49 ta

8. 8`49 you lift: remaining 4`. 4` steps of 2, 8`. 8` from 8`49 you lift:
8,49 nim-*ma ri-ḫi* 4 4 GAM 2 8 8 ta 8,49 nim-*ma*

9. remaining 49. What steps of what may I go
ri-ḫi {erasure} 49 *mi-nu-ú* GAM *mi-ni-i lu*-RÁ-*ma*

10. so that 49? 7 steps of 7, 49. 7 from 23 you lift, remaining 16.
lu 49 7 GAM 7 49 7 ta 23 nim *ri-ḫi* 16

11. 16 steps of 30´ go, 8. 8 the width. 7 to 8 you join: 15. 15 the length.
16 GAM 30 RÁ 8 8 sağ 7 šè 8 *t*[*e-ṭip-p*]*i-ma* 15 15 uš

#12

17. One reed together with a wall I have erected. 3 kùš as much as I have gone down
⟨1⟩^{en} gi ki i.zi *za-qip* 3 kùš *ki ú-ri-du*

18. 9 kùš it has moved away. What the reed? What the wall? Since you do not know,
9 kùš bad en gi en i.zi mu nu.zu^{ú}

19. ⌊3 steps of 3,⌋ 9. 9 steps of 9, 1`21. 9 to 1`21 y[ou jo]in[:]
⌊3 GAM 3⌋ 9 9 GAM 9 1,21 9 šè 1,21 *t*[*e-ṭip-p*]*i-m*[*a*]

20. 1`3[0 steps of 3]0´ you go: 45. Igi 3, 20´. 20´ [steps of 45 you go:]
1,3[0 GAM 3]0 RÁ-*ma* 45 igi 3 ğál.bi 20 20 [GAM 45 RÁ-*ma*]

21. 15 the reed. Wh[at the wa]ll? 15 steps of 15, 3`45. 9 [steps of 9],
15 gi e[n i.]zi 15 GAM 15 3,45 9 [GAM 9]

22. 1`21. 1`21 from 3`45 you lift, remaining [2`24].
1,21 1,21 ta 3,45 nim-*ma ri-ḫi* [2,24]

23. What steps of what may I go so that 2`[24?]
mi-nu-ú GAM *mi-ni-i lu*-RÁ-*ma lu* 2,[24]

24. 12 steps of 12, 2`24. 12 the wall.
12 GAM 12 2,24 12 i.zi

#13

25. The diagonal and the length I have accumulated: 9. The diagonal and the width I have accumulated[:8.]
bar.NUN *ù* uš ğar-*ma* 9 bar.NUN *ù* sağ ğar-*m*[*a* 8]

26. What the width? Since you do not know, 9 steps of 9, 1`21. 8 [steps of 8, 1`4].
en sağ mu nu.zu^{ú} 9 GAM 9 1,21 8 [GAM 8 1,4]

27. 1`21 and 1`4 you accumulate: 2`25. 1 from [2`25]
1,21 *ù* 1,4 ğar.ğar-*ma* 2,25 1 ta [2,25]

28. you lift: remaining 2`24. Wha[t] steps of wha[t may I go:]
nim-*ma ri-ḫi* 2,24 *mi-nu*[*-ù*] GAM *mi-n*[*i-i lu*-RÁ-*ma*]

29. [so] that 2`24? 12 steps of 12, 2`24. 9 from 12 [you lift:]
[*l*]*u* 2,24 12 GAM 12 2,24 9 ta 12 [nim-*ma*]

30. remaining 3, the width. 3 from 8 [you l]ift: remaining [!]5, the diagonal.]
ri-ḫi 3 sağ 3 ta 8 [n]im-*ma ri-ḫi* [!]5 bar.NUN]

31. 5 from 9 you lift: remaining 4. 4 the length.
5 ta 9 nim-*ma ri-ḫi* 4 4 uš

#14

32. The length, the width, and the diagonal I have accumulated: 1`10, and [7` the surface].
uš sağ *ù* bar.NUN ğar-*ma* 1,10 *ù* [7 a.šà]

Rev. I

1. [what, as much as the length, the width, and the diago]nal? Since you do n[ot kno]w
[*mi-nu-ú ki-i* us sağ *ù* bar.NU]N mu n[u.zu][ú]

2. [1`10 steps of 1`10, 1``21`4]0. 7` step[s of 2 you go, 14`. 14` from]
[1,10 GAM 1,10 1,21,4]0 7 a.r[á 2 RÁ-*ma* 14 14 ta]

3. [1``21`40 you lift: re]maining 1``7`4[0. 1``7`40 steps of 30` you go:]
[1,21,40 nim-*ma r*]*i-ḫi* 1,7,4[0 1,7,40 GAM 30 RÁ-*ma*]

4. [33`50. 1`10 steps] of w[h]at ma[y I go]: so [that 33`50?]
[33,50 1,10 GAM] *m*[*i-n*]*i-i l*[*u*-RÁ-*ma*] *l*[*u* 33,50]

5. [1`10 steps of 29, 3]3`50. 29 the diagonal.
[1,10 GAM 29 3]3,50 29 bar.NUN

#15

6. [The length over the wid]th 7 kù[š] goes beyo[nd. 2`⌉ the surface. Wha[t]
[uš ugu sa]ğ 7 kù[š] *ú-a-ti*[*r* 2⌉ a.šà *mi-n*[*u-ú*]

7. [the length and[!] the width? 2` steps of 4, 8`. 7 steps of 7, 49. 8` ⌊and]
[uš *ù*[!] sağ 2 GAM 4 8 7 GAM 7 49 8 ⌊*ù*]

8. [49 (with) each] other you join: 8`49. What [ste]ps of wha[t]
[49 *a-ḫa*]-*míš* tab-*ma* 8,49 *mi-nu-ú* [GA]M *mi-n*[*i-i*]

9. [may I go: so that 8`4]9? 23 steps of 23, 8`49. ⌊7⌋ from
[*lu*-RÁ-*ma lu* 8,4]9 23 GAM 23 8,49 ⌊7⌋ ta

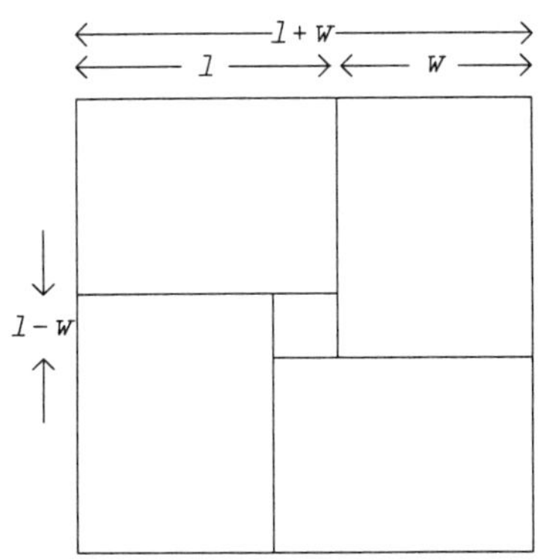

Figure 84. The diagram underlying BM 34568 #9 and #15.

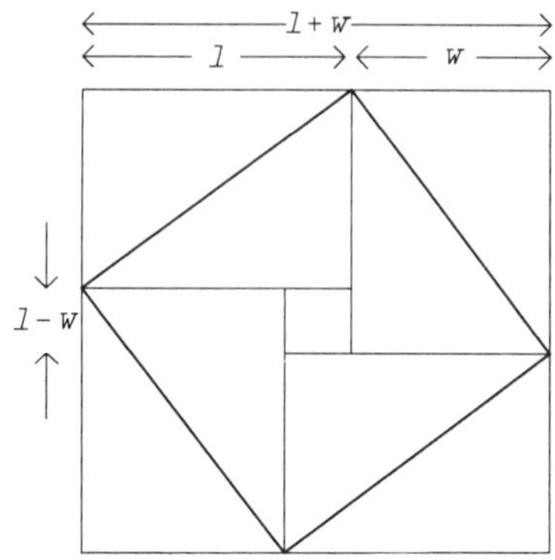

Figure 85. The diagram underlying Bm 34568 #10.

10. [23 you lif]t [: re]maining 16. 16 steps of 30′ you go, 8. 8 the wid[th].
[23 ni]m-[*ma r*]*i-ḫi* 16 16 GAM 30 RÁ 8 8 sa[g̃]

11. [7 and$_1$ 8 join: 15. 15 the length.
[7 *ù*$_1$ 8 tab-*ma* 15 15 uš

Of the problems omitted above, #4 repeats #3, switching only the roles of length and width (and changing the numbers). #7–8 repeat #5–6 with other numbers. #11 repeats #3 with other numbers. #16 is a simple alloying problem. #17 and #18 repeat #14 with other numbers, #18 in general formulation. Finally, # 19 repeats #13 with other numbers, also in general formulation. After this follow a number of problems in rev. II, almost completely destroyed. Apart from the alligation problem, the above excerpt thus contains all problem types from the text that have survived.

The first observation to be made is that *everything is new except the trivial determination of the area and the diagonal of a rectangle from the sides and of the width from length and diagonal.*

#1 is not particularly interesting from our present point of view but only as an early instance of a mathematically trivial interest in the rectangle or right triangle where *w*, *l*, and *d* form an arithmetical progression, an interest which surfaces again in various later practical geometries.[487]

#9 and #15 are not new as problems, but their method is innovative. The procedure is based on the principle shown in Figures 84 and 85 – which we already know from Figure 12 (etc.), but used in a new way. In #9, the complete square □(*l*+*w*) is found to be 3`16, from which 4 times the area ⊏⊐(*l*,*w*) is removed, that is, 3`12; this leaves □(*l*–*w*) = 4, whence *l*–*w* = 2. Subtracting this difference from *l*+*w* = 14 leaves twice the width. The width is therefore 6, and addition of the difference gives the length. Not the faintest

487 Thus in *Liber mensurationum*, #49–50 [ed. Busard 1968: 97*f*], and in Fibonacci's *Pratica geometrie* [ed. Boncompagni 1862: 70*f*].

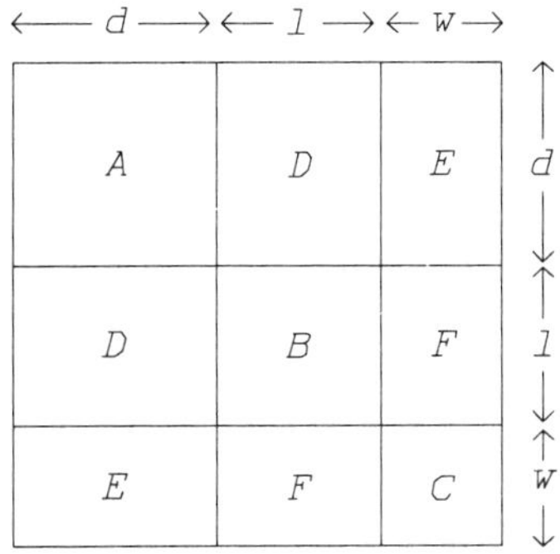

Figure 86. The diagram underlying BM 34568 #3, etc.

trace of the method of average and deviation is left, nor of any cut-and-paste procedure.

#15 is strictly analogous, and #10 is closely related, using the same configuration but with inserted diagonals (see Figure 85), as demonstrated by the apparently roundabout calculation of $23^2-2\cdot(23^2-17^2)$ instead of $2\cdot17^2-23^2$. At first the square on the diagonal $\square(d)$ = 4`49 is lifted from $\square(l+w)$ = 8`49, which leaves 4` as the area of 4 half-rectangles. This remainder is doubled, which tells us that 4 times the rectangular area is 8` – and lifting this again from 8`49 leaves 49 for $\square(l-w)$. Thereby we are brought to the situation of #9, line 3, and the rest follows exactly the same pattern.

#3 and its analogues seem to be based on something like the diagram shown in Figure 86 – in which, we should keep in mind, $\square(d) = \square(l)+\square(w)$, whence $A = B+C$. At first $\square(l+d) = A+2D+B$ is found to be 1`21. Subtracting $\square(w) = C = 9$ leaves $(A-C)+2D+B = 2(B+D) = 2\sqsubset\!\sqsupset(l,l+d)$ = 1`12; this is first divided by 2 (outcome 36), whence $l = \sqsubset\!\sqsupset(l,l+d)\div(l+d) = 36\div9 = 4$. That division by 2 is made separately and before the division by $l+d$ fits the assumption that 1`12 is really found as two rectangles and not as a mere number, and hence that a diagram akin to Figure 86 is used.

The same diagram is apparently used in #14 (and #17–18):[488]

$$\square(l+w+d) = A+B+C+2D+2E+2F,$$
$$\sqsubset\!\sqsupset(l,w) = F$$

Therefore, since $A = B+C$,

$$\square(l+w+d)-2\sqsubset\!\sqsupset(l,w) = 2A+2D+2E = 2\sqsubset\!\sqsupset(d,l+w+d)\ ,$$

whence d is found by halving followed by a division by $l+w+d$ = 1`10.

In #13, the number 1 which is subtracted represents $\square(l-w) = \square(9-8)$. We may try to follow the procedure in Figure 86, although it is not certain that

488 This problem was also present in AO 6484, we remember, but without description of the procedure.

exactly this configuration was thought of:[489]

$$\begin{aligned}\square(d+l)+\square(d+w) &= (A+2D+B)+(A+2E+C)\\ &= (A+B+C+2D+2E+2F)+(A-2F)\end{aligned}$$

Now, since $A = B+C$,

$$A-2F = B+C-2F = \square(l)+\square(w)-2\sqsubset\!\sqsupset(l,w) = \square(l-w)\ ,$$

according to the rule which was shown in Figure 73. Therefore,

$$\square(d+l)+\square(d+w)-\square(l-w) = A+B+C+2D+2E+2F = \square(l+w+d)\ ,$$

as used in the text.

#12 corresponds to the rectangle problem $d-l = 3$, $w = 9$. An easy solution follows from the observation that

$$\square(w) = \square(d)-\square(l) = \sqsubset\!\sqsupset(d+l,d-l) = \sqsubset\!\sqsupset(4\frac{d+l}{2},\frac{d-l}{2})$$

– quite in the spirit of the method of average and deviation (cf. pp. 172 and 267). It is probably significant that the text makes a different calculation, which we may follow on Figure 87 (again without certainty that exactly this simple configuration was thought of – as long as no constructional prescriptions are given, different diagrams may lead to the same conclusions): As usual, we should use that

$$\square(d) = \square(l)+\square(w)\ .$$

Since $\square(l) = Q$, this implies that

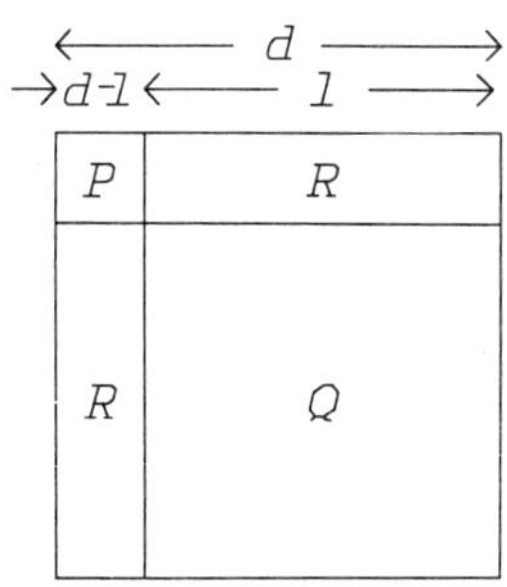

Figure 87. A diagram on which the procedure of BM 34568 #12 can be followed.

489 Fibonacci [ed. Boncompagni 1862: 68] uses the diagram of Figure 86 when solving #14 of the present text with different numbers. For #13 (still with different numbers) he only cites the rule $\square(d+l)+\square(d+w)-\square(l-w) = \square(l+w+d)$ (p. 66); the reason could be that the proof of this rule from Figure 86 is more intricate and therefore not transmitted to him or too laborious to explain: but is might also be that a different proof was used.

We notice that the diagram of Figure 70 – also from Fibonacci's *Pratica* – is a special case of that shown in Figure 86.

$$\square(w) = P+2R\ ,$$

whence, since $\square(d-l) = P$,

$$\square(w)+\square(d-l) = P+2R+P = 2\cdot(P+R) = 2\sqsubset\!\sqsupset(d,d-l)\ .$$

Therefore, halving the sum gives $\sqsubset\!\sqsupset(d,d-l)$; and dividing afterwards by $d-l$ gives the diagonal, that is, the length of the reed. Afterwards, the length l of the rectangle (that is, the height of the wall) follows from the Pythagorean rule.

The dress of this problem goes back to the Old Babylonian age, as we have seen (p. 275). Whether the actual problem was equally old we do not know, only that it could have been solved by what was probably the method of TMS I, the transformation of the border between two concentric squares into a rectangle.

The *method* that is used in our Seleucid text, on the other hand, resembles of nothing we have encountered in the Old Babylonian corpus (or the pre-Seleucid Late Babylonian texts), and is therefore highly unlikely to be old. It looks as if the new approach of which the other problems of the tablet bear witness was consciously applied, either to an old problem or to an old problem type for which it allowed the solution of a more complex variant than what had circulated so far. In either case, new wine poured into a venerated old Bottle.

We have no evidence as to the time or the place where the new methods and problems were created; as we shall see, indirect evidence speaks vaguely in favour of a non-Mesopotamian cradle, and more distinctly of a recent date.

Chapter XI
Repercussions and Influences

Above (pp. 368*ff*), the traces of the lay surveyors' tradition in later epochs were used as a means to establish the existence of this tradition and to identify and characterize its stock of proto-algebraic riddles. In this final chapter we shall take up the theme again in order to outline the impact of the tradition in later epochs. Since the topic is vast, I shall restrict myself to a brief sketch.

Greek Theoretical Mathematics

Since Neugebauer set forth his thesis that Greek "geometric algebra" would be a translation of Babylonian knowledge, it has been clear that *Elements* II is a central piece – more precisely, *Elements* II.1–10. In symbolic translation, these propositions say the following:

1. $\sqsubset\!\sqsupset(e,p+q+...+t) = \sqsubset\!\sqsupset(e,p) + \sqsubset\!\sqsupset(e,q) +...+ \sqsubset\!\sqsupset(e,t)$
2. $\square(e) = \sqsubset\!\sqsupset(e,p) + \sqsubset\!\sqsupset(e,e-p)$
3. $\sqsubset\!\sqsupset(e,e+p) = \square(e) + \sqsubset\!\sqsupset(e,p)$
4. $\square(e+f) = \square(e) + \square(f) + 2\sqsubset\!\sqsupset(e,f)$
5. $\sqsubset\!\sqsupset(a+d,a-d) + \square(d) = \square(a)$
6. $\sqsubset\!\sqsupset(e,e+2d) + \square(d) = \square(e+d)$
7. $\square(e+p) + \square(e) = 2\sqsubset\!\sqsupset(e+p,e) + \square(p)$
8. $4\sqsubset\!\sqsupset(a,d) + \square(a-d) = \square(a+d)$
9. $\square(a+d) + \square(a-d) = 2[\square(a) + \square(d)]$
10. $\square(e) + \square(e+2d) = 2[\square(d) + \square(e+d)]$

In more detail:

Prop. 1 shows that rectangles can be cut (or, the other way round, pasted if possessing a common side); prop. 2 and 3, actually nothing but corollaries of prop. 1, show that sides (transformed into rectangles by being provided with a projecting breadth *1*) may be removed from or joined to a square.

Prop. 7 is nothing but the rule $\square(R\text{-}r) = \square(R)\text{-}2\sqsubset\sqsupset(R,r)+\square(r)$ which appears to be presupposed in the Sargonic square problems referred to in note 304 with preceding text ($R = e+p$, $r = e$); and prop. 4 is its additive counterpart (which is not likely not to have been discovered together with the other rule).

Prop. 5 and 6 were always used in later times when the algebraic solutions to mixed second-degree equations were argued by means of geometric theory (only a few Renaissance writers would use II.4); if a is the average between e and $e+2d$, they are easily seen to be algebraically equivalent. Geometrically, however, they correspond to different situations: In prop. 5, the lines $a+d$ and $a–d$ are distinct parts of a total $2a$, in prop. 6 the line e is part of the line $e+2d$. The proofs are made correspondingly, and agree with the Old Babylonian respective ways of resolving the two rectangle problems $\sqsubset\sqsupset(l,w) = \alpha$, $l+w = \beta$ and $\sqsubset\sqsupset(l,w) = \alpha$, $l–w = \beta$. The former also conforms to the solution of problems of the type $\alpha s–Q = \beta$, the latter to those of problems $Q\pm\alpha s = \beta$. Cf. the above confrontation of *Elements* II.6 with YBC 6967, p. 97.

Prop. 8 corresponds to the rule that the border between concentric squares $\square(a+d)$ and $\square(a–d)$ is equal to four rectangles with length equal to the average side a and width equal to the width d of the border, that is, to a rectangle whose length is the average length of the border and whose width is the distance between the squares (cf. Figures 10, 11 and 74); this serves, as we have seen, for two-square problems $Q_1–Q_2 = \alpha$, $s_1\pm s_2 = \beta$. Prop. 9 and 10 correspond to the two-square problems $Q_1+Q_2 = \alpha$, $s_1\pm s_2 = \beta$. We may also think of the "thrusting" problem UET V, 864, where the four rectangles turn up separately as in the Euclidean proposition.

Prop. 4–7 serve later in the *Elements* (in particular in Book X), but the others are never referred to again: their substance seems to have been so familiar that it needed not be mentioned explicitly once its reliability had been validated – cf. [Mueller 1981: 301]. Clearly, *Elements* II.1–10 establish nothing new. As argued on p. 97, they are meant to consolidate the well-known – to be a "critique of mensurational reason", showing why and under which conditions (e.g., genuine right angles) the traditional ways could be accepted, and formulating the outcome in a general form.

Being critiques, Euclid's proofs are of course different from the "naive" procedures; but as pointed out on p. 98 in connection with prop. 6, most of them fall in two parts, of which the second repeats the naive procedures but on a foundation which causes them no longer to be naive. For propositions 1–7 the *proof ideas* are thus the same; for prop. 8 a slight rearrangement is likely to have taken place, corresponding to Figure 10 instead of Figure 11. Only the proofs of propositions 9–10 are incongruous with anything we know from the second millennium, and therefore almost certainly a late development.[490]

It is noteworthy that the proofs of the single propositions are independent, though some of them could easily be demonstrated from others. That each proposition gets its own proof shows that not only the knowledge contained in the theorems but also the traditional heuristic proofs were meant (if not by Euclid then by a source which he follows faithfully) to be consolidated by theoretical critique.

We observe that all propositions where the distinction can be made (5, 6, 8, 9, 10) are based on average and deviation, not on sum and difference.

Euclid is certainly not responsible for the adoption. What little we know about the work of Hippocrates of Chios and Theodoros of Cyrene shows them to have used some kind of metric geometry. The step from *use* to *critique* could but need not be slightly later: coins from Aegina, which in the fifth century had carried a "naive" geometric diagram, exhibit the diagram of II.4 (including the diagonal that makes the proof "critical") from 404 BCE onward [Artmann 1990: 47], which could mean that the topic was hot at that date. All in all it seems plausible that the theory presented in *Elements* II was created in the mid- to late fifth century B.C.

One of the interesting features that connect the Hippocratic fragment on the lunules and Plato's oblique account of Theodoros's work on irrationals with each other and with "algebra" and later metric geometry is the use of the term *dýnamis*.

In Diophantos's *Arithmetic* we are told that "it has been approved" to designate the second power of the unknown number as *dýnamis*, thus making it an "element of arithmetical theory", i.e., of *algebra* as treated by Diophantos [ed. Tannery 1893: I, 4]. This, and various other agreements, shows us that a number of passages in the *Republic* and other Platonic works refer to a second- and third-degree algebra used by calculators (λογιστικόι), and that even this early algebra used the term *dýnamis* for the second power – cf. [Høyrup 1990b: 368*f*].

But as it is evident from other passages in Plato and Aristotle; from the Hippocratic fragment; and from the *Elements* and certain Archimedean writings, the term was *also* part of the geometers' idiom. In this function, its interpretation has provoked much discussion – at times is seems to mean a square, at times it seems to stand for the side of a square or a square root.

Closer analysis of all occurrences shows that the term is no more ambiguous than mathematical terms in general, only unfamiliar [Høyrup 1990b]. As the Old Babylonian *mitḫartum*, "confrontation" (see p. 25), it stands for the quadratic figure parametrized by its side, that is, for a square

490 It is still possible, however, that they were invented in the practitioners' environment. This is indeed suggested by the use of the characteristic configuration from II.10 for a wholly different purpose in *Geometrica*/AC [ed. Heiberg 1912: 331] – a striking parallel to the recycling of the bisected trapezium as an elementary problem in UET V, 858.

identified by – and hence potentially *with* – its side. A *dýnamis* is a square that *is* its side and *possesses* an area, whereas our square (and the Greek *tetrágōnon*) *has* as side of, e.g., 2 m, and *is* 4 m^2.

The *dýnamis* did not correspond to the Greek conceptualization of *a figure* as "that which is contained by any boundary or boundaries" (*Elements* I, def. 14 [trans. Heath 1926: I, 153]), and already in Euclid's and Archimedes's times the term tended to vanish from geometry. This incongruity is striking, given the central importance of the term in the corpus of early references to geometry, and it makes the structural agreement with the *mitḫartum* interesting.

Árpád Szabó [1969: 46*f*], points out that both *dýnamis* and the verb *dýnasthai*[491] have connotations of equivalence and commercial value, together with the basic denotation of physical strength; exactly the same range of de- and connotations belongs with the verb *maḫārum*, from which *mitḫartum* is derived.

All this does not prove that the Greek *dýnamis* is a calque of the Babylonian *mitḫartum* (or, rather, of some Aramaic term with a corresponding meaning and semantic range and used by Near Eastern mathematical practitioners around 500 BCE). But taken together it must at least be counted as circumstantial evidence pointing in that direction. It certainly fits other types of evidence that the adoption of the Near Eastern proto-algebra took place in the fifth century.

Elements II is not the only place in the Euclidean corpus where influence from the Near Eastern tradition is directly visible. *Data*, propositions 84 and 85 show that if the area of a rectangle and either the difference between the sides or their sum are given in magnitude, then the sides are also given in magnitude. Whereas *Elements* II.5 and 6 are critiques of the procedures by which the two most prominent rectangle problems were solved, *Data* 84–85 may thus be considered as *solvability theory* for the same problem types. The treatise *On the Division of Figures* contains the bisection of the trapezium by a parallel transversal as one of its problems. But both works go far beyond anything that can be imagined to have belonged to any group of lay practitioners, and take whatever was borrowed from them as the inspiration and starting point for far-ranging independent developments.

A similar though more modest generalization starting from but leaving behind the adopted material is found in *Elements* VI.28–29, the application of an area with defect or excess, where the defect or excess is not required to be square (as in II.5–6) but similar to a given parallelogram.

---

491 To "master" or to "be worth", used in geometry to say that a line "masters" that square of which it is the side; in Aristotle's formulation of the Pythagorean theorem (*De incessu animalium* 708^{b}33-709^{a}2, ed. E. S. Forster in [Peck and Forster 1937: 510]), the hypotenuse "is worth"/"masters" the sides containing the right angle.

On a larger scale, the process characterizes the development of Greek metric geometry as a whole. *Elements* II.1–10 may prove no results that were not known since times (by then) immemorial; but without the consolidation of the well-known – including, for instance, the *definition* of a right angle and the discovery that this definition is useless unless the equality of all right angles is made *a postulate* – neither the theory of irrationals in *Elements* X nor Apollonios's *Conics* could have been produced. In mathematics as elsewhere, the expansion of knowledge is a dialectical process where neither naive creativity nor critical consolidation can function alone, each of them establishing the possibility that the other may become operative.[492]

Beyond the Euclidean works, the proto-algebraic inheritance has also left its traces in Diophantos's *Arithmetic* I.27–30 – see above, p. 373. Diophantos's immediate aim coincides with that of the practitioners: to find the solution – neither to construct a critique nor to formulate solvability theory. In contrast to what we find in sources that reflect the culture or teaching of practitioners, however, theoretical reflection is made explicit in two different ways: firstly, each problem is formulated in general terms, even though the solution is demonstrated on a paradigmatic example; secondly, by the formulation of diorisms telling the conditions for solvability when such conditions exist. These conditions are said to be *plasmatikós*, which may (but need not) mean that they can be seen in a diagram, a *plásma* (which indeed they can, namely, the traditional "naive" diagrams) – cf. the discussion in [Høyrup 1990: 349*f*].

All of this – *Elements* II.1–10 and VI.28–29, *Data* 84–85, the bisected trapezium from the treatise *On the Division of Figures*, *Arithmetic* I.27–30 – refers to the stock of problems that seems to have belonged to the lay surveyors' tradition already before the Old Babylonian scribe school adopted

---

492 Without pursuing the matter one might claim that analysis, presupposing the existence and properties of the objects it deals with without having proved either – always starts "naively" but then may need to pass through phases of synthetical critique. With respect to early medieval *al-jabr* algebra, al-Khayyāmī's geometrical existence proofs and Jordanus of Nemore's arithmetical ditto are early instances of such synthetical critique – none of them without noteworthy impact; Viète's *algebra speciosa* and its elaboration by Descartes and Wallis constitute a second phase, which changed the discipline thoroughly and made possible a new stage of naive expansion, including the creation of the *analysis infinitorum*. Both post-Viète algebra and *analysis infinitorum* were then recast completely in a new process of critique in the nineteenth century – becoming so unrecognizable in the process that most applied practice sticks to the eighteenth-century approach to the two fields even when using the results and concepts of later centuries (no engineer ever needed to know that π is transcendental, and none ever needed to refer to the Lebesgue integral).

As illustrated by this last observation, the "need" for critique is always relative to a particular setting – often, as in the case of the Old Babylonian critique of the naive surveyors' style, by location within a teaching institution. In other cases (al-Khayyāmī, Jordanus, Viète), privately held views of how mathematics *should* look may have played a larger role.

its riddles. With explainable exceptions it also exhausts this stock: rectangle problems $A+(l\pm w) = \alpha$, $l\mp w = \beta$ were always solved by an elegant "change of variable" which allowed reduction to the types $A = \alpha$, $l\pm w = \beta$, and they were thus uninteresting on their own from a theoretical point of view – the only place where they *might* at a pinch have fitted in is in the *Arithmetica*. The circle problem $A+d+p = \alpha$ was of course inaccessible to treatment inasmuch as the ratio between the circular diameter and perimeter was inexpressible. The rectangle with given area and diagonal is not present in itself, but by means of the Pythagorean theorem it is solved via either *Elements* II.4 or II.7 (as explained by Fibonacci and Savasorda). The equality of perimeter and area for squares and rectangles has no place within a geometry not based on a unit length.

In all cases where the alternative presents itself, these texts make use of average and deviation; the new ways of BM 34568 leave no trace in the works of the Greek theoreticians. This does not prove but speaks in favour of the assumption that these new ways were not at hand at the time and place when the Greek geometers borrowed from the Near Eastern tradition and started to develop their critique – that is, as argued, in the fifth century, perhaps in Phenicia/Syria.

Demotic Egypt

Apart from the family likeness between the filling problems in IM 53957 and Rhind Mathematical Papyrus #37 (see p. 373), no evidence suggests the slightest connection between Old Babylonian mathematics and "classical" (Pharaonic) Egyptian mathematics as found in Middle and New Kingdom papyri – nor between the surveyors' tradition and classical Egyptian mathematics. In Demotic mathematical papyri from the Ptolemaic and Roman period, however, the presence of material with roots in Mesopotamia is indubitable.

Of some interest is the (heavily damaged) Demotic Papyrus Carlsberg 30 [ed., trans. Parker 1972: 74], probably from the second century CE. It contains a diagram showing the 10×10 square with diagonal $14\frac{1}{7}$, and another with side $14\frac{1}{7}$ and indicated area 200. As we remember from p. 370, Savasorda's version of the quadratic "surface added to its four sides" presupposes $s = 7$; a problem in AO 6484 asks for the square side when the diagonal is 10 kùš. This papyrus thus shares the interest in "cascades of squares" which manifests itself both in the Seleucid material and in the variations of the problem of "the four sides and the area".[493]

493 Two types of "cascades" are involved: the one starting with a square $\square(s)$ and its diagonal d, which takes d or an approximate value (and, downwards, $\frac{d}{2}$) as the

More important is Papyrus Cairo J.E. 89127–30, 89137–43 [ed., trans. Parker 1972: 41–43] from the third c. BCE. Its problems #33–34 are two variants of the rectangle problem where the area and the diagonal are given (the problem of Db_2-146, p. 257). In contrast to the Eshnunna solution (and the solutions found with Savasorda, Abū Bakr, and Fibonacci), it finds both $l+w$ and $l-w$, as $\sqrt{\square(d)+2A}$ and $\sqrt{\square(d)-2A}$, respectively, and next (as in BM 34568 #9) w as the half of $(l+w)-(l-w)$. Finally, l is found as $(l+w)-w$. The same papyrus contains material with indubitable Babylonian affinities: the "pole against a wall" is found both in elementary variants (to determine w from d and $d-l$, or $d-l$ from d and w) and in the sophisticated variant of BM 34568, where $d-l$ and w are given (together, no less than 8 out of 40 problems). The solution to the sophisticated variant coincides with that of BM 34568.

Greek Underbrush

Less famous and less studied than Greek theoretical mathematics is the underbrush of writings derived more directly from mathematical practice – so much less studied indeed that the *Geometrica*, which Heiberg tells explicitly to have put together from two independent treatises of very different character, *not* written by Hero and used in different environments, are mostly cited as a Heronian work.[494]

---

next side; and those which connect a pair (s,d) to another pair (S,D) via the condition ${}_4s+\square(s) = \square(S)-{}_4S$. The former contains the sequence 7–10–14 (or $7\frac{1}{14}$–10–$14\frac{1}{7}$), the latter condition reduces to $S-s = 4$, which gives rise to the sequence 6–10–14.

494 In the introduction to vol. V of Hero's *Opera quae supersunt omnia* [Heiberg 1914], we read that *Geometrica* (a plural from the substantial as well as the grammatical point of view) "were not made by Hero, nor can a Heronian work be reconstructed by removing a larger or smaller number of interpolations" (p. xxi); that mss AC represent a book which was not meant to serve field mensuration directly but was for use "in [commercial and legal] life" and in general education (p. xxi); that ms S, with the closely related ms V, was intended to serve youth studying "architecture, mechanics, and field mensuration" in the "University of Constantinople" and thus "more familiar with theoretical mathematics" – a use which in Heiberg's view agrees with the presence of Hero's (more or less) genuine *Metrica* in the same manuscript (p. xxiii); and that both families (each in its own way) merge "various problem collections together with Heronian and Euclidean excerpts" (p. xxiv).

In detail, Heiberg's judgments can certainly be challenged – that the "surveyors' formula" is present in ms S but not in mss A+C, and that only the latter manuscripts give the rule for finding the height in a scalene triangle (in a form that derives from *Elements* II.12–14) does not exactly make ms S appear

The surviving corpus is small, and not much in it is relevant for our present concern – but a small part *is* highly relevant. Various components of the *Geometrica* conglomerate were cited above: on p. 368 the independent treatise from ms. S that has become Heiberg's Chapter 24 – namely, because of the problem "square area together with the perimeter given"; on p. 371 both this treatise and mss A+C because of the problem "circular diameter plus perimeter plus area given"; and p. 382 mss A+C (together with Hero's *Metrica*) because of the predilection for 10 as the side of regular polygons.

In the square problem, two features of the phrasing are significant. Firstly, the task is not to find the side (or the area or the perimeter) but to get the area and the perimeter separated (διαχωρίζω); secondly, when a step is independent of the actual numerical parameter (the sum 896), it is to be performed καθολικῶς, "in general".

The idea of "separating" is not common in the Old Babylonian texts, but it does occur as *bêrum*, "to single out". Least characteristic is the appearance in TMS VII, line 4, "30′ and 5′ single out" (p. 181); quite analogous to the *Geometrica*-24-occurrence are BM 10822 #1, which asks for the separation of three brick types, and AO 8862 #7, which asks for the separation of men, days, and bricks – cf. [Friberg 2001: 143–148].

Other problems in the *Geometrica* material also ask for separation with varying terms: διαχωρίζω, διαστέλλω, ἀποδιαστέλλω.[495] Statements that an operation is independent of the specific parameters are also common, and may occur as καθολικῶς, "in general", παντός or ἀεί, "always", or as πάντοτε, "at all times". However, the distribution of the terms is very far from being random. Explicitations of general validity are mostly found in problems that ask for the separation of a sum, and all separation problems but one contain such an explicitation. In spite of terminological differentiation, these two ideas clearly belong with a particular strain in the material; the extant occurrences of "separation" in the Babylonian material are too few (and too specifically linked to brick problems) to allow us to connect this strain to the Babylonian or the surveyors' tradition; as we shall see, other sources establish the bond.

Also of interest are the Greco-Egyptian Papyrus Genève 259 and a *Liber podismi* known only in Latin but obviously built on Greek material if not directly a translation from the Greek (ποδισμός meaning "measuring by feet", that is, "area"). Papyrus Genève 259,[496] contains three problems on right

---

more "theoretical" than mss A+C – cf. [Høyrup 1997]; but it is symptomatic of the lack of interest in these works that these passages from Heiberg's introduction were neither read nor criticized. It is true that they are found in the Latin introduction to vol. V, whereas the *Geometrica* themselves are in vol. IV. But if work on the writings in question had been minimally serious, their presence in a volume where Hero's name appears on the title page would not have been enough to transform them into a Heronian treatise.

495 Details on the distribution are given in [Høyrup 1997: 92*f*].

496 [Ed., trans. Rudhardt 1978], much improved edition with acute analysis in [Sesiano

triangles:

#1. $w = 3$, $d = 5$

#2. $w+d = 8$, $l = 4$

#3. $l+w = 17$, $d = 13$

The first, of course, is uninformative. The second and third, however, belong to types that only turn up in the Seleucid and Demotic material (in BM 34568 #10 and #4, respectively). #3, moreover, is one of the types which, when adopted into the tradition reported by Abū Bakr, was solved by means of average and deviation (below, p. 410). The Geneva Papyrus does nothing similar; even though its exact procedures for #2–3 are not the same, their general tenor is that of the Seleucid text.

The *Liber podismi* is a short collection of problems on right triangle, most of them too simple to tell us much. One of them, however, repeats the old rectangle problem where the diagonal and the area are given [ed. Bubnov 1899: 511*f*]. The solution follows the same pattern as the Demotic Cairo papyrus (without referring to average and deviation), and is thus in the new Seleucid style.

Without digging any deeper, I shall recall two facets of the low level of Greek mathematics that were mentioned above: Theon of Smyrna's presentation of the side-and-diagonal number algorithm (see note 297), and the Neopythagorean (?) interest in squares and rectangles whose area equals the perimeter.

Even the underbrush of Greek mathematics was thus in debt to the Near Eastern tradition. As in the case of theoretical mathematics, nothing suggests specific inspiration from what was brought forth by the Old Babylonian scribe school – but the connections to the "new" approach of which the Seleucid texts bear witness are unmistakeable.

India

The Indian Śulbasūtras, rules for altar construction and transformation, are roughly contemporary with the beginnings of Greek metric geometry. The general design of some of their constructions is quite similar to the naive cut-and-paste procedures of the surveyors' tradition; but arguments solely from the "general design" of a mathematical procedure are dangerous unless this design is very complex or presupposes a fallacy that is not likely to occur often – both the number system and plane geometry have their inner constraints that may easily give rise to parallel developments. It is not to be excluded that closer analysis of the Śulbasūtras will substantiate links to the

1999]; probably 2nd c. CE.

Near Eastern surveyors; but for the moment this hypothesis is *not* substantiated, and I shall leave it aside.

That Indian astronomy of the early first millennium CE received influences from both Babylonian and Hellenistic astronomy is a well-established fact; the presence of arithmetical "recreational" problems in India which are also known from the Mediterranean and Near Eastern region (the "hundred fowls", and many others) is also incontrovertible. But again this does not prove that Indian *algebra* from the period received inspirations from the same regions. *If* so, it soon transformed the material it had received beyond recognition: the algebraic schemes of Brahmagupta and the Bakhshālī manuscript[497] are so different from anything Babylonian, Greek, or Arabic that Léon Rodet's rejection [1878] of an Indian inspiration for al-Khwārizmī on this account would seem to hold no less certainly the other way.

After the recurrent references to Mahāvīra in preceding chapters it will come as no surprise that his early ninth-century *Gaṇita-sāra-sangraha* [ed., trans. Raṅgācārya 1912] is the Indian work where the link is established (but still not any link to the Śulbasūtras or Brahmagupta etc.). Beyond the circle problem $c+d+S = \alpha$ (p. 371) and the problems on the square and the rectangle whose perimeter equals the area (p. 372), we may also note the rectangle problem $\sqsubset\!\sqsupset(l,w) = \alpha,\ {}_2l+{}_2w = \beta$ (VII.129), where Mahāvīra finds $l+w$ and then goes on as in BM 34568 #9; and the rectangle with given area and given diagonal, where the method is that of the Cairo papyrus (VII.127). In VII.125, where $2l+2w$ and d is given, $\square(l-w)$ is found as $2\square(d)-\square(l+w)$; the same trick is used in the Greek Papyrus Genève 259 #3 (see note 496). In all cases the final steps are slightly different in Mahāvīra's version, l and w being found as average and deviation from $l+w$ and $l-w$

These similarities go far beyond what could arise as accidental parallel developments. Most of them connect Mahāvīra's treatise to what was characterized as the Greek underbrush, and to the Seleucid style; only the circle problem is in the style of the pre-Old-Babylonian lay tradition. Interestingly, Mahāvīra knows that there is a difference. The circle problem belongs (together with the surveyors' formula and the formula for the circle ring as mid-length times width) in the section "approximate measurement (of areas)", VII.7–48; the others are found in the section "Devilishly difficult problems" (VII.112–232½), after "Minutely accurate calculation of the measure of areas" (VII.49–111½).

Mahāvīra wrote in the early ninth century CE, and he is thus a contemporary of al-Khwārizmī. It is obvious, however, that his material is not a recent borrowing – many points are referred to "learned teachers", and other formulations too show that his material had been fully naturalized within the Jaina community. If the circle problem had been imported in recent centuries, it would certainly have gone together with $\pi = 3\frac{1}{7}$, which was already used by practitioners in Hero's times (*Metrica* I.xxxi [ed. Schöne 1903: 74]). Most

497 See [Datta 1929: 28*f*]; and [Datta and Singh 1962: II, 28–32].

probable is a borrowing in the later first millennium BCE,[498] the bloom of Jaina mathematics: a date much after 200 CE seems impossible.

Impact in Islamic and Post-Islamic Mathematics: Towards Early Modern Algebra

Of all the later sources that have been quoted, Abū Bakr's *Liber mensurationum* is certainly the one which conserves the heritage most faithfully; beyond the shared problems and techniques it also conserves many aspects of the characteristic Old Babylonian format (see p. 370 and note 456) – so many indeed that a purely oral transmission seems unlikely. Even though the cultural type is oral (natural coefficients, riddle formulation), we may perhaps assume that the advent of alphabetic Aramaic literacy around 1000 BCE allowed the stabilization of the tradition in the Syro-Mesopotamian region among practitioners whose literacy was not carried by a genuine school.[499]

The *Liber mensurationum* contains many of the original proto-algebraic riddles (the circle problem and the two-square problems are absent), and solve all of these by the method of average and deviation – including the rectangle problem with given area and diagonal and a problem that coincides numerically with BM 34568 #9. It also contains many of the "new" problems, but when these can be solved by the traditional method (as, for instance, the rectangle problem $d = 10$, $l+w = 14$), this is preferred; when it cannot be done, the Seleucid method is used[500].

This observation suggests that the Seleucid innovations did not originate in the Syro-Mesopotamian region, where Abū Bakr can be presumed to have worked, given his linguistic fidelity to the ancient model; though almost certainly descending from the ancient tradition they are more likely to have been created elsewhere in deviant form; to have reached Mesopotamia during the Seleucid era;[501] and to have undergone there a gradual assimilation to

---

498 It is not *a priori* excluded that the "Seleucid" innovations originated in the Jaina school and radiated from there; since they never go together with anything characteristically Indian, what is "not *a priori* excluded" remains very improbable.

499 The colophon of W 23291-x states that it was copied from a wax tablet [ed. Friberg, Hunger and al-Rawi 1990a: 545]. Wax tablets might certainly be used for cuneiform writing, but was also a normal medium for alphabetic Aramaic; the left-right inversion of the format of Late Babylonian metrological tables [Friberg 1993] may also have been brought about by the habit of writing from right to left, that is, in Aramaic, and not from left to right (the direction of cuneiform writing).

500 With the variation that calculations $\frac{a^2+b^2}{2} \div b$ are replaced by the algebraically equivalent $(\frac{a^2}{b} + b) \div 2$. The reappearance of the same problems in Fibonacci's *Pratica* confirms, however, that the underlying idea is the same.

501 Probably not much earlier, since the approach has left no traces in Greek

the ancient tradition during the millennium that separates its arrival into the region from Abū Bakr.

Abū Bakr's handbook contains many other problems, including reworking of the traditional rectangle problems as problems about rhombs. They need not concern us here. Nor do we need go into details with the other works from the Islamic world and that late medieval Europe which in this respect is "post-Islamic" – works that betray the wide diffusion of the tradition: ibn Thabāt's *Reckoners' Wealth* (see p. 371), and Savasorda, Fibonacci, Piero della Francesca, Pacioli (p. 370).

Instead we shall turn to the beginning of *our* algebra: al-Khwārizmī's short introduction to the topic *al-jabr wa'l-muqābalah*, written in the early ninth century on the exhortation of the caliph al-Maʿmūn. It contains a chapter on practical geometry and one on inheritance computations, but none of them have much relevance for the present discussion.

As I have argued elsewhere [Høyrup 1998], the best extant witness of al-Khwārizmī's original chapter on *al-jabr* proper is Gerard of Cremona's twelfth-century Latin translation [ed. Hughes 1986]; the best translation of the published Arabic text (whose branch of the stemma has undergone at least three successive revisions after its separation from the version used by Gerard) was made by Boris Rozenfeld [1983]. Since Rosen's English translation [1831] remains more widely diffused than both of these, I shall point out when appropriate its insufficiencies with respect to the present discussion.

The *al-jabr* technique, we are told, is based on three kinds of numbers: [square] roots, possessions,[502] and simple numbers. Fundamentally, it thus deals with [unknown] amounts of money, square roots of these amounts, and numbers; al-Khwārizmī explains, however, that the possession is produced as the product of the root with itself, in agreement with the identification of the "root" with the "thing" and of the "possession" with the product of the "thing" with itself when the technique is applied (see presently).

The three kinds of numbers are combined in 6 equation types:

Possession is made equal to roots[503]
Possession is made equal to number

---

theoretical mathematics.

502 "Squares" in Rosen's translation; but the Arabic word is *māl*, meaning a [monetary] property, adequately translated by Gerard as *census*. In Rosen's defense it should be said that he takes care to translate the Arabic term for a square (*murabbaʿ*) as "quadrate" – a point too fine to be observed by most users of the text.

Rozenfeld uses *kvadrat* for both *māl* and *murabbaʿ*.

503 The revised Arabic text, followed by Rosen and Rozenfeld, speaks of "possessions", in agreement with the habit of later Arabic algebra; since all the basic examples are normalized, we may be confident that Gerard's singular form corresponds to the original text.

The rules as quoted by Thābit (see note 505) also refer to a single possession.

Roots are made equal to number
Possession and roots are made equal to number[504]
Possession and number is made equal to roots
Roots and number are made equal to possession

For each of these, a numerical example is given, and a rule for solving it (followed by non-normalized examples, whose normalization is explained). In the fourth case, for instance, the example is "A possession and 10 roots are made equal to 39 dirhams", and the rule [ed. Hughes 1986: 234] that

> you halve the roots, which in this question are 5. You then multiply them with themselves, from which arises 25; add them to 39, and they will be 64. You should take the root of this, which is 8. Next remove from it the half of the roots, which is 5. Then 3 remains, which is the root of the possession. And the possession is 9.

Thus, if the equation is $y+a\sqrt{y} = b$, then

$$\sqrt{y} = \sqrt{b+(\frac{a}{2})^2} - \frac{a}{2} \text{ and } y = (\sqrt{y})^2 .$$

Working in the ambience of the House of Wisdom, al-Khwārizmī was not satisfied with a list of unexplained rules (though this may have been all the caliph had asked for), and he therefore added a set of geometrical proofs.[505] Later on he makes it clear that the geometrical proofs he constructs in order to illustrate the calculation with binomials (and which he tries, though without being satisfied with the outcome, to construct for trinomials) are of his own making; he says nothing similar about the present proofs, and already for this reason we may therefore assume that he borrowed them from somewhere.

"Somewhere" turns out to be familiar. The case whose rule was quoted above is furnished with two proofs. The diagram of the first is shown in Figure 88. At first the square *AB* with unknown side is drawn, which represents the possession; each of its sides is thus equal to the root. The 10 roots are distributed equally, as the four rectangles *G*, *H*, *T*, and *K*, each of which has the width $2\frac{1}{2}$. In each corner a square $2\frac{1}{2}\times2\frac{1}{2}$ is missing; adding these we get for the larger square a total area $39+4\cdot6\frac{1}{4} = 39+25 = 64$; etc. All in all we get a proof of the rule

---

504 For obscure reasons, Rosen changes the order into "Roots and squares are equal to number", against his own Arabic text.

505 That only the rules and not the proofs belonged with the extant *al-jabr* technique (that which al-Maʿmūn had asked al-Khwārizmī to expound in handy form) follows from Thābit ibn Qurrah's slightly later Euclidean justification of the rules of *al-jabr* [ed., trans. Luckey 1941].

As we shall see, later representatives of the "low" variety of *al-jabr* are also devoid of proofs.

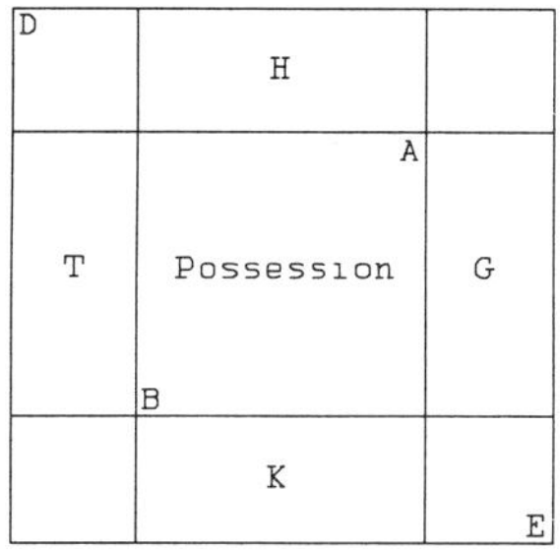

Figure 88. The first proof of the case "Possession and roots made equal to number".

$$y = \sqrt{b+4\cdot(\frac{a}{4})^2} - 2\cdot\frac{a}{4},$$

– not of the rule that was to be proved.

Afterwards comes a second proof, the diagram of which is shown in Figure 89. This time the proof fits the procedure to be proved perfectly. Both for grammatical reasons and because the proof style is more concise and formal (also compared with the proofs of the following cases), there are reasons to believe that this second proof was added by al-Khwārizmī in a later revision of the text. Even if this is not the case it comes after the first, and since it is obviously more adequate, al-Khwārizmī must have had some particular, conscious or subconscious reason to put it in this position – and to include the first proof at all. This reason can be one of two: it may have been the one which came first to al-Khwārizmī's mind, and which he found simpler; or it may be the one with which he expected the reader to be more familiar (or both). Since the first proof is evidently derived from the solution of the problem of "the four fronts and the surface" with which we are familiar from BM 13901 #23 (without the quadripartition, which is likely to be a scribe school innovation), al-Khwārizmī's source for the proof in question (and then by association also the others, which are all of the habitual types) is clearly our familiar surveyors' tradition and its riddle collection. These must hence have been around, and apparently more familiar than *al-jabr* itself.

The origin of the *al-jabr* technique is not known. Al-Maʿmūn's request suggests that it was not stock knowledge in early ninth-century Iraq. It has been suggested at times that it came from Central Asia, since not only al-Khwārizmī but also ibn Turk, another early writer on the topic, had their family roots there. A medieval story reported by David King [1988] relates that the technique was adopted from the Iranian Fars province already under the caliph ʿUmar (634–644), transmitted orally for a while and then lost, and only restored by al-Khwārizmī. Though hardly reliable in its details, the story supports a descent from a location somewhere to the east or north-east of Iraq.

Nor is it known whether *al-jabr* is somehow a descendant of earlier "algebras" – Babylonian, Greek "geometric" or Diophantine, Indian, or the

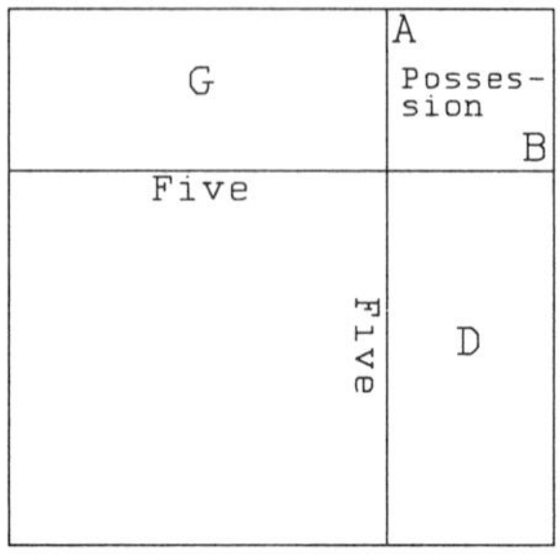

Figure 89. The second proof of the case "Possession and roots made equal to number".

surveyors' riddles. Once the geometric proofs have been seen to be secondary grafts, both Greek possible sources become unlikely;[506] since *al-jabr* is a numerical technique and not geometrical, having the possession and not the root as its basic unknown, any assumption of a link is gratuitous unless supported by arguments that are independent of the agreement between numerical steps (arguments which nobody has produced so far). As to a possible descent from Indian algebra as known, e.g., from Brahmagupta, Léon Rodet already argued in [1878] that the sophisticated algebraic schemes of the Indians and their free use of negative entities make al-Khwārizmī's work look much too primitive for this hypothesis to seem plausible.

Thābit does not mention al-Khwārizmī's proofs when presenting his own, based on *Elements* II.5–6. Further on in the "high" tradition of Islamic algebra – from Abū Kāmil onward – al-Khwārizmī was the recognized founder of the discipline as it had come to look, and the geometric proofs were accepted as an integral part of the subject (though from Thābit and Abū Kāmil onward mostly formulated with reference to *Elements* II.5–6).[507] They were evidently taken over in the Latin translations of al-Khwārizmī and Abū Kāmil, and also in Fibonacci's *Liber abbaci* and *Pratica geometrie*. With Fibonacci, however, the development is taken one step further: when paraphrasing Gerard's translation of al-Khwārizmī in the *Pratica* [ed. Boncompagni 1862:

---

506 Al-Khwārizmī's use of *šay*ʾ, "thing", could remind of the Greek αριθμός (in Diophantos's *Arithmetic*, but also in humble Greco-Egyptian papyri, see [Robbins 1929] and [Vogel 1930]). However, even the *šay*ʾ (*res* in the Latin translations, *cosa* and *coss* in Italian and German Renaissance algebra) is likely to be a secondary graft. This kind of computation is called *regula recta* by Fibonacci (*Liber abbaci* [ed. Boncompagni 1857: 191 and *passim*]) and introduced long before his presentation of *al-jabr wa'l-muqābalah*, namely, when he solves the dressed version of Diophantos' *Arithmetica* I.15 exactly as Diophantos solves it by means of an unknown *arithmós*, "number". Unless we are deceived by a highly improbable coincidence, Arabic *šay*ʾ- and Greek αριθμός-algebra belong to the same family, and the *šay*ʾ-technique only coalesced with *al-jabr* at a late moment (though certainly well before al-Khwārizmī's time).

507 [Dold-Samplonius 1987] is a convenient survey.

56] he corrects the statement that numbers are "roots, possessions, and simple numbers". Now they are *aut radices quadratorum, aut quadrati, aut numeri semplices*, "either roots of squares, squares, or simple numbers" – and the squares are real geometric squares, whereas the roots are strips as long as the side of the square, and with width *1*. The formulations in *Liber abbaci* are similar [ed. Boncompagni 1857: 406]. Here, the *census* only appears when the first problem type is introduced (p. 407): "The first mode is, when the square, which is called *census*, is made equal to roots".

Liber abbaci gives two versions of the geometric proof of the first mixed case; one is similar to al-Khwārizmī's second proof (but also to *Elements* II.4), the other to Thābit's and Abū Kāmil's (that is, it refers implicitly to *Elements* II.6). In the *Pratica*, the paradigmatic example for the same case, here defined as "number is made equal to quadrate and roots", is nothing but the problem "a [quadratic] area and its four sides make 140" (solved, however, with another implicit reference to *Elements* II.6). Regarding geometric proofs as the gist of the discipline, Fibonacci in fact reconstructs it, deriving it in part indirectly (through Euclid and those who had used him), in part directly from the old sides-and-area riddles (which he knew from Gerard's translation of Abū Bakr, from Savasorda, and from unidentified sources[508]).

Not all Islamic *al-jabr* after al-Khwārizmī and Thābit belongs to the "high" division, even though the "low" register has attracted the attention of modern scholars and medieval translators into Latin much less. The "low" register seems to be characterized by an ordering of cases that differs from al-Khwārizmī's; by defining the cases in non-normalized form, as in the revised al-Khwārizmī text; *and by having no geometric proofs*. This is exemplified in al-Karajī's *Kāfī* [ed., trans. Hochheim 1878];[509] ibn al-Bannā''s *Talkhīṣ* [ed., trans. Souissi 1969: 92]; and ibn al-Yāsamīn's *'Urjuza fi'l-jabr wa'l-muqābalah* (paraphrase in symbols in [Souissi 1983: 220–223]).

Common prejudice notwithstanding, this type – and not Fibonacci's *Liber abbaci* – was the kind of algebra that inspired the beginnings of Italian vernacular algebra in the earliest fourteenth century. This can be seen from the

---

508 One trace of the latter is Fibonacci's replacement of a corrupt problem from the *Liber mensurationum* with a problem which is certainly not of his own making – cf. [Høyrup 1996a: 56]. Among other things, the problem in question has the sides before the area, whereas Fibonacci's own preference (strong enough to make him correct Abū Bakr/Gerard on this account) is to have the area first.

509 It may astonish to see al-Karajī listed in the "low" category, but the surprise may serve to show that the categorization has nothing to do with mathematical competence and incompetence; as shown by Saliba [1972], al-Karajī's terminology in the *Kāfī* demonstrates its algebra to be derived from a pre–al-Khwārizmīan model; similarly, much in its geometry turns out to be close to the traditional practitioners' model – cf. [Høyrup 1997].

The *Fakhrī* shows al-Karajī to be wholly conversant with the geometric proofs and ready to present them when he thinks they fit the context. But even here his definition of key terms is pre–al-Khwārizmīan.

earliest specimens: the Vatican manuscript of Jacopo da Firenze's *Tractatus algorismi* (dated 1307);[510] Paolo Gherardi's *Libro di ragioni* from 1328 [ed. Arrighi 1987]; and a composite *abbaco* book from Lucca from c. 1330 [ed. Arrighi 1973]. But even vernacular European algebra succumbed to the spell of geometrical reasoning, and *not* in the first instance because of the influence of Fibonacci or al-Khwārizmī. Like the former in the *Pratica*, but in wholly independent and fully "naive" and non-Euclidean terms, Piero della Francesca uses the problem of "a square whose surface, joined to its four sides, makes 140" as the paradigmatic example explaining the rule for the case "*censo* and things are equal to number" (cf. p. 370). His proof runs thus (see Figure 90): *TE* is explained to be the square or *censo* (like Fibonacci, Piero takes *censo* to be another term for the square figure), *AI* to be 4, and *G* the mid-point of *AI*; *GF* is then drawn so as to exceed *BE* with as much as *IG*; etc.[511]

As a rule, Piero's algebra problems are derived from the preceding vernacular tradition, not from the *Liber abbaci*, as his geometrical riddles are generally derived from some unidentified (but certainly indirect) link to the Islamic world; but his use of geometry shows that times were ripe for Pacioli's reintroduction of Fibonacci's version in *Summa de arithmetica*. In part in fairly original "naive" shape, in part in versions more or less touched by the

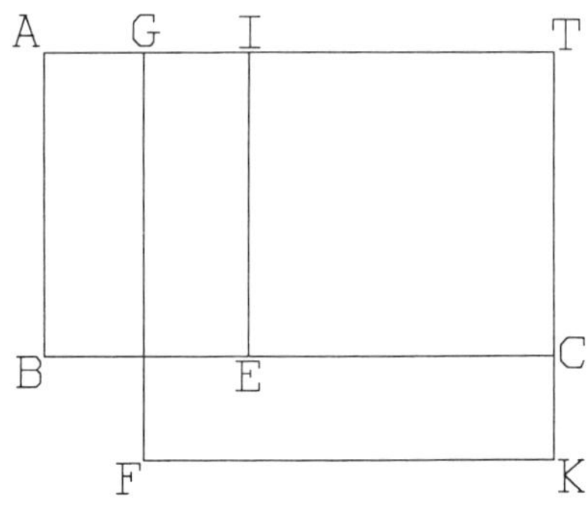

Figure 90. Piero's diagram for his "surface joined to its four sides".

510 [ed. Høyrup 2000]. Jacopo's algebra differs from the Latin tradition by lacking all geometric proofs; by sharing no single example or problem with *Liber abbaci* or with the Latin translations of al-Khwārizmī and Abū Kāmil; and by containing no examples in the abstract form involving *censo*, *cosa*, and number.

511 Much later in the treatise comes a whole collection of problems derived from the riddle tradition: $Q = 2\cdot{}_4s$; $Q = {}_4s+60$; $d-s = 6$; $A = \alpha$, $l = w+2$; etc., in versions which (when numerical parameters were not fixed by tradition) are shared neither with Abū Bakr nor (with a single exception) with Fibonacci. Piero has obviously borrowed the area-with-sides problem from this group and put it in the place where he needed it for pedagogical reasons. The two other mixed cases [ed. Arrighi 1970: 133, 136] are illustrated by the problems ${}_4s-Q = 3$ and $Q-{}_4s = 77$, similarly borrowed from the geometric collection.

Euclidean "critical" form, the riddle tradition had reconquered all levels of European algebra when Cardano was to encounter it.[512] Even though Old Babylonian school algebra, much too refined to survive without its institutional support, ended up as one of the glorious blind alleys of history, Modern algebra turns out to owe at least as much to the lay predecessors of the Babylonians as it owes to that *al-jabr* technique from which it took its name. Al-Khwārizmī remains the "father of algebra" – not because he created *al-jabr* (which he did not) but by bringing about the synthesis.

512 The introductory passage [trans. Witmer 1968: 7*f*] of his *Ars magna* runs as follows:

> This art originated with Mahomet the son of Moses the Arab [i.e., al-Khwārizmī / JH]. Leonardo of Pisa is a trustworthy source for this statement. There remain, moreover, four propositions of his with their demonstrations, which we will ascribe to him in their proper places. After a long time, three derivative propositions were added to these. They are of uncertain authorship, though they were placed with the principal ones by Luca Paccioli. [...].

Cardano obviously knows some of the particular types of the *abbaco* tradition – but the names of the discipline are the geometrizers: al-Khwārizmī, Fibonacci, and Pacioli. Quite appropriate as a background for his own decomposition of the cube.

Abbreviations and Bibliography

ABZ: Rykle Borger, *Assyrisch-babylonische Zeichenliste*. (Alter Orient und Altes Testament, 33). Kevelaer: Butzon & Bercker / Neukirchen-Vluyn: Neukirchener Verlag, 1978.

AHw: Wolfram von Soden, *Akkadisches Handwörterbuch*. Wiesbaden: Otto Harrassowitz, 1965–1981.

al-Rawi, Farouk N. H., and Michael Roaf, 1984. "Ten Old Babylonian Mathematical Problem Texts from Tell Haddad, Himrin". *Sumer* **43** (1984, printed 1987), 195–218.

Allotte de la Fuÿe, François-Maurice, 1915. "Mesures agraires et formules d'arpentage à l'époque présargonique". *Revue d'Assyriologie* **12**, 117–146.

Amiet, Pierre, 1979. "Archaeological Discontinuity and Ethnic Duality in Elam". *Antiquity* **53**, 195–204, pl. xx–xxi.

Arrighi, Gino (ed.), 1970. Piero della Francesca, *Trattato d'abaco*. Dal codice ashburnhamiano 280 (359*–291*) della Biblioteca Medicea Laurenziana di Firenze. A cura e con introduzione di Gino Arrighi. (Testimonianze di storia della scienza, 6). Pisa: Domus Galileana.

Arrighi, Gino (ed.), 1973. *Libro d'abaco*. Dal Codice 1754 (sec. XIV) della Biblioteca Statale di Lucca. Lucca: Cassa di Risparmio di Lucca.

Arrighi, Gino (ed.), 1987. Paolo Gherardi, *Opera mathematica: Libro di ragioni – Liber habaci*. Codici Magliabechiani Classe XI, nn. 87 e 88 (sec. XIV) della Biblioteca Nazionale di Firenze. Lucca: Pacini-Fazzi.

Artmann, Benno, 1990. "Mathematical Motifs on Greek Coins". *Mathematical Intelligencer* **12**:4, 43–51.

Baqir, Ṭaha, 1950. "An Important Mathematical Problem Text from Tell Harmal". *Sumer* **6**, 39–54.

Baqir, Taha, 1950a. "Another Important Mathematical Text from Tell Harmal". *Sumer* **6**, 130–148.

Baqir, Taha, 1951. "Some More Mathematical Texts from Tell Harmal". *Sumer* **7**, 28–45.

Baqir, Taha, 1962. "Tell Dhiba'i: New Mathematical Texts". *Sumer* **18**, 11–14, pl. 1–3.

Bauer, Joseph, 1967. "Altsumerische Wirtschaftstexte aus Lagasch". *Inauguraldissertation*, Philosophische Fakultät der Julius-Maximilian-Universität, Würzburg.

Berger, Hermann, 1992. "Modern Indo-Aryan [Numerals]", pp. 243–287 *in* Jadranka Gvozdanović (ed.), *Indo-European Numerals*. (Trends in Linguistics. Studies and Monographs, 57). Berlin and New York: Mouton de Gruyter.

Boncompagni, Baldassare (ed.), 1857. *Scritti* di Leonardo Pisano matematico del secolo decimoterzo. I. Il *Liber abbaci* di Leonardo Pisano. Roma: Tipografia delle Scienze Matematiche e Fisiche, 1857.

Boncompagni, Baldassare (ed.), 1862. *Scritti* di Leonardo Pisano matematico del secolo decimoterzo. II. *Practica geometriae* et *Opusculi*. Roma: Tipografia delle Scienze Matematiche e Fisiche, 1862.

Borger, Rykle, 1967. "SU-GU$_7$ = *sugû* 'Hungersnot' und GU$_7$ 'essen'". *Orientalia*, NS **36**, 429–431.

Brack-Bernsen, Lis, and Olaf Schmidt, 1990. "Bisectable Trapezia in Babylonian Mathematics". *Centaurus* **33**, 1–38.

Brentjes, Sonja, and Manfred Müller, 1982. "Eine neue Interpretation der ersten Aufgabe des altbabylonischen mathematischen Textes AO 6770". *NTM. Schriftenreihe für Geschichte der Naturwissenschaften, Technik und Medizin* **19**:2, 21–26.

Bruins, Evert M., 1966. "Fermat Problems in Babylonian Mathematics". *Janus* **53**, 194–211.

Bruins, Evert M., 1971. "Computation in the Old Babylonian Period". *Janus* **58**,

222–267.

Bubnov, Nicolaus (ed.), 1899. Gerberti postea Silvestri II papae *Opera mathematica* (972—1003). Berlin: Friedländer.

Busard, H. L. L., 1968. "L'algèbre au moyen âge: Le 'Liber mensurationum' d'Abû Bekr". *Journal des Savants*, Avril-Juin 1968, 65–125.

Busard, H. L. L. (ed.), 1998. Johannes de Muris, *De arte mensurandi*. A Geometrical Handbook of the Fourteenth Century. (Boethius, 41). Stuttgart: Franz Steiner.

Cantor, Moritz, 1875. *Die römischen Agrimensoren und ihre Stellung in der Geschichte der Feldmesskunst. Eine historisch-mathematische Untersuchung*. Leipzig: Teubner.

Cantor, Moritz, 1907. *Vorlesungen über Geschichte der Mathematik*. Erster Band, *von den ältesten Zeiten bis zum Jahre 1200 n. Chr.* Dritte Auflage. Leipzig: Teubner.

Castellino, G. R., 1972. *Two Šulgi Hymns (BC)*. (Studi semitici, 42). Roma: Istituto di studi del Vicino Oriente.

Chace, Arnold Buffum, with the assistance of Henry Parker Manning, 1927. *The Rhind Mathematical Papyrus. British Museum 10057 and 10058*. I. Free Translation and Commentary. *Bibliography of Egyptian Mathematics* by R.C. Archibald. Oberlin, Ohio: Mathematical Association of America.

Chace, Arnold Buffum, Ludlow Bull and Henry Parker Manning, 1929. *The Rhind Mathematical Papyrus*. II. Photographs, Transcription, Transliteration, Literal Translation. *Bibliography of Egyptian and Babylonian Mathematics* (Supplement), by R. C. Archibald. *The Mathematical Leather Roll in the British Museum*, by S. R. K. Glanville. Oberlin, Ohio: Mathematical Association of America.

Chemla, Karine, 1991. "Theoretical Aspects of the Chinese Algorithmic Tradition (First to Third Centuries)". *Historia Scientiarum* **42**, 75–98.

Chemla, Karine, 1996. "Relations between Procedure and Demonstration. Measuring the Circle in the *Nine Chapters on Mathematical Procedures* and Their Commentary by Liu Hui (3rd Century)", pp. 69–112 *in* Hans Niels Jahnke, Norbert Knoche and Michael Otte (eds), *History of Mathematics and Education: Ideas and Experiences*. Göttingen: Vandenhoeck & Ruprecht.

Chemla, Karine, 1997. "Reflection on the World-wide History of the Rule of Double False Position, or: How a Loop Was Closed". *Centaurus* **39**, 97–120.

Curtze, Maximilian (ed.), 1902. *Urkunden zur Geschichte der Mathematik im Mittelalter und der Renaissance*. (Abhandlungen zur Geschichte der mathematischen Wissenschaften, vol. 12–13). Leipzig: Teubner.

Damerow, Peter, and Robert K. Englund, 1987. "Die Zahlzeichensysteme der Archaischen Texte aus Uruk", Kapitel 3 (pp. 117–166) *in* M. W. Green and Hans J. Nissen, *Zeichenliste der Archaischen Texte aus Uruk*, Band II (ATU 2). Berlin: Gebr. Mann, 1987.

Damerow, Peter, 1993. "Buchhalter erfanden die Schrift". [Anmerkungen zu dem Buch *Before Writing* von Denise Schmandt-Besserat]. *Rechthistorisches Journal* **12**, 9–35.

Datta, Bibhutibhusan, 1929. "The Jaina School of Mathematics". *Bulletin of the Calcutta Mathematical Society* **21**, 115–145.

Datta, Bibhutibhusan, and Avadhesh Narayan Singh, 1962. *History of Hindu Mathematics. A Source Book*. Parts I and II. Bombay: Asia Publishing House. [1]Lahore: Motilal Banarsidass, 1935–38.

Dauben, Joseph W. (ed.), 1984. *The History of Mathematics from Antiquity to the Present: A Selective Bibliography*. (Bibliographies of the History of Science and Technology, vol. 6). New York: Garland.

Dijksterhuis, Eduard J., 1956. *Archimedes*. (Acta Historica Scientiarum Naturalium et Medicinalium, 12). København: Munksgaard.

Dold-Samplonius, Yvonne, 1987. "Developments in the Solution to the Equation $cx^2+bx=a$ from al-Khwārizmī to Fibonacci", pp. 71–87 *in* D. A. King and G. Saliba (eds), *From Deferent to Equant: A Volume of Studies in the History of Science in the Ancient and Medieval Near East in Honor of E. S. Kennedy*. (Annals of the New York Academy of Sciences, Volume 500). New York: New York Academy of Sciences.

Dupuis, J. (ed., trans.), 1892. Théon de Smyrne, philosophe platonicien, *Exposition des connaissances mathématiques utiles pour la lecture de Platon*. Traduite pour la première fois du grec en français. Paris: Hachette.

Edzard, Dietz Otto, 1969. "Eine altsumerische Rechentafel (OIP 14, 70)", pp. 101–104 *in* W. Röllig (ed.), *Lišān mitḫurti. Festschrift Wolfram Freiherr von Soden zum 19.VI.1968 gewidmet*. (Altes Orient und Altes Testament, Band 1). Kevelaer: Butzon & Bercker / Neukirchen-Vluyn: Neukirchener Verlag des Erziehungsvereins, 1969.

Edzard, Dietz Otto, 1980. "Keilschrift", pp. 544–568 *in Reallexikon der Assyriologie und Vorderasiatischen Archäologie*, Band V. Berlin: Walter de Gruyter.

Englund, Robert K., 1988. "Administrative Timekeeping in Ancient Mesopotamia". *Journal of the Economic and Social History of the Orient* **31**, 121–185.

Folkerts, Menso, 1978. "Die älteste mathematische Aufgabensammlung in lateinischer Sprache: Die Alkuin zugeschriebenen *Propositiones ad acuendos iuvenes*. Überlieferung, Inhalt, Kritische Edition". *Österreichische Akademie der Wissenschaften, Mathematisch-Naturwissenschaftliche Klasse. Denkschriften*, 116. Band, 6. Abhandlung.

Fowler, David, and Eleanor Robson, 1998. "Square Root Appproximations in Old Babylonian Mathematics: YBC 7289 in Context". *Historia Mathematica* **25**, 366–378.

Frank, Carl, 1928. *Straßburger Keilschrifttexte in sumerischer und babylonischer Sprache*. (Schriften der Straßburger Wissenschaftlichen Gesellschaft in Heidelberg, Neue Folge, Heft 9). Berlin and Leipzig: Walter de Gruyter.

Friberg, Jöran, 1981. "Methods and Traditions of Babylonian Mathematics. Plimpton 322, Pythagorean Triples, and the Babylonian Triangle Parameter Equations". *Historia Mathematica* **8**, 277–318.

Friberg, Jöran, 1981a. "Methods and Traditions of Babylonian Mathematics, II: An Old Babylonian Catalogue Text with Equations for Squares and Circles". *Journal of Cuneiform Studies* **33**, 57–64.

Friberg, Jöran, 1982. "A Survey of Publications on Sumero-Akkadian Mathematics, Metrology and Related Matters (1854–1982)". *Department of Mathematics, Chalmers University of Technology and the University of Göteborg* No. 1982–17.

Friberg, Jöran, 1986. "The Early Roots of Babylonian Mathematics. III: Three Remarkable Texts from Ancient Ebla". *Vicino Oriente* **6**, 3–25.

Friberg, Jöran, 1990. "Mathematik". *Reallexikon der Assyriologie und Vorderasiatischen Archäologie* VII, 531–585. Berlin and New York: de Gruyter.

Friberg, Jöran, Hermann Hunger, and Farouk al-Rawi, 1990a. "'Seeds and Reeds': A Metro-Mathematical Topic Text from Late Babylonian Uruk". *Baghdader Mitteilungen* **21**, 483–557, Tafel 46–48.

Friberg, Jöran, 1993. "On the Structure of Cuneiform Metrological Table Texts from the -1st Millennium". *Grazer Morgenländische Studien* **3**, 383–405.

Friberg, Jöran, 1994. "Preliterate Counting and Accounting in the Middle East. A Constructively Critical Review of Schmandt-Besserat's Before Writing". *Orientalistische Literaturzeitung* **89**, 477–502.

Friberg, Jöran, and Farouk al-Rawi, 1994a. "Equations for Rectangles and Methods of False Value in Old Babylonian Geometric-Algebraic Problem texts". *Manuscript in progress*, courtesy of the authors.

Friberg, Jöran, 1997. "'Seed and Reeds Continued'. Another Metro-Mathematical Topic Text from Late Babylonian Uruk". *Baghdader Mitteilungen* **28**, 251–365, pl. 45–46.

Friberg, Jöran, 2000. "Mathematics at Ur in the Old Babylonian Period". *Department of Mathematics, Chalmers University of Technology, The University of Göteborg. Preprint* 2000/25.

Friberg, Jöran, 2001. "Bricks and Mud in Metro-Mathematical Cuneiform Texts", pp. 61–154 *in* Jens Høyrup and Peter Damerow (eds), *Changing Views on Ancient Near Eastern Mathematics*. (Berliner Beiträge zum Vorderen Orient, 19). Berlin: Dietrich Reimer.

Froidefond, Christian (ed., trans.), 1988. Plutarque, *Isis et Osiris*. (Oeuvres morales V.2). Paris: Les Belles Lettres.

GAG: Wolfram von Soden, *Grundriss der Akkadischen Grammatik*. (Analecta

Orientalia, 33). Roma: Pontificium Institutum Biblicum.
Gandz, Solomon, 1937. "The Origin and Development of the Quadratic Equations in Babylonian, Greek, and Early Arabic Algebra". *Osiris* **3**, 405–557.
Gandz, Solomon, 1939. "Studies in Babylonian Mathematics II. Conflicting Interpretations of Babylonian Mathematics". *Isis* **31**, 405–425.
Gandz, Solomon, 1948. "Studies in Babylonian Mathematics I. Indeterminate Analysis in Babylonian Mathematics". *Osiris* **8**, 12–40.
Gardiner, Alan H., 1911. *Egyptian Hieratic Texts*. Series I: *Literary Texts from the New Kingdom*. Part I: *The Papyrus Anastasi I* and the *Papyrus Koller*, together with Parallel Texts. Leipzig: J. C. Hinrichs'sche Buchhandlung.
Gericke, Helmuth, 1984. *Mathematik in Antike und Orient*. Berlin etc.: Springer.
Glassner, Jean-Jacques, 2000. *Écrire à Sumer. L'invention du cunéiforme*. Paris: Seuil, 2000.
Goetze, Albrecht, 1951. "A Mathematical Compendium from Tell Harmal". *Sumer* **7**, 126–155.
Goody, Jack, 1987. *The Interface between the Oral and the Written*. Cambridge: Cambridge University Press.
Granger, Frank (ed., trans.), 1931. Vitruvius, *De Architectura*. 2 vols. (Loeb Classical Library). London: Heinemannj / Cambridge, Mass.: Harvard University Press, 1931, 1934.
Gundlach, Karl-Bernhard, and Wolfram von Soden, 1963. "Einige altbabylonische Texte zur Lösung 'quadratischer Gleichungen'". *Abhandlungen aus dem mathematischen Seminar der Universität Hamburg* **26**, 248–263.
Guttman, Miquel, and J. Millas y Vallicrosa (eds, trans.), 1931. Abraam bar Hiia. *Llibre de geometria. Hibbur hameixihà uebatixbòret*. (Biblioteca Hebraico-Catalana, 3). Barcelona: Alpha.
Gvozdanović, Jadranka (ed.), 1992. *Indo-European Numerals*. (Trends in Linguistics. Studies and Monographs, 57). Berlin & New York: Mouton de Gruyter.
HAHw: W. Gesenius, *Hebräisches und Aramäisches Handwörterbuch über das Alte Testament*. Bearbeitet von F. Buhl. [16]Leipzig: F. C. W. Vogel, 1915.
Heath, Thomas L., 1921. *A History of Greek Mathematics*. 2 vols. Oxford: The Clarendon Press.
Heath, Thomas L. (ed., trans.), 1926. *The Thirteen Books of Euclid's Elements*, Translated with Introduction and Commentary. 2nd revised edition. 3 vols. Cambridge: Cambridge University Press / New York: Macmillan. [1]1908.
Heiberg, J. L. (ed., trans.), 1912. Heronis *Definitiones* cum variis collectionibus. Heronis quae feruntur *Geometrica*. (Heronis Alexandrini Opera quae supersunt omnia, IV). Leipzig: Teubner.
Heiberg, J. L. (ed., trans.), 1914. Heronis quae feruntur *Stereometrica* et *De mensuris*. (Heronis Alexandrini Opera quae supersunt omnia, V). Leipzig: Teubner.
Herz-Fischler, Roger, 1987. *A Mathematical History of Division in Extreme and Mean ratio*. Waterloo, Ontario: Wilfrid Laurier University Press.
Hochheim, Adolph (ed., trans.), 1878. *Kâfî fîl Hisâb (Genügendes über Arithmetik)* des Abu Bekr Muhammed ben Alhusein Alkarkhi. I-III. Halle: Louis Nebert, 1878.
Hofmann, Joseph E. (ed.), 1970. François Viète, *Opera mathematica* recognita à Francisci à Schooten. Hildesheim & New York: Georg Olms Verlag.
Høyrup, Jens, 1982. "Investigations of an Early Sumerian Division Problem, c. 2500 B.C.". *Historia Mathematica* **9**, 19–36.
Høyrup, Jens, 1990. "Algebra and Naive Geometry. An Investigation of Some Basic Aspects of Old Babylonian Mathematical Thought". *Altorientalische Forschungen* **17**, 27–69, 262–354.
Høyrup, Jens, 1990a. "Sub-Scientific Mathematics. Observations on a Pre-Modern Phenomenon". *History of Science* **28**, 63–86.
Høyrup, Jens, 1990b. "*Dýnamis*, the Babylonians, and Theaetetus 147c7—148d7". *Historia Mathematica* **17**, 201–222.
Høyrup, Jens, 1990c. "On Parts of Parts and Ascending Continued Fractions". *Centaurus* **33**, 293–324.
Høyrup, Jens, 1992. "The Babylonian Cellar Text BM 85200 + VAT 6599. Retranslation and Analysis", pp. 315–358 *in* Demidov et al. (eds), *Amphora.*

Festschrift für Hans Wussing zu seinem 65. Geburtstag. Basel etc.: Birkhäuser.

Høyrup, Jens, 1992a. "Sumerian: The Descendant of a Proto-Historical Creole? An Alternative Approach to the 'Sumerian Problem'". AIΩN. *Annali del Dipartimento di Studi del Mondo Classico e del Mediterraneo Antico. Sezione linguistica. Istituto Universitario Orientale, Napoli* **14** (1992; publ. 1994), 21–72, Figs. 1–3.

Høyrup, Jens, 1993. "Mathematical Susa Texts VII and VIII. A Reinterpretation". *Altorientalische Forschungen* **20**, 245–260.

Høyrup, Jens, 1993a. "'Remarkable Numbers' in Old Babylonian Mathematical Texts: A Note on the Psychology of Numbers". *Journal of Near Eastern Studies* **52**, 281–286.

Høyrup, Jens, 1993b. "On Subtractive Operations, Subtractive Numbers, and Purportedly Negative Numbers in Old Babylonian Mathematics". *Zeitschrift für Assyriologie und Vorderasiatische Archäologie* **83**, 42–60.

Høyrup, Jens, 1994. *In Measure, Number, and Weight. Studies in Mathematics and Culture*. New York: State University of New York Press.

Høyrup, Jens, 1995. "A Note on an Anomalous Area Measurement from Ur III (with a Remark about Confidence in Experts)". *Altorientalische Forschungen* **22**, 84–86.

Høyrup, Jens, 1995a. "Linee larghe. Un'ambiguità geometrica dimenticata". *Bollettino di Storia delle Scienze Matematiche* **15**, 3–14.

Høyrup, Jens, 1995b. [Review of Kazuo Muroi, "Reexamination of the first problem of the Susa mathematical text No. 9 (English)". *Sūgakushi Kenkyū* **140** (1994), 50–56]. *Zentralblatt für Mathematik und ihre Grenzgebiete* **795**, #01001.

Høyrup, Jens, 1996. "Changing Trends in the Historiography of Mesopotamian Mathematics: An Insider's View". *History of Science* **34**, 1–32.

Høyrup, Jens, 1996a. "'The Four Sides and the Area'. Oblique Light on the Prehistory of Algebra", pp. 45–65 *in* Ronald Calinger (ed.), *Vita mathematica. Historical Research and Integration with Teaching*. Washington, DC: Mathematical Association of America. Marred by printing errors, due to the publisher's omission of a proof reading stage.

Høyrup, Jens, 1996b. [Review of Kazuo Muroi, "Reexamination of Susa Mathematical Text No. 8". *Sūgakushi Kenkyū* **140** (1994), 50–56]. *Zentralblatt für Mathematik und ihre Grenzgebiete* **838**, #01001.

Høyrup, Jens, 1997. "Hero, Ps.-Hero, and Near Eastern Practical Geometry. An Investigation of *Metrica*, *Geometrica*, and other Treatises", pp. 67–93 *in* Klaus Döring, Bernhard Herzhoff and Georg Wöhrle (eds), *Antike Naturwissenschaft und ihre Rezeption*, Band 7. Trier: Wissenschaftlicher Verlag Trier. (For obscure reasons, the publisher has changed "□" into "~" and "⊏⊐" into "¤§" on p. 83 after having supplied correct proof sheets).

Høyrup, Jens, 1998. "'Oxford' and 'Cremona': On the Relations between two Versions of al-Khwārizmī's *Algebra*", pp. 159–178 *in Actes du 3me Colloque Maghrébin sur l'Histoire des Mathématiques Arabes, Tipaza (Alger, Algérie), 1–3 Décembre 1990*, vol. II. Alger: Association Algérienne d'Histoire des Mathématiques.

Høyrup, Jens (ed., trans.), 2000. Jacobus de Florentia, *Tractatus algorismi* (1307), the chapter on algebra (Vat. Lat. 4826, fols 36[v]–45[v]). *Centaurus* **42**, 21–69.

Høyrup, Jens, 2000a. "The Finer Structure of the Old Babylonian Mathematical Corpus. Elements of Classification, with some Results", pp. 117–177 *in* Joachim Marzahn & Hans Neumann (eds), *Assyriologica et Semitica*. Festschrift für Joachim Oelsner anläßlich seines 65. Geburtstages am 18. Februar 1997. (Altes Orient und Altes Testament, 252). Münster: Ugarit Verlag.

Høyrup, Jens, 2001. "The Old Babylonian Square Texts BM 13901 and YBC 4714. Retranslation and Analysis", pp. 155–218 *in* Jens Høyrup and Peter Damerow (eds), *Changing Views on Ancient Near Eastern Mathematics*. (Berliner Beiträge zum Vorderen Orient, 19). Berlin: Dietrich Reimer.

Huber, Peter, 1955. "Zu einem mathematischen Keilschrifttext (VAT 8512)", *Isis* **46**, 104–106.

Hughes, Barnabas, O.F.M., 1986. "Gerard of Cremona's Translation of al-Khwārizmī's *Al-Jabr*: A Critical Edition". *Mediaeval Studies* **48**, 211–263.

Hultsch, Friedrich, 1900. "Die pythagoreischen Reihen der Seiten und Diagonalen von Quadraten und ihre Umbildung zu einer Doppelreihe ganzer Zahlen". *Bibliotheca mathematica*, 3. Folge **1** (1900–1901), 8–12.

Hunger, Hermann, 1968. *Babylonische und assyrische Kolophone*. (Alter Orient und Altes Testament, 2). Kevelaer: Butzon & Bercker / Neukirchen-Vluyn: Neukirchener Verlag.

Jacobsen, Thorkild, 1988. "The Sumerian Verbal Core". *Zeitschrift für Assyriologie und Vorderasiatische Archäologie* **78**, 161–220.

Jones, Alexander (ed., trans.), 1986. Pappus of Alexandria, Book 7 of the *Collection*. 2 vols. New York etc.: Springer, 1986.

Kilmer, Anne Draffkorn, 1960. "Two New Lists of Key Numbers for Mathematical Operations". *Orientalia*, NS **29**, 273–308.

Kilmer, Anne Draffkorn, 1964. "The Use of Akkadian *DKŠ* in Old Babylonian Geometry Texts", pp. 140–146 *in Studies Presented to A. Leo Oppenheim, June 7, 1964*. Chicago: The Oriental Institute of the University of Chicago.

King, David, 1988. "A Medieval Arabic Report on Algebra Before al-Khwārizmī". *al-Masāq* **1**, 25–32.

Knuth, Donald E., 1972. "Ancient Babylonian Algorithms". *Communications of the ACM* **15**, 671–677 (+**29** (1976), 108).

Krasnova, S. A. (ed., trans.), 1966. "Abu-l-Vafa al-Buzdžani, *Kniga o tom, čto neobxodimo remeslenniku iz geometričeskix postroenij*", pp. 42–140 *in* A. T. Grigor'jan and A. P. Juškevič, *Fiziko-matematičeskie nauki v stranax vostoka.* Sbornik statej i publikacij. Vol. I (of IV). Moskva: Izdatel'stvo 'Nauka'.

Krecher, Joachim, 1973. "Neue sumerische Rechtsurkunden des 3. Jahrtausends". *Zeitschrift für Assyriologie und Vorderasiatische Archäologie* **63** (1973–74), 145–271.

Kroll, Wilhelm (ed.), 1899. Procli Diadochi *In Platonis Rem publicam commentarii*. 2 vols. Leipzig: Teubner, 1899, 1901.

Lambert, Maurice, 1953, "Textes commerciaux de Lagash". *Revue d'Assyriologie* **47**, 57–69, 1+5–120.

Liverani, Mario, 1990. "The Shape of Neo-Sumerian Fields". *Bulletin of Sumerian Agriculture* **5**, 147–186.

Lord, Albert B. 1960. *The Singer of Tales*. Cambridge, Mass.: Harvard University Press.

Luckey, Paul, 1941. "Ṯābit b. Qurra über den geometrischen Richtigkeitsnachweis der Auflösung der quadratischen Gleichungen". *Berichte der Sächsischen Akademie der Wissenschaften zu Leipzig. Mathematisch-physische Klasse* **93**, 93–114.

Ma Li, 1993. "The Rule of False: Early Applications and Conjectured Transmissions" *Department of Mathematics, Chalmers University of Technology & The University of Göteborg, Preprint* 1993-15.

Mahoney, Michael S., 1971. "Babylonian Algebra: Form *vs.* Content". [Essay Review of a reprint edition of O. Neugebauer 1934]. *Studies in History and Philosophy of Science* **1** (1970–71), 369–380.

MCT: O. Neugebauer and A. Sachs, *Mathematical Cuneiform Texts*. (American Oriental Series, vol. 29). New Haven, Connecticut: American Oriental Society, 1945.

MEA: 1963. René Labat, *Manuel d'épigraphie akkadienne (signes, syllabaire, idéogrammes)*. ^{4}Paris: Imprimerie Nationale. ^{1}1948.

Menge, Heinrich (ed.), 1896. Euclidis *Data* cum *Commentario* Marini et scholiis antiquis. (Euclidis Opera Omnia vol. VI). Leipzig: Teubner.

MKT: O. Neugebauer, *Mathematische Keilschrift-Texte*. I-III. (Quellen und Studien zur Geschichte der Mathematik, Astronomie und Physik. Abteilung A: Quellen. 3. Band, erster-dritter Teil). Berlin: Julius Springer, 1935, 1935, 1937.

Mueller, Ian, 1981. *Philosophy of Mathematics and Deductive Structure in Euclid's Elements*. Cambridge, Mass., & London: MIT Press.

Muroi, Kazuo, 1991. "The Expressions of Zero and of Squaring in the Babylonian Mathematical Text VAT 7537". *Historia Scientiarum*, Second Series **1**, 59–62.

Muroi, Kazuo, 1992. "Reexamination of Susa Mathematical Text No. 3: Alleged Value $\pi \fallingdotseq 3^1/_8$" *Historia Scientiarum*, Second Series **2**, 45–49.

Muroi, Kazuo, 1994. "Reexamination of the First Problem of the Susa Mathematical Text No. 9 (English)". *Historia Scientiarum*, Second Series **3**, 231–233.

Muroi, Kazuo, 1994a. "Reexamination of Susa Mathematical Text No. 8". *Sūgakushi Kenkyū* **140**, 50–56.
Muroi, Kazuo, 1998. "Early Old Babylonian Mathematical Problems Written in Sumerian". *Historia Scientiarum*, Second Series **7**, 199–203.
Muroi, Kazuo, 1999. "Extraction of Square Roots in Babylonian Mathematics". *Historia Scientiarum*, Second Series **9**, 127–133.
Muroi, Kazuo, 2001. "Inheritance Problems in the Susa Mathematical Text No. 26". *Historia Scientiarum*, Second Series **10** (2001), 226–234.
Muroi, Kazuo, 2001a. "Reexamination of the Susa Mathematical Text No. 12: A System of Quartic Equations". *SCIAMUS* **2**, 3–8.
Nesselmann, G. H. F., 1842. *Versuch einer kritischen Geschichte der Algebra*. Nach den Quellen bearbeitet. Erster Theil, *Die Algebra der Griechen*. Berlin: G. Reimer.
Neugebauer, O., 1929. "Zur Geschichte der babylonischen Mathematik". *Quellen und Studien zur Geschichte der Mathematik, Astronomie und Physik.* Abteilung B: *Studien* **1** (1929–31), 67–80.
Neugebauer, O., and W. Struve, 1929a. "Über die Geometrie des Kreises in Babylonien". *Quellen und Studien zur Geschichte der Mathematik, Astronomie und Physik.* Abteilung B: *Studien* **1** (1929–1931), 81–92.
Neugebauer, Otto, 1932. "Zur Transkription mathematischer und astronomischer Keilschrifttexte". *Archiv für Orientforschung* **8** (1932–33), 221–223.
Neugebauer, Otto, 1932a. "Studien zur Geschichte der antiken Algebra I". *Quellen und Studien zur Geschichte der Mathematik, Astronomie und Physik.* Abteilung B: *Studien* **2** (1932–33), 1–27.
Neugebauer, Otto, 1934. *Vorlesungen über Geschichte der antiken mathematischen Wissenschaften.* I: *Vorgriechische Mathematik*. (Die Grundlehren der mathematischen Wissenschaften in Einzeldarstellungen, Bd. XLIII). Berlin: Julius Springer.
Neugebauer, O., 1936. "Zur geometrischen Algebra (Studien zur Geschichte der antiken Algebra III)". *Quellen und Studien zur Geschichte der Mathematik, Astronomie und Physik.* Abteilung B: *Studien* **3** (1934–36), 245–259.
Nissen , Hans J., 1974. "Zur Frage der Arbeitsorganisation in Babylonien während der Späturuk-Zeit". *Acta Antiqua Academiae Scientiarum Hungaricae* **22**, 5–14.
Nissen, Hans J., Peter Damerow, and Robert Englund, 1993. *Archaic Bookkeeping: Writing and Techniques of Economic Administration in the Ancient Near East*. Chicago: Chicago University Press.
Nuñez, Pedro, 1567. *Libro de Algebra en Arithmetica y Geometria*. Anvers: En casa de los herederos d'Arnaldo Birckman.
Ong, Walter J., 1982. *Orality and Literacy. The Technologizing of the Word*. London and New York: Methuen.
Orwell, George, 1954. *Nineteen Eighty-Four*. Harmondsworth, Middlesex: Penguin.
Pacioli, Luca, 1523. *Summa de Arithmetica geometria Proportioni: et proportionalita*. Novamente impressa. Toscolano: Paganinus de Paganino.
Parker, Richard A., 1972. *Demotic Mathematical Papyri*. Providence and London: Brown University Press.
Paton, W. R. (ed., trans.), 1979. *The Greek Anthology*. In Five Volumes. (Loeb Classical Library) Cambridge, Mass.: Harvard University Pressj / London: Heinemann. [1]1918.
Peck, A. L., and E. S. Forster (eds, trans.), 1937. Aristotle, *Parts of Animals. Movement of Animals. Progression of Animals*. (Loeb Classical Library). Cambridge, Mass.: Harvard University Libraryj / London: Heinemann.
Petersson, John Sören, 1991. *Critique of Evolutionary Accounts of Writing*. (Reports from Uppsala University Linguistics 21/1991). Uppsala:Department of Linguistics, Uppsala University.
Picutti, Ettore, 1989. "Sui plagi matematici di frate Luca Pacioli". *Le Scienze* **40**:246 (febbraio 1989), 72–79.
Powell, Marvin A., 1990. "Maße und Gewichte". *Reallexikon der Assyriologie und Vorderasiatischen Archäologie* VII, 457–516. Berlin and New York: de Gruyter.
Raṅgācārya, M. (ed., trans.), 1912. The *Gaṇita-sāra-sangraha* of Mahāvīrācārya with English Translation and Notes. Madras: Government Press.
Rebstock, Ulrich, 1993. *Die Reichtümer der Rechner (Ġunyat al-Ḥussāb) von Aḥmad b.*

Ṯabāt (gest. 631/1234). Die Araber – Vorläufer der Rechenkunst. (Beiträge zur Sprach- und Kulturgeschichte des Orients, 32). Walldorf-Hessen: Verlag für Orientkunde Dr. H. Vorndran.

Ritter, Jim, forthcoming. "Reading Strasbourg 368: A Thrice-Told Tale".

Robbins, Frank Egleston, 1929. "P. Mich. 620: A Series of Arithmetical Problems". *Classical Philology* **24**, 321–329.

Robson, Eleanor, 1995. "Old Babylonian Coefficient Lists and the Wider Context of Mathematics in Ancient Mesopotamia, 2100–1600 BC". *Dissertation*, submitted for D.Phil in Oriental Studies. Wolfson College, Oxford. Published with revisions as [Robson 1999].

Robson, Eleanor, 1996. "Building with Bricks and Mortar. Quantity Surveying in the Ur III and Old Babylonian Periods", pp. 181–190 *in* Klaas R. Veenhof (ed.), *Houses and Households in Ancient Mesopotamia.* Papers Read at the 40e Rencontre Assyriologique Internationale, Leiden, July 5–8, 1993. Leiden and Istanbul: Nederlands Historisch-Archaeologisch Instituut te Istanbul.

Robson, Eleanor, 1999. *Mesopotamian Mathematics 2100–1600 BC. Technical Constants in Bureaucracy and Education.* (Oxford Editions of Cuneiform Texts, 14). Oxford: Clarendon Press.

Robson, Eleanor, 2000. "Mathematical Cuneiform Tablets in Philadelphia. Part 1: Problems and Calculations". *SCIAMUS* **1**, 11–48.

Robson, Eleanor, 2001. "Neither Sherlock Holmes nor Babylon: a Reassessment of Plimpton 322". *Historia Mathematica* **28**, 167–206.

Rodet, Léon, 1878. "L'algèbre d'al-Khârizmi et les méthodes indienne et grecque". *Journal Asiatique*, 7e série **11**, 5–98.

Rosen, Frederic (ed., trans.), 1831. The *Algebra* of Muhammad ben Musa, Edited and Translated. London: The Oriental Translation Fund, 1831.

Rozenfeld, Boris A. (trans.), 1983. Muhammad ibn Musa al-Xorezmi, *Kratkaja kniga ob isčiclenii algebry i almukabaly*, pp. 20–81, notes 118–142 *in* S. X. Siraždinov, (ed.), Muxammad ibn Musa al-Xorezmi *Matematičeskie traktaty.* Taškent: Izdatel'stvo 'FAN' Uzbekskoj CCP.

Rudhardt, Jean, 1978. "Trois problèmes de géométrie, conservés par un papyrus genevois". *Museum Helveticum* **35**, 233–240.

Rudolff, Christoff, 1540. *Künstliche rechnung mit der ziffer und mit den zalpfenningē/ sampft der Wellischen Practica/ und allerley forteyl auff die Regel de tri. Item vergleichung mancherley Land uñ Stet/ gewicht/ Einmas/ Müntz etc.* [2]Wien.

Sachs, Abraham J., 1952. "Babylonian Mathematical Texts. II-III". *Journal of Cuneiform Studies* **6**, 151–156.

Saggs, H. W. F. 1960. "A Babylonian Geometrical Text". *Revue d'Assyriologie* **54**, 131–146.

Saliba, George A., 1972. "The Meaning of al-jabr wa'l-muqābalah". *Centaurus* **17** (1972-73), 189–204.

Schmandt-Besserat, Denise, 1992. *Before Writing.* I. *From Counting to Cuneiform.* Austin: University of Texas Press.

Schöne, Hermann (ed., trans.), 1903. Herons von Alexandria *Vermessungslehre* und *Dioptra.* Griechisch und deutsch. (Heronis Alexandrini Opera quae supersunt omnia, vol. III). Leipzig: Teubner.

Schuster, H. S., 1930. "Quadratische Gleichungen der Seleukidenzeit aus Uruk". *Quellen und Studien zur Geschichte der Mathematik, Astronomie und Physik.* Abteilung B: *Studien* **1** (1929–31), 194–200.

Sesiano, Jacques, 1987. "Survivance médiévale en Hispanie d'un problème né en Mésopotamie". *Centaurus* **30**, 18–61.

Sesiano, Jacques, 1999. "Sur le Papyrus graecus genevensis 259". *Museum Helveticum* **56**, 26–32.

Shelby, Lon R. (ed.), 1977. *Gothic Design Techniques. The Fifteenth-Century Design Booklets of Mathes Roriczer and Hanns Schmuttermayer.* Carbondale and Edwardsville: Southern Illinois University Press.

SL: Anton Deimel, S.I. *Šumerisches Lexikon.* I. *Vollständiges Syllabar mit den wichtigsten Zeichenformen.* II:1–4. *Vollständige Ideogramm-Sammlung.* III:1. *Šumerisch-Akkadisches Glossar.* II:2.*Akkadisch-Šumerisches Glossar.* (Scripta

Pontificii Instituti Biblici). Rom: Verlag des Päpstlichen Bibelinstituts, 1925–1937.
SLa: Marie-Louise Thomsen, *The Sumerian Language. An Introduction to its History and Grammatical Structure*. (Mesopotamia, 10). København: Akademisk Forlag, 1984.
Souissi, Mohamed (ed., trans.), 1969. Ibn al-Bannā', *Talkhīṣ a'māl al-ḥisāb*. Texte établi, annoté et traduit. Tunis: L'Université de Tunis.
Souissi, Mohammed, 1983. "Ibn al-Yāsamīn, savant mathématicien du Maghrib", pp. 217–225 *in Actas del IV Coloquio Hispano-Tunecino* (Palma, Majorca, 1979). Madrid: Instituto Hispano-Arabe de Cultura.
Steinkeller, Piotr, 1979. "Alleged GUR.DA = ugula-géš-da and the Reading of the Sumerian Numeral 60". *Zeitschrift für Assyriologie und Vorderasiatische Archäologie* **69**, 176–187.
Suter, Heinrich, 1910. "Das Buch der Seltenheiten der Rechenkunst von Abū Kāmil al-Miṣrī". *Bibliotheca Mathematica*, 3. Folge **11** (1910–1911), 100–120.
Szabó, Arpád, 1969. *Anfänge der griechischen Mathematik*. München and Wien: R. Oldenbourg/Budapest: Akadémiai Kiadó.
Tannery, Paul (ed., trans.), 1893. Diophanti Alexandrini *Opera omnia* cum graecis commentariis. 2 vols. Leipzig: Teubner, 1893–1895.
Tanret, M., 1982. "Les tablettes 'scolaires' découvertes à Tell ed-Dēr". *Akkadica* **27**, 46–49.
Thompson, Reginald Campbell (ed.), 1930. *The Epic of Gilgamish*. Text, Transliteration and Notes. Oxford: Clarendon Press.
Thureau-Dangin, François, 1897. "Un cadastre chaldéen". *Revue d'Assyriologie* **4**, 13–27.
Thureau-Dangin, F., 1936. [Review of MKT1 I–II]. *Revue d'Assyriologie* **33**, 55–61.
Thureau-Dangin, F., 1936a. "L'Équation du deuxième degré dans la mathématique babylonienne d'après une tablette inédite du British Museum". *Revue d'Assyriologie* **33**, 27–48.
Thureau-Dangin, F., 1937. "Notes sur la mathématique babylonienne". *Revue d'Assyriologie* **34**, 9–28.
Thureau-Dangin, F., 1940. "Notes sur la mathématique babylonienne". *Revue d'Assyriologie* **37**, 1–10.
TMB: F. Thureau-Dangin, *Textes mathématiques babyloniens*. (Ex Oriente Lux, Deel 1). Leiden: Brill, 1938.
TMS: E. M. Bruins and M. Rutten, *Textes mathématiques de Suse*. (Mémoires de la Mission Archéologique en Iran, XXXIV). Paris: Paul Geuthner, 1961.
Tredennick, Hugh (ed., trans.), 1933. Aristotle, *The Metaphysics*. 2 vols. (Loeb Classical Library, 271, 287). Cambridge, Mass.: Harvard University Press / London: Heinemann, 1933, 1935.
Vajman, A. A., 1961. *Šumero-vavilonskaja matematika. III-I Tysjačeletija do n. e.* Moskva: Izdatel'stvo Vostočnoj Literatury.
van der Waerden, B. L., 1962. *Science Awakening*. [2]Groningen: Noordhoff. [1]1954.
Van Egmond, Warren, 1986. "The Contributions of the Italian Renaissance to European Mathematics". *Symposia Mathematica* **27**, 51–67.
Vogel, Kurt, 1930. "Die algebraischen Probleme des P. Mich. 620". *Classical Philology* **25**, 373–375.
Vogel, Kurt, 1933. "Zur Berechnung der quadratischen Gleichungen bei den Babyloniern". *Unterrichtsblätter für Mathematik und Physik* **39**, 76–81.
Vogel, Kurt, 1934. "Kubische Gleichungen bei den Babyloniern?" *Sitzungsberichte der Bayerischen Akademie der Wissenschaften zu München. Mathematisch-Naturwissenschaftliche Abteilung* 1934, 87–94.
Vogel, Kurt, 1958. "Ist die babylonische Mathematik sumerisch oder akkadisch?" *Mathematische Nachrichten* **18**, 377–382.
Vogel, Kurt, 1959. *Vorgriechische Mathematik*. II. *Die Mathematik der Babylonier*. (Mathematische Studienhefte, 2). Hannover: Hermann Schroedelj / Paderborn: Ferdinand Schöningh.
Vogel, Kurt, 1960. "Der 'falsche Ansatz' in der babylonischen Mathematik". *Mathematisch-Physikalische Semesterberichte* **7**, 89–95.
von Soden, Wolfram, 1939. [Review of TMB]. *Zeitschrift der Deutschen Morgenländi-*

schen Gesellschaft **93**, 143–152.

von Soden, Wolfram, 1952. "Zu den mathematischen Aufgabentexten vom Tell Harmal". *Sumer* **8**, 49–56.

von Soden, Wolfram, 1964. [Review of TMS]. *Bibliotheca Orientalis* **21**, 44–50.

von Weiher, Egbert, 1993. *Uruk: Spätbabylonische Texte aus dem Planquadrat U 18*. Teil IV. (Ausgrabungen der Deutschen Forschungsgemeinschaft in Uruk-Warke, Endberichte, 12). Mainz: Philip von Zabern.

Weil, André, 1978. "Who Betrayed Euclid?" *Archive for History of Exact Sciences* **19**, 91–93.

Whiting, Robert M., 1984. "More Evidence for Sexagesimal Calculations in the Third Millennium B.C." *Zeitschrift für Assyriologie und Vorderasiatische Archäologie* **74**, 59–66.

Witmer, T. Richard (ed., trans.), 1968. Girolamo Cardano, *The Great Art or The Rules of Algebra*. Cambridge, Mass., and London: M.I.T. Press.

Wright, Henry T., and Gregory A. Johnson, 1975. "Population, Exchange, and Early State Formation in Southwestern Iran". *American Anthropologist* **77**, 267–289.

Zeuthen, Hieronimus Georg, 1886. *Die Lehre von den Kegelschnitten im Altertum*. København: Höst & Sohn.

Index of Tablets

Index of Akkadian and Sumerian Terms and Key Phrases

Only discussions of terms and phrases are indexed, not their appearance in the texts; metrological units are indexed only where they are discussed or referred to as such, not when they merely serve as units. In cases where a declinated form or a whole phrase is listed separately (e.g., *tammar*, "you see", *atta ina epēšika*, "you, by you procedure"), a cross-reference is given under the relevant lexical form or forms (in case *amārum*, "to see", and *epēšum*, "proceed"/"proceeding"), unless the two are immediate neighbours.

Name Index

Subject Index

Sources and Studies in the History of Mathematics and Physical Sciences

Continued from page ii

J. Høyrup
Length, Widths, Surfaces: A Portrait of Old Babylonian Alegbra and Its Kin

A. Jones (Ed.)
Pappus of Alexandria, Book 7 of the *Collection*

E. Kheirandish
The Arabic Version of Euclid's *Optics*, Volumes I and II

J. Lützen
Joseph Liouville 1809–1882: Master of Pure and Applied Mathematics

J. Lützen
The Prehistory of the Theory of Distributions

G.H. Moore
Zermelo's Axiom of Choice

O. Neugebauer
A History of Ancient Mathematical Astronomy

O. Neugebauer
Astronomical Cuneiform Texts

F.J. Ragep
Naṣīr al-Dīn al-Ṭūsī's *Memoir on Astronomy*
(al-Tadhkira fī ʿilm al-hay'a)

B.A. Rosenfeld
A History of Non-Euclidean Geometry

J. Sesiano
Books IV to VII of Diophantus' *Arithmetica*: In the Arabic Translation Attributed to Quṣṭā ibn Lūqā

B. Stephenson
Kepler's Physical Astronomy

N.M. Swerdlow/O. Neugebauer
Mathematical Astronomy in Copernicus's De Revolutionibus

G.J. Toomer (Ed.)
Appolonius *Conics* Books V to VII: The Arabic Translation of the Lost Greek Original in the Version of the Banū Mūsā, Edited, with English Translation and Commentary by G.J. Toomer

G.J. Toomer (Ed.)
Diocles on Burning Mirrors: The Arabic Translation of the Lost Greek Original, Edited, with English Translation and Commentary by G.J. Toomer

Continued from the previous page

C. Truesdell
The Tragicomical History of Thermodynamics, 1822–1854

K. von Meyenn, A. Hermann, V.F. Weisskopf (Eds.)
Wolfgang Pauli: Scientific Correspondence II: 1930–1939

K. von Meyenn (Ed.)
Wolfgang Pauli: Scientific Correspondence III: 1940–1949

K. von Meyenn (Ed.)
Wolfgang Pauli: Scientific Correspondence IV, Part I: 1950–1952

K. von Meyenn (Ed.)
Wolfgang Pauli: Scientific Correspondence IV, Part II: 1953–1954